Edited by
Jürgen Stampfl, Robert Liska, and
Aleksandr Ovsianikov

Multiphoton Lithography

Edited by Jürgen Stampfl, Robert Liska, and Aleksandr Ovsianikov

Multiphoton Lithography

Techniques, Materials and Applications

WILEY-VCH
Verlag GmbH & Co. KGaA

Editors

Dr. Jürgen Stampfl
TU Wien
Institute of Materials Science and Technology
Getreidemarkt 9 E308
1060 Wien
Austria

Dr. Robert Liska
TU Wien
Institute of Applied Synthetic Chemistry
Getreidemarkt 9
1060 Wien
Austria

Dr. Aleksandr Ovsianikov
TU Wien
Institute of Materials Science and Technology
Getreidemarkt 9 E308
1060 Wien
Austria

Cover
The cover image was kindly provided by the Additive Manufacturing Technologies group of the TU Wien (Aleksandr Ovsianikov, Peter Gruber) and MäID – FutureRetrospectiveNarrative (Daniela Mitterberger, Tiziano Derme)

Library of Congress Card No.: applied for

British Library Cataloguing-in-Publication Data
A catalogue record for this book is available from the British Library.

Bibliographic information published by the Deutsche Nationalbibliothek
The Deutsche Nationalbibliothek lists this publication in the Deutsche Nationalbibliografie; detailed bibliographic data are available on the Internet at <http://dnb.d-nb.de>.

Print ISBN: 978-3-527-33717-0
ePDF ISBN: 978-3-527-68268-3
ePub ISBN: 978-3-527-68269-0
Mobi ISBN: 978-3-527-68266-9
oBook ISBN: 978-3-527-68267-6

Typesetting SPi Global, Chennai, India
Printing and Binding Markono Print Media Pte Ltd, Singapore

Printed on acid-free paper

Contents

List of Contributors

Nikita Bityurin
Institute of Applied Physics of Russian Academy of Sciences
46, Ul'yanov Street
603950 Nizhniy Novgorod
Russia

Yaoyu Cao
Swinburne University of Technology
Centre for Micro-Photonics (H34)
John Street
Hawthorn
VIC 3122
Australia

Boris Chichkov
Laser Zentrum Hannover e.V.
Hollerithallee 8
D-30419 Hannover
Germany

Daniel S. Correa
National Laboratory for Nanotechnology in Agribusiness (LNNA)
Embrapa Instrumentation
13560-970 São Carlos, SP
Brazil

Cole A. DeForest
University of Washington
Department of Chemical Engineering
4000 15th Ave NE
Seattle, WA 98195
USA

and

University of Washington
Department of Bioengineering
3720 15th Ave NE
Seattle, WA 98105
USA

and

University of Washington
Institute for Stem Cell and Regenerative Medicine
850 Republican Street
Seattle, WA 98109
USA

Jennefir L. Digaum
University of Central Florida
CREOL, The College of Optics and Photonics
4000 Central Florida Blvd.
Orlando, FL 32816
USA

Xuan-Ming Duan
Chinese Academy of Sciences
Technical Institute of Physics
and Chemistry
Laboratory of Organic
NanoPhotonics and Key
Laboratory of Functional Crystals
and Laser Technology
No. 29, Zhongguancun East Road
Beijing 100190
PR China

and

Chinese Academy of Sciences
Chongqing Institutes of Green
and Intelligent Technology
No. 266 Fangzheng Ave
Shuitu Technology Development
Zone
Beibei District
Chongqing 400714
PR China

Maria Farsari
Institute of Electronic Structure
and Laser (IESL)
Foundation for Research and
Technology-Hellas (FORTH)
N. Plastira 100
Vassilika Vouton
70013 Heraklion
Crete
Greece

John T. Fourkas
University of Maryland
Department of Chemistry
and Biochemistry
Institute for Physical Science and
Technology
College Park, MD 20742
USA

Zongsong Gan
Swinburne University of
Technology
Centre for Micro-Photonics (H34)
John Street
Hawthorn
VIC 3122
Australia

Christopher N. Grabill
University of Central Florida
Chemistry Department
4000 Central Florida Blvd.
Orlando, FL 32816
USA

Min Gu
Swinburne University of
Technology
Centre for Micro-Photonics (H34)
John Street
Hawthorn
VIC 3122
Australia

and

Swinburne University of
Technology
Centre for Ultrahigh-Bandwidth
Devices for Optical Systems
(CUDOS)
John Street
Hawthorn
VIC 3122
Australia

David J. Hagan
University of Central Florida
CREOL, The College of Optics
and Photonics
4304 Scorpius Street
Orlando, FL 32816-2700
USA

Ulf Hinze
Laser Zentrum Hannover e.V.
Hollerithallee 8
D-30419 Hannover
Germany

Saulius Juodkazis
Swinburne University of
Technology
Centre for Micro-Photonics
Hawthorn
VIC 3122
Australia

and

Melbourne Centre for
Nanofabrication (MCN)
Australian National Fabrication
Facility (ANFF)
Clayton
VIC 3168
Australia

Stephen M. Kuebler
University of Central Florida
Chemistry Department
4000 Central Florida Blvd.
Orlando, FL 32816
USA

and

University of Central Florida
CREOL, The College of Optics
and Photonics
4000 Central Florida Blvd.
Orlando, FL 32816
USA

and

University of Central Florida
Physics Department
4000 Central Florida Blvd.
Orlando, FL 32816
USA

Josef Kumpfmüller
TU Wien
Institute of Applied Synthetic
Chemistry
Getreidemarkt 9/163
1060 Vienna
Austria

Christopher N. LaFratta
Bard College
Department of Chemistry
Annandale-on-Hudson
New York, NY 12504-5000
USA

Samuel Clark Ligon
TU Wien
Institute of Applied Synthetic
Chemistry
Getreidemarkt 9/163
1060 Vienna
Austria

Robert Liska
TU Wien
Institute of Applied Synthetic
Chemistry
Getreidemarkt 9/163
1060 Vienna
Austria

Mangirdas Malinauskas
Vilnius University
Laser Research Center
Department of Quantum Electronics
Saulėtekio Ave. 10
10223 Vilnius
Lithuania

Shoji Maruo
Yokohama National University
Department of Mechanical Engineering and Materials Science
79-5 Tokiwadai, Hodogaya-ku
Yokohama 240-8501
Japan

Cleber R. Mendonça
Universidade de São Paulo
Departamento de Física e Ciência dos Materiais
Instituto de Física de São Carlos
Av. Trabalhador Sãocarlense
400 Centro
13566-590 São Carlos, SP
Brazil

Adriano J. G. Otuka
Universidade de São Paulo
Departamento de Física e Ciência dos Materiais
Instituto de Física de São Carlos
Av. Trabalhador Sãocarlense
400 Centro
13566-590 São Carlos, SP
Brazil

Alexander Pikulin
Institute of Applied Physics of Russian Academy of Sciences
46, Ul'yanov Street
603950 Nizhniy Novgorod
Russia

Niklas Pucher
TU Wien
Institute of Applied Synthetic Chemistry
Getreidemarkt 9/163
1060 Vienna
Austria

Casey M. Schwarz
University of Central Florida
Chemistry Department
4000 Central Florida Blvd.
Orlando, FL 32816
USA

and

University of Central Florida
CREOL, The College of Optics and Photonics
4000 Central Florida Blvd.
Orlando, FL 32816
USA

Alexandros Selimis
Institute of Electronic Structure and Laser (IESL)
Foundation for Research and Technology-Hellas (FORTH)
N. Plastira 100
Vassilika Vouton
70013 Heraklion
Crete
Greece

Gediminas Seniutinas
Swinburne University of Technology
Centre for Micro-Photonics
Hawthorn
VIC 3122
Australia

and

Melbourne Centre for Nanofabrication (MCN)
Australian National Fabrication Facility (ANFF)
Clayton
VIC 3168
Australia

Jared A. Shadish
University of Washington
Department of Chemical Engineering
4000 15th Ave NE
Seattle, WA 98195
USA

Jürgen Stampfl
Tu Wien
Institute of Materials Science and Technology
Getreidemarkt 9 E 308
1060 Vienna
Austria

Mark W. Tibbitt
David H. Koch Institute for Integrative Cancer Research
Massachusetts Institute of Technology
500 Main Street
Cambridge, MA 02139
USA

Vinicius Tribuzi
Universidade de São Paulo
Departamento de Física e Ciência dos Materiais
Instituto de Física de São Carlos
Av. Trabalhador Sãocarlense
400 Centro
13566-590 São Carlos, SP
Brazil

Eric W. Van Stryland
University of Central Florida
CREOL, The College of Optics and Photonics
4304 Scorpius Street
Orlando, FL 32816-2700
USA

Henry E. Williams
NanoSpective
12565 Research Parkway
Suite 390
Orlando, FL 32826
USA

Mei-Ling Zheng
Chinese Academy of Sciences
Technical Institute of Physics and Chemistry
Laboratory of Organic NanoPhotonics and Key Laboratory of Functional Crystals and Laser Technology
No. 29, Zhongguancun East Road
Beijing 100190
PR China

Albertas Žukauskas
Vilnius University
Laser Research Center
Department of Quantum Electronics
Saulėtekio Ave. 10
10223 Vilnius
Lithuania

Foreword

Nanotechnology has great potential in manufacturing, manipulating, observing, and functioning of materials, and understanding the physics and chemistry at the nanoscale, but it is limited to surface science and technology in two dimensions, with use of, for example, e-beam, ion beam, light beam, and atomic force through the processes of illumination, deposition, etching, and modification of the surface. Available nanotechnological devices such as nanoelectronic circuits, high-density data storage, liquid-crystal displays, and micro and nanoelectromechanical systems (MEMSs and NEMSs) are all basically of two dimensions. However, future nanotechnology, in particular nanomanufacturing, should be made in three dimensions similar to the existing ordinary machinery. In the book *Engines of Creation: The Coming Era of Nanotechnology* (Doubleday, 1986), the author Eric Drexler discusses future nanotechnologies such as molecular assemblers, atom-by-atom building, and self-replicating machines. In the novel *Prey* (Harper Collins, 2002), Michael Crichton describes a machine that makes nanobots from bacteria, which turn into a nano-swarm attacking humans. The nanomachines appearing in those books are all three-dimensional machines.

I had been frustrated by the fact that there was no method available or being developed (except in the novels) for three-dimensional nanotechnology, and in 1997 we published a paper in *Opt. Lett.* on the method of laser fabrication of three-dimensional plastic micro/nano structures with use of two-photon polymerization. The method was applied to make a movable micromachine (Kawata *et al.*, *Nature*, 2001). This work fortunately interested and stimulated scientists and engineers who wanted three-dimensional nanotechnology to be used in a variety of fields in science, industry, and medicine.

This book thoroughly reviews the principle, materials, instrumentation, and applications of two-photon and multiphoton polymerization with the latest developments. The key device of two-photon nanostructuring including two-photon polymerization is a mode-locked femtosecond laser. In addition to mode-locked Ti: Sapphire lasers, fiber lasers are also used. The laser power should be high enough (as high as 1 kW) at peak to produce multiphoton processes in the medium, while the pulse width should be short enough (shorter than a picosecond) to avoid any thermal effect in the medium. The wavelength of the excitation of laser should be in the transparency (no absorption) range of materials for single-photon process. Plastic, glass, biocells, and photoresists are nearly transparent in the near-infrared range but they all absorb in the UV. Objective lenses with high numerical aperture should be used for exciting the nonlinear absorption process (multiphoton process) only in the confined volume at the focus. Two-photon absorption has been also applied to two-photon photoisomerization of photochromic molecules, two-photon photorefraction of ferroelectric crystals, and two-photon photoreduction of metals for the purposes of three-dimensional nanostructuring.

One of the remaining practical issues for future studies would be the limit in the total size and time to manufacture large-scale structures with fine details with the use of a focused beam to draw the structures. There have been some reports on the mass-production of a number of identical structures in parallel by focusing multiple spots with a microlens array (Kato *et al.*, *Appl. Phys. Lett.*, 2004) and with holographic illumination (Shoji and Kawata, *Appl. Phys. Lett.*, 2000), although still higher peak power is required for the pulses laser, and the design of the structure is not arbitrary. Two-photon absorption cross-section of available materials (dye molecules) is still not large enough in comparison to that of their single-photon counterpart. Hence, further development of materials is expected. The most important issue is the lack of applications. We need a killer application for three-dimensional nanostructuring. Unlike the recent enthusiasm for 3D printing with lower resolution but on a large scale, two- and multiphoton nanostructuring has not found a big market yet. The other way of 3D nanostructuring would be based on the bottom-up self-assembly or self-growth approach. Self-assembly and self-growth can result in a large structure in a short time at a low cost. We recently published a work on the self-growth of fractal metamaterials of silver with 20 nm resolution with plasmonic top-down illumination (Takeyasu *et al.*, *APL Photonics*, 2016). Such a combination of top-down and bottom-up methods may open up a number of nanostructuring applications.

Osaka, Japan
June 2016

Satoshi Kawata

Introduction

When Maria Göppert investigated the possibility of two-photon absorption in her PhD thesis in 1930, this phenomenon was considered a purely academic effort. It took many decades until lasers became available, which facilitated the experimental assessment of two-photon-activated processes in 1961 by Kaiser and Garrett. From this point on, it was still a long journey to the practical use of multiphoton processes. Again, it was the development of easy-to-use and reliable laser systems, which helped to initiate many developments in this field.

Due to the nonlinearity of multiphoton processes, the size of the active zone can be smaller than the diffraction limit. This is certainly of special interest for microscopy, and it is therefore no surprise that biologists were the first to make use of multiphoton processes for 3D imaging. Within a few years, two-photon microscopy has emerged as a standard technique and is now employed for many applications in cell biology.

Two features of multiphoton processes make their use appealing in microscopy: (i) The outstanding feature resolution and (ii) the possibility to arbitrarily position the active focus volume within a 3D construct (e.g., a living cell). These two features are also highly beneficial for 3D structuring, and in 1997 a group around Satoshi Kawata (see also his foreword in this book) was able to show that two-photon-induced photopolymerization can be used for high-resolution lithography. This pioneering work initiated numerous research activities worldwide, and a literature search for the terms "two-photon lithography" or "multiphoton lithography" yields thousands of publications, indicating the academic relevance of the topic in current research. Different research groups have also been referring to this technique as two-photon-absorbed photopolymerization, two-photon laser scanning lithography, multiphoton-excited microfabrication, 3D laser lithography, or simply direct laser writing. In addition, the number of related publications, describing photochemical processes other than polymerization or cross-linking, is growing rapidly.

Multiphoton lithography is also getting increasing attention from the industrial side. Several commercial suppliers offer systems for multiphoton lithography, and the World Technology Award 2015 has acknowledged these commercialization efforts. Nevertheless, there is still a lack of industrial applications, or in other words: There are plenty of opportunities to bring the technology forward. With this book the authors aim to provide an overview of the state of the art in the field and also give indications where the journey is heading toward. One of the key challenges for future applications is to substantially increase throughput and build volume while maintaining the excellent feature resolution of multiphoton lithography. Therefore, several sections of this book are dedicated to design of multiphoton lithography systems and available light sources. Of equal importance are suitable materials, including high-performance photoinitiators developed toward improved multiphoton cross-section, but also monomer systems that fulfil the requirements of the targeted applications (microfluidics, applications in photonics, hydrogels for cell biology and tissue engineering, etc.). A chapter on modeling of polymerization processes helps to gain insight into the chemical mechanisms occurring in multiphoton active materials during exposure.

By inviting authors from Asia, Europe, and the Americas, we aim at providing a comprehensive overview of the global activities in this dynamic field of research. Last but not least we would like to thank all the contributors for their great enthusiasm and commitment to this book project, which was a great pleasure to work on.

Hopefully the readers can enjoy the text and gain new insights and ideas for their own research.

Vienna, July 2016

R. Liska
A. Ovsianikov
J. Stampfl

Part I
Principles of Multiphoton Absorption

Multiphoton Lithography: Techniques, Materials and Applications, First Edition.
Edited by Jürgen Stampfl, Robert Liska, and Aleksandr Ovsianikov.

1
Rapid Laser Optical Printing in 3D at a Nanoscale

Albertas Žukauskas, Mangirdas Malinauskas, Gediminas Seniutinas, and Saulius Juodkazis

1.1
Introduction

Three-dimensional (3D) laser structuring of materials is widely used in photopolymer prototyping applications including micro-optical elements, parts of optically actuated micromachines in microfluidics, scaffolds for cell growth, templates for plasmonic and metamaterials, and photonic crystals (PhCs) [1–7]. The technique originated from nonlinear microscopy, providing 3D confined recording inside UV-sensitive polymers [8]. Now it is emerging as the most precise and true 3D printing technology in both scientific and industrial fields [9–12]. In order to achieve the resolution of structuring required for PhC structures operational at the visible spectral range, the feature sizes of 3D structures should be smaller than ~100 nm in all cross sections. The surface quality has to be even higher in terms of surface roughness, and in some applications the volume fraction of polymer has to be only 30% in a 3D PhC, which are very demanding requirements [13–15]. For this aim, the optical means of light beam delivery as well as the material's response should be precisely engineered to control the resolution, but without compromising the fabrication throughput in cubic micrometers per hour (in three dimensions) or square micrometers per hour(in two dimensions). The mask writing by electron-beam lithography (EBL) has a throughput of $\sim 10^5 \mu m^2 h^{-1}$ for a current 22-nm node in microelectronics [16].

The most popular materials in the case of 3D laser photopolymerization are acrylate- and epoxy-based resins, which were developed decades ago before the era of tabletop femtosecond lasers for one-photon stereolithography. The photopolymers are photosensitized for the wavelength of an eximer laser emitting at 308 nm or the i-line of a Hg light source at 360 nm. The photosensitization and initiation of nonlinear photopolymerization are fundamentally different from the one-photon processes in the case of ultrashort laser pulses at longer wavelengths. The excitation of the electronic subsystem occurs faster than the absorption of energy into ionic subsystem proceeding via electron–ion equilibration, recombination, and thermal diffusion in the case of ultrashort sub-picosecond laser pulses.

Multiphoton Lithography: Techniques, Materials and Applications, First Edition.
Edited by Jürgen Stampfl, Robert Liska, and Aleksandr Ovsianikov.

Since thermal processes are very efficient in chemical modifications of materials, by tuning photoexcitation with controlled thermal activation, the processing of materials is done by acquiring new functionalities when femtosecond laser pulses with repetition rates of tens of megahertz are used [17–24].

In this chapter, we demonstrate how optical and thermal resist exposure protocols at high scanning speeds can be used to ensure high resolution as well as a high throughput of fabrication. Discussion is focused on 3D microfabrication using two of the most popular resists, SU8 and SZ2080. We show that avalanche ionization plays an important part in photopolymerization at tight focusing conditions even though a seeding stage of bond-breaking and radical generation might occur via a nonlinear two-photon absorption (TPA). The role of photoinitiators in the achievable resolution in 3D structuring is revealed based on two-photon nonlinearities [25]. In resists photosensitized by TPA, there is a wider processing window when 3D structures of high quality and fidelity can be prototyped. Thermal control of conditions at the focus are of paramount importance and are explicitly discussed.

1.2
3D (Nano)polymerization: Linear Properties

As technology evolves, the increasing complexity of devices requires 3D packaging in electronics, and now in fluidics, with integrated electrical, optical, and chemical functionalities. Another trend is the reduction in size/volume with an increase in the operational speed and detection sensitivity. These trends are leading modern technology, currently still not achievable, to a molecular level of integration, precision, and 3D control that we find in nature: DNA code from nanoscale drives the production of a 3D mass.

The analysis provided here is focused on the 3D resists SU8 and SZ2080 suitable for the fabrication of 3D nano/microstructures.[1)] Figure 1.1 illustrates the flexibility of usage for these resists, which can be spin-coated with a controlled thickness or drop-cast on complex substrates such as black-Si [26] for mask projection and DLW exposure. This shows that resists that have properties comparable to those of glass (a transparent dielectric) can be formed into a 3D shape in the mesoscale spanning from 20 nm [27] to 1 mm reaching nanoscale precision when required.

Polymerization is usually achieved by the addition of photoinitiators up to a few weight percent (wt%) to absorb at the wavelength of exposure and to promote polymerization, which occurs via opening of chemical bonds (formation of radicals) and subsequent crosslinking. In SU8, each crosslinking involves the release of molecules, which opens the bonds and promotes further polymerization in as in a chain reaction. DLW with ultrashort laser pulses at $\lambda_l = 800$ and 1030–1060

1) Here, “nano” addresses the targeted dimension but not limited to the exact level of spatial dimensions.

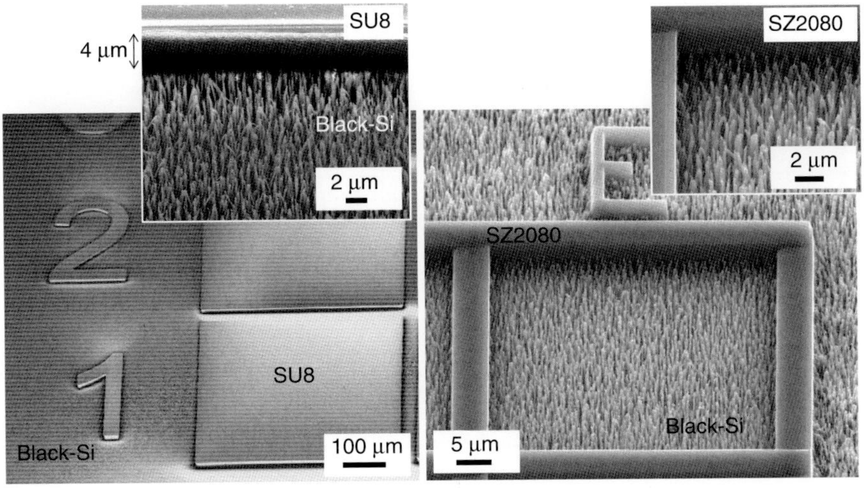

(a) (b)

Figure 1.1 Popular 3D resists SU8 and SZ2080 exposed to (a) mask projection and (b) direct laser writing. The substrate is black-Si needles with ~2.5 μm length and aspect ratio ~10. Laser writing was done with 1030 nm/280 fs pulses with an objective lens of numerical aperture (NA) = 1.4. The resist was spin-coated in (a) and drop-cast in (b).

nm is mostly used. At these wavelengths, TPA of photoinitiators which are optimized for absorption at $\lambda_a = 308$ nm (excimer) and 360 nm (i-line of Hg lamp) is not efficient. For example, if a photoinitiator absorbs at $\lambda_a = 400$ nm, then for the most efficient TPA the excitation wavelength should be $\lambda_a/0.7 = 571$ nm rather than the usually considered $\lambda_a/0.5 = 800$ nm [28]; hence, visible (rather near-IR) wavelengths of ultrashort laser pulses are required for efficient TPA in most of the popular photoresists and resins.

Apart from the TPA being the most efficient nonlinear process, thermal effects have to be considered since DLW polymerization takes place under conditions that are close to the dielectric breakdown of ~ 1 TW cm^{-2} per pulse at the focus. The breakdown is an avalanche ionization-driven process with strong generation of free carriers and localized heating as electrons couple energy to the ions over a few picoseconds (already after sub-picosecond optical pulses).

1.2.1 Photocure and Thermal Cure of Photoresists

Stereolithography started the field of 3D printing with single-photon (direct) absorption on the surface of a liquid resin [29]. Recent advances and the availability of high-quality microscopes and optics allow imaging microscale volumes, and one can now deliver light energy (absorption) and effect 3D printing on a (nano)microscale in a very similar manner as thermal extrusion of larger (millimeter scale) 3D structures [30].

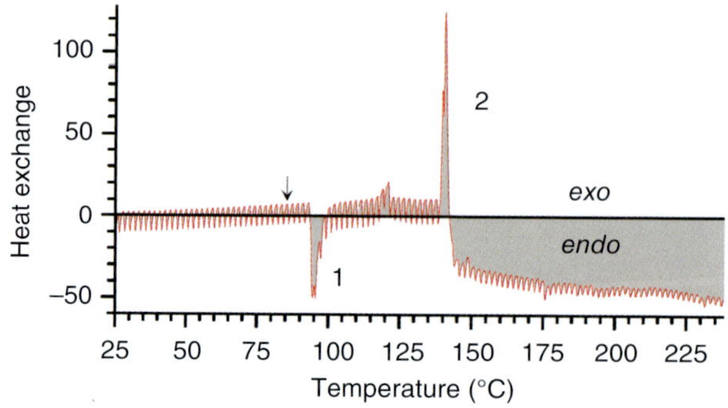

Figure 1.2 Calorimetry measurement of heat exchange during polymerization of SZ2080 carried out simultaneously under Raman scattering detection [36]. Arrow represents the thermal consequence of a single laser exposure for Raman scattering detection. Peak 1 corresponds to the final drying of the resist, and the exothermic peak 2 at ∼140 °C is due to polymerization.

Understanding light-matter interaction under tight focusing, high irradiance/intensity, radiative and thermal energy transfer in sub-wavelength volumes, and photochemistry will help create a technology that either defines high-resolution fabrication, or is optimized for high throughput, or both at the same time. DLW and photocuring of SU8 without any usual post-exposure annealing required for chemical reactions of crosslinking were shown to deliver sub-wavelength resolution of 3D polymerization [31]. There is no disagreement that both direct heating and thermal curing contribute to the polymerization, although, currently, the significance of each is being debated [32–34].

Figure 1.2 shows that the sol–gel organic–inorganic SZ2080 resist [35] is polymerized at 140 °C; a sharp exothermic peak is followed by an endothermic region that indicates the removal of water, which is a result of gelification [36]. Local heating at the relatively low temperature (140°C) provides efficient polymerization. How the optical energy is delivered and absorbed depends on the optical properties of the focal region, which are discussed later. Scanning with a hot spot is the method to write 3D polymerized structures.

1.2.2
Tight Light Focusing

Very tight focusing is used in DLW using oil-immersion objective lenses with numerical apertures $NA = n \sin \alpha > 1$, where $n = 1.515$ is the refractive index of immersion oil at vis–IR wavelengths, and α is the half-angle of the focusing cone. This is required for high-resolution fabrication. Modification of the material follows energy delivery (the optical penetration depth) or/and the thermal penetration depth. Both are determined by light delivery and material properties, which change during excitation.

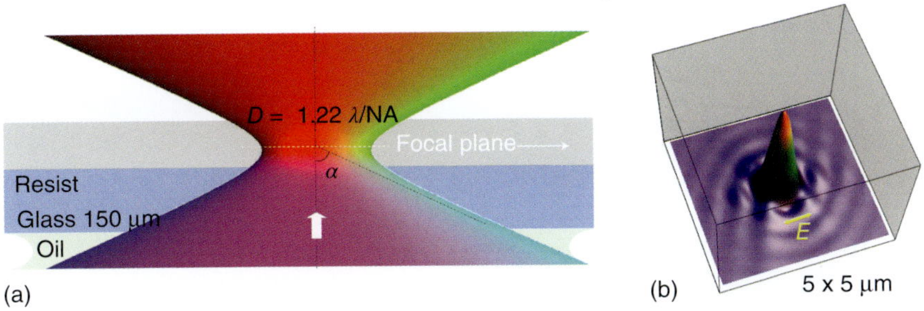

Figure 1.3 Schematic presentation of tight focusing. (a) Focusing of a Gaussian beam with NA = 1.4 (or $\alpha = 67.5°$ angle) inside the resist. (b) Focal intensity distribution of linearly polarized (E-field) beam of NA = 1.4 lens without spherical aberrations. Calculations were done using the scalar Debye theory.

Figure 1.3 shows tight focusing, which is widely used in 3D laser polymerization, with objective lenses with numerical aperture NA = 1.2–1.4. Close matching of indices allows avoiding spherical aberration, and light intensity is delivered to the focal volume without strong distortions. Because of tight focusing, polarization effects become pronounced; for example, the focal spot becomes slightly elongated along the direction of linear polarization (Figure 1.3b).

Vectorial Debye's theory predicts that focusing using a high numerical aperture lens (NA > 0.7) breaks the cylindrical symmetry of the focal electric field distribution (assuming that input beam is linearly polarized along the x-direction) and a corresponding elongation takes place with alteration of the E_y and E_z (longitudinal) components [37]. The electric field at the focal point is expressed as [38]

$$\mathbf{E}(r,\psi,z) = \frac{\pi i}{\lambda}\{[I_0 + \cos(2\psi)I_2]\mathbf{i} + \sin\,(2\psi)I_2\mathbf{j} + 2i\cos(\psi)I_1\mathbf{k}\}, \tag{1.1}$$

where **i**, **j**, and **k** are the unit vectors in the x-, y-, and z-directions, respectively, and the variables r, ψ, and z are the cylindrical coordinates of an observation point. The integrals I_0, I_1, and I_2 are cumbersome to handle but can be straightforwardly calculated by integration over the focusing angle θ for a chosen apodization function [37–39]. So, light is depolarized at the focal region with a preferentially elongated focal spot along the direction of the linear polarization of light. Figure 1.3b shows a numerical simulation for the sin-apodization function $P(\theta) = \sqrt{\cos(\theta)}$, where θ is the focusing cone covering $0-\alpha$ with NA $= n\sin(\alpha)$.

For a circularly polarized beam, scalar focusing is used to calculate the focal volume [37, 38]:

$$E_{\mathrm{sc}}(r,\psi,z) = \frac{2\pi i}{\lambda}\int_0^{a} P(\theta)\sin(\theta)J_0(kr\sin(\theta))e^{-ikz\cos(\theta)}d\theta. \tag{1.2}$$

Figure 1.3b shows spherical-aberration-free focal intensity distribution calculated by the scalar Debye expression with the terms accounting for Fresnel coefficients

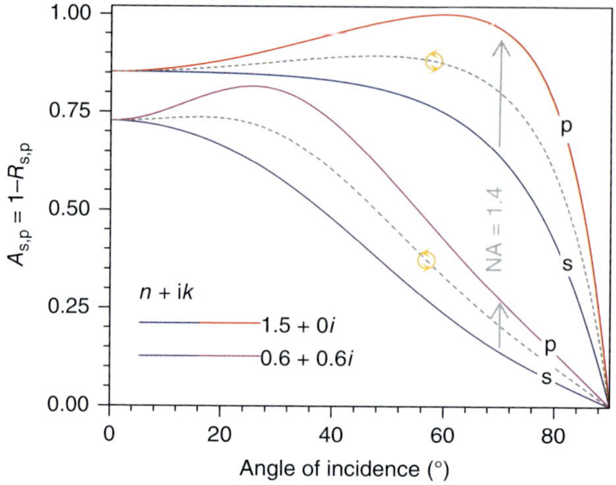

Figure 1.4 Dependence of the Fresnel absorption coefficient $A_{s,p} = 1 - R_{s,p}$ on the angle of incidence for s- and p-polarizations, respectively, at different excitation levels (see Section 1.2.3 for details). The arrows mark the 67.5° angle for focusing with NA = 1.4 objective lens inside a material of refractive index 1.5. The dashed lines are for unpolarized and circularly polarized light.

and the aberration function which depends on refractive index mismatch and focal depth (Figure 1.4) (for details see Refs [40, 41]).

1.2.3
Optical Properties at High Excitation: From Solid to Plasma

The SZ2080 resist has a refractive index close to that of glass $n + ik = 1.5 + 0i$ at the laser fabrication wavelength $\lambda = 1030$ nm. Fresnel reflection coefficients for intensity are [42][2)]

$$R_s(\theta) = \frac{(a(\theta) - \cos(\theta))^2 + b(\theta)^2}{(a(\theta) + \cos(\theta))^2 + b(\theta)^2}, \tag{1.3}$$

$$R_p(\theta) = R_s(\theta)\frac{(a(\theta) - \sin(\theta)\tan(\theta))^2 + b(\theta)^2}{(a(\theta) + \sin(\theta)\tan(\theta))^2 + b(\theta)^2}, \tag{1.4}$$

where

$$a(\theta) = \frac{1}{2}\left(\sqrt{(n^2 - k^2 - \sin(\theta)^2)^2 + 4n^2k^2} + (n^2 - k^2 - \sin(\theta)^2)\right), \tag{1.5}$$

$$b(\theta) = \frac{1}{2}\left(\sqrt{(n^2 - k^2 - \sin(\theta)^2)^2 + 4n^2k^2} - (n^2 - k^2 - \sin(\theta)^2)\right). \tag{1.6}$$

2) The corrected expression are shown here, courtesy Prof. Andrei Rode.

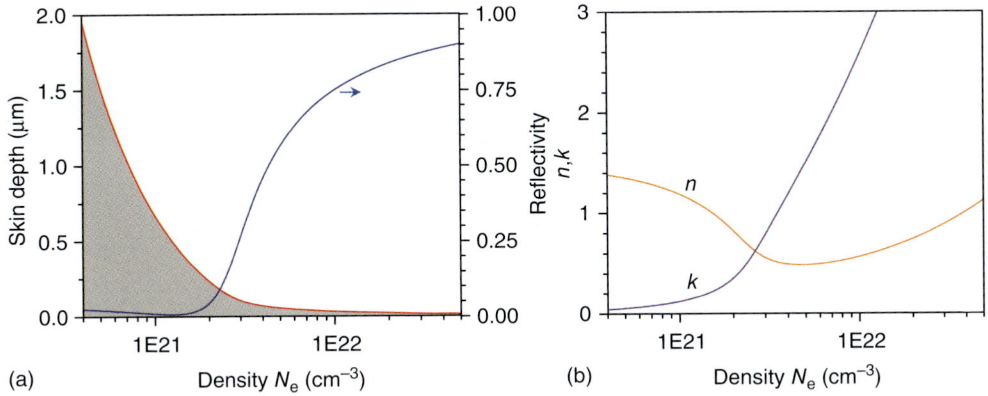

Figure 1.5 (a) Skin depth $\delta = c/(2\omega \mathrm{Im}(\sqrt{\varepsilon_D})) \equiv \lambda/(4\pi \mathrm{Im}(\sqrt{\varepsilon_D}))$ (for intensity) and reflection coefficient $R = \frac{(\mathrm{Re}(\sqrt{\varepsilon_D})-1)^2+\mathrm{Im}(\sqrt{\varepsilon_D})^2}{(\mathrm{Re}(\sqrt{\varepsilon_D})+1)^2+\mathrm{Im}(\sqrt{\varepsilon_D})^2}$ for $\lambda = 1030$ nm wavelength at different plasma densities N_e; the critical plasma density is $N_{cr}(\lambda) = 1.05 \times 10^{21}$ cm^{-3}. (b) Real (Re) and imaginary (Im) parts of refractive index calculated by formulae given in Section 1.2.3.

Dielectric permittivity of excited material can be considered as that of unexcited ε_r with the Drude contribution [43]:

$$\varepsilon_{\mathrm{D}}(\lambda) = \varepsilon_{\mathrm{r}} - \frac{\omega_p^2}{\omega(\omega + i/\tau_{\mathrm{d}})}, \tag{1.7}$$

where the plasma cyclic frequency $\omega_{\mathrm{p}} = \sqrt{N_e e^2/(m\varepsilon_0)}$ with the electron density N_e; the electron charge and mass are, respectively, e, m; $\tau_{\mathrm{d}} = 1.6$ fs is the electron–ion impulse relaxation time similar to that of glass [44, 45]; ε_0 is the dielectric constant; c is the speed of light; and $\varepsilon_{\mathrm{r}} = n^2 - k^2$ is the real part of dielectric permittivity of the material under strong excitation and presence of free carrier plasma, which is defined by refractive index $n + ik$.

Figure 1.5 shows the absorption skin depth, reflectivity, and refractive index evolution of the resist, which is initially defined by $n + ik = 1.5 + 0i$ for $\lambda = 1030$ nm excitation wavelength at different free carrier concentrations (corresponding to different excitation levels).

The above-mentioned formula allows us to retrieve the real and imaginary parts of the dielectric function (refractive index) from the reflectivity transients in pump-two-probes experiment. Backreflection has to be measured at two different angles of incidence (two probe beams introduced at different radial positions at the entrance of the pupil of the objective lens). Simultaneous reflectivity at two angles is required to obtain both the real and imaginary parts of the dielectric function. Such an experiment has never been accomplished and could bring out the temporal evolution of the material parameters as they change upon excitation (Figure 1.5). Pump–probe imaging of the interaction regions is challenging to obtain down to spatial resolution comparable with the wavelength, as was attempted in glasses [44, 46–48].

1.2.4 Heat Accumulation

The temperature diffusion for 3D localized energy deposition is considered next, assuming a Gaussian laser beam intensity distribution at the focal spot of diameter d_f:

$$I = I_0 \exp\left(-2\frac{r^2}{(d_f/2)^2}\right), \tag{1.8}$$

where I_0 is the amplitude, and the intensity is I_0/e^2 at the point of $r = d_f/2$.

During irradiation with an ultrashort laser pulse, electrons can be heated to temperatures up to $T_e = 1000\,\text{K}$, which is close to the dielectric breakdown conditions $\sim 1\ \text{TW cm}^{-2}$. From Wien's law, the maximum emission according to the black body radiation law is at $\lambda_{max} = 2.9 \times 10^9\ \text{nm K}$; hence $\lambda_{max}(1000\ \text{K}) = 2.9\ \ \mu\text{m}$. This falls in the absorption band of most of organic compounds including photoresists. The thermodynamic limit of radiative energy transfer efficiency for the emission of hot electrons at $T_e = 1000\ \text{K}$ with absorbers at $T_c = 300\ \text{K}$ (cold atoms at room temperature) would have $\eta = (1 - T_c^4/T_e^4) \simeq 99.2\%$ [49]. Such a process could cause reduction in resolution if absorption is taking place outside the focal volume. In usual laser fabrication, super-resolution (sub-diffraction-sized features) is achieved, indicating that radiative processes are not efficient for increasing absorbed energy in the focal region.

The focal spot diameter (at $1/e^2$) is equal to $d_f = 1.22\lambda/\text{NA} \simeq 900\,\text{nm}$, assuming Gaussian intensity profile for simplicity. The actual distribution of the temperature increase due to a 1-fs pulse depends on the absorption energy. Considering the case of linear absorption valid for free carrier absorption at the high-irradiance condition results in the temperature profile closely following the spatial distribution of the the excitation (Eq. 1.8), hence [50]

$$T = T_{max} \exp\left(-8\frac{r^2}{d_f^2}\right), \tag{1.9}$$

where T_{max} is the temperature maximum at the center of the irradiation spot. Solution of a 3D heat diffusion equation of spherical symmetry can be found from

$$\frac{\partial T}{\partial t} = D\left(\frac{\partial^2 T}{\partial r^2} + \frac{2\partial T}{r\partial r}\right), \tag{1.10}$$

where D is the temperature diffusion coefficient, t is the time, and

$$T(r,t) = T_{max}\left(\frac{d_f^2}{d_f^2 + 32Dt}\right)^{3/2} \exp\left(\frac{8r^2}{d_f^2 + 32Dt}\right), \tag{1.11}$$

for the initial temperature deposition given by Eq. (1.9). Between consecutive pulses, the temperature elevation from a previous pulse at the center decreases by a factor

$$a = \left(\frac{1}{1 + 32DR_{rep}^{-1}d_{\mathrm{f}}^{-2}} \right)^{3/2}. \tag{1.12}$$

It is to be noted that the exponent 3/2 is pertinent for the 3D heat diffusion and defines the temperature decrease at the center while the size of heated region is spreading as $r \propto t^{1/2}$ due to an isotropic (or 1D) heat diffusion [51]. After n pulses, the contribution of a single pulse is decreased by a factor

$$a_n = \left(\frac{1}{1 + 32nDR_{rep}^{-1}d_{\mathrm{f}}^{-2}} \right)^{3/2}. \tag{1.13}$$

The total accumulated temperature reaches the maximum at the end of the last pulse of the N pulse train:

$$T_N = T_1 \cdot \sum_{n=0}^{N-1} a_n, \tag{1.14}$$

where T_1 is the temperature jump after the first pulse. In general, $a_n \neq a^n$ and only if $32R_{\mathrm{rep}}^{-1} \ll t_{th}$ where the cooling time $t_{\mathrm{th}} = d_{\mathrm{f}}^2/D$, $a_n \approx a^2$. Estimating the temperature accumulation for the above-mentioned Parameters, one gets $N = N_{\mathrm{spot}} = 1.8 \times 10^3$, $D = 10^{-3}\mathrm{cm^2\,s^{-1}}$, $d_{\mathrm{f}} = 900$ nm, and $R_{\mathrm{rep}} = 2 \times 10^5\ \mathrm{s^{-1}}$. The summation results in $T_N \approx 1.03T_1$. This can be a negligibly small value for the studied case. On the contrary, for a common femtosecond oscillator operating at the 82 MHz regime, the heat accumulation factor can reach a value of $F \approx 42$. Thus, employing amplified laser systems and oscillators covers two different heat accumulation estimations, yielding a the resultant temperature increase by more than an order of magnitude difference. Furthermore, at tight focusing, the beam of the complex shape is absorbed, resulting in complex cooling evolution in time. In other words, in the classical case of material ablation from the surface, it is just a thin layer of a relatively large area that is heated, but when absorption and the heated volume are elongated along the beam in a cylindrical shape (voxel) distribution, the 1D cooling and 3D cooling become the two limiting cases of the real temperature evolutions at the focus. There is scope for further theoretical studies matching the exact case of applied experimental conditions.

In actual 3D laser fabrication at tight focusing, the 3D temperature diffusivity after a point-like energy deposition is affected by the 3D spot cross sections, which are polarization- and focusing-dependent. At tight focusing, the vectorial Debye theory (Section 1.2.2) has to be used to estimate two lateral cross sections of ellipsoidal focal spot; one would find $W_{\mathrm{l}} = 481$ nm and $W_{\mathrm{s}} = 332$ nm for long and short cross sections, respectively, calculated for the full width at half-maximum (FWHM), which is smaller than the cross section at the $1/e^2$ level by a factor $2\sqrt{\ln\ 2} = 1.665$ (for Gaussian).

Thermal accumulation effects account for the laser-ablated volume using picosecond and femtosecond pulses semiquantitatively [51]. Here, we adopt the same treatment as for polymerization. The dwell time of each pulse at the focal spot of diameter d_f equals $t_{dw} = d_f/v_{scan} \simeq 9$ ms for a scan speed $v_s = 100\ \ \mu m\,s^{-1}$. Thus, the number of pulses per spot at the repetition rate of the typically used industrial fabrication laser Pharos (Light Conversion Ltd.) $R_{rep} = 2 \times 10^5$ pulses s^{-1} is equal to $N_{spot} = t_{dw} \times R_{rep} \simeq 1.8 \times 10^3$ pulses. The heat diffusion coefficient for a cold resist is similar to that of silica, $D_{diff} = 10^{-3}$ $cm^2\,s^{-1}$ [45, 52]. Thus, the cooling time for the heated area, $t_{th} = d_f^2/D$, has to be compared with the time gap between subsequent pulses arriving at 5 μs. The heat transfer to the surrounding cold material between the pulses results in the average temperature drop at the arrival of the next pulse, and the temperature accumulation can be explicitly calculated for the N pulses as [51]

$$T_N = T_1(1 + a + a^2 + \ldots + a^N) = T_1 \frac{1 - a^N}{1 - a}, \tag{1.15}$$

where $a = \sqrt{\frac{t_{th}}{t_{th}+1/R_{rep}}}$; as $a \to 1$, the temperature accumulation becomes larger. Assuming a temperature jump $T_1 = constant$, for one pulse regardless of polarization heat accumulation is $T_N \simeq 4.67T_1$ with $a = 0.786$. For SZ2080, it means that if one pulse can increase the local temperature by $\Delta T_1 = 140°/4.67 = 30\,°C$, this would be sufficient for thermal polymerization during continuous scan (Figures 1.2 and 1.6) solely due to thermal accumulation. Figure 1.6 shows how temperature accumulation is affected at different laser repetition rates. The main factor causing accumulation is the low temperature diffusivity of the resist. Industrial-scale fabrication is carried out at ~200 kHz repetition rate; after the first 15 pulses, thermal conditions become stable (Figure 1.6). For a repetition

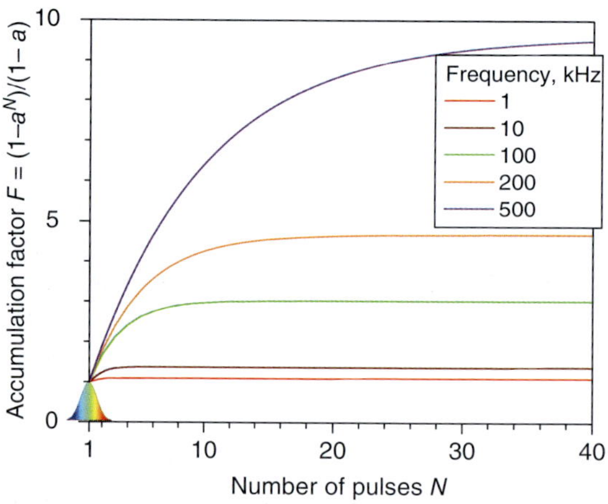

Figure 1.6 Visualization of temperature accumulation after N laser pulses $T_N = T_1 \times F$, where the factor $F = \frac{1-a^N}{1-a}$ (Eq. 1.15) at different repetition rates (or the number of pulses per focal spot of diameter $d_f = 0.9\ \mu m$) at a fixed scan speed $v_s = 100\ \mu m$ and for temperature diffusion coefficient $D = 10^{-3}$ $cm^2\,s^{-1}$; $a = \sqrt{\frac{t_{th}}{t_{th}+1/R_{rep}}}$ (see Section 1.2.4). The Gaussian chirped pulse marker depicts a T_1 temperature jump due to one (first) pulse.

rate of $R_{rep} = 82$ MHz (typical for a femtosecond oscillator), $F \simeq 1300$ at the same conditions as those estimated in Figure 2.4. This shows how efficient the thermal polymerization can be at multi-megahertz repetition rates.

The maximum temperature jump due to a single pulse can be estimated from the absorbed energy as for the ablation studies [53]:

$$T_1^{\max} = \frac{AF_{\mathrm{p}}}{l_{\mathrm{s}} C_{\mathrm{L}} n_{\mathrm{a}}}, \quad (1.16)$$

where A is the Fresnel absorption coefficient (Section 1.2.3); $F_{\mathrm{p}} = E_{\mathrm{p}}/S$ is the fluence per pulse energy E_{p} over a surface area S; $l_{\mathrm{s}} = c/(\omega k) \equiv \delta$ is the absorption depth, which is the skin depth in plasma at high excitation (Section 1.2.3); $n_{\mathrm{a}} = \rho N_{\mathrm{A}}/M$ is the atomic density of the material with mass density ρ and molar mass M; N_{A} is the Avogadro number; and C_{L} [J/K] is the lattice specific heat.

1.3 3D (Nano)polymerization: Nonlinear Properties

The most probable optically nonlinear effects occurring at the lowest irradiance/intensity are TPA and refraction change via the Kerr effect. The direct measurements of these nonlinear parameters are carried out by the Z-scan method [54] using ultrashort laser pulses; for example, the best fit of an open aperture Z-scan data for the Gaussian spatial distribution of a laser pulse provides a measure of the TPA cross section. It is noteworthy that Z-scan measurements at intensities approaching $\sim 1\ \mathrm{TW\,cm^{-2}}$ become invalid due to strong Raman scattering. Other methods such as transmission measurements and photoluminescence usually overestimate the actual TPA cross section [55, 56]. The intensity-dependent TPA coefficient β and nonlinear refractive index n_2 have been measured for many crystalline and amorphous materials [25, 57]. The same characteristic dependencies are also valid for molecular solutions of the highly efficient TPA absorber MBAPB [58], which is similar to the photoinitiators used in resists [59, 60]. Optical nonlinearities of dyes and photoinitiators measured by femtosecond Z-scan corroborated the expected spectral dependencies established in the case of inorganic solid-state materials [55, 56]. Based on these premises, wavelength of excitation for the most efficient TPA was determined for 3D polymerization of SZ2080 with different photoinitiators [28], whiich is discussed later.

1.3.1 Strongest Optical Nonlinearities

The nonlinear refractive coefficient n_2 ($n = n_0 + n_2 I$) is given by Boyd [25, 57]

$$n_2[\mathrm{cm^2/W}] = K \frac{\hbar c \sqrt{E_{\mathrm{p}}}}{2 n_0^2 E_{\mathrm{g}}^4} G_2(\hbar\omega/E_{\mathrm{g}}), \quad (1.17)$$

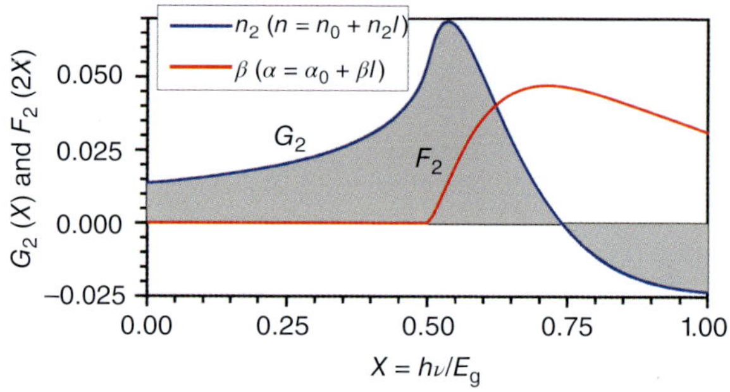

Figure 1.7 Functional dependencies of the TPA coefficient β and the nonlinear refractive index n_2 on the normalized photon energy determined by the polynomial functions G_2 and F_2, respectively [25]. The bandgap energy is E_g (corresponding to the direct absorption), $h\nu$ is the photon energy at the irradiation wavelength.

where $E_p = 21$ eV is an empirical constant and G_2 is the universal function plotted in Figure 1.7. The nonlinear TPA coefficient $\beta(\alpha = \alpha_0 + \beta I)$ is given by Boyd [25, 57]

$$\beta[\mathrm{cm/W}] = K\frac{\sqrt{E_p}}{n_0^2 E_g^3} F_2(2\hbar\omega/E_g), \tag{1.18}$$

where F_2 is the function plotted in Figure 1.7. The functions G and F are defined by the following polynomial expressions: $G_2(x) = (2 + 6x - 3x^2 - x^3 - 3x^4/4 - 3x^5/4 + 2(1 - 2x)^{3/2}\Theta(1 - 2x))/(64x^4)$, where the Heaviside function $\Theta(y) = 0$ for $y < 0$ and $\Theta(y) = 1$ for $y \geq 0$; $F_2(2x) = (2x - 1)^{3/2}/(2x)^5$ for $2x > 1$. Figure 1.7 shows the functional dependencies for β and n_2. The abscissa value $x = 1$ corresponds to the one-photon (or fundamental) absorption when the photon energy is equal to the bandgap E_g. For our analysis, we use the wavelength corresponding to the absorption maximum, λ_{abs}^{max}, as the wavelength of the most efficient 3D structuring [28]. This wavelength defines the central wavelength of overlap between the absorption and emission spectra of the photoinitiator or the resist matrix.

Justification of this wavelength choice for $x = 1$ condition was tested on Z-scan data of the dye MBAPB [60]. The $\lambda_{abs}^{max} = 470$ nm was considered corresponding to $x = 1$, while the strongest TPA was at 680–710 nm, which is close to the expected location of the most efficient TPA [25, 57] (see Figure 1.7) at 671 nm (the $x = 0.7$ condition).

In terms of 3D laser structuring, the spectral dependence of the functions G and F would suggest that the irradiation wavelength for the most efficient photoinitiation of absorption should be carefully chosen. The optimal case for two-photon structuring is at a photon energy of $0.7E_g$ when β is the largest and n_2 is the smallest. One can also recognize an expected tendency; that is, closer to the fundamental absorption $x = 1$, the n_2 values become negative, as would be expected, because of free carrier absorption. Free carriers cause defocusing and alter light beam delivery to the focus. Free carrier effects are accounted for by including Drude contribution to the dielectric permittivity, as discussed in Section 1.2.3. In terms of linear optical properties, the presence of free carriers increases absorption (imaginary part of refractive index in not 0) and reduces the light penetration depth, both of which contribute to light localization and absorption at the focus.

1.3.2
Avalanche Versus Multiphoton Excitation

The rates of multiphoton absorption and avalanche multiplication of electrons are estimated for the polymerization at pre-breakdown conditions in photoresists. The number density of electrons n_e created at the end of the pulse by avalanche and multiphoton processes can be obtained from a rate equation [53, 61]:

$$\frac{dn_\mathrm{e}}{dt} = n_\mathrm{e} w_\mathrm{imp} + n_\mathrm{a} w_\mathrm{mpi}, \tag{1.19}$$

where n_a denotes the molecular density, that is, available electron donors, which provide free electrons and free radicals by photocleavage of the chemical bonds in the resist. If the laser exposure intensity is constant during the laser pulse (corresponding to a flat-top intensity distribution in the time domain), and the recombination during the pulse is negligible, then the solution of Eq. (1.19) with the initial condition $n_\mathrm{e}(t = 0) = n_\mathrm{e0}$ and w_imp and w_mpi is straightforward [51]:

$$n_\mathrm{e}(I, \lambda, t) = \left\{ n_\mathrm{e0} + \frac{n_\mathrm{a} w_\mathrm{mpi}}{w_\mathrm{imp}} \left[1 - \exp\left(-w_\mathrm{imp} t\right)\right] \right\} \exp(w_\mathrm{imp} t). \tag{1.20}$$

It is commonly accepted that breakdown of a dielectric occurs when the plasma frequency of excited electrons equals the frequency of laser light [52]. For example, the critical electron density—the breakdown threshold—for 800 nm wavelength or $\omega = 2.35 \times 10^{15}\ \mathrm{s}^{-1}$ frequency is $n_\mathrm{c} = \frac{m_\mathrm{e}\omega^2}{4\pi e^2} = 1.735 \times 10^{21}\ \mathrm{cm}^{-3}$.

Free electrons oscillate in the electromagnetic field of the laser pulse. These electrons gain a net energy by multiple electron–lattice/atom collisions and eventually can be accelerated to reach the excess energy of the ionization potential J_i. Energetic electrons create an avalanche, which has the estimated ionization rate as follows [62]:

$$w_\mathrm{imp} \approx \frac{\varepsilon_\mathrm{osc}}{J_\mathrm{i}} \frac{2\omega^2 v_\mathrm{e-ph}}{(v_\mathrm{e-ph}^2 + \omega^2)}, \tag{1.21}$$

where $v_\mathrm{e-ph}$ and ω are the electron–phonon momentum exchange rate and laser frequency, respectively. Based on the electron–phonon momentum exchange rate, the rate of collisions at the breakdown can be estimated as $v_\mathrm{e-ph} = 6 \times 10^{14}$ $1\ \mathrm{s}^{-1}$ [62]. The oscillation energy of the electron in a scalar form reads [62]

$$\varepsilon_\mathrm{osc}[eV] = (1 + \alpha^2) 9.3 \left(\frac{I}{10^{14}\ [\mathrm{W}/cm^2]} \right) \lambda_{\mu m}^2, \tag{1.22}$$

from which, for 1 TW cm^{-2} linearly polarized ($\alpha = 0$) irradiance at 800 nm wavelength, we get $\varepsilon_{osc} = 30\,\text{meV}$ and twice that value for circularly polarized ($\alpha = 1$) light.

The multiphoton ionization rate, the probability of ionization per atom per second, can be calculated according to [62]

$$w_{mpi} \approx \omega\, n_{ph}^{3/2} \left(\frac{\varepsilon_{osc}}{2J_i} \right)^{n_{ph}}, \tag{1.23}$$

where n_{ph} is the integer part of $(J_i/\hbar\omega + 1)$ and defines the number of photons required for ionization [63]. The electron production rates via nonlinear TPA and avalanche can be estimated for the typical experimental conditions of 3D polymerization. This analysis is valid only for an order-of-magnitude estimation; however, it is insightful to determine the dominant mechanism of ionization, bond-breaking, radical generation, and, hence, crosslinking.

The light intensity threshold of dielectric breakdown in the case of single-pulse irradiation of SZ2080 with Irg.2wt/% by 150 fs pulses at 800 nm is ~ 8.6 TW cm^{-2} (6 nJ per pulse at tight focusing). This intensity is comparable to the ionization threshold for the dielectrics [64].

The multiphoton and impact ionization (avalanche) rates, w_{mpi} and w_{imp}, respectively, are calculated from the formulas given earlier. The wavelength of direct excitation is $\lambda_{abs}^{max} \simeq 390$ nm for SZ2080 with Irg [28], corresponding to the potential $J_i[\text{eV}] = 1.24/\lambda[\mu\text{m}] \simeq 3.18$ eV, at which free carriers are created. These free carriers undergo avalanche multiplication. The dielectric breakdown at which the avalanche becomes a dominant mechanism of excitation was observed at the pulse energy $E_p = 6$ nJ; the corresponding intensity = 8.6 TW cm^{-2}, $w_{mpi} = 0.81$ THz, and $w_{imp} = 91.1$ THz. The avalanche generation of electrons is more efficient than the multiphoton process, which might be required when the electrons seeding the avalanche cannot be provided by defects and Urbach states. However, there are always defects in glasses and polymers, which provide seeding electrons by one-photon absorption, and hence for photopolymers this is expected to be the major electron seeding mechanism.

The typical pulse energy used for recording high-resolution 3D PhC structures with 800 nm/150 fs pulses in SU8 resist [13, 65–67] was ~ 1 nJ or $I_0 = 1.4$ TW cm^{-2} and $w_{mpi} = 0.0035$ THz with $w_{imp} = 14.8$ THz; for 3D structuring of SZ2080, conditions were very similar for same focusing. The avalanche is even more dominant at lower pulse energies. When the pulse energy is considerably higher than the breakdown threshold, the multiphoton rate catches up with the avalanche and material experiences the breakdown at a very early stage of the laser pulse. The rest of pulse interacts with plasma and creates micro-explosion. Table 1.1 shows the comparison of laser fabrication windows with another popular longer wavelength 1030 nm, where many femtosecond fiber lasers are also available. Very narrow spread of pulse energies is available for the fabrication of 3D structures in undoped SZ2080 resist, almost limited by the laser stability. The window is much wider in the resist with a photoinitiator. However, the

Table 1.1 Qualitative comparison [28] of different laser structuring regimes for 1030 nm/300 fs pulses in SZ2080 with different photoinitiators Irg and Bis (1wt%); focusing $NA = 1.4$; and scanning speed 100 μm s^{-1}.

Condition at focus	SZ2080	SZ2080+Irg	SZ2080+Bis
Polymerization threshold, nJ (TW cm^{-2})	14 (12.5)	5 (4.5)	4 (3.6)
Structuring (middle of the range), nJ (TW cm^{-2})	16 (14.4)	7 (6.3)	5 (4.5)
Explosion threshold, nJ (TW cm^{-2})	17 (15.2)	12 (10.8)	12 (10.8)
Uncontrolled burning, nJ (TW cm^{-2})	18 (16.1)	18 (16.1)	17 (15.2)

achievable resolution becomes worse in the case of photoinitiator-doped resist due to stronger absorption, avalanche, and heating at the focus [28].

1.4 Discussion

Light delivery into micrometer-sized focal spots is also possible with dry objective lenses still maintaining a high $NA = 0.9$ focusing. This allows higher scan speeds since there is no oil immersion, which makes the fabrication less practical. Speeds of 10 cm min^{-1} (1.7 mm s^{-1}) on large-scale industrial laser welding lines are easily exceeded for the fabrication on the microscale due to small inertia of the workpiece. Mechanical constraints of fabrication speed are defined by the mass and acceleration of stages at high velocity of scan and direction changes. With the use of light-weight scanning mirrors combined with θ-lenses that maintain the same focal plane height when a beam is scanned, in combination with a collinear stage scan, is a promising direction for fabrication approaching 1 m s^{-1} linear scan speeds and is practical for fabrication over larger areas. This makes even raster scan a productive approach in 3D additive manufacturing on the $resolution \propto throughput$ scaling [16].

Relying on linear and thermal absorption rather than optically nonlinear energy delivery DLW becomes appealing for new applications. As is shown here, thermal accumulation at high laser repetition rate and fast scan can precisely control thermal conditions and heat accumulation at the focus inside the resists. Heat localization in photoresists, glasses, and polymers is facilitated by low-temperature conductivity, while precise energy delivery via light absorption can be tailored using pulse duration and spectral or spatial chirp of ultrashort laser pulses. Reaching temperatures of modification (polymerization, glass transition, etc.) that are in the range of 150–250 °C and localized on scale of few micrometers inside polymers is a promising fabrication tool. When photoinitiators are not

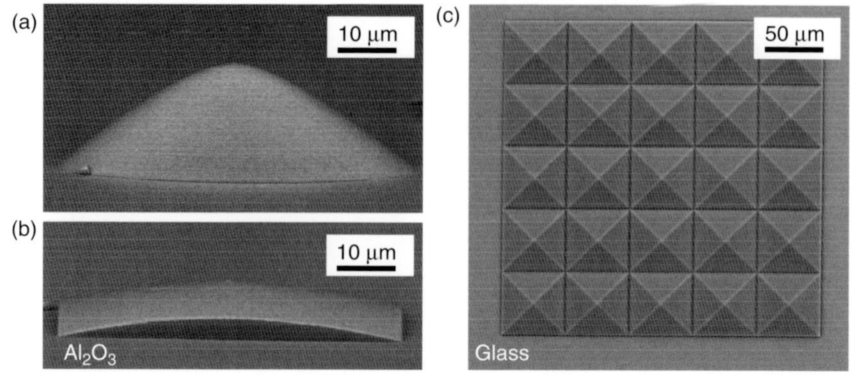

Figure 1.8 Micro-optical elements made out of SZ2080 by DLW with 1030 nm/300 fs pulses without surface post-processing. (a,b) Single 0.1-mm-diameter lenses for light extraction from blue GaN-based LEDs were polymerized on the backside of the sapphire substrate at 0.2 mm/s using a surface definition with 75 nm hatching with subsequent homogenizing exposure after development (total time 9 min per lens). (c) Array of 5×5 lenses made by surface definition at 0.25 mm/s with internal raster exposure at 5 mm/s (array formation time 37 min).

used inside photopolymers such as silicones popular in biomedical applications, a wider range of applications can be opened. Feasibility to tailor thermal curing that is launched by avalanche ionization of undoped pure silicone for its 3D structuring has been demonstrated [68].

Figure 1.8 shows an example of microfabrication that produces micro-optical elements in a single processing step. In the field of micro-optical devices, DLW can be used to make lenses, prisms, gratings, and their hybrids with surface quality, which does not require any post-processing and can be incorporated into microfluidic chips and fabricated on waveguides [69].

1.5 Conclusions and Outlook

Analysis of the light delivery at tight focusing and thermal accumulation presented here shows the importance of avalanche ionization and thermal processes in 3D polymerization by DLW employing high-repetition, femtosecond pulses. By precisely adjusting the temperature at the focal region at a high speed $\gg 1$ mm s^{-1} scanning, high-resolution patterns can be fabricated over areas useful for micro-optical and microfluidic research and practical applications. The proposed nanostructuring approach is not only limited to cross-linkable polymers (resins) but can also be further implemented for pure optical printing of glasses, metals, or even viable proteins [70–73]. It extends dramatically emerging fabrication accuracy of 3D printing by three orders to the realm of nanoscale, offering unmatched, rapid prototyping and additive manufacturing possibilities. Lastly, tightly focused femtosecond pulses enable tunable light–matter interaction level at ultralocalized dimensions. By varying the exposure conditions, one can achieve control in the fourth dimension based on mechanical shape memory [74] or degree of chemical modification [75], thus "unlocking" the possibility of 4D additive manufacturing.

Acknowledgments

We acknowledge the discussions with Dr. Kenji Kamada on *Z*-scan. This work was partially supported via the ARC Discovery 130101205 project. AŽ and MM are thankful to EC's Seventh Framework Programme (LASERLAB-EUROPE, Grant Agreement No. 228334, OPTOBIO) for the financial support and to the multifunctional laser facility "NAGLIS" for providing laboratory access. SJ acknowledges start-up funding of the Nano Lab by Swinburne University and Dr. Gintas Šlekys for joint research project with Window-on-Photonics, Altechna Ltd., on femtosecond laser fabrication.

References

1. Gansel, J.K., Justyna, K., Thiel, M., Rill, M.S., Decker, M., Bade, K., Saile, V., von Freymann, G., Linden, S., and Wegener, M. (2009) Gold helix photonic metamaterial as broadband circular polarizer. *Science*, **325** (5947), 1513–1515.
2. Li, L., Gattass, R.R., Gershgoren, E., Hwang, H., and Fourkas, J.T. (2009) Achieving lambda/20 resolution by one-color initiation and deactivation of polymerization. *Science*, **324** (5929), 910–913.
3. Gattass, R.R. and Mazur, E. (2008) Femtosecond laser micromachining in transparent materials. *Nat. Photon.*, **2** (4), 219–222.
4. Park, S.H., Lim, T.W., Yang, D.Y., Kim, R.H., and Lee, K.S. (2006) Improvement of spatial resolution in nano-stereolithography using radical quencher. *Macromol. Res.*, **14** (5), 559–564.
5. Tanaka, T., Ishikawa, A., and Kawata, S. (2006) Two-photon-induced reduction of metal ions for fabricating three-dimensional electrically conductive metallic microstructure. *Appl. Phys. Lett.*, **88**, 081107.
6. Sundaram, S.K. and Mazur, E. (2002) Inducing and probing non-thermal transitions in semiconductors using femtosecond laser pulses. *Nat. Mater.*, **1**, 217–224.
7. Kawata, S., Sun, H.-B., Tanaka, T., and Takada, K. (2001) Finer features for functional microdevices. *Nature*, **412** (6848), 697–698.
8. Strickler, J. and Webb, W. (1991) Two-photon excitation in laser scanning fluorescence microscopy. *Proc. SPIE*, **1398**, 107–118.
9. Grilli, S., Coppola, S., Vespini, V., Merola, F., Finizio, A., and Ferraro, P. (2011) 3D lithography by rapid curing of the liquid instabilities at nanoscale. *Proc. Natl. Acad. Sci. U.S.A.*, **108** (37), 15106–15111.
10. Symes, M., Kitson, P., Yan, J., Richmond, C., Cooper, G., Bowman, R., Vilbrandt, T., and Cronin, L. (2012) Integrated 3D-printed reactionware for chemical synthesis and analysis. *Nat. Nanotechnol.*, **4**, 349–354.
11. Comina, G., Suska, A., and Filippini, D. (2014) PDMS lab-on-a-chip fabrication using 3D printed templates. *Lab Chip*, **14**, 424–430.
12. Malinauskas, M., Rekštytė, S., Lukoševičius, L., Butkus, S., Balčiunas, E., Pečiukaitytė, M., Baltriukienė, D., Bukelskienė, V., Butkevičius, A., Kucevičius, P., Rutkūnas, V., and

Juodkazis, S. (2014) 3D microporous scaffolds manufactured via combination of fused filament fabrication and direct laser writing ablation. *Micromachines*, **5** (4), 839–858.

13. Seet, K.K., Mizeikis, V., Matsuo, S., Juodkazis, S., and Misawa, H. (2005) Three-dimensional spiral - architecture photonic crystals obtained by direct laser writing. *Adv. Mater.*, **17** (5), 541–545.
14. Fischer, J. and Wegener, M. (2011) Three-dimensional direct laser writing inspired by stimulated-emission-depletion microscopy. *Opt. Mater. Express*, **2**, 614–624.
15. Maigyte, L., Purlys, V., Trull, J., Peckus, M., Cojocaru, C., Gailevičius, D., Malinauskas, M., and Staliunas, K. (2013) Flat lensing in the visible frequency range by woodpile photonic crystals. *Opt. Lett.*, **38** (14), 2376–2378.
16. Marrian, C.R.K. and Tennant, D.M. (2003) Nanofabrication. *J. Vac. Sci. Technol.*, **21**, S207–S215.
17. Maruo, S., Nakamura, O., and Kawata, S. (1997) Three-dimensional microfabrication with two-photon-absorbed photopolymerization. *Opt. Lett.*, **2** (22), 132–134.
18. Borisov, R.A., Dorojkina, G.N., Koroteev, N.I., Kozenkov, V.M., Magnitskii, S.A., Malakhov, D.V., Tarasishin, A.V., and Zheltikov, A.M. (1998) Femtosecond two-photon photopolymerization: a method to fabricate optical photonic crystals with controllable parameters. *Laser Phys.*, **8** (5), 1105–1105.
19. Serbin, J., Egbert, A., Ostendorf, A., Chichkov, B.N., Houbertz, R., Domann, G., Schulz, J., Cronauer, C., Frohlich, L., and Popall, M. (2003) Femtosecond laser-induced two-photon polymerization of inorganic-organic hybrid materials for applications in photonics. *Opt. Lett.*, **28** (5), 301–303.
20. Straub, M. and Gu, M. (2002) Near-infrared photonic crystals with higher-order bandgaps generated by two-photon photopolymerization. *Opt. Lett.*, **27** (20), 1824–1826.
21. Park, S.H., Lee, S.H., Yang, D.-Y., Kong, H.J., and Lee, K.-S. (2005) Subregional slicing method to increase three-dimensional nanofabrication efficiency in two-photon polymerization. *Appl. Phys. Lett.*, **87**, 154108.
22. Qi, F., Li, Y., Tan, D., Yang, H., and Gong, Q. (2007) Polymerized nanotips via two-photon photopolymerization. *Opt. Express*, **15** (3), 971–976.
23. Malinauskas, M., Purlys, V., Rutkauskas, M., and Gadonas, R. (2009) Two-photon polymerization for fabrication of three-dimensional micro- and nanostructures over a large area. *Proc. SPIE*, **7204** (72040C), 72040C–72040C–11.
24. Malinauskas, M., Farsari, M., Piskarskas, A., and Juodkazis, S. (2013) Ultrafast-laser micro/nano-structuring of photopolymers: a decade of advances. *Phys. Rep.*, **533**, 1–31.
25. Boyd, R.W. (2003) *Nonlinear Optics*, 2nd edn, Academic Press, London.
26. Žukauskas, A., Malinauskas, M., Kadys, A., Gervinskas, G., Seniutinas, G., Kandasamy, S., and Juodkazis, S. (2013) Black silicon: substrate for laser 3D micro/nano-polymerization. *Opt. Express*, **21** (6), 6901–6909.
27. Juodkazis, S., Mizeikis, V., Seet, K.K., Miwa, M., and Misawa, H. (2005) Two-photon lithography of nanorods in SU-8 photoresist. *Nanotechnology*, **16**, 846–849.
28. Malinauskas, M., Žukauskas, A., Bičkauskaitė, G., Gadonas, R., and Juodkazis, S. (2010) Mechanisms of three-dimensional structuring of photo-polymers by tightly focussed femtosecond laser pulses. *Opt. Express*, **18** (10), 10209–10221.
29. Jacobs, P. (1992) *Rapid Prototyping and Manufacturing: Fundamental of Stereolithography*, Society of Manufacturing Engineers.
30. Fused filament fabrication (fused deposition modeling), http://reprap.org/wiki/Fused_filament_fabrication (accessed 3 May 2016).
31. Seet, K.K., Juodkazis, S., Jarutis, V., and Misawa, H. (2006) Feature-size reduction of photopolymerized structures by femtosecond optical curing of SU-8. *Appl. Phys. Lett.*, **89**, 024106.
32. Malinauskas, M., Danilevičius, P., and Juodkazis, S. (2011) Three-dimensional

micro-/nano-structuring via direct write polymerization with picosecond laser pulses. *Opt. Express*, **19** (6), 5602–5610.
33. Baldacchini, T., Snider, S., and Zadoyan, R. (2012) Two-photon polymerization with variable repetition rate bursts of femtosecond laser pulses. *Opt. Express*, **20** (28), 29890–29899.
34. Mueller, J., Fischer, J., Mange, Y., Nann, T., and Wegener, M. (2013) *In-situ* local temperature measurement during three-dimensional direct laser writing. *Appl. Phys. Lett.*, **103** (12), 123107.
35. Ovsianikov, A., Viertl, J., Chichkov, B., Oubaha, M., MacCraith, B., Sakellari, I., Giakoumaki, A., Gray, D., Vamvakaki, M., Farsari, M., and Fotakis, C. (2008) Ultra-low shrinkage hybrid photo-sensitive material for two-photon polymerization microfabrication. *ACS Nano*, **2** (11), 2257–2262.
36. Suzuki, T., Morikawa, J., Hashimoto, T., Buividas, R., Gervinskas, G., Paipulas, D., Malinauskas, M., Mizeikis, V., and Juodkazis, S. (2012) Thermal and optical properties of sol-gel and SU-8 resists, in *Advanced Fabrication Technologies for Micro/Nano Optics and Photonics V*, Vol. **8249** (eds W.V. Schoenfeld, R.C. Rumpf, and G. von Freymann), Proceedings of SPIE, p. 82490K.
37. Li, J.L., Jia, B., Zhou, G., and Gu, M. (2006) Fabrication of three-dimensional woodpile photonic crystals in a PbSe quantum dot composite material. *Opt. Express*, **14** (22), 10740–10745.
38. Chon, J.W.M., Gan, X., and Gu, M. (2002) Splitting of the focal spot of a high numerical-aperture objective in free space. *Appl. Phys. Lett.*, **81** (9), 1576–1578.
39. Mizeikis, V., Matsuo, S., Juodkazis, S., and Misawa, H. (2006) Three-dimensional laser microfabrication: fundamentals and applications, *Femtosecond Laser Microfabrication of Photonic Crystals*. John Wiley & Sons, Weinheim.
40. Gamaly, E.E., Juodkazis, S., Nishimura, K., Misawa, H., Luther-Davies, B., Hallo, L., Nicolai, P., and Tikhonchuk, V. (2006) Laser-matter interaction in a bulk of a transparent solid: confined micro-explosion and void formation. *Phys. Rev. B*, **73**, 214101.
41. Marcinkevicius, A., Mizeikis, V., Juodkazis, S., Matsuo, S., and Misawa, H. (2003) Effect of refractive index-mismatch on laser microfabrication in silica glass. *Appl. Phys. A*, **76**, 257–260.
42. Hunter, W.R. (1998) *Measurement of Optical Constants in the Vacuum Ultraviolet Spectral Region*, Chapter 4, Academic Press, San Diego, CA, pp. 69–88.
43. Sokolowski-Tinten, K. and von der Linde, D. (2000) Generation of dense electron-hole plasmas in silicon. *Phys. Rev. B*, **61** (4), 2643–2650.
44. Hayasaki, Y., Iwata, K., Hasegawa, S., Takita, A., and Juodkazis, S. (2011) Time-resolved axial-view of the dielectric breakdown under tight focusing in glass. *Opt. Mater. Express*, **1**, 1399–1408.
45. Sun, Q., Jiang, H.-B., Liu, Y., Wu, Z.-X., Yang, H., and Gong, Q.-H. (2006) Diagnose parameters of plasma induced by femtosecond laser pulse in quartz and glasses. *Front. Phys. China*, **1**, 67–71.
46. Hayasaki, Y., Isaka, M., Takita, A., Hasegawa, S., and Juodkazis, S. (2012) Photo-acoustic sub-micrometer modifications of glass by pair of femtosecond laser pulses. *Opt. Mater. Express*, **2** (5), 691–699.
47. Morikawa, J., Hayakawa, E., Hashimoto, T., Buividas, R., and Juodkazis, S. (2011) Thermal imaging of a heat transport in regions structured by femtosecond laser. *Opt. Express*, **19** (21), 20542–20550.
48. Hayasaki, Y., Isaka, M., Takita, A., and Juodkazis, S. (2011) Time-resolved interferometry of femtosecond-laser induced processes under tight focusing and close-to optical breakdown inside borosilicate glass. *Opt. Express*, **19** (7), 5725–5734.
49. Hardel, P. and Wurfel, P. (2003) Theoretical limits of thermophotovoltaic solar energy conversion. *Semicond. Sci. Technol.*, **18**, S151–S157.
50. Tikhonov, A.N. and Samarski, A.A. (2013) *Equations of Mathematical Physics*, Courier Corp.
51. Luther-Davies, B., Rode, A., Madsen, N., and Gamaly, E.G. (2005) Picosecond

high-repitition-rate pulsed laser ablation of dielectrics: the effect of energy accumulation between pulses. *Opt. Eng.*, **44**, 051102.

52. Stuart, B.C., Feit, M.D., Herman, S., Rubenchik, A.M., Shore, B.W., and Perry, M.D. (1996) Nanosecond-to-femtosecond laser-induced breakdown in dielectrics. *Phys. Rev. B*, **53** (4), 1749–1761.
53. Gamaly, E.G. and Rode, A.V. (2013) Physics of ultra-short laser interaction with matter: from phonon excitation to ultimate transformations. *Progr. Quant. Electron.*, **37** (5), 215–323.
54. Sheik-bahae, M., Said, A.A., Wei, T.H., Hagan, D.J., and van Stryland, E.W. (1990) Sensitive measurement of optical nonlinearities using a single beam. *IEEE J. Quantum Electron.*, **26** (4), 760–769.
55. Kamada, K. (2004) Characterization of two-photon absorption and its resonance enhancement by Z-scan method. *Proc. SPIE*, **5516**, 97–105.
56. Kamada, K., Matsunaga, K., Yoshino, A., and Ohta, K. (2003) Two-photon-absorption-induced accumulated thermal effect on femtosecond Z-scan experiments studied with time-resolved thermal-lens spectrometry and its simulation. *J. Opt. Soc. Am. B*, **20**, 529–537.
57. DeSalvo, R., Said, A.A., Hagan, D., Van Stryland, E.W., and Sheik-Bahae, M. (1996) Infrared to ultraviolet measurements of two-photon absorption and n(2) in wide bandgap solids. *IEEE J. Quantum Electron.*, **32** (8), 1324–1333.
58. Rumi, M., Ehrlich, J., Heikal, A., Perry, J., Barlow, S., Hu, Z., McCord-Maughon, D., Parker, T.C., Rockel, H., Thayumanavan, S., Marder, S.R., Beljonne, D., and Bredas, J.-L. (2000) Structure-property relationships for two-photon absorbing chromophores: bis-donor diphenylpolyene and bis(styryl)benzene derivatives. *J. Am. Chem. Soc.*, **122** (39), 9500–9510.
59. Murazawa, N., Juodkazis, S., Misawa, H., and Kamada, K. (2008) Two-photon excitation of dye-doped liquid crystal by a cw-laser irradiation. *Mol. Cryst. Liq. Cryst.*, **489**, 310–319.
60. Juodkazis, S., Mizeikis, V., and Misawa, H. (2008) Three-dimensional structuring of resists and resins by direct laser writing and holographic recording. *Adv. Polym. Sci.*, **213**, 157–206.
61. Gamaly, E.G. (2011) The physics of ultra-short laser interaction with solids at non-relativistic intensities. *Phys. Rep.*, **508**, 91–243.
62. Raizer, Y.P. (1977) *Laser-induced Discharge Phenomena*, Consultant Bureau, New York.
63. Vogel, A., Noack, J., Huttman, G., and Paltauf, G. (2005) Mechanisms of femtosecond laser nanosurgery of cells and tissues. *Appl. Phys. B*, **81**, 1015–1047.
64. Juodkazis, S., Rode, A.V., Gamaly, E.G., Matsuo, S., and Misawa, H. (2003) Recording and reading of three-dimensional optical memory in glasses. *Appl. Phys. B*, **77**, 361–368.
65. Mizeikis, V., Seet, K.K., Juodkazis, S., and Misawa, H. (2004) Three-dimensional woodpile photonic crystal templates for infrared spectral range. *Opt. Lett.*, **29** (17), 2061–2063.
66. Seet, K.K., Mizeikis, V., Juodkazis, S., and Misawa, H. (2005) Spiral three-dimensional photonic crystals for telecomunications spectral range. *Appl. Phys. A*, **82** (4), 683–688.
67. Seet, K.K., Mizeikis, V., Juodkazis, S., and Misawa, H. (2006) Three-dimentional circular spiral potonic crystalls structures recorded by femtosecond pulses. *J. Non-Cryst. Solids*, **352** (23-25), 2390–2394.
68. Rekštytė, S., Malinauskas, M., and Juodkazis, S. (2013) Three-dimensional laser micro-sculpturing of silicone: towards bio-compatible scaffolds. *Opt. Express*, **21** (14), 17028–17041.
69. Malinauskas, M., Žukauskas, A., Purlys, V., Belazaras, K., Momot, A., Paipulas, D., Gadonas, R., Piskarskas, A., Gilbergs, H., Gaidukeviciute, A., Sakellari, I., Farsari, M., and Juodkazis, S. (2010) Femtosecond laser polymerization of hybrid/integrated micro-optical elements and their characterization. *J. Opt.*, **12** (12), 124010.
70. Cao, Y.-Y., Takeyasu, N., Tanaka, T., Duan, X.-M., and Kawata, S. (2009) 3d metallic nanostructure fabrication by

surfactant-assisted multiphoton-induced reduction. *Small*, **5** (10), 1144–1148.

71. Buividas, R., Rekštytė, S., Malinauskas, M., and Juodkazis, S. (2013) Nanogroove and 3D fabrication by controlled avalanche using femtosecond laser pulses. *Opt. Mater. Express*, **3** (10), 1674–1686.

72. Sugioka, K. and Cheng, Y. (2014) Femtosecond laser three-dimensional micro- and nanofabrication. *Appl. Phys. Rev.*, **1** (4), 041303.

73. Spivey, E., Ritschdorff, E., Connell, J., McLennon, C., Schmidt, C., and Shear, J. (0013) Multiphoton lithography of unconstrained three-dimensional protein microstructures. *Adv. Funct. Mater.*, **23**, 333–339.

74. Liu, D.-X., Sun, Y.-L., Dong, W., Yang, R.-Z., Chen, Q.-D., and Sun, H.-B. (2014) Dynamic laser prototyping for biomimetic nanofabrication. *Laser Photon. Rev.*, **8** (6), 882–888.

75. Žukauskas, A., Matulaitienė, I., Paipulas, D., Niaura, G., Malinauskas, M., and Gadonas, R. (2015) Tuning the refractive index in 3D direct laser writing lithography: towards GRIN microoptics. *Laser Photon. Rev.*, **9** (6), 706–712. doi: 10.1002/lpor.201500170.

2
Characterization of 2PA Chromophores

Eric W. Van Stryland and David J. Hagan

2.1
Introduction

In this chapter, we describe several techniques for measuring nonlinear absorption (NLA) with special emphasis on the instantaneous nonlinear response of two-photon absorption (2PA) [1] due to its extensive use in microscopic imaging and 3D stereolithography/additive manufacturing [2]. We also give some information on nonlinear refraction (NLR) measurements since NLR almost always accompanies NLA signals. In general, multiple experimental techniques are needed to fully unravel the various physical processes involved in the NLA for a given material, as it is rare to find only a single dominant nonlinear response. For example, 2PA not only causes a change in transmission but also results in the generation of excited states. These excited states can also change the absorption spectrum of the material, which can easily be confused with the 2PA itself.

Additionally, NLR accompanies NLA and transforms into a spatial redistribution of energy upon propagation. Thus, in any experiment to monitor transmittance changes, considerable caution needs to be taken to ensure that the entire beam is collected by the detector for all input irradiances. Otherwise, the NLR may be easily mistaken for NLA. Also, in order to be able to separate NLA and NLR, the conditions for external self-action, that is, the thin-sample approximation, need to be satisfied, placing stringent requirements on the experimental parameters [3]. This will be discussed in more detail later. Even when NLA and NLR can be separated, obtaining reliable values for nonlinear coefficients can be difficult. For example, if NLA is irradiance-dependent (as for ultrafast nonlinearities), errors come from energy calibration (e.g., ±5%), temporal measurements (e.g., ±10%), and spatial energy distribution measurements (e.g., ±10% for Gaussian beams but for two dimensions). These errors add along with fitting errors (e.g., ±10%) to give an overall confidence error for the third-order nonlinear response. The example errors given in the previous sentence add to give a total error for the NLA or NLR coefficient of $\pm\sqrt{0.05^2 + 0.1^2 + 2 \times 0.1^2 + 0.1^2} = \pm 21\%$. Measurement of higher order responses would result in larger errors. All of these difficulties, as well as

Multiphoton Lithography: Techniques, Materials and Applications, First Edition.
Edited by Jürgen Stampfl, Robert Liska, and Aleksandr Ovsianikov.

light simply missing the detector and incorrect identification of the physical processes, have led to reported values of nonlinear coefficients that vary over orders of magnitude from one publication to the next [4–7]. Although such large discrepancies are becoming rarer due in part to more reliable, stable, and well-characterized optical sources and to our more complete understanding of the nonlinearities themselves along with the potential measurement problems, obtaining reliable coefficients with small confidence errors is still challenging.

2.2
Description of Nonlinear Absorption and Refraction Processes

2.2.1
Two-Photon Absorption and Bound-Electronic Nonlinear Refraction

The second term after the linear optics term in Bloembergen's [8] expansion of the electric polarization

$$P \sim \varepsilon_0(\chi^{(1)}E + \chi^{(2)}E^2 + \chi^{(3)}E^3 + \dots), \tag{2.1}$$

is the third-order polarization, which is characterized by the third-order nonlinear susceptibility $\chi^{(3)}$. This is the lowest order term in the expansion that leads to "self-action" nonlinearities, that is, NLA, which is proportional to the imaginary part of $\chi^{(3)}$ and NLR proportional to the real part of $\chi^{(3)}$.

Assuming that the sample is thin, that is, the sample thickness is less than the Rayleigh length, and the nonlinear phase does not propagate far enough to cause an irradiance change within the sample (i.e., external self-action [3]), the resulting equations for the propagation of the light irradiance I and nonlinear phase shift φ are

$$\frac{dI(z,t)}{dz} = -\alpha I(z,t) - \alpha_2 I^2(z,t)$$

and

$$\frac{d\phi(z,t)}{dz} = kn_2 I(z,t), \tag{2.2}$$

where z is the propagation distance, α is the linear absorption coefficient, α_2 is the 2PA coefficient, k is the wave number, and n_2 is the nonlinear refractive index from $n = n_0 + n_2 I$. Here, in MKS units, which are used throughout this chapter unless noted otherwise, we have

$$\alpha_2(\omega;\omega) = \frac{\omega\mu_0}{n_0}\mathrm{Im}\chi_{\mathrm{eff}}^{(3)}$$

and

$$n_2(\omega;\omega) = \frac{\mu_0 c}{2n_0^2}\mathrm{Re}\chi_{\mathrm{eff}}^{(3)}, \tag{2.3}$$

where $\chi_{\mathrm{eff}}^{(3)} = \frac{3}{2}\chi_{\Re,xxxx}^{(3)}(\omega;-\omega,\omega,\omega)$ for a single, linearly polarized input frequency, where $E(z,t) = E_\omega \cos(kz - \omega t)$, and $E_\omega(z)$ is the slowly varying field amplitude with irradiance $I = \frac{1}{2}c\varepsilon_0 n_0|E_\omega|^2$ [9]. We must mention here that this definition of α_2 in terms of Im $\chi^{(3)}$ assumes that we are not within a region of linear loss where other effects, including absorption saturation, would contribute. Equation (2.3) assumes that the linear absorption $\alpha \ll 2\omega n_0/c$; otherwise α_2 becomes dependent on both the real and the imaginary parts of $\chi^{(3)}$, as does n_2. This condition is usually satisfied except very near the linear resonance [10]. For example, in the visible region, for an index of ~1.5, this becomes $\alpha \ll 2\times10^5\ \text{cm}^{-1}$; however, for metallic nanoparticles, this can be important.

The solution of Eq. (2.2) for the irradiance propagation is

$$I(L,t) = \frac{e^{-\alpha L}(1-R)^2 I(0,t)}{1+\alpha_2(1-R)I(0,t)L_{\text{eff}}}, \tag{2.4}$$

with $L_{\text{eff}} = (1-e^{-\alpha L})/\alpha$, where L is the sample length, $I(0,t)$ is the irradiance incident on the sample, and R is the surface reflectance. If the thin-sample approximation is not valid, then amplitude and phase become intertwined with propagation within the sample, and the full wave equation must be solved [11]. In the following sections, we discuss specific methods for measuring the 2PA coefficient along with some complementary methods for determining n_2. NLR always occurs when α_2 is being measured, and sometimes it can be important. In addition, because 2PA and bound-electronic n_2 are related through Kramers–Kronig relations via causality, careful and complete measurements of nonlinear spectra can be used to derive spectra/dispersion of the other. One should note that these Kramers–Kronig relations are strictly valid only for the frequency-nondegenerate forms of the nonlinearities [9].

When making measurements of organic dyes, solutions are often used, since growing crystals can be very research-intensive, and making good optical quality solid-state films can also be difficult. In such measurements, nonlinear cross sections are the more meaningful parameters to use. These cross sections are usually given in GM [12] units ($1\ \text{GM} = 1\times10^{-50}\ \text{cm}^4\ \text{s}$). For NLA, the cross section in terms of the 2PA coefficient is

$$\delta = \frac{\alpha_2 \hbar\omega}{N}$$

and

$$\delta_r = \frac{k_0 n_2 \hbar\omega}{N}, \tag{2.5}$$

where we have defined the NLR coefficient in an analogous manner, with N being the molecular density. The nonlinear coefficient per molecule may also be defined in terms of the second hyperpolarizability γ, which is defined by

$$\gamma = \frac{\varepsilon_0 \chi^{(3)}}{N f^4}, \tag{2.6}$$

where $f = (n+2)/3$ is the local field correction factor [13]. The molecular second hyperpolarizability is usually given as $\gamma_{\text{esu}} = \frac{\chi^{(3)}_{\text{esu}}}{N_{\text{cgs}} f^4}$ in esu. For noninteracting solute and solvent molecules, assuming a nonabsorbing solvent, the nonlinear susceptibility of the solution is given by

$$|\chi^{(3)}| = \frac{f^4}{\varepsilon_0}\left[\left(N_{\text{solvent}}\text{Re}\gamma_{\text{solvent}} + N\text{Re}\gamma\right)^2 + (N\text{Im}\gamma)^2\right]^{1/2}, \quad (2.7)$$

where the index used in the local field correction factor is the index of the solution, that is, solvent plus solute, and the γ's are the orientationally averaged molecular hyperpolarizabilities. If $N_{\text{solvent}}\gamma_{\text{solvent}} = -N_{\text{solute}}\gamma_{\text{solute}}$ (the contribution of the real parts due to solute and solvent cancel each other), the absolute value of $\chi^{(3)}$ of the solution has a minimum value given by [14]:

$$\left|\chi^{(3)}_{\text{solution}}\right|_{\text{min}} = \varepsilon_0 f^4 N_{\text{solute}}\text{Im}\gamma_{\text{solute}} = \left|\text{Im}\chi^{(3)}_{\text{solution}}\right|. \quad (2.8)$$

This allows the separation and determination of the real and imaginary parts of the nonlinear susceptibilities by varying the concentration to find the minimum.

2.2.2 Excited-State Absorption and Refraction

The above description of 2PA and n_2 is strictly for bound-electronic nonlinearities, which are ultrafast (sub-femtosecond). There are many other nonlinear responses that are pulsewidth-dependent, and we describe just a few of these in the following. Many of these, such as excited-state nonlinearities, involve resonant, that is, real, absorption processes [15]. Organic dyes often display excited-state nonlinearities both in absorption, ESA (excited-state absorption), and in refraction, ESR (excited-state refraction). These effects are due to the redistribution of level/state populations. Figure 2.1 shows the "instantaneous" 2PA (a), where the intermediate off-resonance state is labeled "e" and the final upper state is "u." Transitions are of

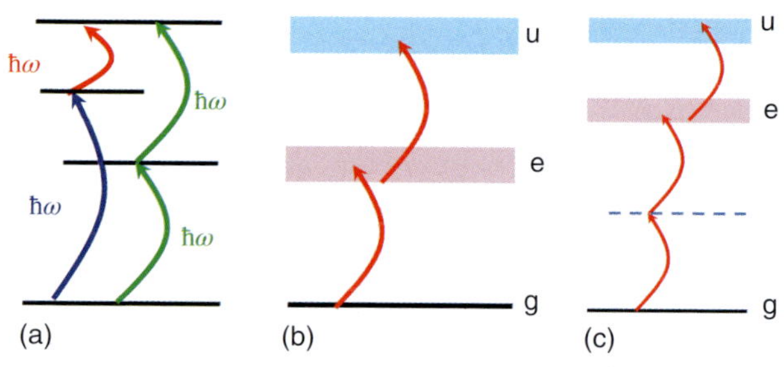

Figure 2.1 (a) Schematic energy-level diagram showing two-photon absorption from the ground state to the excited state with equal photons (solid green lines) of energy $\hbar\omega$ or unequal photons (dotted blue and red lines) of energies $\hbar\omega_1$ and $\hbar\omega_2$ where $2\hbar\omega = \hbar\omega_1 + \hbar\omega_2$. The dashed horizontal lines indicate what are commonly referred to as *virtual states*. (b) Resonant two-step excitation of the upper state, u, with linear absorption cross sections between the states of σ_{ge} and σ_{eu}, respectively. The rectangles are meant to indicate rovibronic transition absorption bands. (c) Two-photon absorption to a band e, followed by excited-state absorption to band u.

one photon allowed from g to e and from e to u but not allowed by parity from g to u. It is instructive to look at the specific case of one-photon-induced ESA in the case where saturation for the g to e transition is relatively small [15]. Figure 2.1b shows resonant 2PA, that is, the intermediate state resonantly absorbs $\hbar\omega$ and is populated. The equations describing this phenomenon in terms of the absorption cross sections σ are

$$\frac{dI}{dz} = -\sigma_{\text{ge}} N_{\text{g}} I - \sigma_{\text{eu}} N_{\text{e}} I$$

and

$$\frac{dN_{\text{e}}}{dt} = \frac{\sigma_{\text{ge}} N_{\text{g}} I}{\hbar\omega}, \tag{2.9}$$

where N_{g} is the ground-state population density, and N_{e} is the population density of excited state in Figure 2.1b, and the cross sections are between the labeled one-photon allowed transitions. As the absorption has not been written in terms of a population difference, we have included saturation of the linear absorption only as a ground-state depletion to a lower order, and we have assumed that, once the excited state has been populated, it does not decay back to the ground state within the duration of the pulse. This is often a good approximation since the initial absorption in molecules is usually into a rovibronic band, which rapidly relaxes to a lower energy from which ESA occurs, and this approximation is especially good when considering triplet states (not shown).

As 2PA and ESA are typically measured using short laser pulses, for which the measured quantities are the transmitted pulse energy or fluence (energy per unit area), we need to integrate Eq. (2.9) over time. Under our approximations, these two equations can be integrated over time to give a new propagation equation:

$$\frac{dF}{dz} = -\sigma_{\text{ge}} N_{\text{g}} F - \sigma_{\text{ge}} N_{\text{g}} \frac{(\sigma_{\text{eu}} - \sigma_{\text{ge}})}{2\hbar\omega} F^2, \tag{2.10}$$

where F is the time-integrated irradiance or fluence, $F = \int_{-\infty}^{\infty} I(t)dt$ [15, 16]. Equation 2.10 is good for the second order in F. The last term, proportional to the square of σ_{ge}, comes from the ground-state depletion, which can sometimes be ignored, as is true for the experimental case shown in Section 2.3.1.3. This equation is analogous to the equation for 2PA except that the fluence has replaced the irradiance, and α_2 is replaced by products of one-photon cross sections. When performing perturbation theory to calculate 2PA, it is the product of one-photon absorption (1PA) matrix elements (proportional to the linear absorption cross sections) that enters the calculation. Thus, this equation shows the transition from nonresonant 2PA to resonant, two-step NLA. There is a limited applicability of this solution as, for example, when the irradiance/fluence is increased, the ground-state depletion becomes important [16, 17]. Looking at the similarity between Eqs. (2.10) and (2.2), it is not surprising that this ESA has been mistaken for 2PA in some experiments. We show an example in Section 2.3.1.3. It is also the case where some authors define an effective 2PA coefficient to describe this ESA process; however, this coefficient is pulse-width-dependent, while the true 2PA coefficient is independent of the pulse width.

Associated with this ESA is ESR, since a change in absorption also results in a change in index, both due to the redistribution of populations. The sign of the NLR will depend on the relative spectral positions of the resonances, that is, above or below resonance. In this case, the nonlinear phase shift is given by

$$\frac{d\varphi(z,t)}{dz} = kn_2 I(z,t) + (\sigma_{\mathrm{euR}} + \sigma_{\mathrm{geR}})N_{\mathrm{e}}(z,t), \tag{2.11}$$

where σ_{geR} and σ_{euR} are the nonlinear refractive cross sections for the ground and excited states, respectively, and we have kept the instantaneous NLR. Here, the integration over time involves averaging of the overall phase shift, which results in a similar integral as for ESA that does not depend on pulse shape, giving an averaged phase shift of one-half the phase shift at the peak of the pulse:

$$\left.\frac{d\langle\varphi(z,t)\rangle_t}{dz}\right|_{\mathrm{ESR}} = (\sigma_{\mathrm{geR}} + \sigma_{\mathrm{exR}})\frac{\alpha}{\hbar\omega}\frac{\int_{-\infty}^{\infty} dt I(z,t)\int_{-\infty}^{t} dt' I(z,t')}{\int_{-\infty}^{\infty} dt I(z,t)}$$
$$= \frac{1}{2}(\sigma_{geR} + \sigma_{exR})\frac{\alpha}{\hbar\omega}F. \tag{2.12}$$

On the other hand, the instantaneous term gives an average phase shift that depends on pulse shape and is $1/\sqrt{2}$ for Gaussian temporal pulses, that is

$$\left.\frac{d\langle\varphi(z,t)\rangle_t}{dz}\right|_{n_2} = kn_2\frac{\int_{-\infty}^{\infty} dt I^2(z,t)}{\int_{-\infty}^{\infty} dt I(z,t)} = \frac{kn_2}{\sqrt{2}}F. \tag{2.13}$$

Measurements of 2PA in semiconductors and many organic dyes are also accompanied by ESA but when the excited state is populated by 2PA. In this case, the irradiance propagation equation becomes

$$\frac{dI(z,t)}{dz} = -\alpha I(z,t) - \alpha_2 I^2(z,t) - \sigma_{\mathrm{ex}} N_{\mathrm{e}} I, \ \text{with} \ \frac{dN_{\mathrm{e}}}{dt} = \frac{\alpha_2 I^2}{2\hbar\omega} \tag{2.14}$$

as discussed in more detail in the experimental section where the data are shown.

This is a higher order process (Im $\chi^{(3)}{:}\chi^{(1)}$), that is, 2PA followed by linear absorption (Im $\chi^{(1)}$), or 2PA followed by linear refraction (Re $\chi^{(1)}$), as indicated in Figure 2.1c. Thus, these are effectively fifth-order nonlinearities, so that this ESA can easily be confused with three-photon absorption (3PA), and many authors have defined effective 3PA coefficients; but again, these will be pulse-width-dependent by necessity [18]. These nonlinearities are discussed in more detail in the sections describing specific experimental arrangements, for example, Sections 2.3.1.3 and 2.4.2.

The similarities of the propagation equations for ESA and 2PA, as well as similar issues for higher order nonlinearities, present challenges for characterization techniques to uniquely determine the underlying physical processes. Unique determination often requires multiple methods and parametric studies, for example, pulse-width dependence. In the following sections, we discuss several techniques for measuring these nonlinearities, the limitations of these methods, and the possible combinations of methodologies to eliminate ambiguities.

2.3 Methods for Measurements of NLA and NLR

Measurements of NLA and NLR fall into two broad categories: *direct methods*, where we measure the self-induced change in transmission of a beam, and *indirect methods*, where an excitation is induced by a beam, to be sensed by some other means (transmittance of a probe, heating of materials, fluorescence, etc.) [19].

2.3.1 Direct Methods

2.3.1.1 Nonlinear Transmission

We start with perhaps the most straightforward and direct measurement of transmission as a function of irradiance. From Eq. (2.4), we find

$$\frac{1}{T} = \frac{I(0,t)}{I(L,t)} = \frac{e^{\alpha L}}{(1-R)^2}[1 + \alpha_2 I(0,t)(1-R)L_{\text{eff}}] \tag{2.15}$$

where T is the transmittance [20, 21]. The equation is written in this way to point out a simple method for determining the 2PA coefficient by measuring the transmittance and plotting inverse transmittance versus irradiance. The result should be a straight line whose slope determines α_2 and the intercept determines α. Of course, using Gaussian spatial and temporal profiles (or other shapes) requires integration over space and time. These integrals reduce the slope for Gaussian shapes by a factor of $2\sqrt{2}$ (root 2 for each dimension of space and time); and for higher irradiance, they produce a downward curvature owing to spatial and temporal beam reshaping, thus deviating from a simple straight line [22].

Similarly, for 3PA

$$\frac{1}{T^2} = \left[\frac{I(0,t)}{I(L,t)}\right]^2 = \frac{e^{2\alpha L}}{(1-R)^4}\left(1 + \alpha_3 I^2(0,t)(1-R)^2\left(\frac{1+e^{-2\alpha L}}{\alpha}\right)\right), \tag{2.16}$$

where it is assumed that there is no 2PA. Figure 2.2 shows a plot of inverse transmittance for a two-photon absorbing organic compound in thin-film form using 810-nm femtosecond,pulses [23].

Only the low irradiance inputs are plotted in Figure 2.2 because at higher inputs two-photon excited states linearly absorb leading to a higher order nonlinear response. The slight upward turn as opposed to a slight downward curvature predicted as a result of beam broadening by NLA is due to this higher order process of two-photon ESA. The straight line in Figure 2.2 gives a slope that yields a value of $\alpha_2 \cong 30\,\text{cm GW}^{-1}$, which agrees with the values reported in Ref. [23].

Although the experimental procedure appears to be simple and straightforward, one must be aware of the difficulties that result from the presence of NLR in these

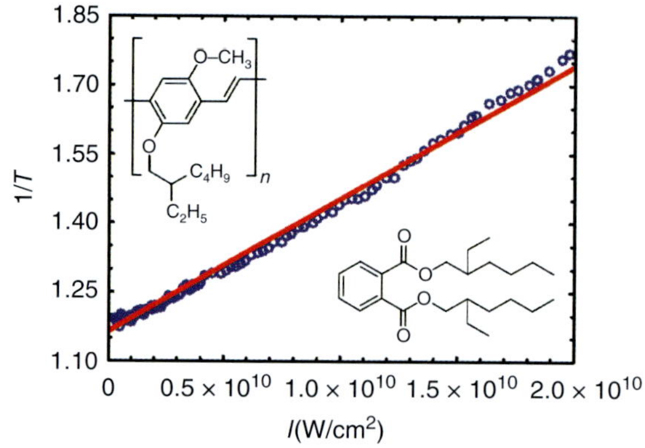

Figure 2.2 Data on a 25-μm-thick film of MEH-PPV; DOP (50 : 50 melt-processed blend, where DOP is dioctyl phthalate, a plasticizer) as measured in Ref. [23]. Data provided by J. Hales, and replotted as inverse transmission versus irradiance. The upper left is MEH-PPV and the lower right is DOP. The wavelength used was 810 nm with 110-fs Full-Width at Half Maximum (FWHM) pulses.

experiments. If this NLR is defocusing, it is quite easy for some of the beam to miss the detector at higher irradiances, giving the appearance of a larger nonlinear loss. It is therefore not surprising that most inconsistencies in the literature involve reported values for the nonlinear coefficients being too large. It is also important that the detector response is spatially uniform. Additionally, as for any nonlinear measurement, the irradiance needs to be carefully calibrated, necessitating measurements of the spatial profile as well as the temporal profile. Gaussian spatial profile beams are desirable since their propagation is particularly simple.

The literature is filled with reports of high 2PA coefficients that are either the result of missing the beam due to NLR or the use of pulses that have rapid temporal modulation that was not noticed [22, 24]. Similar problems can occur if the spatial profile is not smoothly varying at the sample. These comments can be applied to nearly all of the measurement techniques described in this chapter; however, there are many ways to mitigate such problems, and some of them involve using known nonlinear samples as references or test samples.

Another comment about single beam direct transmission measurements that is self-evident is that they are sensitive to all NLA processes and they do not differentiate between the different mechanisms that give rise to them; thus, other techniques must usually be used to determine the process or processes involved.

2.3.1.2 *Z*-Scan

Introduced in 1989 [25], the original Z-scan consists of translating a thin sample through the focus of a laser beam while monitoring the light transmitted through an aperture [26]. This gives a measure of the NLR if there is no NLA present. Gaussian spatial beams are preferred since modeling the results is greatly simplified, but not necessary. For measuring ultrafast nonlinearities, pulsed sources are required, and the signal is proportional to the temporally integrated nonlinearity. Figure 2.3a

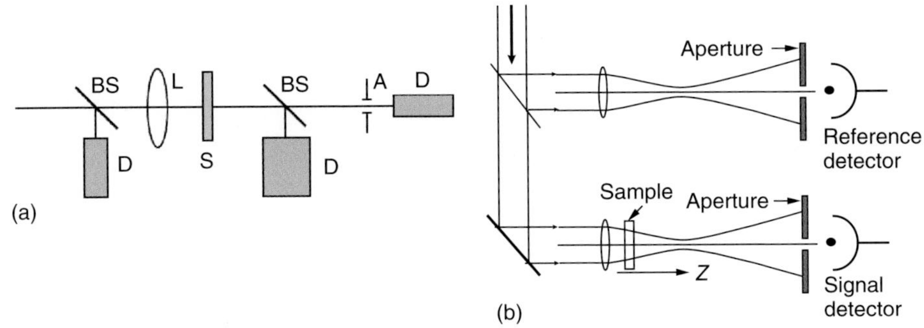

Figure 2.3 (a) *Z*-scan setup allowing for simultaneous open- and closed-aperture *Z*-scans. (b) Closed-aperture Z-scan with identical reference arm increasing the signal-to-noise ratio [27].

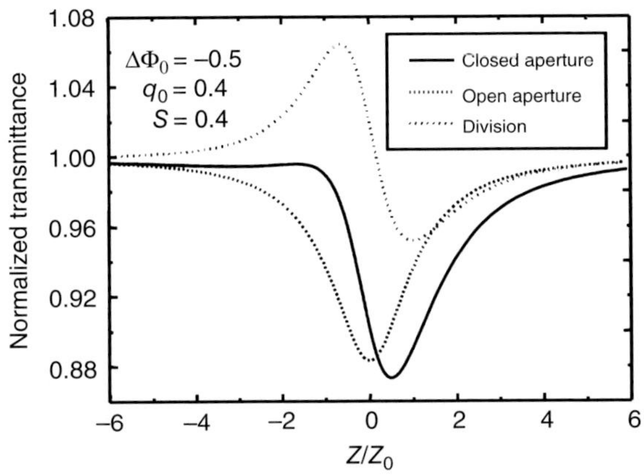

Figure 2.4 Calculated *Z*-scan signals using open, closed, and divided signals as described for third-order nonlinearities (2PA and self-defocusing) with the parameters shown in the figure using the analysis described in Section titled "Simple *Z*-Scan Analysis". The separation of the transmission "peak" and "valley" for the divided signal is typically labeled ΔZ_{pv}, while the difference between peak and valley transmittances is commonly labeled ΔT_{pv}. (Ferdinandus 2012 [28].)

shows the usual *Z*-scan setup for simultaneously performing open- and closed-aperture *Z*-scans, while Figure 2.3b shows the configuration for a closed-aperture *Z*-scan where a reference arm is utilized to increase the signal-to-noise ratio (S/N) [27]. When NLA is to be measured, care is taken to open the aperture and collect all of the transmitted light. Typical signal profiles for *Z*-scans are shown in Figure 2.4.

A beam splitter is often placed after the sample, as shown in Figure 2.3, to split the transmitted beam to fall on two detectors. The first detector measures the total transmitted light and is referred to as an *open-aperture Z-scan*. The second beam passes through the aperture in the far field that typically transmits ~40%

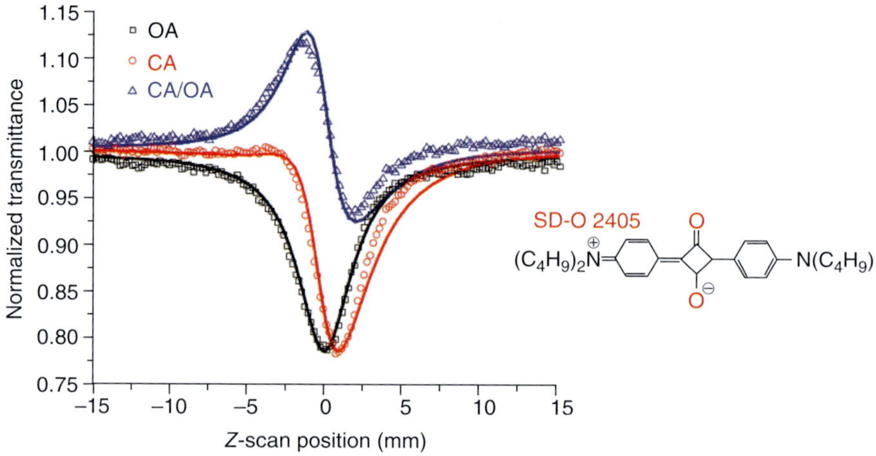

Figure 2.5 *Z*-scans of open (black squares), closed (red circles), and divided (blue triangles) of squaraine molecule shown along with fits using only ultrafast α_2 and n_2. (Ferdinandus 2012 [28].)

of the light in the linear regime, which is collected onto a second detector. This is referred to as a *closed-aperture Z-scan*, that is, partially closed. The open-aperture *Z*-scan is sensitive only to NLA as long as the sample is thin as previously defined, and the curve can be fittted with a parameter describing the NLA (assuming there is a single NLA process!). The closed-aperture *Z*-scan is affected by both NLR and NLA and, given that the NLA has previously been fitted, can be fitted with a single parameter for the NLR (again under the assumption that there is a single NLR process!). However, if the closed-aperture signal is divided by the open-aperture signal, the resulting curve is nearly identical to the signal that would have been obtained if the NLA were absent (see Figure 2.5). This can greatly simplify the data analysis, although the curve may also be fitted with a two-parameter fitting procedure for the NLA and NLR coefficients of the third-order response materials, but with the usual increased errors associated with a two-parameter fit. The popularity of the *Z*-scan technique is due in large part to the simplicity of separating absorptive and refractive nonlinearities; however, it is sensitive to *all* NLA and NLR mechanisms.

Figure 2.5 shows an example of *Z*-scans (actually dual-arm *Z*-scans described in Section titled "Dual-Arm Z-Scan") using femtosecond pulses where only 2PA and n_2 are observed. The NLR is from the solute only, where the NLR from the solvent has been subtracted, as described in Section titled "Dual-Arm Z-Scan". In the usual *Z*-scan, in order to determine the n_2 of the solute, the NLR of the solvent must be subtracted by taking two sequential *Z*-scans, one with only the solvent and the other with the solution.

Simple Z-Scan Analysis A simple empirical relation exists in the transmission difference between the peak and valley measured in a *Z*-scan with Gaussian beams for the closed-aperture Z-scan signal and the nonlinearly induced phase distortion [26]:

$$\Delta T_{\mathrm{pv}} \cong 0.406(1-S)^{0.27}|\Delta\Phi_0|, \tag{2.17}$$

where S is the linear transmittance of the aperture placed in the far field and

$$\Delta\Phi_0(t) = \frac{2\pi}{\lambda} n_2 I_0(t) L_{\mathrm{eff}}, \tag{2.18}$$

where $I_0(t)$ is the axis irradiance at focus and the effective length $L_{\mathrm{eff}} = (1-e^{-\alpha L})/\alpha$. This empirical relation is accurate for both cases without NLA and with NLA after division by the open-aperture Z-scan data. Since the temporally averaged transmittance is measured in a Z-scan, the pulse shape needs to be included. This is relatively straightforward given the linear relation between transmittance change and phase distortion of Eq. (2.17). If the irradiance-dependent pulse shape in time is described by the function $f(t)$, then the temporal integral given by

$$A_\tau = \int_{-\infty}^{\infty} f^2(t)dt \bigg/ \int_{-\infty}^{\infty} f(t)dt$$

reduces the measured ΔT_{pv} by A_τ. This has the value of $1/\sqrt{2}$ for Gaussian pulses with nonlinearities much faster than the pulse width used. For pulses much shorter than the nonlinear response, for example, excited-state nonlinearities, $A_\tau = 1/2$ independent of the pulse shape. In addition, the separation in Z between the position of the peak and valley is given by

$$|\Delta Z_{\mathrm{pv}}| \approx 1.71 Z_0, \tag{2.19}$$

where $Z_0 = n\pi w_0{}^2/\lambda$, with w_0 the beam HW$1/e^2M$ in irradiance. This allows an independent determination of the focused spot size if the sample is known to exhibit a third-order nonlinear response.

For the open-aperture Z-scan, the peak energy transmittance change is given by

$$\Delta T(Z) \approx -\frac{q_0}{2\sqrt{2}} \frac{1}{[1+Z^2/Z_0^2]}, \tag{2.20}$$

where $q_0 = \alpha_2 I_0(0) L_{\mathrm{eff}}$ ($|q_0| \ll 1$). Here, no propagation to the far field is required in the calculation since total energy is detected. The irradiance at the exit of the sample I_{e} is

$$I_{\mathrm{e}}(Z,r,t) = \frac{I(Z,r,t)\exp^{-\alpha L}}{1+q(Z,r,t)}, \tag{2.21}$$

while the phase change at the exit is

$$\Delta\phi(Z,r,t) = \frac{kn_2}{\alpha_2} \ln[1+q(Z,r,t)], \tag{2.22}$$

where $q(Z,r,t) = \alpha_2 I(Z,r,t) L_{\mathrm{eff}}$. This phase along with the irradiance distribution must be propagated to the detector in the far field to model the closed-aperture Z-scan data. There are multiple ways to perform this propagation, and two of these are described in Ref. [26]. Thus, while the above-mentioned empirical results are useful, the full propagation results can be fitted for all Z, which is good, for example, for checking the order of the nonlinearity. For example, a 3PA signal will be narrower in Z than a 2PA signal, and $|\Delta Z_{pv}| \cong 1.5Z_0$ for a fifth-order nonlinearity.

Combining Eqs. (2.21) and (2.22), we obtain the complex field at the exit surface of the sample as

$$E_e(Z,r,t) = E(Z,r,t)e^{-\alpha L/2}(1+q(Z,r,t))^{(ikn_2/\alpha_2-1/2)}, \tag{2.23}$$

while the field at the aperture is found by performing a Fresnel diffraction calculation:

$$E_a(Z,r,t,d) = \frac{2\pi}{i\lambda d'}\exp\left(\frac{i\pi r^2}{\lambda d'}\right)\int_0^\infty r'dr'E_e(Z,r',t-d'/c)\exp\left(\frac{i\pi r'^2}{\lambda d'}\right)J_0\left(\frac{2\pi rr'}{\lambda d'}\right), \tag{2.24}$$

where $d' = d - Z$ is the distance from the sample to the aperture plane. The measured quantity is the pulse energy or the average power transmitted through the far-field aperture having a radius of r_a. The normalized transmittance is then obtained as

$$T(Z) = \frac{\int_{-\infty}^{\infty} dt \int_0^{r_a} |E_a(Z,r,t,d)|^2 r dr}{U}, \tag{2.25}$$

where U is the same as the numerator but in the linear regime (i.e., for $\Delta\phi = 0$). In the case of an *EZ*-scan, the limits of the spatial integral in Eq. (2.25) must be replaced by r_d to ∞, where r_d is the radius of the obscuration disk [29, 30]. It is generally more convenient to represent the aperture (or disk) size by the normalized transmittance (or rejection) S in the linear regime.

The formalism thus far presented is generally applicable to any radially symmetric beam. Here, however, we assume a $TEM_{0,0}$ Gaussian distribution for the incident beam by

$$E(Z,r,t) = E_0(t)\frac{w_0}{w(Z)}\exp\left(-\frac{r^2}{w^2(Z)} + i\frac{\pi r^2}{\lambda R(Z)} + i\phi\right), \tag{2.26}$$

where $w(Z) = w_0(1 + Z^2/Z_0{}^2)^{1/2}$ and $R(Z) = Z + Z_0{}^2/Z$. The radially invariant phase terms contained in ϕ are immaterial to our calculations and hence will be ignored.

The integral in Eq. (2.24) can be analytically evaluated if we assume that $|q| < 1$ (in Eq. (2.23)). We can then perform a binomial series expansion of E_e in powers of q. Recalling that $q \propto I \propto \exp(-r^2/w^2)$, this expansion effectively decomposes E_e into a sum of Gaussian beams with varying beam parameters. This method of beam propagation, known as *Gaussian decomposition*, was first given by Weaire *et al.* [31]. Following the expansion, we obtain

$$E_e = E(Z,r,t)e^{-\alpha L/2}\sum_{m=0}^{\infty} F_m \exp(2mr^2/w^2(Z)), \tag{2.27}$$

where F_m, the factor containing the nonlinear optical coefficients, is given by

$$F_m = \frac{(i\Delta\phi_0(Z,t))^m}{m!}\prod_{j=1}^{m}\left[1 + i\left(j - \frac{1}{2}\right)\frac{\lambda\beta}{2\pi n_2}\right]. \tag{2.28}$$

Remarkably, this single-beam technique has a demonstrated sensitivity to induce phase distortion of $\sim\lambda/10^3$, while a simple variation, the "*EZ*-scan," which replaces the aperture with a stop, has a demonstrated sensitivity of $\lambda/10^4$ [29]. There have been multiple other modifications of the basic *Z*-scan reported in the literature [27, 32–36]. This interferometric-like sensitivity comes about by the sample serving as a phase mask, which upon propagation to the far field is transformed into a spatial amplitude redistribution, that is, via diffraction. That is, diffraction is an interference phenomenon between the wings of the beam that are weak and propagate nearly linearly,, and the center of the beam that undergoes a nonlinear phase shift, either advanced as for self-defocusing ($n_2 < 0$) or delayed as for self-focusing ($n_2 > 0$). Thus, the *Z*-scan serves as a single-beam interference technique without the alignment complexities of interferometers.

Dual-Arm Z-Scan Figure 2.5 shows an example of *Z*-scans where only 2PA and n_2 are observed. Mostly, solvents do not exhibit 2PA in the visible or near-infrared, so any observed 2PA can be attributed to the solute. However, the same is not true of NLR, and any measured NLR comes from a combination of the NLR of the solvent and the solute. In order to determine the n_2 of the solute, the NLR of the solvent must be subtracted. Since sometimes the solubility of molecules is limited, the solvent n_2 can overwhelm the solute NLR and make its determination problematic. One method to alleviate this problem is the "dual-arm" (DA) *Z*-scan [28].

In the DA *Z*-scan (experiment shown in Figure 2.6), a second *Z*-scan arm is arranged to have an optical path identical to that of the first arm. When initially

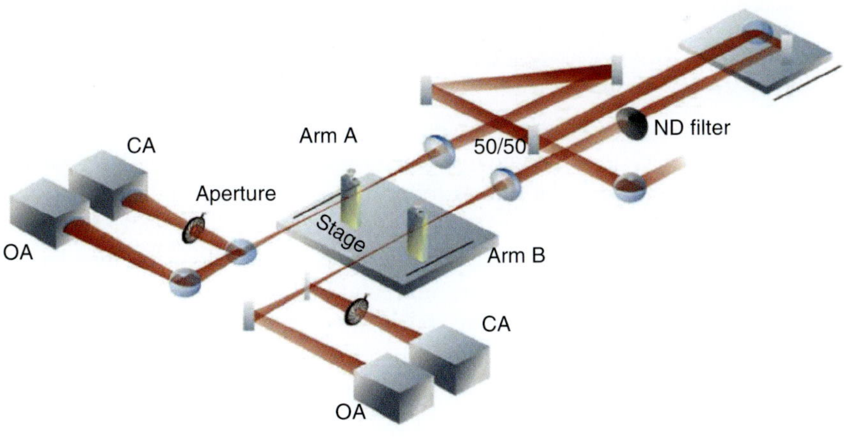

Figure 2.6 Schematic of dual-arm *Z*-scan. The items labeled CA and OA represent the closed-aperture and open-aperture detectors for each arm, respectively. The reference beam used for energy monitoring is not shown. (Ferdinandus 2012 [28].)

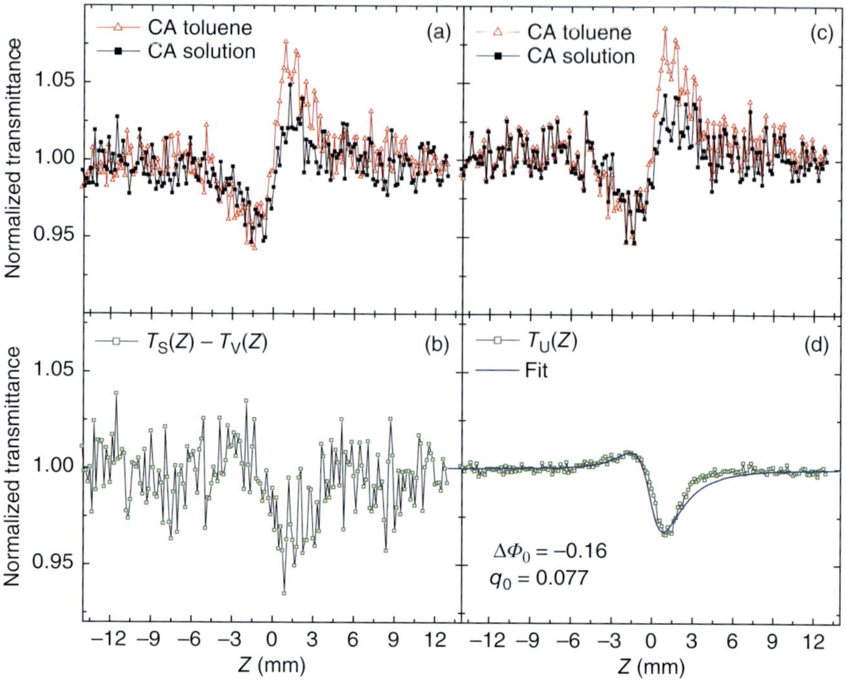

Figure 2.7 (a) Sequential CA single-arm *Z*-scans of the solvent toluene (open red triangles) and the solution of SD-O 2405 in toluene (closed black squares) at 695 nm where the concentration $C = 47\,\mu m$ and the pulse energy $E = 31$ nJ ($I_0 = 51$ GW cm^{-2}). (b) Subtraction of the solvent CA signal from the solution CA signal (open green squares). Note that this is for the same molecule as shown in Figure 2.5. (c) Simultaneous CA dual-arm *Z*-scans of the solvent toluene (open red triangles) and the solution SD-O 2405 in toluene (closed black squares) at 695 nm using the same pulse. (d) the subtraction of the solvent CA signal from the solution CA signal after low-energy background signal (LEB(*Z*)) Subtraction (open green squares) and corresponding fit of both 2PA and NLR (solid blue line) with $\Delta\phi_0 = -0.16$ and $q_0 = 0.077$ using $S = 0.33$. (Ferdinandus 2012 [28].)

aligned with identical cuvettes and the same solvent in each arm, the signal, which is the difference between the signals from each arm, can be nulled. When the solute is added to one arm, the resultant difference signal is then only due to the solute. In practice, we find that this increases the S/N for NLR by a factor of ~10× (see Figure 2.7 of a squaraine molecule showing the increased S/N). The S/N increase for NLA is less dramatic. This increase in S/N for NLR comes about because much of the noise in each arm correlates with each laser firing. This noise is effectively subtracted on each laser shot. For example, as closed-aperture *Z*-scans are sensitive to the beam-pointing instabilities, the dual arms provide an excellent way to reduce their effect because both arms move synchronously, thus giving the same change in signal from this beam movement. This technique also works well for thin-film samples on substrates to subtract the substrate signal and correlate noise [37]. Figure 2.7 shows the effectiveness of using this method for a squaraine dye in toluene showing both NLA and NLR.

2.3.1.3 Determining Nonlinear Response from Pulse-width Dependence of Z-Scans

Z-scans of materials exhibiting 1PA-induced ESA (see Eqs. (2.9)–(2.12) cannot easily be distinguished from *Z*-scans performed on materials with only ultrafast nonlinearities. One way to distinguish these is to perform the *Z*-scan at two different pulse widths. If the *Z*-scans are identical for the same irradiance but with different pulse widths, the response time of the nonlinearity is shorter than the temporal width of the pulses used. For very short pulses (e.g., femtosecond), this would indicate that the mechanism is 2PA. If the open-aperture *Z*-scans are identical for the same fluence, then they may be due to ESA. Similarly, if the closed-aperture *Z*-scans are the same for the same fluence rather than irradiance, then the NLR may be due to the redistribution of the populations.

In order to illustrate this, we show the *Z*-scan data for a sample of chloroaluminum phthalocyanine (CAP) in Figure 2.8 using a fixed energy but with pulse widths differing by a factor of 2 [15]. Thus, the irradiance changes by a factor of 2 while the signals for both NLA and NLR are nearly identical. Both NLA and NLR are due to excited states rather than to 2PA and bound-electronic n_2.

Creating new linear absorption resonances by creating excited states changes the index, and the sign depends on which side of the resonance the light is; additionally, removing absorbers from the ground state also changes the index, and, again, the sign depends on the resonance frequency, and these contributions add to determine the net NLR. These can be determined by the usual Kramers–Kronig relations for the materials with the changed absorption profile, that is, including the new ESA or saturated absorption profile. It is important to note that this is not n_2 because it depends on the density of excited states created, which is energy-dependent.

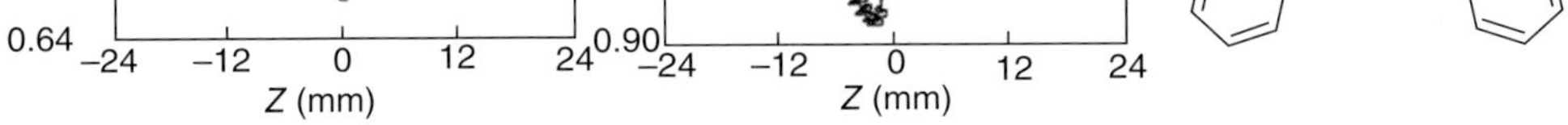

Figure 2.8 (a) Open-aperture *Z*-scan data at 532 nm using (open triangles) 30-ps pulses (FWHM) and (open squares) 62-ps pulses (FWHM) on chloroaluminum phthalocyanine and (b) closed-aperture *Z*-scan data (divided by open-aperture data) for the same pulse widths. The molecule is shown at (c). (Wei 1992 [15]. Reproduced with permission of Springer.)

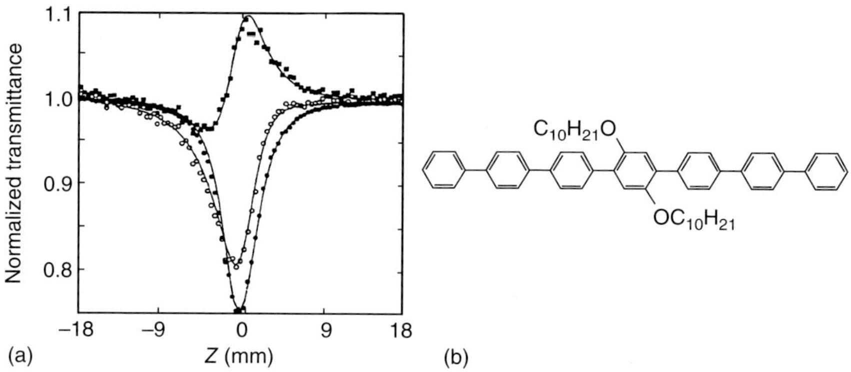

Figure 2.9 Open (open circles), closed (closed circles), and divided (closed squares) Z-scan data of the organic molecule shown on the right. The solid lines are fits to the data. (Said 1994 [38]. Reproduced with permission of Elsevier.)

When the excited states are created by 2PA, as described in Eq. (2.14), the NLA can mimic a fifth-order 3PA while the NLR also appears as a fifth-order NLR signal [18, 38]. Figure 2.9 shows a molecule in which ESA and ESR occur because of 2PA.

Here, the equations are modified in a straightforward way following Eq. (2.14) but ignoring linear absorption. The irradiance varies as

$$\frac{dI(t)}{dz} = -\alpha_2 I^2(t) - \sigma_{ex} N_e(t) I(t) \tag{2.29}$$

while the excited-state generation varies as

$$\frac{dN_e}{dt} = \frac{\alpha_2 I^2(t)}{2\hbar\omega}. \tag{2.30}$$

Although these do not have a simple solution as in the 1PA-generated excited-state case, they can be numerically evaluated for Z-scans, as shown in the fits of Figure 2.9. Again, for a fixed irradiance, longer pulses contain more energy, and the role of ESA and ESR is increased. Some researchers have quoted effective 3PA coefficients for the NLA portion of these equations by temporally integrating the excited-state density. For example, assuming Gaussian temporal pulses of HW1/eMt_0 yields

$$N_e(t) = \frac{\alpha_2}{2\hbar\omega}\int_{-\infty}^{t} I^2(t')dt' = \frac{\alpha_2 I_0^2}{2\hbar\omega}\int_{-\infty}^{t} e^{-2(t'/t_0)^2}dt' \tag{2.31}$$

which can be used in Eq. (2.29). If the pulses are long enough, the second term in Eq. (2.29) dominates, and since $N_e(t) \propto I^2$, the open-aperture Z-scan signal looks like 3PA with $N_e(\infty) = \sqrt{\frac{\pi}{2}}\frac{\alpha_2 I_0^2 t_0}{2\hbar\omega}$, which when used in Eq. (2.29) yields an I^3 dependence for the spatial derivative of the irradiance. This provides a quick estimate of the size of the NLA. However, it also points out how easily this cascaded 2PA + ESA (i.e., Im $\chi^{(3)}$:Im $\chi^{(1)}$) effect can be misinterpreted in an experiment as 3PA [18]. The NLR from creating these excited states also looks like a fifth-order

"n_3" using similar math (Im $\chi^{(3)}$:Re $\chi^{(1)}$); however, again this does not give the essential physics behind the NLR.

2.3.1.4 White-Light-Continuum *Z*-Scan (WLC Z-Scan)

Performing nonlinear spectroscopy using *Z*-scan can be time consuming because most broadly tunable sources, for example, optical parametric sources, require lengthy characterization each time they are tuned since the temporal and, in particular, the spatial profiles are not reproducible. Usually, spatial filtering is required if Gaussian profile pulses are to be used. Again, this is not a necessary condition for performing *Z*-scans but it greatly simplifies the analysis [39–41].

Nonlinear spectroscopy using *Z*-scan is possible, as shown in Figure 2.10, where the 2PA spectrum of a squaraine molecule shown in the figure was measured by performing open-aperture *Z*-scans at many wavelengths using an optical parametric amplifier (OPA). Such a spectrum can take several days to perform because the *Z*-scan needs to be realigned for each wavelength, as the OPA beam alignment changes upon tuning [42].

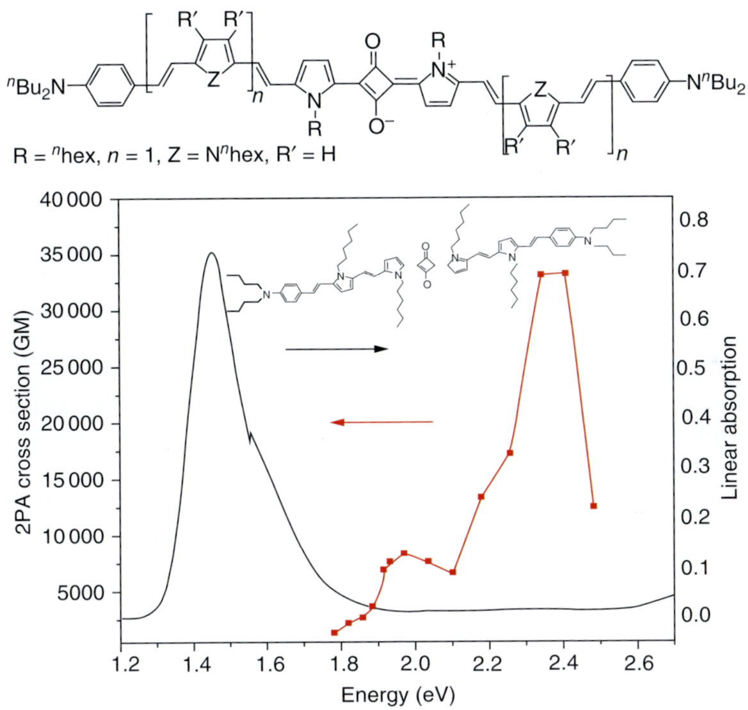

Figure 2.10 Linear absorption (right arbitrary axis) along with 2PA cross section (left axis) determined by open-aperture *Z*-scans versus photon energy for the molecule shown. The energy corresponds to the energy of final excitation for both linear and two-photon absorption; for example, the 2PA peak is near an input photon energy of 2.35/2 eV. (Jchung 2008 [42]. Reproduced with permission of American Chemical Society.)

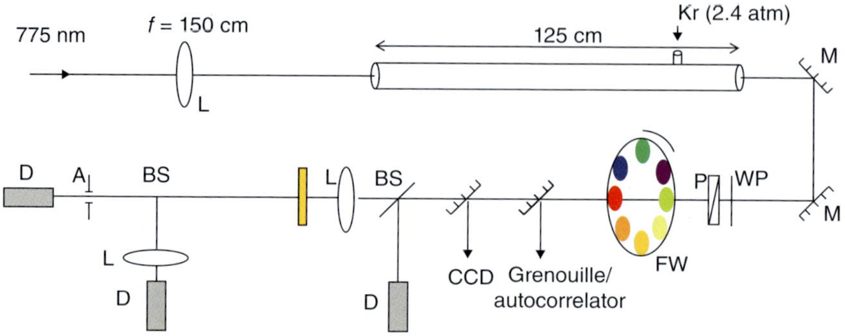

Figure 2.11 WLC Z-scan experimental setup: L, lens; M, mirror; WP, half-wave plate; P, polarizer; FW, filter wheel; BS, beam splitter; D, detector; A, aperture; S, sample; and dotted M's, removable mirrors for beam characterization. (Balu 2008 [40]. Reproduced with permission of The Optical Society.)

A solution that allows rapid Z-scan characterization of materials over a broad spectral range is the white-light-continuum, WLC Z-scan [39–41, 43]. Here, the Z-scan methodology is the same but we replace the usual "single" frequency source in the Z-scan with a spectrally filtered WLC [44].

The apparatus is shown in Figure 2.11. The WLC is produced by weak focusing in a long (~1.5 m) cell filled with Kr gas at a pressure of ~2 atm. Using Gaussian spatial profile, femtosecond excitation pulses of ~1 mJ at ~800 nm produce a clean spatial profile pulse of spectral content from <400 to >800 nm, that is, greater than one octave.

The entire WLC cannot be used in a single Z-scan because, in addition to the degenerate 2PA, there will be a strong nondegenerate 2PA, and these two processes cannot be simply separated. The simplest method to apply is to spectrally filter the WLC prior to the sample and perform a normal single-frequency Z-scan. Then the spectral filter can be changed (e.g., using a computer-controlled rotating platform holding many spectral filters as shown in Figure 2.11, or using a linear variable filter) and another Z-scan is performed.

The filters must be spectrally broad enough to support the pulses used, which is ~100 fs in this case. The difference between using the WLC and, for example, tuning an optical parametric generator/amplifier (OPG/A) is in the length of time it takes to measure a nonlinear spectrum (or nonlinear dispersion curve). As it turns out, once characterized in terms of the spatial properties and temporal profiles at all wavelengths, the WLC is stable from day to day and week to week. This cannot be said of optical parametric sources. We have found that OPG/As when tuned change their spatial profile, and usually spatial filtering is needed to obtain Gaussian profile beams. Figure 2.12 shows the characterization of a WLC demonstrating that near-Gaussian profile beams are obtained at all the wavelengths over an octave span. New research is extending the useful range to greater than two octaves using different excitation wavelengths. Thus, once characterized, the WLC

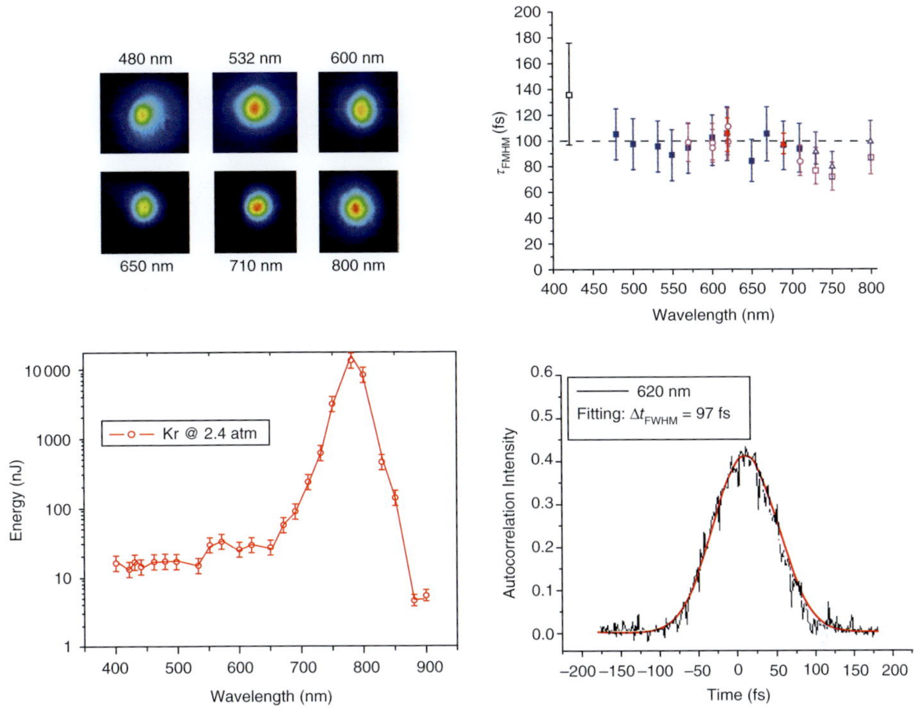

Figure 2.12 Characterization of the WLC. (a) Spatial profiles, (b) temporal pulsewidth, (c) available energy after narrow band filters, and (d) example of an autocorrelation after a 620-nm narrow band filter (~10 nm bandwidth). (Balu 2008 [40]. Reproduced with permission of The Optical Society.)

can be used to perform nonlinear spectroscopy, while the OPG/A needs to be recharacterized after each tuning cycle.

This WLC *Z*-scan has the potential of serving as the nonlinear equivalent of a linear spectrophotometer. When automated, a researcher can insert a sample, initiate a measurement, and then after sometime (perhaps <1 h for kilohertz sources) retrieve the degenerate NLA spectrum; however, with this nonlinear spectrophotometer, there is an added benefit not available with a linear spectrometer. The dispersion of the NLR is measured simultaneously with the closed-aperture *Z*-scan, as shown in Figure 2.11. The data of Figure 2.13 were taken using this configuration.

2.3.1.5 Other Variants of the Z-Scan Method

Determining the physical mechanism or mechanisms responsible for the signal is seldom revealed by a single *Z*-scan experiment, although ultrashort femtosecond pulses can sometimes limit the significant responses to bound-electronic nonlinearities [45, 46]. Figure 2.8 shows that either one of the *Z*-scans could have been used to extract erroneous values for a 2PA coefficient and n_2. There have been

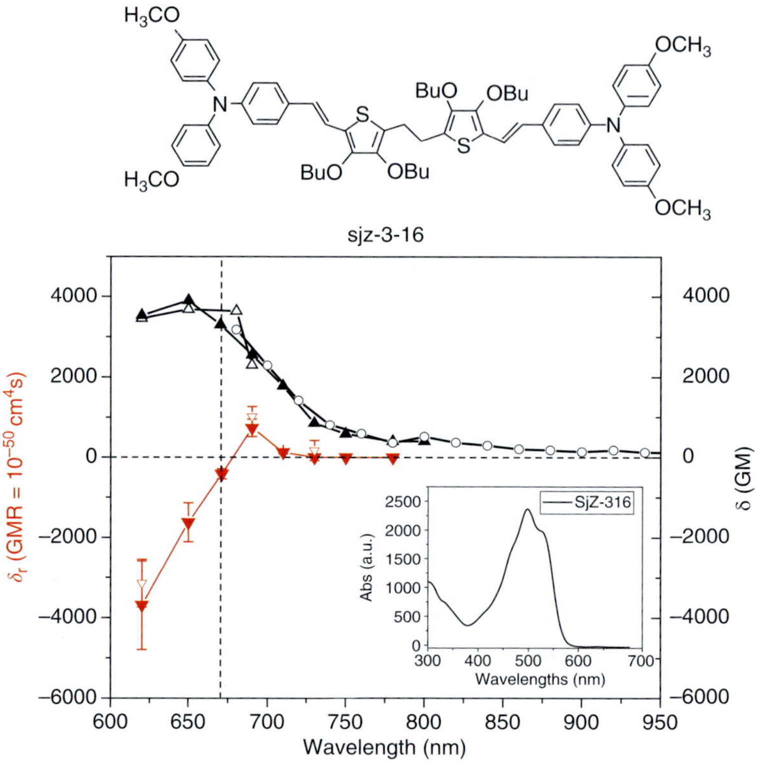

Figure 2.13 2PA cross section (right axis) and nonlinear refractive cross section (left axis) for the molecule shown to the right. Data taken using the WLC Z-scan, except for wavelengths longer than 800 nm where two-photon fluorescence is used (open circles). (Balu 2008 [40]. Reproduced with permission of The Optical Society.)

many variations of the Z-scan introduced over the past years. Many of these have shown the potential to further increase the sensitivity, for example, the EZ-scan [29], while others have introduced multiple beams, which have the potential to allow the measurement of nondegenerate nonlinearities [27, 35, 47]. However, all these "improvements" come at the cost of complicating the methodology, and the simplicity of this technique is the likely reason why it became so popular.

The 4f coherent imaging method [48] and many other techniques are closely related to the Z-scan. Here, for example, a beam profile is focused through a nonlinear sample and imaged in the linear regime. Then the irradiance is increased and the change in the imaged irradiance distribution is observed. The nonlinearly induced phase distortion can be calculated and the nonlinear index determined. NLA can also be analyzed. In the particular setup given in [48], a circular flat-top beam is imaged and the light going into the originally dark area in the image plane is monitored. When it is fully aligned, the nonlinear index (and/or absorption) can be determined in a single shot.

As previously stated, the Z-scan technique is sensitive to all NLA processes and all NLR mechanisms. Unraveling the various physical processes that contribute to these nonlinearities cannot be unambiguously determined with single Z-scans.

2.3.2 Indirect Methods

By "indirect" we mean that a material is excited optically to be sensed by some other means than the transmission of the excitation beam itself. These "other means" could include measuring the transmittance of a probe, heating of materials, fluorescence, and so on. For example, an excitation beam induces excited states via 2PA whose presence is sensed by their fluorescence in which the intensity is proportional to how many excited states were produced via 2PA. We give examples of these indirect methods in the following.

2.3.2.1 Excitation–Probe Methods

Figure 2.14 shows the standard excite–probe experimental setup where an excitation pulse is incident on the system at time $\tau_d = 0$ and the probe pulse examines the change in transmission properties of the sample at later times, that is, $\tau_d > 0$. For example, see articles in Ref. [49].The excitation and probe pulses could be at the same (degenerate) or different (nondegenerate) wavelengths, but in order that they are synchronized in time, they must be derived from the same pulsed laser source. The time delay τ_d of the probe is controlled by a delay line, as shown in Figure 2.14.

2PA can easily be differentiated from ESA using excite–probe techniques since, for times after a short pulse excitation (pulse width shorter than the decay time of the system τ), ESA will be present after the excitation pulse is over, that is, for τ greater than the pulse width. For example, for 1PA-generated excited states, the dynamics are described by a modification of Eq. (2.9), which includes simple decay:

$$\frac{dN_e}{dt} = \frac{\sigma_{ge} N_g I_e}{\hbar\omega} - \frac{N_e}{\tau}, \tag{2.32}$$

where the definitions are as given in Section 2.2.1, I_e is the excitation irradiance, and τ is the decay time. It is assumed that the probe irradiance is weak enough so that it generates no significant excited-state population. The transmission change is sensed as a function of time delay by the probe pulse according to

$$\frac{dI_p}{dz} = -(\sigma_{ge} N_g + \sigma_{eu} N_e) I_p. \tag{2.33}$$

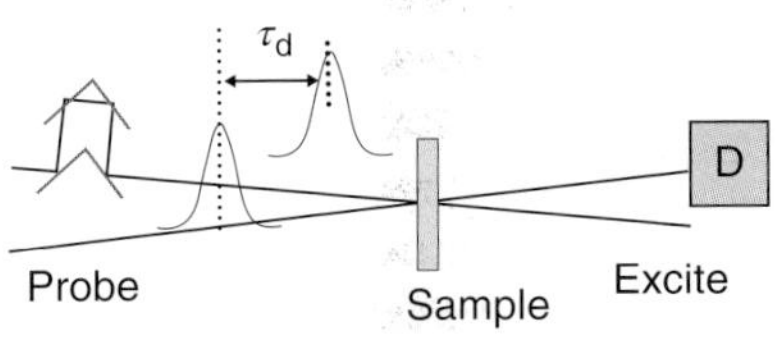

Figure 2.14 Excite–probe experimental setup showing the optical delay line.

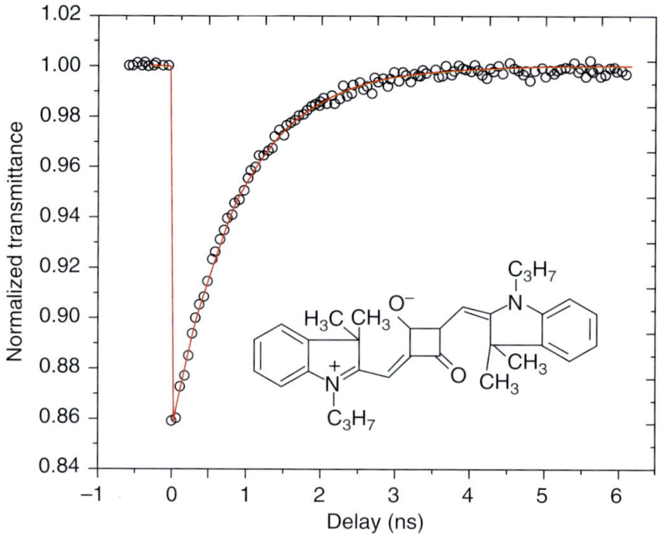

Figure 2.15 Excite–probe on the squaraine molecule shown in the inset (SD 2577) at a wavelength of 532 nm, showing ESA with a decay time of 0.9 ns. (Peceli 2013 [50]. Reproduced with permission of American Chemical Society.)

Provided the excitation and probe pulses are much shorter than the decay time τ, integration over time to find the probe fluence is straightforward. Defining the probe fluence transmittance as $T_{\text{probe}}(\tau_d) = F(\tau_d, L)/F(\tau_d, 0)$ and the normalized transmittance as $T_{\text{NL}}(\tau_d) = T_{\text{probe}}(\tau_d)/T_L$, where T_L is the linear transmittance of the probe, we find

$$T_{\text{NL}}(\tau) = [T_{\text{NL}}(0)]^{(\exp(-\tau_d/\tau))} \tag{2.34}$$

where $T_{\text{NL}}(0) = \left[1 + \frac{(\sigma_{eu} - \sigma_{ge})F_e}{2\hbar\omega}(1 - e^{-\alpha_0 L})\right]^2$ is the probe fluence transmission for the probe delay immediately after the excitation pulse. Here, we have assumed that $\alpha_0 = \sigma_{ge} N_g$ is constant, that is, small depletion of the ground state. Provided the changes in normalized transmittance are small, that is, $\Delta T = 1 - T_{\text{NL}} \ll 1$, then Eq. (2.34) becomes

$$\Delta T_{\text{NL}}(\tau) \cong \Delta T_{\text{NL}}(0) e^{-\tau_d/\tau} \tag{2.35}$$

so that the transmittance change can simply be fitted with an exponential decay to determine the excited-state lifetime. Meantime, the ESA cross section σ_{eu} can be determined from $T_{\text{NL}}(0)$. Even if we do not satisfy the conditions for Eq. (2.35), we may still fit the decay with Eq. (2.34); however, provided we remain in this low-excitation regime where there is small depletion of the ground-state population, we do not need to know the molecular density to obtain σ_{eu}. Figure 2.15 shows an example of an excite–probe experiment on a squaraine molecule (SD 2577) at a wavelength of 532 nm with 30-ps pulses [51].

This single excite–probe experiment can often determine the singlet manifold molecular parameters, but it is usually incapable of determining the triplet yield, cross section, and lifetime, which are often important for organic dyes. This can be done using a double excite–probe geometry, as discussed in detail in Ref. [52]. In these experiments, a first excitation creates a population of the triplet state, which is excited again after allowing sufficient time to create a triplet population and then probed to determine these molecular constants.

Although degenerate excite–probe methods are probably best for fully understanding the mechanisms involved in a single-beam experiment such as Z-scan, it is often useful to perform nondegenerate excite–probe measurements in which the excitation and probe are at different wavelengths. Experimentally, this has the advantage that the weak probe can be easily distinguished from the strong excitation beam using spectral filters; also, using different excitation and probe wavelengths allows other measurements to be performed. This allows, for example, the entire ESA spectrum to be measured by tuning the probe while keeping the excitation wavelength fixed [53]. Also, the excitation beam can be used to generate excited states via 2PA, which has the advantage of creating a more uniform excitation through the depth of the sample. Finally, there may be nondegenerate 2PA involving the absorption of one photon from each of the excitation and probe beams. In this case, the propagation equations for the probe and excitation beams are

$$\begin{aligned}\frac{dI_{\mathrm{p}}}{dz} &= -2\alpha_2(\omega_{\mathrm{p}};\omega_{\mathrm{e}})I_{\mathrm{e}}I_{\mathrm{p}} - \sigma_{\mathrm{ex}}(\omega_{\mathrm{p}})N_{\mathrm{e}}I_{\mathrm{p}} \\ \frac{dI_{\mathrm{e}}}{dz} &= -\alpha_2(\omega_{\mathrm{e}};\omega_{\mathrm{e}})I_{\mathrm{e}}^2,\end{aligned} \tag{2.36}$$

where $\alpha_2(\omega_{\mathrm{p}};\omega_{\mathrm{e}})$ is the nondegenerate 2PA coefficient, and it is assumed that there is no ESA experienced by the excitation beam. In the event that there is degenerate 2PA at the excitation wavelength, the excited-state population is given by

$$\frac{dN_{\mathrm{e}}}{dt} = \frac{\alpha_2(\omega_{\mathrm{e}};\omega_{\mathrm{e}})I_{\mathrm{e}}^2}{2\hbar\omega}. \tag{2.37}$$

As illustrated in Figure 2.1c, the effects of these excited states may be seen by the presence of ESA after the excitation pulse, and since the excited-state density is proportional to I_{e}^2, the ESA experienced by the probe will appear as a higher order nonlinearity than the nondegenerate 2PA, which is proportional to I_{e}. With femtosecond pulses, the population densities generated by 2PA are usually too small for significant ESA to be observed. However, Figure 2.16 shows a lead bis(ethynyl)porphyrin polymer, where 2PA-generated ESA is large using femtosecond excitation [54]. Figure 2.16 shows excite–probe data for two excitation energies differing by a factor of 2 along with fits using 2PA excitation of state e followed by ESA to state u. In this case, there is actually a multilevel excited-state system where the 2PA-excited state decays to a low-lying state in ~240 fs, which leads to the delay of the onset of ESA. The ESA scales as the input energy squared, while at zero delay there are contributions from both 2PA, which scales linearly with the energy, and ESA.

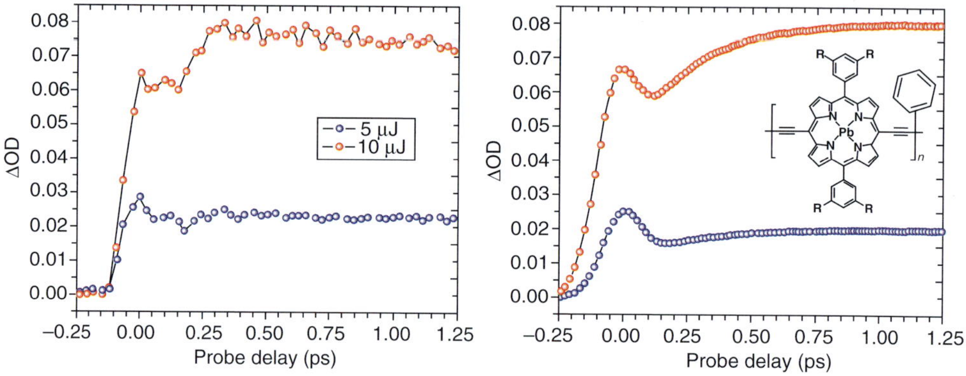

Figure 2.16 (a) Excite–probe data showing the change in optical density on the lead bis(ethynyl)porphyrin molecule shown in the inset (right) versus temporal delay for excitation energies differing by a factor of 2. (b) Fits for the two energies on the left using a combination of 2PA and ESA. (Hales 2009 [54].)

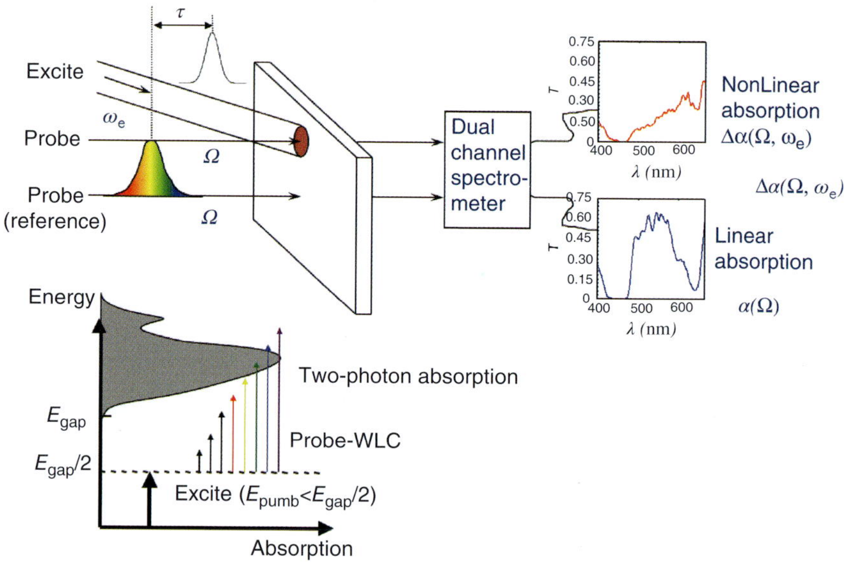

Figure 2.17 Femtosecond excitation, white-light-continuum probe spectroscopy setup showing a 2PA measurement. (Negres 2002 [55]. Reproduced with permission of The Optical Society.)

2.3.2.2 White-Light-Continuum (WLC) Excite–Probe Spectroscopy

Excitation–probe experiments using a WLC probe, as shown in Figure 2.17, have greatly speeded data acquisition of nondegenerate NLA spectra since tuning of the probe becomes unnecessary [55–57]. A commercial product based on WLC excite–probe spectroscopy has been developed by Helios Inc.

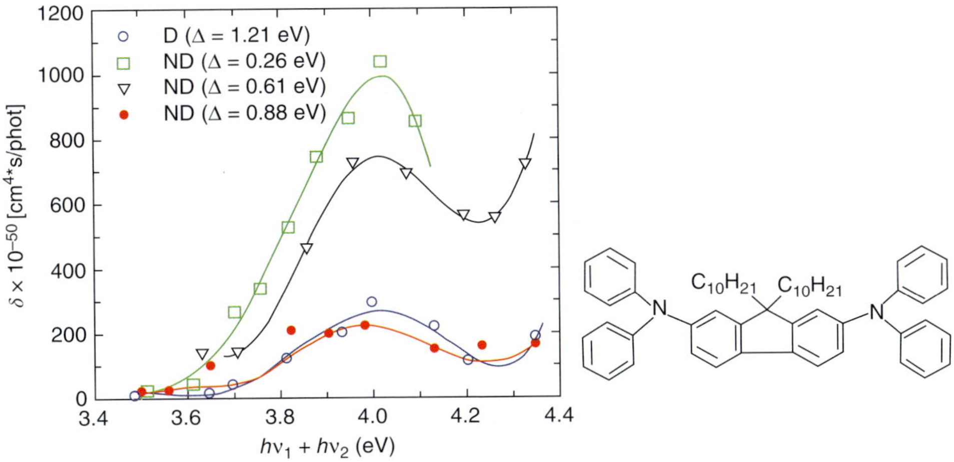

Figure 2.18 Two-photon absorption spectrum of the molecule on the right as a function of the photon energy sum for different excitation photon energies from degenerate D (blue circles) to very nondegenerate ND (green squares).

Usually, femtosecond excitation pulse is used, which is also used to create a femtosecond WLC. Here, the probe (WLC) irradiance is kept weak to avoid any nonlinear effects induced by the probe itself. Then, in most cases, the entire WLC can be incident on the sample at the same time; however, care still needs to be applied because linear absorption of the WLC can produce excited states. Thus, the femtosecond excitation prepares the sample to be probed by the WLC, which can be temporally delayed to give the spectral dynamics of the NLA. A WLC reference can also be used to give the linear transmittance spectrum, which can be used to determine the nonlinear change in transmittance at all WLC wavelengths as a function of temporal delay [55].

The WLC probe may be used to measure the nondegenerate 2PA spectrum $\alpha_2(\omega_p; \omega_e)$, as described by Eq. (2.36). Figure 2.18 shows a nondegenerate NLA spectrum of the molecule shown for different excitation wavelengths giving different levels of nondegeneracy. For the largest nondegeneracy, the nonlinear response is enhanced the most because of intermediate-state resonance enhancement (ISRE), that is, 2PA becomes larger if the intermediate state becomes closer to being one-photon absorbed.

Additionally, by tuning the excitation wavelength to a linear absorption band and delaying the WLC probe by a few picoseconds with respect to the excitation pulse, we can measure the induced absorption of the probe over a broad wavelength range to obtain the ESA spectrum. Figure 2.19 shows the ground-state and ESA spectrum of a thiacarbocyanine dye, which was obtained with a 120-fs (FWHM) excitation pulse at 900 nm and a 117-fs (FWHM) white-light probe pulse, delayed by 13 ps with respect to the excitation pulse over the spectral range of 400–800 nm. In this case, we observe an excited-state cross section that is

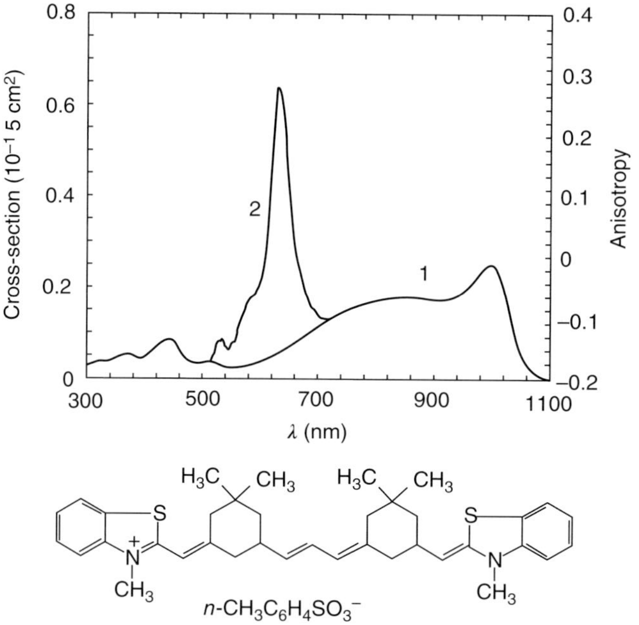

Figure 2.19 Ground-state absorption (1) and ESA spectrum (2) for the molecule shown at the bottom [58].

approximately 3 times larger than that of the ground-state cross section at the peak spectral position, while the integrated areas of their absorption bands are nearly the same.

Unfortunately, there are other experimental artifact signals that appear in this technique, which require subtraction to obtain the signals of interest. References [56] and [57] describe show how to do this. The resultant excite–probe data can be fitted using a single equation that depends on the WLC chirp, the group velocity dispersion (GVD), the excitation irradiance, and the sample's nondegenerate NLA. The GVD and the WLC chirp can be measured using the same apparatus of Figure 2.17 by adding appropriate polarizers as in the optical Kerr effect measurement [59]. This excite–probe spectroscopic method is useful for measuring both 2PA and ESA and can distinguish between these nonlinearities since two-beam experiments show the temporal dynamics.

If the nondegenerate NLA spectrum obtained is broad enough to cover all the spectral features, Kramers–Kronig relations can be applied to calculate the nondegenerate NLR dispersion [9, 60, 61]. In addition to the WLC probe for measuring NLA, a narrower spectrum, ultrashort probe pulse can be spectrally resolved to give remarkably sensitive measurements of NLR in addition to giving the NLA. Here, the reported sensitivity to a nonlinearly induced phase change is $\sim 10^{-7}$ waves [62, 63]. This method was first used by Mysyrowicz to measure the NLR of air and temporally resolve the various nuclear components coming from N_2

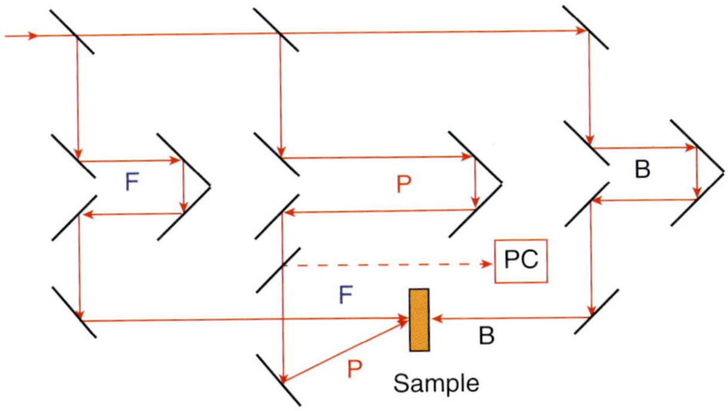

Figure 2.20 DFWM setup for pulsed measurements. F, B, and P refer to the forward excitation, backward excitation, and probe beams, respectively. PC is the phase conjugate beam or signal that is detected.

and O_2. By spectrally resolving the transmitted probe pulse, the cross-phase modulation (XPM) can be seen as a time-dependent shift of the frequency [63].

2.3.2.3 **Degenerate Four-Wave Mixing (DFWM)**

Another excite–probe measurement is degenerate four-wave mixing (DFWM), but in this case two excitation beams are used. For short-pulse, nonlinear measurements, a standard geometry is shown in Figure 2.20, where all beams can be temporally delayed. The excitation beams are usually the forward and backward excitation beams that are counterpropagating, while the probe comes in at a small angle with respect to the forward excitation. The analysis usually involves discussions of the multiple gratings produced that scatter the various beams in the phase-conjugate or signal direction, which is backward propagating from the probe. In addition to the usual grating terms, there is a so-called two-photon coherence term, which also results in signal [64–66]. The PC signal using this geometry in the low-depletion (small signal) limit is

$$\left|\chi_{\text{eff}}^{(3)}\right|^2 \cong \frac{n_0^4 c^4 \varepsilon_0^2}{\omega^2 L^2} \frac{I_c}{I_f I_b I_p}, \tag{2.38}$$

and the spot sizes and pulsewidths of the PC signal are $w_{\text{PC}}^{-2} = w_f^{-2} + w_b^{-2} + w_p^{-2}$ and $\tau_{\text{PC}}^{-2} = \tau_f^{-2} + \tau_b^{-2} + \tau_p^{-2}$, respectively. The important result to note is that the nonlinear susceptibility phase is not given and, thus, both real and imaginary parts contribute to the PC signal and they cannot be separated without determining this phase. Various homodyne and heterodyne techniques have been developed, but they often add a great deal of complication [67, 68]. The geometry discussed next can simplify this problem.

Another standard geometry is the so-called boxcar geometry shown in Figure 2.21, where three input beams (four where homodyne detection is used to determine the phase) are arranged to give a PC/signal beam, which forms the fourth corner of the "box" [69]. If the phase of the signal with respect to the excitations can be measured, this allows separation of the absorptive and

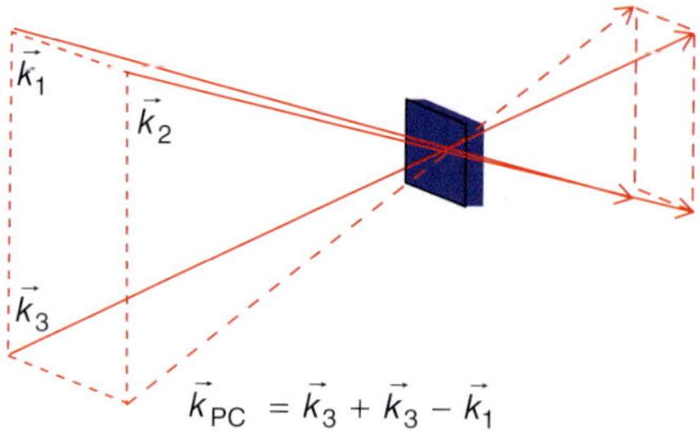

Figure 2.21 Boxcar geometry for DFWM. The dashed line is the phase-matched signal/PC. For homodyne detection, a weak beam is also incident along the direction to obtain the phase information.

refractive nonlinearities. These beams can be generated using diffraction gratings (2D gratings can produce all the needed beams from a single input and essentially eliminate the need for phase-stabilized beams) [69, 70].

The magnitude of the nonlinear susceptibility can be determined with the use of a known reference sample from

$$\left|\chi_s^{(3)}\right| = \left|\chi_r^{(3)}\right| \sqrt{\frac{I_s}{I_r}} \left(\frac{n_s}{n_r}\right)^2 \frac{L_s}{L_r} \left[\frac{\ln T}{\sqrt{T}\,[1-T]}\right], \tag{2.39}$$

where L is the sample length, the subscripts s and r refer to the sample and reference, respectively, and T is the linear transmittance of the samples. Here, the reference sample is assumed to be nonabsorbing and the experimental conditions are identical. The signals from the reference and sample are I_r and I_s, which are the maxima in the DFWM signals with time delay. Figure 2.22 shows an example of a DFWM signal in the boxcar geometry without homodyne detection on a sample

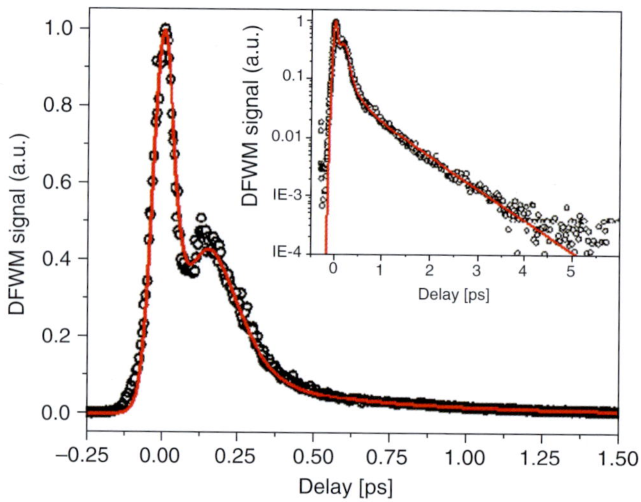

Figure 2.22 DFWM data as a function of the probe time delay using the boxcar geometries on CS_2 for 88 and 42 fs pulse widths of the excitation and probe pulses, respectively, at 700 nm. The solid line is a fit using the bound-electronic and nuclear responses. Inset uses a logarithmic vertical axis. (Reichert 2014 [70].)

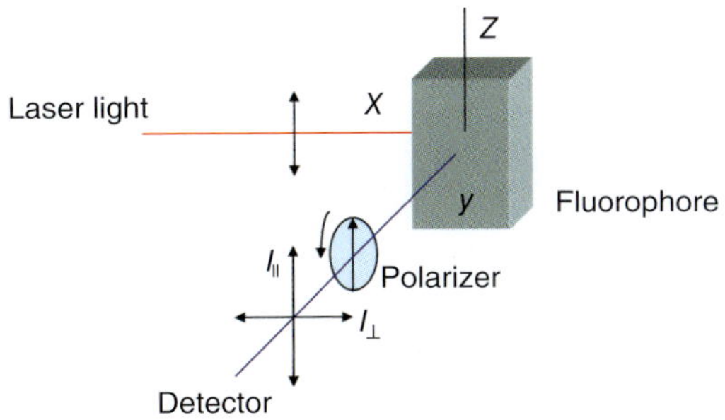

Figure 2.23 Experiment schematic for the two-photon fluorescence measurements.

of carbon disulfide (CS_2), where the solid line comes from a calculation of the expected signal from the known response function determined in Ref. [70].

2.3.2.4 Two-Photon-Absorption-Induced Fluorescence Spectroscopy

Many organic molecules show significant fluorescence quantum yields. This allows measurements of both one-photon absorption-induced fluorescence (1PF) and two-photon absorption-induced fluorescence (2PF) to obtain a degenerate 2PA spectrum. A typical experimental setup is shown in Figure 2.23.

After simultaneous absorption of two photons, the excited state rapidly relaxes to the zero vibrational level of the first excited state. The fluorescence intensity is a measure of the population of the first excited state, which is proportional to the 2PA coefficient. Since the 2PA is proportional to the square of the input irradiance I, the resulting fluorescence signal is also proportional to I^2. This dependence is thus a standard criterion to distinguish 1PF from 2PF, which was first used by Kaiser and Garrett in 1961 [71], just 1 month after second harmonic generation was reported [72]. This technology was further developed to measure 2PA spectra [73, 74] and for two-photon microscopy [75, 76]. Due to the uncertainty of various experimental parameters – for example, collection geometry and efficiency – absolute calibration of the fluorescence signal to determine the 2PA is problematic. Standard reference molecules with known 2PA spectra are often used to measure relative optical fluorescence signals [73].When using 2PF to determine 2PA spectra, as shown in Figure 2.23, one needs to minimize strong reabsorption of the fluorescence signal for high sample solution concentrations by focusing the excitation beam near the wall of the cuvette close to the collecting lens. The fluorescence spectrum also needs to be corrected for the spectral sensitivity of the detector and for any residual reabsorption by the solution. To determine the 2PA spectrum of the sample solution, the integrated fluorescence signal is then compared to a reference molecule with a known 2PA cross section under identical measuring conditions. The 2PA cross section for the sample can then be calculated from

$$\delta_{\mathrm{2PA\,sample}} = \frac{\langle F\rangle_{\mathrm{sample}} n^2_{\mathrm{sample}} \Phi_{\mathrm{F\,ref}} C_{\mathrm{M\,ref}} \langle P\rangle^2_{\mathrm{ref}}}{\langle F\rangle_{\mathrm{ref}} n^2_{\mathrm{ref}} \Phi_{\mathrm{F\,sample}} C_{\mathrm{M\,sample}} \langle P\rangle^2_{\mathrm{sample}}} \delta_{\mathrm{2PA\,ref}} \tag{2.40}$$

where $\langle F \rangle$ is the integrated fluorescence and $\langle P \rangle$ is the average excitation power which is proportional to pulse energy, C_M is the concentration of the solution obtained by measuring the 1PA spectrum, and Φ_F is the fluorescence quantum yield measured by 1PF.

There are several other factors contributing to accurate 2PF measurements. The concentration of the solution should be as high as possible (but below the aggregation threshold) to increase the probability of 2PA; thus, reabsorption can be a significant issue. Assuming the fluorescence spectrum and quantum yield are independent of the concentration of the solution and the type of excitation (i.e., 1PA or 2PA), the 2PF spectrum can be fitted to the 1PF spectrum measured at a lower concentration (with peak OD < 0.1) to obtain a correct spectral contour. The selection of a reference molecule is based on the spectral range of the fluorescence of the sample. A large overlap of the fluorescence spectra will help minimize the measurement error induced by the spectral sensitivity of the detector. Several 2PF standards have been published with emission wavelengths ranging from 390 to 1000 nm and excitation wavelength from 500 to 1600 nm [73, 77, 78]. However, there are still spectral ranges where new/more standards are needed.

Figure 2.24 shows the 2PF spectrum of an organic dye dissolved in chloroform calibrated against a known standard [78]. *Z*-scans of the 2PA spectrum are also shown, which are in agreement with the 2PF data.

2PF spectroscopy is, in principle, a rapid experimental technique to measure 2PA spectra in comparison to other commonly adopted techniques, for example, *Z*-scan. However, verifying the quadratic dependence on irradiance at each wavelength considerably lengthens the experimental time. It is also not sensitive to the

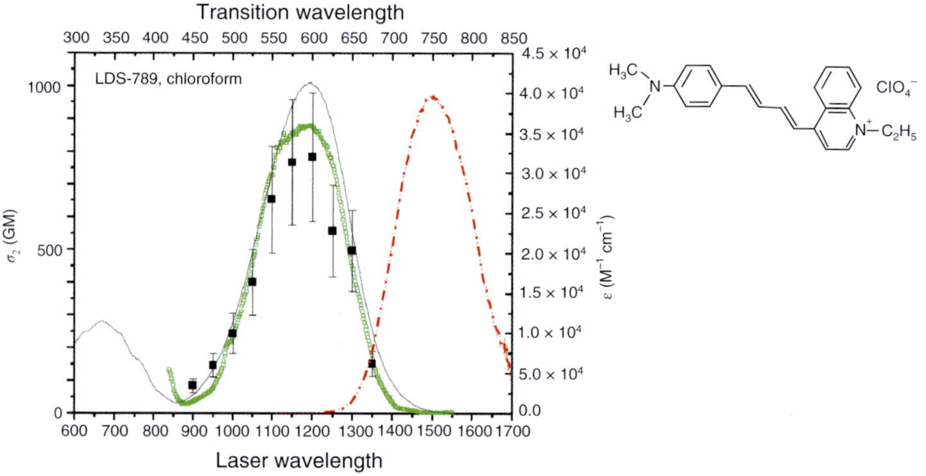

Figure 2.24 Linear absorption spectrum (black line, right vertical and top horizontal scales), normalized fluorescence spectrum (red dash-dot line, top horizontal scale), 2PA spectrum using reference calibrated 2PF (green symbols, left and bottom scales), and 2PA spectra by *Z*-scan (black squares) of the molecule shown to the right. (Makarov 2011 [78]. Reproduced with permission of The Optical Society.)

ESA, which can follow 2PA because ESA does not contribute to the fluorescence. The sensitivity of 2PF can be very high, thanks to the high responsivity of photomultiplier tube (PMT) detectors and the fact that 2PF is a zero-background signal technique. A 2PA cross section of ~1 GM can be measured given a sample solution with large fluorescence quantum yield at high concentration.

However, there are molecules whose fluorescent quantum yield is too low to measure by 2PF. It is also difficult to measure 2PA spectra of solid-state samples because reference material geometries will be difficult to match although the relative spectral shape of 2PA spectra can be determined. Finally, the 2PF technique can only measure 2PA, that is, imaginary part of $\chi^{(3)}$. The nonlinear refractive index n_2 has to be measured by other experimental techniques.

2.3.2.5 Fluorescence Anisotropy

One-photon excitation fluorescence anisotropy measurements, especially when linked to quantum chemical calculations, can reveal the spectral positions and orientations of the transition dipole moments from the ground state S_0 to the first excited state $S_1(\mu_{01})$ and to higher excited states $S_n(\mu_{0n})$ relative to the orientation of the emission dipole moment (μ_{10}). This information cannot be fully obtained from linear absorption spectra. The excitation anisotropy spectrum $r(\lambda)$ is defined as

$$r(\lambda) = [I_{||}(\lambda) - I_{\perp}(\lambda)]/[I_{||}(\lambda) + 2I_{\perp}(\lambda)], \tag{2.41}$$

where $I_{||}$ and $I|_{\perp}$ are the fluorescence intensities polarized parallel and perpendicular to the excitation light, respectively [79]. The angle β between the absorption transition moment and the emission transition moment can be determined from [74]

$$r = \frac{2}{5}\left(\frac{\cos^2\beta - 1}{2}\right). \tag{2.42}$$

Similarly, 2PA anisotropy can be measured by two-photon excitation and measuring the ratio of parallel and perpendicular polarizations, which for a three-level system gives r_{2PA} as [80]

$$r_{2PA} = \frac{18\cos(\gamma/2 - \beta_{01})\cos\beta_{01}\cos\gamma/2 - 7\cos^2\gamma/2 + 1}{7(2\cos^2\gamma/2 + 1)}. \tag{2.43}$$

Here, γ is the angle between dipole moments participating in the 2PA process, and β_{01} is the angle between the bisector of γ and the emission transition moment, which can be found from one-photon anisotropy measurements in Eq. (2.42) [81].

Figure 2.25 shows the fluorescence anisotropy spectra for both 1PA and 2PA for a squaraine dye along with other data described in the next section [82].

2.4 Examples of Use of Multiple Techniques

In this section, we describe how organic dyes are characterized by the combined use of several methods. Figure 2.24 already showed a dye where both 2PF and

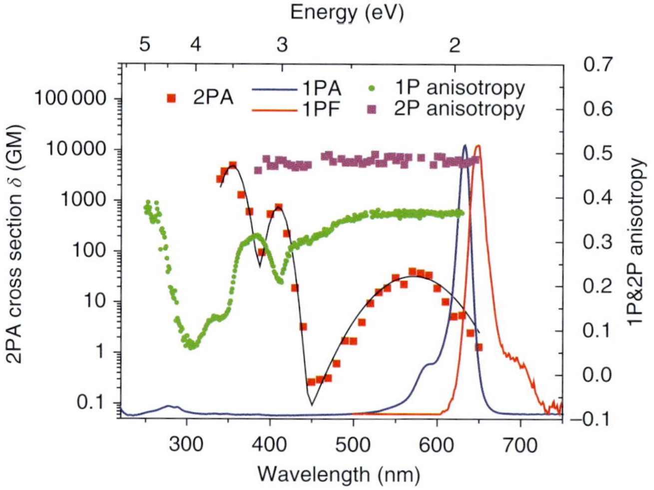

Figure 2.25 Linear absorption spectrum (blue line) of the squaraine molecule (SD 2577 shown in Figure 2.15) with the one-photon fluorescence, 1PF (red line), 1PF anisotropy (green circles), 2PF anisotropy (purple squares), and the 2PF and *Z*-scan measurements (red squares) of the 2PA spectrum. Black line is a model of the 2PA spectrum based on a single intermediate state and three 2PA states. (Fu 2007 [82]. Reproduced with permission of The Optical Society.)

Z-scan were used to measure the 2PA spectrum, and Figure 2.25 shows multiple data for a squaraine dye.

2.4.1 Squaraine Dye

Figure 2.25 shows the results of both linear and nonlinear spectroscopy of a squaraine dye first reported in Ref. [82]. The linear absorption was measured in a solution of ethanol using a spectrophotometer, revealing a peak absorption wavelength at 632 nm with a peak extinction coefficient of $3.1 \times 10^{-5}\,M^{-1}\,cm^{-1}$. The fluorescence spectrum and fluorescence anisotropy were measured in a dilute solution using a spectrofluorometer from PTI, Inc. As described earlier, the one-photon excited fluorescence anisotropy reveals the position of the one- and two-photon transitions, which usually cannot be determined from the linear absorption spectrum [79]. Also shown in Figure 2.25 is the two-photon excited anisotropy, which shows a constant value at all wavelengths measured. The 2PA spectrum shown is obtained by a combination of 2PF and *Z*-scan measurements, where the two-photon fluorescence is measured at every wavelength; in each case the square-law dependence of fluorescence on excitation energy was verified, while the *Z*-scan was measured at three wavelengths to provide an absolute calibration of the magnitude of the 2PA cross section.

The experimentally observed 2PA spectrum for SD 2577 includes three bands: a weakly allowed band at the vibronic shoulder of the $S_0 \rightarrow S_1$ transition; a more intense band at $\lambda = 408$ nm with $\delta_{2PA} = 760$ GM; and a much more intense band at $\lambda = 350$ nm with $\delta_{2PA} = 5200$ GM. The 2PA spectrum for this molecule can be directly mapped onto the anisotropy function $r(\lambda)$. The observed dips in anisotropy at 410 and at 345 nm correspond to the peak positions of the 2PA bands. The last dip at 309 nm corresponds to a position nearly twice the energy of the main $S_0 \rightarrow S_1$ absorption band. As a result, the 2PA cross section at this energy (corresponding excitation wavelength is 618 nm) cannot be experimentally determined because of linear absorption.

2.4.2 Tetraone Dye

The characterization of this dye, labeled TD2765 and shown in Figure 2.26, is similar to the squaraine dye described earlier. The methodologies are similar to those described in Section 2.4.1, and the results are shown in Figure 2.26.

The 2PA spectrum in this case was also taken using a combination of 2PF and Z-scan methods. However, close to the linear absorption edge ($\lambda \approx 700$ nm), the Z-scan gives 2PA cross sections as large as 35 000 GM. Since this is very

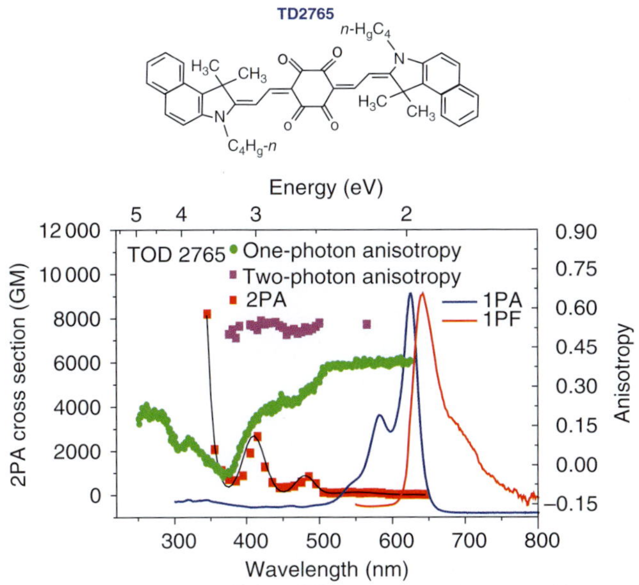

Figure 2.26 Linear absorption spectrum (blue line) of the tetraone molecule (TD2765) shown on the left along with the one-photon fluorescence, 1PF (red line), 1PF anisotropy (green circles), 2PF anisotropy (purple squares), and the 2PF and Z-scan measurements (red squares) of the 2PA spectrum. Solid black line is a model of the 2PA spectrum based on a single intermediate state and three 2PA states. Structure of TD2765 is shown above the plot. (Webster 2008 [83]. Reproduced with permission of Elsevier.)

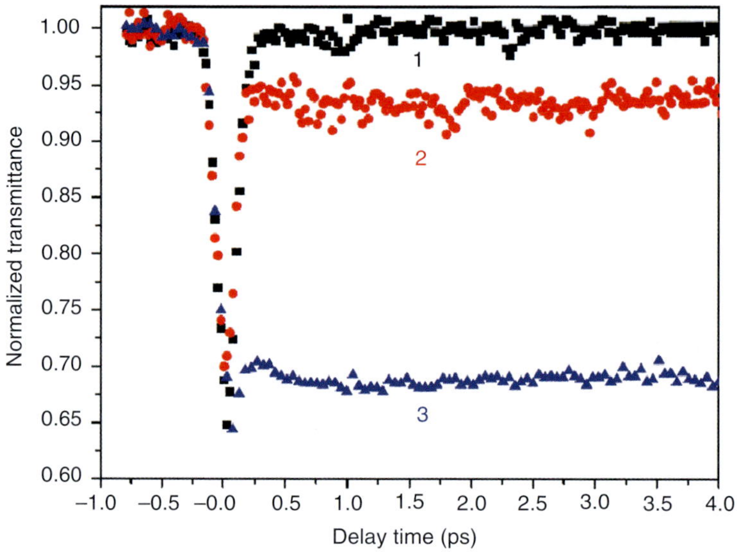

Figure 2.27 Normalized transmittance of TD2765 as a function of probe temporal delay with a probe wavelength of 670 nm and three different input excitation wavelengths getting closer to the linear absorption edge from (1) 710 nm and (2) 700 nm to (3) 690 nm for the tetraone molecule. The absorption peaks at 625 nm as shown in Figure 2.26. This shows how resonant excitation of excited states grows to nearly dominate the absorption as the wavelength is reduced. (Webster 2008 [83]. Reproduced with permission of Elsevier.)

close to the linear absorption region, one must be cognizant that ESA from excited states produced by 1PA can easily be mistaken for 2PA, as described in Section 2.2. The best means of determining the mechanisms for the NLA in such cases is to perform excite–probe measurements, since the 2PA has an ultrafast response while the ESA will have a decay time equal to the fluorescence lifetime, typically hundreds of picoseconds to nanoseconds. Figure 2.27 shows the results of excite–probe measurements of TD2765 with a probe wavelength of 670 nm and three different excitation wavelengths. Exciting at 710 nm (curve 1), no ESA is observed as verified by the full recovery of the probe signal within the pulse temporal length. At 700 nm (curve 2), excite–probe measurements show a small fraction of ESA (signal lasting after the pulse), indicating mostly 2PA evidenced by a recovery of 94% of the normalized transmittance. Significant ESA is observed with an excitation wavelength of 690 nm (curve 3). At 690 nm, the linear absorption cross section is 1.6×10^{-18} cm^2, which is almost three orders of magnitude smaller than the peak absorption cross section, yet there is significant ESA. If the ESA was not accounted for in these measurements, it would have appeared that we measured a 2PA cross section in excess of 35 000 GM, and the errors would probably be significantly larger even closer to the absorption edge. Analysis of the excite–probe data allows us to determine the relative contribution

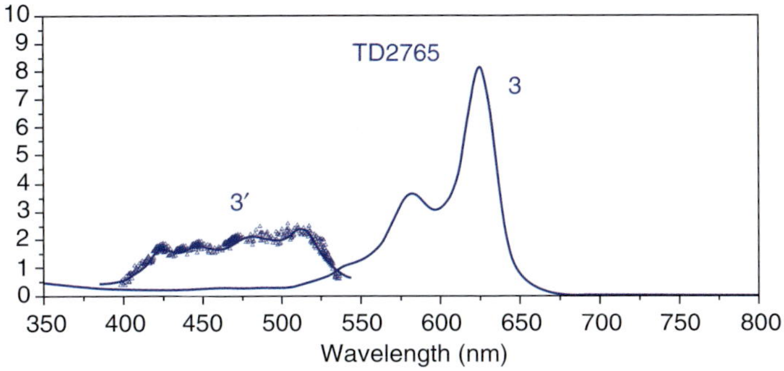

Figure 2.28 Absorption cross section for TD 2765 (3) and the corresponding excited-state absorption spectrum 3′.

of ESA and 2PA and correct the Z-scan data to give the 2PA spectrum shown in Figure 2.26.

To measure the ESA spectrum shown in Figure 2.28, a femtosecond excitation near the linear absorption resonance is used along with a WLC probe. A description of this technique can be found in Refs. [53] and [58] and is discussed in Section 2.3.2.2.

2.5 Other Methods

In this chapter, we have discussed many methods that are used in the characterization of nonlinear optical materials along with other selected techniques. There are several other methods that are not discussed. Here, we mention a few of these and give some references for further investigation. The following are the thermal lensing techniques from 2PA [84–86]:

1) Techniques related to photothermal lensing [87, 88]
   a. Photothermal displacement [89]
   b. Photothermal beam deflection [90]
   c. Photoacoustic (optoacoustic) detection [91–93]
   d. Thermal calorimetry of 2PA [94]
   e. Photothermal interferometry [95].
2) Beam deflection [96]. This is a new excite–probe method that allows a direct measurement of ultrafast or cumulative NLR and NLA to be measured with a demonstrated sensitivity to nonlinear phase distortion of $\lambda/20\,000$ [70].
3) Two-photon-induced phosphorescence to visualize singlet oxygen concentration [97].
4) Lifetime imaging to differentiate phosphorescence from fluorescence [98].
5) Multiphoton ionization spectroscopy [99–101].

2.6 Conclusion

There are many different experimental methodologies for measuring NLA. The associated NLR can often lead to problems of unambiguously determining the nonlinear loss; thus, it can be useful to simultaneously determine the NLR. This can also be valuable for determining the underlying physical process or processes that can often be difficult. Parametric studies of the wavelength dependence, temporal dependence, and sometimes the optical geometry dependence can also be important. The variety of nonlinear processes occurring in different materials are rich and complicated, and it is rarely the case that a single measurement at a single wavelength with a single pulsewidth can determine the physics involved in a nonlinear interaction. Modern nonlinear spectroscopic techniques are essential. Ultrafast nonlinearities often require femtosecond resolution to unambiguously determine their sign and magnitude. Longer pulses display other effects, such as ESA and refraction, which can even dominate the nonlinear response. The nonlinearities occurring in organic materials mimic the nonlinearities in semiconductors, where free-carrier effects take the place of excited-state effects. However, unlike in semiconductors where the band structure determines the response, the individual molecular levels in organics can be engineered, allowing the nonlinear response to be altered. However, the magnitudes of the nonlinearities are still not as large as is needed for many of the applications including all-optical switching and optical limiting.

Acknowledgments

We would like to thank the many former students and postdocs, who worked on several of the methods and materials presented in this chapter, as well as our current students. Special thanks go to our former student Joel Hales who provided data on some of the materials he studied after leaving CREOL. We also thank Mykhailo Bondar for proofreading the manuscript. We acknowledge the many sponsors of our research with recent thanks to the Air Force Office of Scientific Research (AFOSR) (FA9550-10-1-0558) and the National Science Foundation (NSF) (ECCS-1202471, ECCS-1229563).

References

1. Penzkofer, A., Falkenstein, W., and Kaiser, W. (1976) *Appl. Phys. Lett.*, **28** (6), 319–321.
2. Cumpston, B.H., Ananthavel, S.P., Barlow, S., Dyer, D.L., Ehrlich, J.E., Erskine, L.L., Heikal, A.A., Kuebler, S.M., Lee, I.Y.S., McCord-Maughon, D., Qin, J.Q., Rockel, H., Rumi, M., Wu, X.L., Marder, S.R., and Perry, J.W. (1999) *Nature*, **398** (6722), 51–54.
3. Kaplan, A.E. (1969) *Radiophys. Quantum Electron.*, **12**, 692–696.
4. Van Stryland, E.W. and Chase, L.L. (1994) in *Handbook of Laser Science and Technology; Supplement 2: Optical*

Materials, Section 8* (ed. M. Weber), CRC Press, pp. 299–328.

5. Chase, L.L. and Van Stryland, E.W. (1994) in *Handbook of Laser Science and Technology; Supplement 2: Optical Materials, Section 8* (ed. M. Weber), CRC Press, pp. 269–288.
6. Garito, A.F. and Kuzyk, M.G. (1994) in *Handbook of Laser Science and Technology; Supplement 2: Optical Materials* (ed. M. Weber), CRC Press, pp. 329–333.
7. Garito, A.F. and Kuzyk, M.G. (1994) in *Handbook of Laser Science and Technology; Supplement 2: Optical Materials* (ed. M. Weber), CRC Press, pp. 289–298.
8. Armstrong, J.A., Bloembergen, N., Ducuing, J., and Pershan, T.P.S. (1962) *Phys. Rev.*, **127**, 1918–1939.
9. Hutchings, D.C., Sheik-Bahae, M., Hagan, D.J., and Stryland, E.W. (1992) *Opt. Quantum Electron.*, **24** (1), 1–30.
10. Smith, D.D., Yoon, Y., Boyd, R.W., Campbell, J.K., Baker, L.A., Crooks, R., and George, M.M. (1999) *J. Appl. Phys.*, **86**, 6200–6205.
11. Kovsh, D.I., Yang, S., Hagan, D.J., and Van Stryland, E.W. (1999) *Appl. Opt.*, **38** (24), 5168–5180.
12. Göppert-Mayer, M. (1931) *Ann. Phys.*, **401**, 273–294.
13. Prasad, P.N. and Williams, D.J. (1991) *Introduction to Nonlinear Optical Effects in Molecules and Polymers*, John Wiley & Sons, Inc., New York.
14. Aranda, F.J., Rao, D.V.G.L.N., Roach, J.F., and Tayebati, P. (1993) *J. Appl. Phys.*, **73**, 7949–7951.
15. Wei, T.H., Hagan, D.J., Sence, M.J., Van Stryland, E.W., Perry, J.W., and Coulter, D.R. (1992) *Appl. Phys.*, **B54**, 46–51.
16. Xia, T., Hagan, D., Dogariu, A., Said, A., and Van Stryland, E.W. (1997) *Appl. Opt.*, **36**, 4110–4122.
17. Miles, P.A. (1994) *Appl. Opt.*, **33**, 6965–6979.
18. Sutherland, R.L., Brant, M.C., Heinrichs, J., Rogers, J.E., Slagle, J.E., McLean, D.G., and Fleitz, P.A. (2005) *J. Opt. Soc. Am. B-Opt. Phys.*, **22**, 1939–1948.
19. Rumi, M. and Perry, J.W. (2010) *Adva. Opt. Photonics*, **2**, 451–518.
20. Bechtel, J.H. and Smith, W.L. (1976) *Phys. Rev. B*, **13**, 3515.
21. Van Stryland, E.W., Woodall, M.A., Vanherzeele, H., and Soileau, M.J. (1985) *Opt. Lett.*, **10** (10), 490–492.
22. Van Stryland, E.W., Sheik-Bahae, M., Said, A.A., and Hagan, D.J. (1993) *J. Prog. Cryst. Growth Charact.*, **27**, 279–311.
23. Chi, S.-H., Hales, J.M., Cozzuol, M., Ochoa, C., Fitzpatrick, M., and Perry, J.W. (2009) *Opt. Express*, **17**, 22062–22072.
24. Van Stryland, E.W., Sheik-Bahae, M., Said, A.A., Hagan, D.J., and Soileau, M.J. (1994) *Proc. SPIE*, **2114**, 444–468.
25. Sheik-Bahae, M., Said, A.A., and Van Stryland, E.W. (1989) *Opt. Lett.*, **14**, 955–957.
26. Sheik-Bahae, M., Said, A.A., Wei, T.H., Hagan, D.J., and Vanstryland, E.W. (1990) *IEEE J. Quantum Electron.*, **26** (4), 760–769.
27. Ma, H., Gomes, A.S.L., and Dearaujo, C.B. (1991) *Appl. Phys. Lett.*, **59** (21), 2666–2668.
28. Ferdinandus, M.R., Reichert, M., Ensley, T.R., Hu, H., Fishman, D.A., Webster, S., Hagan, D.J., and Van Stryland, E.W. (2012) *Opt. Mater. Express*, **2** (12), 1776–1790.
29. Xia, T., Hagan, D.J., Sheik-Bahae, M., and Van Stryland, E.W. (1994) *Opt. Lett.*, **19**, 317–319.
30. Van Stryland, M.S.-B.E.W., Xia, T., Wamsley, C., Wang, Z., Said, A.A., and Hagan, D.J. (1994) *Int. J. Nonlinear Opt. Phys.*, **3** (4), 489–500.
31. Weaire, D., Wherrett, B.S., Miller, D.A.B., and Smith, S.D. (1979) *Opt. Lett.*, **4** (10), 331–333.
32. Zhao, W., Kim, J.H., and Palffy-Muhoray, P. (1994) *Appl. Phys. Lett.*, **65**, 673–675.
33. Zhao, W. and Pallf-Muhory, P. (1993) *Appl. Phys. Lett.*, **63** (12), 1613–1615.
34. Kozich, V.P., Marcano, A., Hernandez, F., and Castillo, J. (1994) *Appl. Spectrosc.*, **48**, 1506–1512.
35. Wang, J., Sheik-Bahae, M., Said, A.A., Hagan, D.J., and Van Stryland,

E.W. (1994) *J. Opt. Soc. Am. B*, **11**, 1009–1017.
36. Yang, J., Wu, X., Shi, M., Li, Z., Yang, Y., and Son, Y. (2013) *Appl. Phys. B*, **110**, 591–594.
37. Ensley, T.R., Hu, H., Ernst, A.R., Fuentes-Hernandez, C., Dindar, A., Kippelen, B., Hagan, D.J., and Van Stryland, E.W. (2013) Nonlinear Refraction Measurements of Thin Films by the Dual-arm Z-scan Method, in Nonlinear Optics, (eds B. Boulanger, S. Cundiff, M. kauranen, and W. Knox) OSA Technical Digest (online) (Optical Society of America, 2013), paper NTu1B.4.
38. Said, A.A., Hagan, D.J., Van Stryland, E.W., Reinhardt, B.A., Roderer, P., and Dillard, A.G. (1994) *Chem. Phys. Lett.*, **228**, 646–650.
39. Balu, M., Hales, J., Hagan, D.J., and Van Stryland, E.W. (2004) *Opt. Express*, **12** (16), 3820–3826.
40. Balu, M., Padilha, L.A., Hagan, D.J., Van Stryland, E.W., Yao, S., Belfield, K., Zheng, S.J., Barlow, S., and Marder, S. (2008) *J. Opt. Soc. Am. B-Opt. Phys.*, **25** (2), 159–165.
41. Balu, M., Padilha, L.A., Hagan, D.J., Van Stryland, E.W., Yao, S., Belfield, K., Zheng, S.J., Barlow, S., and Marder, S. (2009) *J. Opt. Soc. Am. B-Opt. Phys.*, **26** (8), 1663.
42. JChung, J., Zheng, S.J., Odani, T., Beverina, L., Fu, J., Padilha, L.A., Biesso, A., Hales, J.M., Zhan, X., Schmidt, K., Ye, A., Zojer, E., Barlow, S., Hagan, D.J., Van Stryland, E.W., Yi, Y., Shuai, Z., Pagani, G.A., Bredas, J.-L., Perry, J.W., and Marder, S.W. (2006) *J. Am. Chem. Soc.*, **128**, 14444–14445.
43. Balu, M., Hales, J.M., Hagan, D.J., and Van Stryland, E.W. (2005) *Opt. Express*, **13**, 3594–3599.
44. Corkum, P.B., Rolland, C., and Srinivasanrao, T. (1986) *Phys. Rev. Lett.*, **57** (18), 2268–2271.
45. Van Stryland, E.W. and Sheik-Bahae, M. (1998) in *Characterization Techniques and Tabulations for Organic Nonlinear Optical Materials* (eds M. Kuzyk and C. Dirk), Marcel Decker, pp. 655–692.
46. Van Stryland, E.W. and Hagan, D.J. (2007) in *Self-focusing: Past and Present* (eds R. Boyd, S. Lukishova, and R. Shen), Springer-Verlag.
47. Sheik-Bahae, M., Wang, J., DeSalvo, R., Hagan, D.J., and Van Stryland, E.W. (1992) *Opt. Lett.*, **17** (4), 258–260.
48. Cherukulappurath, S., Boudebs, G., and Monteil, A. (2004) *J. Opt. Soc. Am. B*, **21**, 273–279.
49. Charles, D.R., Shank, V., Erich, Dr., Ippen, P., Stanley, Dr., and Shapiro, L. (eds) (1978) *Picosecond Phenomena Proceedings of the First International Conference on Picosecond Phenomena*, Hilton Head, South Carolina, USA, 24–26.
50. Peceli, D., Hu, H.H., Fishman, D.A., Webster, S., Przhonska, O.V., Kurdyukov, V.V., Slominsky, Y.L., Tolmachev, A.I., Kachkovski, A.D., Gerasov, A.O., Masunov, A.E., Hagan, D.J., and Van Stryland, E.W. (2013) *J. Phys. Chem. A*, **117** (11), 2333–2346.
51. Lepkowicz, R.S., Przhonska, O.V., Hales, J.M., Hagan, D.J., Van Stryland, E.W., Bondar, M.V., Slominsky, Y.L., and Kachkovski, A.D. (2003) *Chem. Phys.*, **286** (2-3), 277–291.
52. Peceli, D., Webster, S., Fishman, D.A., Cirloganu, C.M., Hu, H., Przhonska, O.V., Kurdyukov, V.V., Slominsky, Y.L., Tolmachev, A.I., and Kachkovski, A.D. (2012) *J. Phys. Chem. A*, **116**, 4833–4841.
53. Negres, R., Przhonska, O., Hagan, D., Van Stryland, E., Bondar, M., Slominsky, Y., and Kachkovski, A. (2001) *IEEE J. Sel. Top. Quantum Electron.*, **7** (Organics for Photonics), 849–863.
54. Hales, J.M., Cozzuol, M., Screen, T.E.O., Anderson, H.L., and Perry, J.W. (2009) *Opt. Express*, **17** (21), 18478–18488.
55. Negres, R., Hales, J.M., Kobyakov, A., Hagan, D., and Van Stryland, E.W. (2002) *Opt. Lett.*, **27**, 270–272.
56. Negres, R.A., Hales, J.M., Kobyakov, A.K., Hagan, D.J., and Van Stryland, E.W. (2002) *IEEE J. Quantum Electron.*, **38** (9), 1205–1216.
57. Negres, R.A., Hales, J.M., Kobyakov, A.K., Hagan, D.J., and Van Stryland, E.W. (2003) *IEEE J. Quantum Electron.*, **39** (2), 392.

58. Lepkowicz, R.S., Przhonska, O.V., Hales, J.M., Fu, J., Hagan, D.J., Van Stryland, E.W., Bondar, M.V., Slominsky, Y.L., and Kachkovski, A.D. (2004) *Chem. Phys.*, **305** (1-3), 259–270.
59. Yamaguchi, S. and Hamaguchi, H. (1995) *Appl. Spectrosc.*, **49**, 1513–1515.
60. Sheik-Bahae, M., Hutchings, D.C., Hagan, D.J., and Van Stryland, E.W. (1991) *IEEE J. Quantum Electron.*, **27** (6), 1296–1309.
61. Bassani, F. and Scandolo, S. (1991) *Phys. Rev. B*, **44**, 8446–8453.
62. Kang, I., Krauss, T., and Wise, F. (1997) *Opt. Lett.*, **22**, 1077–1080.
63. Ripoche, J.-F., Grillon, G., Prade, B., France, M., Nibbering, E., Lange, R., and Mysyrowicz, A. (1997) *Opt. Commun.*, **135**, 310–314.
64. Jain, R.K. and Klein, M.B. (1983) in *Optical Phase Conjugation* (ed. R.A. Fisher), Academic Press, New York, pp. 307–455.
65. Smirl, A.L., Bogges, T.F., Wherrett, B.S., Perryman, G.P., and Miller, A. (1982) *Phys. Rev. Lett.*, **49**, 933–936.
66. Sheik-Bahae, M., Wang, J., Canto-Said, E.J., DeSalvo, R., Hagan, D.J., and Van Stryland, E.W. (1995) *IEEE J. Quantum Electron.*, **31**, 1270–1273.
67. 1983) *Optical Phase Conjugation*, Academic Press.
68. Tseng, S.Y., Cao, W., Peng, Y.H., Hales, J.M., Chi, S.H., Perry, J.W., Marder, S.R., Lee, C.H., Herman, W.N., and Goldhar, J. (2006) *Opt. Express*, **14** (19), 8737–8744.
69. Cao, W., Peng, Y.-H., Leng, Y., Lee, C.H., Herman, W.N., and Goldhar, J. (2010) *Organic Thin Films for Photonic Applications*, Washington, DC, pp. 139–157.
70. Reichert, M., Hu, H., Ferdinandus, M.R., Seidel, M., Zhao, P., Ensley, T.R., Peceli, D., Reed, J.M., Fishman, D.A., Webster, S., Hagan, D.J., and Van Stryland, E.W. (2014) *Optica*, **1** (6), 436–445.
71. Kaiser, W. and Garrett, C.G.B. (1961) *Phys. Rev. Lett.*, **7**, 229–231.
72. Franken, P.A., Hill, A.E., Peters, C.W., and Weinreich, G. (1961) *Phys. Rev. Lett.*, **7**, 118–119.
73. Xu, C. and Webb, W.W. (1996) *J. Opt. Soc. Am. B*, **13**, 481–491.
74. 2002) *Topics in Fluorescence Spectroscopy*, Vol. 5: Nonlinear and Two-Photon Induced Fluorescence, Kluwer Academic Publishers, New York.
75. Zipfel, W.R., Williams, R.M., Christie, R., Nikitin, A.Y., Hyman, B.T., and Webb, W.W. (2003) *Proc. Natl. Acad. Sci. U.S.A.*, **100** (12), 7075–7080.
76. Zipfel, W.R., Williams, R.M., and Webb, W.W. (2003) *Nat. Biotechnol.*, **21**, 1369–1377.
77. Makarov, N.S., Drobizhev, M., and Rebane, A. (2008) *Opt. Express*, **16**, 4029–4047.
78. Makarov, N.S., Campo, J., Hales, J.M., and Perry, J.W. (2011) *Opt. Mater. Express*, **1** (4), 551–563.
79. Lakowicz, J.R. (2006) *Principles of Fluorescence Spectroscopy*, 3rd edn, Springer.
80. Callis, P.R. (2002) in *Topics in Fluorescence Spectroscopy*, Nonlinear and Two-Photon Induced Fluorescence, vol. **5** (ed. J.R. Lakowicz), Kluwer Academic Publishers, New York, pp. 1–42.
81. Fu, J., Przhonska, O.V., Padilha, L.A., Hagan, D.J., Van Stryland, E.W., Belfield, K.D., Bondar, M.V., Slominsky, Y.L., and Kachkovski, A.D. (2006) *Chem. Phys.*, **321**, 257–268.
82. Fu, J., Padilha, L.A., Hagan, D.J., Van Stryland, E.W., Przhonska, O.V., Bondar, M.V., Slominsky, Y.L., and Kachkovski, A.D. (2007) *J. Opt. Soc. Am. B*, **24**, 67–76.
83. Webster, S., Fu, J., Padilha, L.A., Przhonska, O.V., Hagan, D.J., Van Stryland, E.W., Bondar, M.V., Slominsky, Y.L., and Kachkovski, A.D. (2008) *Chem. Phys.*, **348**, 143–151.
84. Castillo, J., Kozich, V.P., and Marcano, A.O. (1994) *Opt. Lett.*, **19** (3), 171–173.
85. Kozich, V.P., Marcano, A.O., Hernández, F.E., and Castillo, J. (1994) *Appl. Spectrosc.*, **48** (11), 1419–1422.
86. White, W.T. III,, Henesian, M.A., and Weber, M. (1985) *J. Opt. Soc. Am. B*, **2** (9), 1402–1408.
87. Kurian, A., Lee, S.T., Unnikrishnan, K.P., Sajan, D., Nampoori, V.P.N., and Vallabhan, C.P.G. (2003) *J. Nonlinear Opt. Phys. Mater.*, **9** (12), 75–80.

88. Fang, H.L., Gustafson, T.L., and Swofford, R.L. (1983) *J. Chem. Phys.*, **78**, 1663.
89. Olmstead, M.A. and Amer, N.M. (1983) *J. Vac. Sci. Technol. B*, **1**, 751–753.
90. Jackson, W.B., Amer, N.M., Boccara, A.C., and Fournier, D. (1981) *Appl. Opt.*, **20**, 1333–1344.
91. Badoz, J. and Fournier, D. (1983) *J. Phys. (Paris)*, **44**, C6.
92. Tam, A.C. (1986) *Rev. Mod. Phys.*, **58**, 381.
93. Van Stryland, E.W. and Woodall, M.A. (1980) Presented at the Laser Induced Damage in Optical Materials, Boulder Damage Symposium, Boulder, CO (unpublished).
94. Bass, M., Van Stryland, E.W., and Stewart, A.F. (1979) *Appl. Phys. Lett.*, **34** (2), 142–144.
95. McLean, E.A., Sica, L., and Glass, A.J. (1968) *Appl. Phys. Lett.*, **13**, 369–371.
96. Ferdinandus, M.R., Hu, H., Reichert, M., Hagan, D.J., and Van Stryland, E.W. (2013) *Opt. Lett.*, **38** (18), 3518–3521.
97. Finikova, O.S., Lebedev, A.Y., Aprelev, A., Troxler, T., Gao, F., Garnacho, C., Muro, S., Hochstrasser, R.M., and Vinogradov, S.A. (2008) *ChemPhysChem*, **9** (12), 1673–1679.
98. Howard, S.S., Straub, A., Horton, N.G., Kobat, D., and Xu, C. (2013) *Nat. Photonics*, **7**, 33–37.
99. Zhang, S., Lu, C., Jia, T., Qiu, J., and Sun, Z. (2013) *Phys. Rev. A*, **87**.
100. Johnson, P.M. (1980) *Appl. Opt.*, **19** (23), 3920–3925.
101. Milan, J.B., Buma, W.J., and de Lange, C.A. (1996) *J. Chem. Phys.*, **105** (16), 6688–6712.

3
Modeling of Polymerization Processes

Alexander Pikulin and Nikita Bityurin

3.1
Introduction

The photopolymerization process locally initiated by multiphoton absorption of laser light provides the most versatile modern nanofabrication and nanoprototyping technology [1, 2]. Arbitrarily shaped three-dimensional (3D) solid structures can be made with submicrometer detail by means of the direct writing with a focused ultrashort-pulsed laser beam. Modern polymerizable resins [3, 4] allow the creation of highly transparent and mechanically stable structures for applications in photonics, optoelectronics, plasmonics, as well as microelectromechanical and microfluidic systems. Biocompatibility of some polymers is important for biomedical applications [5, 6].

The structures are written with the focused beam of an ultrashort pulsed laser in the resin transparent to the laser wavelength. This allows penetration of the beam into the sample for the writing of 3D structures. The photopolymerization process is initiated only in the focal spot of the beam due to multiphoton absorption (though other initiation mechanisms are possible, e.g., ionization [7]). After irradiation of the sample, the nonpolymerized resin is dissolved and washed away during the development, leaving only insoluble polymer structures.

This method was initially proposed by Strickler and Webb [8] in 1991 for 3D optical data storage applications. In 1997, Maruo *et al.* [9] performed 3D microstructuring with multiphoton polymerization (MPP). In their works, the characteristic scales of the resulting structures were on the order of micrometers. Currently, the structures of sub-100-nm size are fabricated [10–13].

Photopolymerization is a set of chemical reactions that transform molecules of monomer, which is the main ingredient of the polymerizable resin, into macromolecules of the polymer. The chemistry and physics of the polymerization process constitute the whole branch of modern science and technology. To study this field, the reader must refer to a number of textbooks, including the classical ones [14, 15]. In this chapter, the peculiarities of the polymerization process initiated at a submicrometer scale are theoretically considered. Here,

Multiphoton Lithography: Techniques, Materials and Applications, First Edition.
Edited by Jürgen Stampfl, Robert Liska, and Aleksandr Ovsianikov.

the major part is played by the nonlocal phenomena that are not of primary importance in the case of the bulk polymerization. Among those phenomena are diffusion of species and fluctuations. This chapter does not aim to address particular nanofabrication systems and polymerizable chemicals. Instead, the common traits of the submicrometer-scale polymerization process are analyzed by employing simplistic theoretical models.

3.2
Basic Laser Polymerization Chemistry and Kinetic Equations

The chain-growth photopolymerization process [14] in the simplest case consists of three basic types of chemical reactions: photoinitiation, chain propagation, and chain termination. Light irradiation causes the destruction or excitement of the photoinitiator leading to the generation of molecules with active centers, such as free radicals, ions, and so on. Such active species become the starting elements of the macromolecular chains that further grow by adding many molecules of the monomer, which is the main component of the polymerizable resin. With each of the chain propagation steps, a monomer molecule is attached to the macromolecule's active center in a way that it is transferred to the newly attached monomer. This allows for the growth of linear macromolecules. The lifetime of the active centers is limited by the chain termination reactions. The termination can occur, for example, because of the reaction of two radicals (quadratic termination), recombination of ions, or reaction of the active center with the inhibitor (also called the *radical quencher* for the case of the free-radical polymerization). The described reactions for the case of free-radical polymerization are shown in the following schematic.

Photoinitiation	$\text{Initiator} \xrightarrow{n \times h\nu} A\bullet$
Chain propagation	$A \bullet + M \rightarrow A - M\bullet$
	$A - M \bullet + M \rightarrow A - M - M\bullet$
	$A - (M)_n \bullet + M \rightarrow A - (M)_{n+1}\bullet$
Chain termination by recombination	$A - (M)_n \bullet + \bullet (M)_m - A \rightarrow A - (M)_{n+m} - A$
disproportionation	$A - (M)_n \bullet + \bullet (M)_m - A \rightarrow A - (M)_n + (M)_m - A$
reaction with an inhibitor	$A - (M)_n \bullet + \text{Inhibitor} \rightarrow A - (M)_n - \text{Inhibitor}$ or
	$A - (M)_n \bullet + \text{Inhibitor} \rightarrow A - (M)_n + \ldots$

Here, M is the monomer; the species A are formed as a result of the photoinitiation. The dot ($\bullet$) denotes the radical center.

If the monomer has several moieties that can undergo the chain propagation, then the active species can attach to the already reacted monomer molecules:

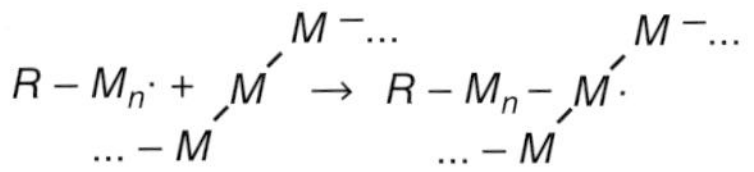

This kind of reaction (also referred to as *crosslinking*) leads to the formation of branched polymer structures whose size significantly exceeds the average length of a single polymer chain. The process finally may result in the aggregation of distinct macromolecules into a single chemically bonded macroscopic network, the gel. Because these gels are more stable relative to only physically connected polymer structures (e.g., by means of entanglement), the multifunctional monomers are very attractive for multiphoton lithography (MPL) applications.

The exact chemical reactions involved in the polymerization depend on the particular polymerizable resin. Typically, the monomer comprises an unsaturated bond or a cycle. The free radical or ion attaches to the bond or opens the cycle, thus propagating the chain. The examples of the former are various acrylate-based resins, including the organic/inorganic hybrids [4]. Examples of the latter are epoxy-based resins, such as SU-8 [6, 16]. Following is the typical chain propagation reaction via double bond:

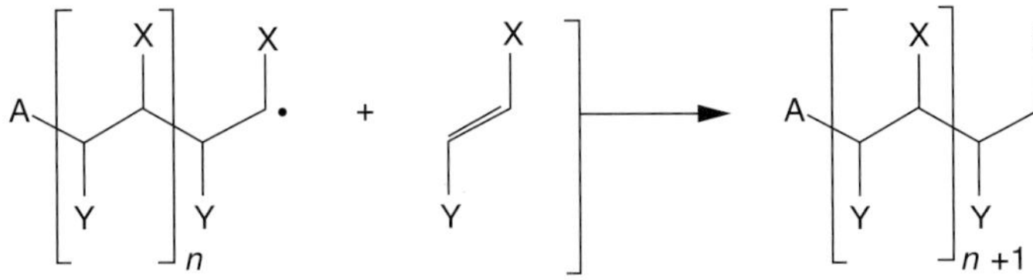

It should be noted that the polymerization of realistic resins may involve other types of reactions [14] in addition to the ones described above, for example, chain transfer or depolymerization.

In the following, we consider the free-radical polymerization. The simplest mathematical model of the polymerization kinetics has to describe the evolution of the number densities of the monomer $M(\vec{r}, t)$ (or non-reacted moieties in multifunctional monomer species) and the active centers (within growing chains and crosslinked clusters) $R(\vec{r}, t)$. Additionally, the inhibitor can be mixed with the resin to prevent occasional polymerization or to add the possibility to manipulate the polymerization process. In those cases, the evolution of the inhibitor number density $Q(\vec{r}, t)$ has to be considered as well. It should be noted that the number densities depends not only on time t but also on the position $\vec{r}$. This reflects the fact that localized initiation of the polymerization by the focused laser beam also results in localized chemical transformations.

The evolution of the number densities of the substances is described by the following equations:

$$\frac{\partial R}{\partial t} = S(\vec{r}, t) - k_{\mathrm{tRR}} R^2 - k_{\mathrm{tRQ}} RQ, \tag{3.1}$$

$$\frac{\partial M}{\partial t} = -k_{\mathrm{p}} MR, \tag{3.2}$$

$$\frac{\partial Q}{\partial t} = -k_{\text{tRQ}} RQ. \tag{3.3}$$

In Eq. (3.1), $S(\vec{r}, t)$ denotes the generation of the active species due to the photoinitiation; other terms describe the termination reactions. The kinetic constants for the chain propagation, the quadratic chain termination, and the chain termination due to inhibition are k_{p}, k_{tRR}, and k_{tRQ}, respectively. When an active center reacts with an inhibitor molecule, both species are destroyed, as reflected in Eqs. (3.1) and (3.3). Equation (3.2) describes the consumption of the monomer due to the chain propagation.

The major parameter of the polymerization process is conversion, which is the fraction of reacted monomer species:

$$p(\vec{r}, t) = \frac{M_0 - M(\vec{r}, t)}{M_0}, \tag{3.4}$$

where M_0 is total number density of the monomer molecules, both free and already included in the polymer chains.

Another important parameter is the molar mass distribution (or molecular weight distribution) of the polymer chains. Equations (3.1)–(3.3) assume that the kinetic constants do not depend on the length of the macromolecules involved in the reactions. This assumption, known as *Flory principle*, is generally satisfied. An important deviation from this principle is concerned with autoacceleration effect (also known as gel effect, Trommsdorff–Norrish effect) [14], which are particularly important in the case of non-crosslinking (linear) radical polymerization. When reaching a certain conversion level, the quadratic termination of long radicals is hampered as a result of entanglements, thus resulting in a steep increase in the polymerization rate. The proper model [17] of this phenomenon considers different termination constants for long and short radicals.

Molar mass distribution of the growing radicals (see Section 3.5) is also important when considering their diffusion [18], because of the dependence of the diffusivity on the size of a molecule.

The initiation kinetics depends on the particular photoinitiator employed in the resin. It may be a multistage process with complex dynamics [19]. Alternately, a photoinitiator may be totally absent, relying on direct photoactivation of the monomer [20].

In the simplest case of direct multiphoton destruction of the photoinitiator

$$S(\vec{r}, t) = \eta \frac{I^n(\vec{r}, t)\sigma^{(n)}}{(h\nu)^n} G(\vec{r}, t), \tag{3.5}$$

where $G(\vec{r}, t)$ is the number density of the photoinitiator, $I(\vec{r}, t)$ is the laser light intensity, $\sigma^{(n)}$ is the n-photon absorption cross section, $h\nu$ is the energy of the photons, and η is the quantum yield.

In the case of irradiation by femtosecond laser pulses delivered at high repetition rate, the intensity distribution can be averaged over the pulse repetition time:

$$S(\vec{r}, t) \approx \eta \tau_{\text{pulse}} R \frac{{I_0}^n(\vec{r}, t)\sigma^{(n)}}{(h\nu)^n} G(\vec{r}, t). \tag{3.6}$$

Here, τ_{pulse} is the pulse length, R is the repetition timer, and $I_0(\vec{r}, t)$ is the peak pulse intensity distribution. The time dependence of $I_0(\vec{r}, t)$ reflects slow (compared to the pulse repetition time) variations of the beam intensity, for example, due to the beam movement with respect to the sample. Equation (3.6) can be used when the repetition time is much less than any characteristic reaction times (and also the times of the transfer of heat and species). This is generally true for the laser systems working at a megahertz-scale repetition rate. In this case, the repetition time R^{-1} is on the order of nanoseconds or less. Typical polymerization reaction and transfer times range from microseconds to milliseconds.

The consumption of the photoinitiator is given by

$$\frac{\partial G(\vec{r}, t)}{\partial t} = -\frac{1}{2k_G} S(\vec{r}, t), \quad (3.7)$$

where coefficient k_G reflects the efficiency of the generation of the active species.

MPP is a photochemical process that does not rely on the temperature rise due to the laser irradiation. The highly energetic irradiation regimes can trigger unnecessary thermal processes, resulting in nanofabrication distortions and damage to the sample. Such regimes are typically avoided and thus are not considered here. In the modest irradiation regimes, the heating can possibly influence the kinetic constants of the polymerization process. In typical direct laser writing (DLW) conditions, however, the emerging heat is quickly removed from the polymerization domain as a result of thermal conductivity. That is, the time in which the heat dissipates from the laser-irradiated spot is estimated as $\tau_{heat} = {d_{focus}}^2 / \chi_{resin}$. Here, χ_{resin} is the thermal diffusivity of the resin, and d_{focus} is the size of the domain heated by the focused laser beam. Typically, for a polymerizable resin, $\chi_{resin} \sim 10^{-3}\,cm^2\,s^{-1}$; d_{focus} can be crudely estimated as a half of the laser wavelength, for example, $d_{focus} \sim 400\,nm$ for a Ti:sapphire laser operating at 800 nm. This results in $\tau_{heat} \sim 1.6\ \mu s$, which is typically much less than the propagation time, that is, the characteristic time needed by a polymer chain to increase its length by one monomer at zero conversion. For instance, in the case of free-radical polymerization of acrylate resins, the propagation time $\tau_p = (k_p M_0)^{-1} \sim 100\,\mu s$ [18]. If τ_{heat} is much smaller than the pulse repetition time, it is pretty safe to consider only the overall uniform temperature of the sample in the basic nanopolymerization model.

3.3 Phenomenological Polymerization Threshold and Spatial Resolution

Writing of 3D structures requires far-field laser beam focusing. Thus, the minimum achievable size of the focal spot is limited by diffraction. To reduce this size, objectives with high numerical aperture as well as the beams with complex polarization (e.g., radially polarized beams [21]) or spatial field structure (e.g., Bessel beams [22, 23]) are employed. For the conventional focusing conditions, the scalar Debye approximation [24] can be used to estimate the lateral and longitudinal peak intensity distributions near the focal spot:

$$I_0(r) = I_{\max}\left[\frac{2J_1\left(2\pi anf^{-1}\lambda^{-1}r\right)}{2\pi anf^{-1}\lambda^{-1}r}\right]^2 \text{ and } I_0(z) = I_{\max}\left[\frac{\sin\left(\pi a^2 f^{-2}\lambda^{-1}nz/2\right)}{\pi a^2 f^{-2}\lambda^{-1}nz/2}\right]^2. \tag{3.8}$$

Here, r and z are the cylindrical coordinates, $I_{\max}$ is the intensity in the maximum, a is the objective aperture radius, f is its focal length, λ is the wavelength in vacuum, J_1 is the Bessel function of the first order of the first kind, and n is the refractive index at the focal spot.

The above formulas are valid for the case of $a \ll f$ and same refractive indices of the immersion oil and the polymerizable resin. Proper calculation of the field intensity distribution in the focal spot must take into account the design of the employed objective, the beam polarization, the interface between the immersion oil and the polymerization resin, and so on.

For convenience, Eq. (3.8) can be approximated by Gaussian (see Figure 3.1)

$$I(r, z) = I_{\max}\exp\left(-2\frac{r^2}{{w_r}^2} - 2\frac{z^2}{{w_z}^2}\right), \tag{3.9}$$

of widths

$$w_r = \frac{\sqrt{2}f}{\pi\cdot n\, a}\lambda \approx \frac{\sqrt{2}}{\pi\cdot \mathrm{NA}}\lambda \text{ and } w_z = w_r\frac{2\sqrt{3}a}{f} \approx w_r\frac{2\sqrt{3}n}{\mathrm{NA}}, \tag{3.10}$$

where numerical aperture of the objective $\mathrm{NA} \approx n\, a/f$.

For a typical MPP setup consisting of the Ti:sapphire laser operating at the wavelength 800 nm and the objective of $\mathrm{NA} = 1.4$ and $n = 1.5$, the minimum focal spot dimensions are estimated as $w_r \approx 257$ nm and $w_z \approx 955$ nm. Direct writing technique is capable of forming 3D structures with detail on the order of 100 nm

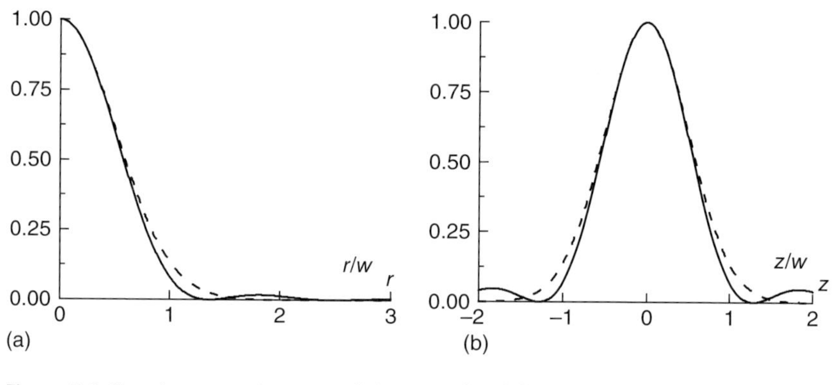

Figure 3.1 Gaussian approximations of the field intensity distributions in the focal spot. (a) Normalized lateral intensity distribution $\left[2J_1\left(2\sqrt{2}r/w_r\right)/(2\sqrt{2}r/w_r)\right]^2$ and its approximation $\exp(-2r^2/{w_r}^2)$ are plotted with solid and dashed lines, respectively. (b) Normalized axial intensity distribution $\left[\sin\left(\sqrt{6}z/w_z\right)/(\sqrt{6}z/w_z)\right]^2$ and its approximation $\exp(-2z^2/{w_z}^2)$ are plotted with solid and dashed lines, respectively.

and smaller with the focused laser beam. According to the above-mentioned estimations, such level of spatial resolution cannot be achieved only by sharp beam focusing. Even taking into account that in the case of the n_{ph}th-order of multiphoton absorption the width of the absorbed intensity distribution is decreased by the factor of $\sqrt{n_{ph}}$, the focusing is definitely not enough to provide sub-100-nm size of a nanostructure.

The high spatial resolution of MPP is provided by the sharp nonlinear response of the polymerizable resin to the laser radiation. The simplest phenomenological model for such a response relies on the polymerization threshold [25]. This is the minimum energy density that must be absorbed for the local transformation of the initial resin into the polymer that can withstand the development process.

Similar to Eq. (3.5), the multiphoton absorbed energy density is given by

$$D^{(n)}(\vec{r}) = \int_0^T \frac{I^n(\vec{r},t)\sigma^{(n)}}{(h\nu)^{n-1}} G(\vec{r},t)dt, \tag{3.11}$$

where T is the irradiation time. For the case of a series of femtosecond pulses at high repetition rate, $I^n(\vec{r},t)$ can be averaged over repetition time. Similar to Eq. (3.6)

$$D^{(n)}(\vec{r}) \approx \frac{\tau_{\text{pulse}} R\sigma^{(n)}}{(h\nu)^{n-1}} \int_0^T I_0{}^n(\vec{r},t)G(\vec{r},t)dt, \tag{3.12}$$

where $I_0(\vec{r},t)$ is the slowly varying peak pulse intensity distribution.

By means of the polymerization threshold, one can estimate the size of a laser-written polymer structure in a way illustrated in Figure 3.2. For example, when writing a polymer voxel with a focused femtosecond laser beam, the two-photon absorbed energy density (from Eqs. (3.9) to (3.12)) is

$$D^{(2)}(r,z) = \frac{\tau_{\text{pulse}} RTI_{\max}{}^2\sigma^{(2)}G_0}{h\nu} \exp\left(-\frac{4r^2}{w_r{}^2} - \frac{4z^2}{w_z{}^2}\right), \tag{3.13}$$

provided the consumption of the photoinitiator can be neglected: $G(\vec{r},t) \equiv G_0$.

A similar absorbed energy density profile is obtained when writing a polymer line with the beam moving along a lateral coordinate with velocity V. The intensity distribution is then

$$I(x,y,z,t) = I_{\max} \exp\left(-2\frac{x^2 + (y - Vt)^2}{w_r{}^2} - 2\frac{z^2}{w_z{}^2}\right), \tag{3.14}$$

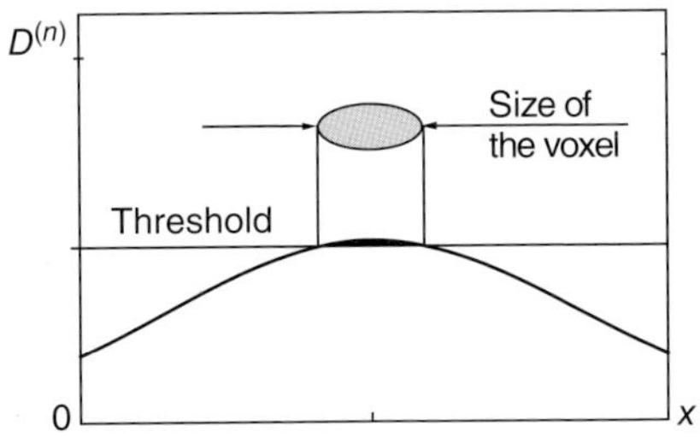

Figure 3.2 Size of a single polymer feature determined by the polymerization threshold. (Pikulin 2012 [26]. Reproduced with permission of Springer.)

Far from the beginning and the end of the line, integration of two-photon absorbed energy density (Eq. (3.12)) results in

$$D^{(2)}(x,z,t) = \frac{w\sqrt{\pi}}{V}\frac{\tau_{\text{pulse}} R I_{\max}{}^{2}\sigma^{(2)} G_0}{h\nu}\exp\left(-\frac{4x^2}{w_r{}^2}-\frac{4z^2}{w_z{}^2}\right). \tag{3.15}$$

Then, in both cases of the voxel and the line, the dimension of the structure along the specified coordinate can be calculated. For instance, the thickness of the line along the coordinate x is

$$d_x = \frac{w_r}{2}\sqrt{\ln\left(\frac{D_{\max}}{D_{\text{th}}}\right)}, \tag{3.16}$$

where $D_{\max}$ is the maximum absorbed energy density, and D_{th} is the polymerization threshold.

It follows from Eq. (3.16), that the size of a single polymer structure (line or voxel) can be made significantly smaller than the beam's focal spot. The ratio w_z/w_r suggests the elongation of the voxels by a factor of about 4, even when using a high-NA objective.

With formula (3.16), however, one cannot calculate the experimentally observed minimal size of a polymer nanofeature. Indeed, when $D_{\max}$ approaches the threshold, the estimated size of the structure tends to zero, which is contrary to common sense. Actually, this means that the polymerization threshold concept fails when approaching a certain minimal feature size.

By means of the polymerization threshold, however, one can explain [27] the spatial resolution limit for the writing of at least two polymer structures close to each other. Let us consider the possibility to write two separate voxels at distance $2L_l$ along a lateral coordinate. Taking into account Eq. (3.13), the absorbed energy density distribution is

$$D^{(2)}(\vec{r}) = D_{\max}\exp\left(-\frac{4x^2}{w_r{}^2}-\frac{4z^2}{w_z{}^2}\right)\left[\exp\left(-4\frac{(y-L_l)^2}{w_r{}^2}\right)+\exp\left(-4\frac{(y+L_l)^2}{w_r{}^2}\right)\right] \tag{3.17}$$

By analyzing Eq. (3.17) (particularly the sign of $\partial^2 D^{(2)}(0)/\partial y^2$), one can find that only if $2L_l > w_r/\sqrt{2}$ has the distribution of $D^{(2)}(\vec{r})$ two separate maxima. In this case, two separate voxels can be obtained by manipulating the maximum absorbed energy density $D_{\max}$ with respect to the threshold (see Figure 3.3a). If $2L_l < w_r/\sqrt{2}$, $D^{(2)}(\vec{r})$ has a single global maximum at $\vec{r} = 0$ (see Figure 3.3b). Thus, regardless of the actual level of the threshold, it is not possible to resolve two voxels at such a distance: a single polymer feature only in the center can be formed instead. Similarly, the longitudinal resolution limit of $2L_z > w_z/\sqrt{2}$ can be derived. For a Ti:sapphire laser operating at 800 nm, in the case of the two-photon absorption, $2L_l \approx 182$ nm and $2L_z \approx 675$ nm, taking into account the above estimations of the dimensions of the focal spot. Because of the diffraction limit, the spatial resolution is different from the minimum feature size.

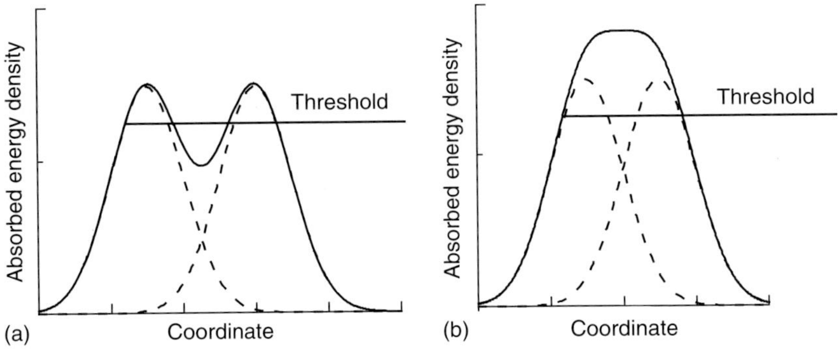

Figure 3.3 Writing of two voxels near each other. (a) When the distance between voxels is longer than the resolution limit, two separate structures can be obtained. (b) When the distance is shorter than the limit, only a single stricture is formed.

It is important to note that the polymerization threshold is a *local* approximation. This means that the polymerization is considered independently in the neighboring subdomains of the medium. Any size-dependent nonlocal phenomena, such as transfer of molecules, finite length of polymer chains, and size-dependent fluctuations, are neglected. Thus, the spatial resolution limit can be possibly overcome when performing the laser writing in a regime that essentially deviates from the local threshold concept.

There are at least two possibilities to improve the spatial resolution in this manner. One of them relies on the concept of super-resolution photoinduced-inhibition nanolithography (SPIN), which is similar to stimulated emission depletion microscopy (STED) [10–13]. Basically, the writing is performed with two beams. One beam with a standard Gaussian-like shape is used for the photoinitiation of the polymerization, that is, generation of free radicals. Another beam with vortex shape is used for the photodepletion of the radicals. The volume in which the polymerization is effectively carried out is narrowed, and thus the spatial resolution is improved. There are several variations of this scheme, which employ femtosecond and continuous wave (CW) lasers, as well as two-color and single-color setups. This technique is described in detail in Chapter 5. The theoretical model of SPIN is given in [28].

Another approach [29, 30] relies on the depletion of the radicals by mobile quencher (inhibitor) molecules specially added in the resin. The irradiation is performed in a regime when the quencher consumption at the beam maximum is compensated by the diffusion from the outside of the irradiated volume. As a result, a nonequilibrium, quasi-stationary spatial distribution of the quencher concentration is formed. The distribution profile has a flat zero minimum around the beam maximum. The polymerization is effectively carried out only where the quencher concentration is near zero. Thus, the effect of the diffusing quencher is similar to that of the vortex beam in STED-inspired writing scheme. This approach is described in detail in Section 3.5.2.

The consideration of the nonlocal phenomena is essential both for the minimization of the size of a single polymer nanofeature and for the improvement of the spatial resolution of multiple nanofeatures at a nanodistance. As a starting point, it is important to understand the possible physical mechanisms behind the phenomenological polymerization threshold. At the larger spatial scale when the influence of the nonlocal phenomena becomes negligible, at least two such mechanisms are possible.

When the inhibitor (radical quencher) is involved in the polymerization, the radicals generated by the photoinitiator are first depleted by the inhibitor. Only if the inhibitor is locally depleted does the growth of the polymer chains occur. Let us consider Eqs. (3.1) and (3.3) assuming that $k_{tRR}R \ll k_{tRQ}Q$, which allows neglecting the quadratic terms:

$$\frac{\partial R}{\partial t} = S(\vec{r}, t) - k_{tRQ}RQ, \tag{3.18}$$

$$\frac{\partial Q}{\partial t} = -k_{tRQ}RQ. \tag{3.19}$$

The initial values are $R(\vec{r}, 0) = 0$ and $Q(\vec{r}, t) = Q_0$. The equations have an integral

$$R(\vec{r}, t) - Q(\vec{r}, t) = \int_0^t S(\vec{r}, t')\mathrm{d}t' - Q_0. \tag{3.20}$$

After the irradiation time T, the following relation holds:

$$R(\vec{r}, t) - Q(\vec{r}, t) = D^{(n)}(\vec{r})\frac{\eta}{h\nu} - Q_0, \tag{3.21}$$

taking into account Eqs. (3.5) and (3.11).

The time evolution of the number densities of the radical and the inhibitor is illustrated by the phase diagram in Figure 3.4. If $D^{(n)}(\vec{r}) < Q_0 \cdot h\nu/\eta$, the system comes to the stable state at $R = 0$ and $Q > 0$. In this case, the number density of the generated radicals is not enough to start the polymerization. When $D^{(n)}(\vec{r}) > Q_0 \cdot h\nu/\eta$, the inhibitor is totally consumed and the polymerization starts.

The applicability of this model is limited by the diffusion of the inhibitor from nonirradiated domains. The diffusion can become a principal factor at the 100-nm

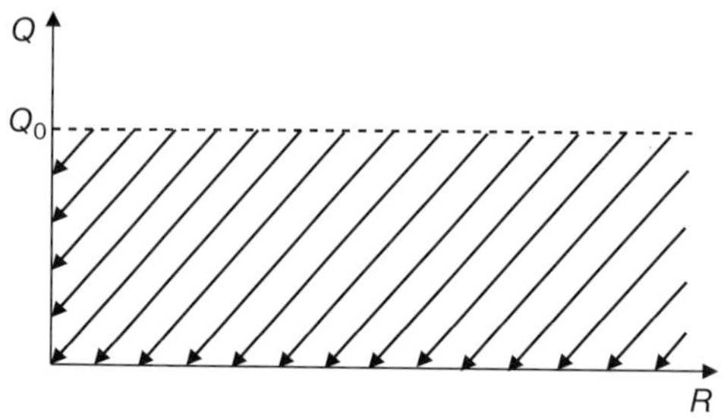

Figure 3.4 Phase diagram of the evolution of the number densities of the quencher (Q) and the radical (R) after the irradiation. The number densities decrease following the arrowed lines and come to the stable states at the axes.

spatial scale, having either positive or negative effect depending on the regime of the laser writing. The diffusion of the inhibitor is one of the factors that determines the duration of the dark polymerization (after the irradiation) at a point in the resin [31]. This factor limits the maximum conversion that can be reached with short-time irradiation schemes. One can expect, however, that the diffusion may be effectively disabled by bonding the inhibitor molecules to a prebuilt matrix, for example, to the inorganic gel of a hybrid polymerizable material. The diffusion of species is considered in Section 3.5.

Another physical mechanism that explains the polymerization threshold is related to gelation. In the case of polymerization of multifunctional monomers, the crosslinking reaction results in random connections of distinct polymer chains by covalent bonds. The basic understanding of the behavior of such a system can be provided by the percolation theory [32]. The growth of the conversion during the polymerization process also means an increase in the probability of the crosslinks. This results in the growth of the average mass of the crosslinked macromolecules containing many linear chains. Percolation theory [33] states that, at a certain conversion level, the average mass diverges and tends to infinity. At this point, the crosslinks result in the formation of the polymer gel, which is a macroscopic-sized network of covalently bonded chains. Such a connectivity transition is a sort of phase transition of the second kind. The gelation point is experimentally observed for bulk crosslinking polymerization at the conversion level of about several percent [34]. In the case of the laser writing, the spatial distribution of the absorbed energy density results in the spatial distribution of the conversion. In the domain where the gelation threshold is exceeded, an insoluble polymer structure is formed. The rest of the sample, where only distinct macromolecules can be found, is dissolved and washed away during development.

As with any phase transition, the formation of the gel suffers from fluctuations in a finite-size domain. Those fluctuations are not caused by the instabilities of the laser beam shape and intensity; instead, they are related to the inhomogeneities in the gel [26, 35]. Such inhomogeneities are one of the factors that determine the minimal feature size. The basic study of this phenomenon is given in the following section.

3.4
Effect of Fluctuations on the Minimum Feature Size

A polymer substance consists of building blocks, namely the monomers, that are connected to each other forming macromolecular chains and crosslinked clusters. Such connections have a certain degree of randomness. Because of the relatively large size of a single monomer, the fluctuations within the polymer substance may become one of the principal factors that limit the minimum feature size in the DLW technique. The aim of this section is to analyze how the properties of a piece of the polymer substance change, in an attempt to minimize its dimensions. For that, we perform Monte Carlo modeling of a voxel, the elementary piece of the polymer made with MPL. Writing of the voxel assumes the polymerization initiated in a localized spot by the focused laser beam. The polymerization results in the formation of the linear chains and branched clusters "glued" from several linear chains through crosslinking reactions. The largest of those clusters is the piece of the polymer gel that is the voxel.

To analyze the properties of the polymer voxel, we employ the simplest numerical model. Cross-linking homopolymerization is assumed. Thus, all the monomer molecules are identical and multifunctional. To speed up the calculations, all the monomer molecules, both individual and already reacted, are placed at the sites of a cubic lattice. Connections between adjacent sites simulate the covalent bonds between monomer units in the macromolecules. The macromolecular chains are simulated with the random walk method. Intersections of the random walks (as well as self-intersections) simulate the crosslinks. The chain lengths are random and obey the distribution

$$p_l \propto \exp(-l/l_0), \tag{3.22}$$

where p_l is the probability that a polymer chain has length l ($l \geq 2$), and l_0 is a parameter of the model that determines the average chain length.

The starting sites of the polymer chains are also random and not correlated. In the simplest case, probability distribution $f(\vec{r})$ to start the random walk at a site at $\vec{r}$ is contributed only by the photoinitiation reaction. The number density of the laser-generated free radicals is determined by the absorbed energy density. Thus (see also Eq. (3.21)) we have

$$f(\vec{r}) = \frac{\int_0^T S(\vec{r}, t)dt}{M_0} = \frac{\eta D^{(n)}(\vec{r})}{h\nu M_0}, \tag{3.23}$$

where M_0 is the starting monomer concentration, $D^{(n)}$ is the distribution of the absorbed energy density, n is the order of multiphoton absorption, $h\nu$ is the photon energy, and η is the quantum yield. In more complex cases, this distribution is affected by radical quenching, diffusion, and so on. For instance, in diffusion-assisted or STED-inspired DLW schemes, a part of the generated active species must be neglected in Eq. (3.23) since they are depleted before making any significant contribution to the conversion.

In this section, the universal properties of polymer voxels are studied rather than a particular MPP scheme. Thus, for simplicity, a model Gaussian-shaped, spherically symmetric distribution

$$f(\vec{r}) = A\exp(-r^2/2w^2) \tag{3.24}$$

is assumed.

In numerical simulation, the ensemble of realizations of similar polymer voxels with same A, w, and l_0 is made. For each realization, the actual voxel is selected as the largest connected cluster of the polymer chains. The parameters of interest are calculated for this cluster and then averaged over the ensemble.

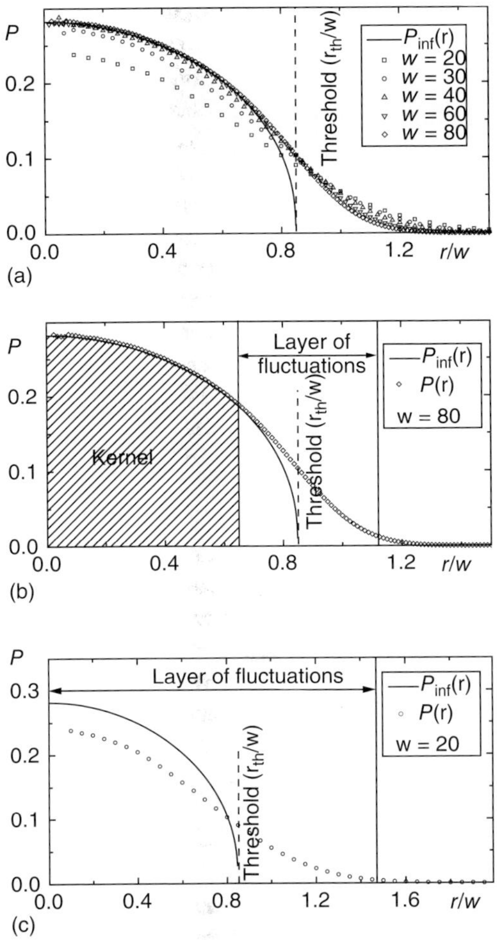

Figure 3.5 (a) Spatial distributions of voxel average density $P(r)$ (dots) for different widths w and fixed amplitude $A = 0.06$ of polymer chain distribution. The macroscopic limit of $w \to \infty$ is shown with a solid line. The cases of the stable voxels (large) and the strongly fluctuating ones (small) are shown separately in (b) and (c), respectively. (Pikulin 2012 [26]. Reproduced with permission of Springer.)

Figure 3.5 shows the spatial profile of the average density of the voxel for different widths of the polymerization spot, w, while keeping A and l_0 fixed. The density is calculated as

$$P(r) = \frac{\langle M|_r \rangle}{N|_r}, \tag{3.25}$$

where $M|_r$ is the number of the monomer units that belong to the largest connected cluster within the layer of $r < |\vec{r}| < r + 1$. $N|_r$ is the total number of monomer molecules (total number of lattice sites) within this layer. The sign "$\langle\ \rangle$" means the ensemble average.

For comparison, uniform bulk samples of the same polymer are simulated. We calculate the density P_{inf} of the gel fraction f for a given spatially uniform chain start probability. The calculated density distribution $P_{\text{inf}}(f(r))$ is plotted in Figure 3.5a–c with a solid line. This distribution corresponds to the macroscopic limit of the voxel, that is, to the case of an infinitely large polymerization spot: $w \to \infty$.

It is seen that the bulk polymer gel is formed only if the threshold is exceeded, that is, if $f > f_{\text{th}}$. Here, the connectivity percolation-like transition occurs, which is a sort of phase transition of the second kind. As discussed earlier, such a transition is one of the physical mechanisms of the phenomenological polymerization threshold. The threshold expectation of the voxel size is calculated as $r_{\text{th}} = w\sqrt{2\ln(A/f_{\text{th}})}$, similar to Eq. (3.16). When the polymerization spot becomes finite, the phase transition becomes spatially confined and thus prone to fluctuations.

It is seen in Figure 3.5 that the average density of the central part of the larger voxels (see Figure 3.5b) does not depend on the voxel size and matches the density of the bulk gel. This part is further referred to as *the kernel*, which possesses the same properties as the bulk polymer. The polymer substance that surrounds the kernel is different. Its average density is size-dependent and significantly deviates from the density of the bulk gel. The observation of the voxel realizations and correlation length analysis [35] suggest pronounced density fluctuations within this zone, which is further referred to as "the layer of fluctuations." The smaller voxels do not have any kernel at all. Their average density is size-dependent everywhere and does not match that of the macroscopic polymer substance. Thus, the properties of the smaller voxels are expected to significantly differ from those of the polymer bulk. Example realizations of the "large" and "small" voxels are shown in Figure 3.6.

For further understanding the transition between the large and the small types of voxels, let us study how the fluctuations of the voxel radius and location depend on w. The position of the center of mass of a voxel is given by $\vec{r}_{\text{cmass}} = \sum_i \vec{r}_i/M$. Its gyration radius is given by $r_{\text{gyr}} = \sqrt{\sum_i (\vec{r}_i - \vec{r}_{\text{cmass}})^2/M}$. Here, $\sum_i$ is the sum of all lattice sites that belong to the largest cluster, and $\vec{r}_i$ are the locations of those sites. The mass M of the voxel refers to the total number of sites in the largest cluster.

To measure the fluctuations of the size and location of the voxel between realizations, we calculate the dispersion of the gyration radius $D[r_{\text{gyr}}] = \langle (r_{gyr} - \langle r_{gyr} \rangle)^2 \rangle$ and the dispersion of a coordinate (e.g., z) of the center of mass, $D[z_{\text{cmass}}] = \langle {z_{cmass}}^2 \rangle$. To indicate the fluctuations relative to the size w of the polymerization spot, the dimensionless normalized dispersions are calculated: $d[r_{\text{gyr}}] = D[r_{\text{gyr}}]/w^2$ and $d[z_{\text{cmass}}] = D[z_{\text{cmass}}]/w^2$.

The dependencies of $D[z_{\text{cmass}}]$, $d[z_{\text{cmass}}]$, $D[r_{\text{gyr}}]$, $d[r_{\text{gyr}}]$, and the average voxel mass $\langle M \rangle$ on w are presented in Figure 3.7. The graphs are given for fixed $l_0 = 20$ and different values of $A = 0.05$ and $A = 0.06$.

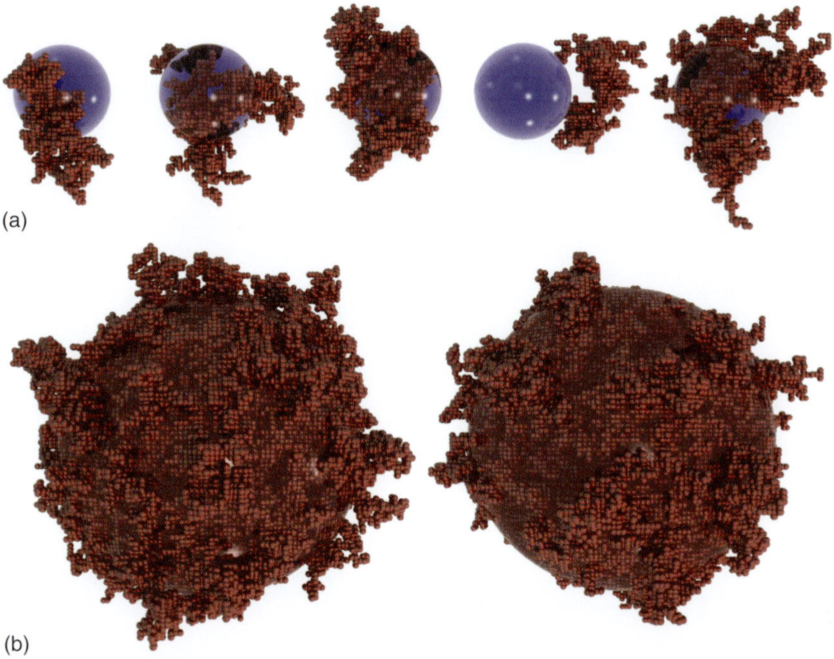

Figure 3.6 Realizations of the small voxels (a) without the kernel and the large voxels (b) with the kernel. The red elements depict the monomer units. The blue spheres are the expectations that result from the gelation threshold. (Pikulin 2010 [35]. Reproduced with permission of American Physical Society.)

It is seen that there is a critical scale $w = w_{cr}$ that separates the large and small voxels, and this scale depends on A. At both $w > w_{cr}$ and $w < w_{cr}$, the dependence of $d[r_{gyr}]$ and $d[z_{cmass}]$ on w obey power laws ($\propto w^{\alpha}$); however, the exponents change at $w = w_{cr}$. For the large voxels, the decrease in the normalized dispersions with w is steeper.

The dependence of the average mass of the large voxel scales as a cube of its size:

$$\langle M \rangle \propto w^3. \tag{3.26}$$

The fractal dimension of the large voxel is 3, as expected for macroscopic solid bodies. The mass of the small voxels deviate from this law. Thus, the small voxels exhibit fractal behavior, which is peculiar to distinct macromolecules.

The critical size of a nanostructure at which the kernel disappears and the structure becomes unstable is calculated in [35] for the case of percolation transitions. The calculation is based on the general scaling relation for the correlation length

$$\xi(f) \propto |f - f_{th}|^{-\nu}, \tag{3.27}$$

which is the characteristic scale of inhomogeneities caused by fluctuations in a bulk sample. Such type of a scaling relation is obeyed by various phase transitions

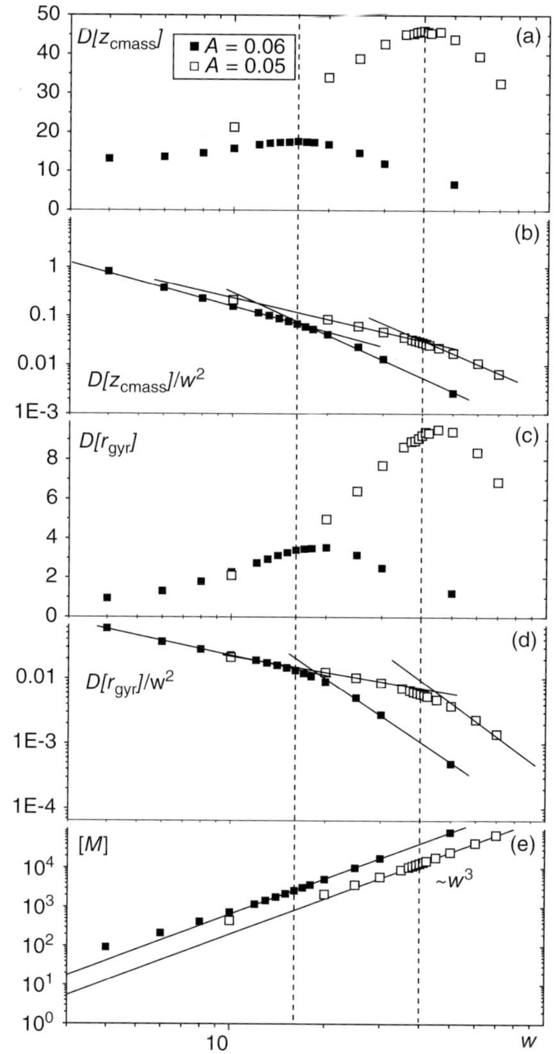

Figure 3.7 Results of the Monte Carlo simulation of polymer voxels. (a) Dispersion of the coordinate of center of mass of the voxel, (b) normalized dispersion of the coordinate of center of mass, (c) dispersion of the radius of gyration of the voxel, (d) normalized dispersion of the radius of gyration, and (e) averaged mass of the voxel plotted against width w of the polymerization spot for distribution amplitudes $A = 0.05$ and 0.06. (Pikulin 2012 [26]. Reproduced with permission of Springer.)

of the second kind, including thermal and connectivity transitions. The exponent ν is the same within the universality class of the transition. For instance, for any 3D percolation transition, $\nu \approx 0.88$. Assuming that the relation (3.27) is valid for our problem as well, one obtains (see [35] for details)

$$w_{cr} = \frac{bK_2^{-1}}{(A - f_{th})^{\nu}\sqrt{2\ln(A/f_{th})}}, \tag{3.28}$$

where $K_2 = B(1/2, \nu + 1)/2 \approx 0.69$, $B(\ )$ is the Euler beta function, and b is a fitting parameter on the order of 1.

Figure 3.8 shows that the results of the numerical experiment fit well the above-mentioned formula with $\nu = 0.88$. This exponent is expected for the 3D percolation-like problem; however, for the real polymerization, the exponent may be different. Significant deviation is observed only when attempting to make w smaller than the average size of a single 20-bond chain (the size is, approximately, $\sqrt{20} \approx 4.47$). Here, the conformational fluctuations of single chains become important; however, they are not considered in Eq. (3.28).

The critical radius of the volume above the threshold is

$$r_{th} = r_{cr} = \frac{bK_2^{-1}}{(A - f_{th})^{\nu}}. \tag{3.29}$$

As it is seen in Figure 3.5b, the radius of the stable voxel is larger than r_{th} estimated within the threshold approximation. The voxel consists of the volume above the threshold and the outer fluctuation layer: $r_{vox} = r_{th} + \sigma_+$, where [35]

$$\sigma_+ = ((1 + \nu)b)^{1/(1+\nu)} \left| \frac{df\,(r_{th})}{dr} \right|^{-\nu/(1+\nu)}. \tag{3.30}$$

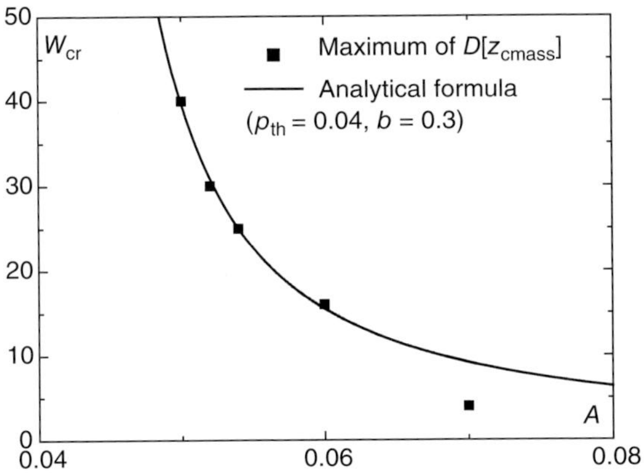

Figure 3.8 Critical width that demarcates stable and unstable voxels. Dots denote the results of the numerical experiment with $l_0 = 20$. The width is determined by the maximum of $D[z_{cmass}]$. The results of analytical calculations by means of Eq. (3.28) are plotted with solid line. (Pikulin 2012 [26]. Reproduced with permission of Springer.)

In what follows, we estimate the minimal voxel radius for the DLW in realistic polymerizable media. Let us expand Eq. (3.24) in Taylor series at the central point up to the second-order term. Then, the threshold and the critical radius are related as

$$f_{\mathrm{th}} = A + \frac{1}{2} f''(0) r_{\mathrm{cr}}^{2} \tag{3.31}$$

A similar relation is obtained for the absorbed energy density, assuming spherical symmetry for simplicity:

$$D_{\mathrm{th}} = D_{\mathrm{max}} + \frac{1}{2} D^{(n)\prime\prime}(0) r_{\mathrm{cr}}^{2}, \tag{3.32}$$

Here, we assume a more general local dependence of $f(D^{(n)})$ than a particular linear case suggested by Eq. (3.23):

$$f''(0) = \frac{df(D_{\mathrm{max}})}{dD^{(n)}} D^{(n)\prime\prime}(0). \tag{3.33}$$

Substitution of Eq. (3.33) into Eq. (3.31) yields

$$A - f_{\mathrm{th}} = -\frac{1}{2} \frac{df(D_{\mathrm{max}})}{dD^{(n)}} D^{(n)\prime\prime}(0) r_{\mathrm{cr}}^{2} \approx -\frac{D_{\mathrm{th}}}{2} \frac{df(D_{\mathrm{max}})}{dD^{(n)}} \frac{D^{(n)\prime\prime}(0)}{D_{\mathrm{max}}} r_{\mathrm{cr}}^{2} \tag{3.34}$$

accurate to r_{cr}^{2}. Then, from Eq. (3.29), we have

$$r_{\mathrm{cr}} = \left(\frac{L_{\mathrm{mat}}}{K_2} \right)^{1/(1+2\nu)} \left(\frac{2D_{\mathrm{max}}}{|D^{(n)\prime\prime}(0)|} \right)^{\nu/(1+2\nu)}. \tag{3.35}$$

Here

$$L_{\mathrm{mat}} = b \left[D_{\mathrm{th}} \frac{df\,(D_{\mathrm{th}})}{dD^{(n)}} \right]^{-\nu} \tag{3.36}$$

is the characteristic scale of the polymerization medium, which is independent of the shape of the laser field distribution. Similarly, we find an expression for thickness σ_+ of the part of the fluctuation layer that is outside the gelation threshold:

$$\sigma_+ = r_{\mathrm{cr}} \cdot \left[\frac{(1+\nu) K_2}{2^{\nu}} \right]^{1/(1+\nu)} \tag{3.37}$$

Substituting two-photon absorbed energy density distribution Eq. (3.15) into formula (3.37) yields

$$r_{\mathrm{cr}} = \left(\frac{L_{\mathrm{mat}}}{K_2} \right)^{1/(1+2\nu)} \left(\frac{\lambda}{\pi \sqrt{2}\mathrm{NA}} \right)^{2\nu/(1+2\nu)} \tag{3.38}$$

Then, let us make crude estimations of L_{mat} and r_{vox} on the basis of the employed numerical model of the polymer. With regard to Eq. (3.23), one has $D_{\mathrm{th}} \cdot (df(D_{\mathrm{th}})/dD^{(n)}) = f_{\mathrm{th}}$. The fitting parameters of Eq. (3.28) to the results of the numerical experiment are $f_{\mathrm{th}} = 0.04$ and $b = 0.30$ (here, b is measured in grid periods). In the absence of solvents, the effective "grid spacing" can be estimated as the length of the edge of a cube, the volume of which is occupied by one monomer molecule. Thus, $L_{mat} = 0.30 \cdot (N_A M_0)^{-1/3} \cdot 0.04^{-0.88} \approx 5.1 \cdot (N_A M_0)^{-1/3}$. For example, for triethylene glycol dimethacrylate (TRGDMA), $L_{mat} \approx 4.1$ nm [26]. For the case of MPP with a Ti:sapphire laser operating at 800 nm and objective of NA = 1.4, one gets $r_{cr} \approx 42$ nm, $\sigma_+ \approx 35$ nm, and $r_{vox} = r_{cr} + \sigma_+ \approx 77$ nm.

It can be shown that the minimum voxel size can be made smaller by adding an inhibitor to a polymerizable resin. Here, we consider the case of a nondiffusing inhibitor. As explained in the previous section, in this case the absorbed energy density is partially spent on the inhibitor depletion; that is, $f \propto D^{(n)} - D_Q$, where $D_Q = Q_0 \cdot h\nu/\eta$ (see Eq. (3.21) and the following text). Then

$$D_{th} \cdot (df(D_{th})/dD^{(n)}) = f_{th} \frac{D_{th}}{D_{th} - D_Q} \tag{3.39}$$

resulting in a smaller

$$L_{mat} = 0.30 \cdot (N_A M_0)^{-1/3} \left(f_{th} \frac{D_{th}}{D_{th} - D_Q} \right)^{-\nu}. \tag{3.40}$$

For instance, for $D_Q = D_{th}/2$, $L_{mat} \approx 2.2$ nm, $r_{cr} \approx 34$ nm, $\sigma_+ \approx 28$ nm, and $r_{vox} \approx 62$ nm [26]. It is seen that the presence of the inhibitor (quencher) results in a smaller feature size, as is observed in the experiment [36, 37].

The above-mentioned simulation shows that an attempt to make polymer voxels smaller than some limit results in statistically unstable, fluctuating ones that have properties of big macromolecules rather than of small pieces of macroscopic polymer substance. The difference between the stable and the fluctuating voxels is indicated by different fractal dimensions and scaling laws of the voxels' size and position fluctuation strength. Such fluctuations can lead to significant distortions of complex polymer structures made with MPL. The proposed analytical expression (3.29) for the limiting feature size suggests the use of a writing regime with narrow polymerization spot where high contrast of conversion/density of polymer chains is reached with respect to the gelation threshold. As we show in the next section, this can be achieved in writing regimes that deliberately employ quencher/inhibitor diffusion (diffusion-assisted DLW).

3.5 Diffusion of Molecules

Polymerization involves many types of molecules, including growing and dead chains of different lengths, crosslinked macromolecules, the inhibitor, the photoinitiator, and so on. Their contribution to the spatial profile of the final conversion is not equal. The diffusion coefficient of a macromolecule in a dilute solution can be evaluated with the Einstein–Stokes formula

$$D_m = \frac{k_B T}{3\pi \eta_m d_m}, \tag{3.41}$$

where k_B is the Boltzmann constant, T is the temperature, η_m is the viscosity, and d_m is the diameter of the molecule. Thus, the contribution of the diffusion of long chains and large crosslinked clusters is smaller than that of the fast and small molecules. Among them, the contribution of the short dead chains to the overall conversion is minimal. On the contrary, the diffusion of the growing chains is important. They diffuse fast during the short-length stage, grow, and then make significant contribution to the conversion.

In this section, the diffusion effects are studied separately for the growing chains and the inhibitor (quencher).

3.5.1
Diffusion of the Growing Chains

In the following model [18], the growing radicals are divided into two groups, small and large. The small radicals are diffusing, whereas the diffusion of the large radicals can be neglected. The total radical number density is

$$R(\vec{r},t) = \sum_{i=1}^{m} R_i(\vec{r},t) + R_G(\vec{r},t), \tag{3.42}$$

where R_i are the number densities of the diffusing radicals comprising i monomers, and R_G denotes the number density of the nondiffusing radical centers. The generation and growth of the small radicals along with their diffusion are described by the following equation set, assuming the absence of the inhibitor:

$$\frac{\partial R_1}{\partial t} = S(\vec{r},t) + D_1 \cdot \Delta R_1 - k_p M R_1, \tag{3.43}$$

$$\frac{\partial R_i}{\partial t} = D_i \cdot \Delta R_i + k_p M R_{i-1} - k_p M R_i, \tag{3.44}$$

We assume that the termination reactions can be neglected at a timescale of the growth of the radical up to the length m. This assumption makes sense when the average length of the polymer chain is much greater than m. The effect of the quadratic termination of small radicals is studied in [18] and is shown to deteriorate the spatial resolution.

The number density of the large radicals obey Eq. (3.1) without the inhibition term and with effective source term $k_p M R_m$:

$$\frac{\partial R_G}{\partial t} = k_p M R_m - k_{tRR} R_G{}^2. \tag{3.45}$$

Since the timescale of the growth up to length m and diffusion is short compared to the overall polymerization time, the effect of the diffusion can be calculated by substituting the actual absorbed energy density distribution

$$D^{(n)}(\vec{r}) = \frac{h\nu}{\eta} \int_0^{\infty} S(t)dt \tag{3.46}$$

by the effective one

$$\tilde{D}(\vec{r}) = \frac{h\nu}{\eta} \int_0^{\infty} k_p M R_m dt, \tag{3.47}$$

which can be calculated by means of Eqs. (3.43) and (3.44). At the beginning of the polymerization process (near the zero conversion level), those equations yield [18]

$$\widetilde{D}(\vec{r}) = \frac{1}{(2\pi)^3} \iiint_{R^3} \left\{ \prod_{i=1}^{m} \left(1 + \frac{D_i}{k_p M_0} u^2 \right) \right\}^{-1} \overline{D}^{(n)}(\vec{u}) \cdot \exp\{j\vec{u}\vec{r}\} d\vec{u} \quad (3.48)$$

where

$$\overline{D}^{(n)}(\vec{u}) = \iiint_{R^3} \vec{D}^{(n)}(\vec{r}) \cdot \exp\{-j\vec{u}\vec{r}\} d\vec{r} \quad (3.49)$$

is the Fourier image of the actual absorbed energy density distribution (3.46).

The geometrical size of macromolecules that are shorter than the persistent length scales as $d_m(i) \propto i$. Thus, the diffusion coefficients are estimated as $D_i = D_1/i$ from Eq. (3.41).

Let us consider again the problem of the spatial resolution of two polymer spherically symmetric voxels at nanodistance. We assume that the voxels are written by a short series of laser pulses; thus, the diffusion of short radicals occurs when the conversion is close to zero and thus formula (3.47) is applicable. The actual absorbed energy density distribution is

$$D^{(n)}(r, z) = a \left[\exp\left\{ -\frac{(z-L)^2 + r^2}{2w^2} \right\} + \exp\left\{ -\frac{(z+L)^2 + r^2}{2w^2} \right\} \right]. \quad (3.50)$$

Furthermore, we study how the minimum resolution length $2L_{min}$ (the minimum length at which two separate nanostructures can be formed, see Section 3.3) is affected by the diffusion. The calculation results are shown in Figure 3.9.

It is seen that the minimum resolution distance is on the order of

$$L_{diffus} = \sqrt{\frac{D_1}{k_p M_0}} \quad (3.51)$$

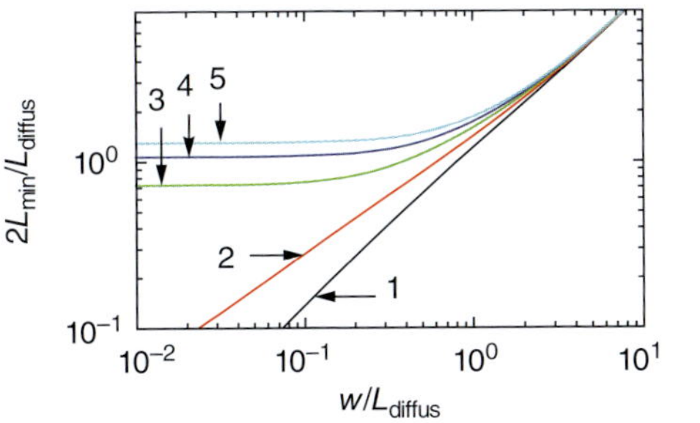

Figure 3.9 Minimum resolution length versus width of absorbed energy density distribution. The numbers indicate the orders of approximation, *m*.

regardless of the actual widths of the absorbed energy density distributions. For $w \gg L_{diffus}$, the diffusion of the growing chains can be neglected. Similar implications can be made [18] for other geometries of polymer structures, including nanorods and nanoplates at nanodistance.

3.5.2
Diffusion of Inhibitor: Diffusion-Assisted Direct Laser Writing

Although the diffusion of small radicals and macromolecules can be detrimental to the spatial resolution, the diffusion of the inhibitor (or the radical quencher) can have a positive effect [29]. The quencher can be added either to the monomer to prevent occasional polymerization or to the resin deliberately. The oxygen dissolved in the resin from the atmosphere can act as an inhibitor as well, by attaching to the growing radicals and producing less active peroxide radicals. The latter can hardly undergo chain propagation; although in some cases they perform chain transfer that effectively reinitiates the polymerization process [38]. Thus, in general, the oxygen cannot be treated as an "ideal" inhibitor that we deal with in our model.

One of the prominent positive effects of the quencher diffusion is the recovery of the initial inhibitor number density after the irradiation. Let us return, again, to the problem of spatial resolution. As shown in Section 3.3, any attempt to make two separate nanofeatures at a distance below the resolution limit fails because the total absorbed energy density in the interstice exceeds the polymerization threshold, thus resulting in a single "glued" structure (see Figure 3.10a). Let us consider

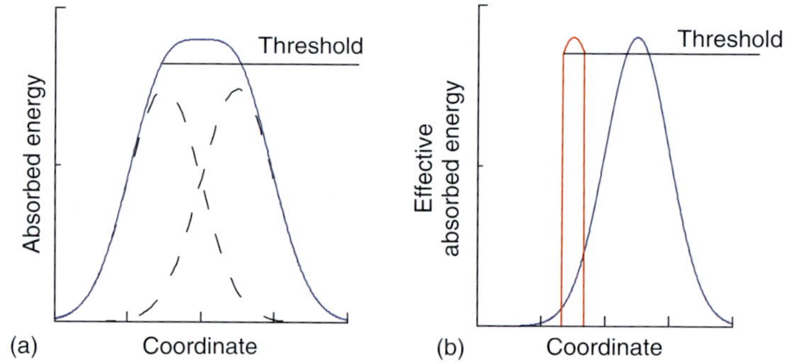

Figure 3.10 (a) Possibility to write two features at a subdiffraction distance is limited by tails of the distributions of the energy density absorbed during scans (black dashed lines). The total absorbed energy (blue line) exceeds the threshold not only where the nanofeatures are expected to form but also in the interstice. (b) Diffusion-assisted writing of the second feature at a subdiffraction distance from the first one. The energy absorbed during the first scan causes both the consumption of the quencher and the formation of the polymer feature. Since the quencher is diffusion-regenerated between scans, the only effect of the irradiation that remains is the formation of the polymer feature (red line). This allows the creation of the second feature at a subdiffraction distance. (Sakellari 2012 [29]. Reproduced with permission of American Chemical Society.)

the writing of, for example, two polymer voxels step by step. When writing the first nanofeature, the inhibitor is totally consumed in the central part of the laser beam, allowing the polymerization process. In the surrounding domain, the consumption of the inhibitor is only partial and the polymerization is prohibited (this is one of the polymerization threshold mechanisms, as was argued earlier). If the second voxel is written immediately after the first, the additional absorbed energy results in the consumption of the rest of the inhibitor in the interstice, starting the polymerization process there as well. If one waits for some time before the writing the second structure, the initial inhibitor concentration will recover in the interstice due to diffusion. When the writing continues with the second voxel, the inhibitor will be consumed again only partially in the interstice, hampering the polymerization there (see Figure 3.10b). Thus, the resolution limit can be overcome through the inhibitor diffusion.

The diffusion of the inhibitor helps in improving the minimum size of a single feature as well. To understand this effect (the detailed theoretical model can be found in [30]), one has to consider the diffusion in Eqs. (3.1)–(3.3):

$$\frac{\partial R}{\partial t} = S(\vec{r}, t) - k_{\mathrm{tRQ}} RQ - k_{\mathrm{tRR}} R^2, \tag{3.52}$$

$$\frac{\partial M}{\partial t} = -k_{\mathrm{p}} MR, \tag{3.53}$$

$$\frac{\partial Q}{\partial t} = D_{\mathrm{Q}} \Delta Q - k_{\mathrm{tRQ}} RQ. \tag{3.54}$$

D_{Q} is the diffusion coefficient of the quencher, and Δ is the Laplacian.

In contrast to the threshold inhibition model, described in Section 3.3, here we consider long-lasting irradiation regimes. Let us assume a constant (in time) or slowly varying source of the radical $S(\vec{r}, t)$. In this case, one can consider quasi-stationary distributions of the number densities of the radical and the quencher: $\partial/\partial t \equiv 0$. The stationary versions of Eqs. (3.52) and (3.54) read

$$R = \frac{\sqrt{k_{\mathrm{tRQ}}{}^2 Q^2 + 4k_{\mathrm{tRR}} S} - k_{\mathrm{tRQ}} Q}{2k_{\mathrm{tRR}}}, \tag{3.55}$$

$$D_{\mathrm{Q}} \Delta Q = \frac{k_{\mathrm{tRQ}}{}^2 Q^2}{2k_{\mathrm{tRR}}} \left(\sqrt{1 + \frac{4k_{\mathrm{tRR}} S}{k_{\mathrm{tRQ}}{}^2 Q^2}} - 1 \right). \tag{3.56}$$

It is seen that the stationary radical number density is either limited by the quadratic termination or by the inhibitor. In domains where the inhibitor is exhausted, that is, where $Q \ll Q_{\mathrm{I}} = \sqrt{k_{\mathrm{tRR}} S} / k_{\mathrm{tRQ}}$

$$R = R_1 = \sqrt{\frac{S}{k_{\mathrm{tRR}}}}, \tag{3.57}$$

as is expected for uninhibited radical polymerization. Alternatively, if $Q \gg Q_{\mathrm{I}}$, then

$$R_2 \approx \frac{S}{k_{\mathrm{Q}} Q}, \tag{3.58}$$

$$D_Q \Delta Q \approx S. \tag{3.59}$$

Here, $R_2 \ll R_1$ (taking into account the expression for Q_I); thus, the polymerization is degraded. The characteristic transition scale between those two types of domains is estimated from Eq. (3.56) at $Q = Q_I$ by substituting ΔQ with $Q_I L_I^{-2}$:

$$L_I \sim \sqrt{\frac{D_Q}{k_{tQR}\sqrt{S/k_{tRR}}}}, \tag{3.60}$$

The main traits of the inhibitor number density distributions that result from Eq. (3.56) can be understood on the basis of a model 1D problem. Let us consider laser-irradiated half-space (coordinate $x > 0$) with $S = S_0$ for $x > 0$ and with the boundary condition $Q = Q_0 \gg Q_I$ at $x = 0$. For this case, Eq. (3.56) is written as

$$\frac{d^2Q}{dx^2} = \frac{{k_{tRQ}}^2 Q^2}{2k_{tRR} D_Q}\left(\sqrt{1 + \frac{4k_{tRR} S_0}{{k_{tRQ}}^2 Q^2}} - 1\right). \tag{3.61}$$

The solution has to obey the boundary condition of $Q \to 0$ at $x \to \infty$. The numerical solution of Eq. (3.61) is presented in Figure 3.11. At the region where $Q \gg Q_I$, Eq. (3.61) transforms into

$$\frac{d^2Q}{dx^2} = \frac{S_0}{D_Q}, \tag{3.62}$$

yielding the solution in the form of a parabolic function. At the region where $Q \ll Q_I$, Eq. (3.61) transforms into

$$\frac{d^2Q}{dx^2} = \frac{Q}{{L_I}^2}. \tag{3.63}$$

Here, the inhibitor number density degrades exponentially to zero with the characteristic scale L_I.

Exact analytical formulas can be obtained for the ideal case of negligible quadratic termination, that is, for $k_{tRR} \to 0$, and thus, $Q_I \to 0$ and $L_I \to 0$. In this case, a sharp transition point $x = x_c$ exists between the regions of inhibited and not inhibited polymerization: $Q(x) > 0$ for $0 \le x < x_c$ and $Q(x) = 0$ for $x \ge x_c$. At $0 \le x \le x_c$, the solution is found from Eq. (3.62) with boundary condition

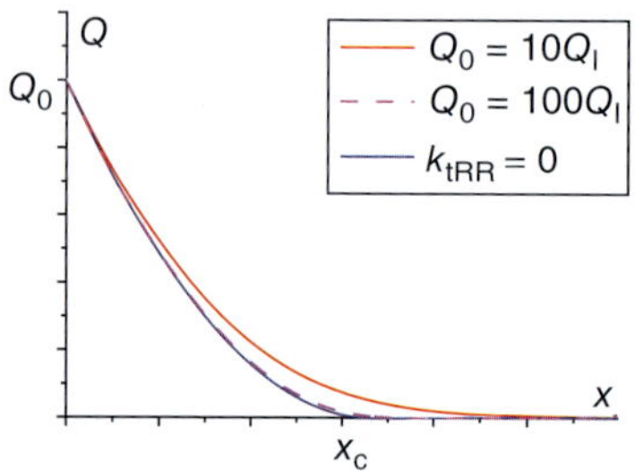

Figure 3.11 Inhibitor number density distributions in a model 1D problem.

$Q(0) = Q_0$. To ensure smoothness and zero number density flux at the transition point, two more boundary conditions

$$Q(x_c) = 0 \text{ and } \partial Q(x_c)/\partial x = 0 \tag{3.64}$$

must be obeyed. There is one extra boundary condition and one additional unknown (x_c) in this problem. The solution is $Q = Q_0(x/x_c - 1)^2$ with $x_c = \sqrt{2D_Q Q_0/S_0}$ (see Figure 3.11).

More complex shapes of the radical source are treated in the similar way. For the Gaussian radical source in the infinite polymerizable sample

$$S = S_0 \exp\left(-\frac{r^2}{2w^2}\right), \tag{3.65}$$

the radius of the inhibitor-free domain was found [29] to be

$$r_1 = w\sqrt{2\ln\left(\frac{S_0 w^2}{Q_0 D_Q}\right)}, \tag{3.66}$$

where Q_0 is the starting number density of the inhibitor. The distributions of the inhibitor number density are shown in Figure 3.12.

In the stationary case, the consumption of the inhibitor in the laser-irradiated spot is balanced by the diffusion from the surrounding volume. The polymerization process is localized within the radius r_1, where the inhibitor is being exhausted during the irradiation. Here, the role of the inhibitor is similar to that of the vortex beam in STED-inspired laser writing. The radius of the polymerization volume depends on the laser intensity and is not affected by the irradiation time. An increase in the irradiation time (or a decrease in the scan speed) only causes an increase in the conversion within the quencher-free domain (Figure 3.13a). This is different from the threshold mode (without pronounced inhibitor diffusion effect) discussed in Section 3.3. Here, the maximum conversion of a nanofeature can be increased by applying a higher radiation exposure by increasing either the beam intensity or the irradiation time (Figure 3.13b). However, the higher exposure inevitably results in increased size of the polymer feature. By employing the diffusion of the inhibitor, one can handle both the size and the maximum conversion of the polymer feature independently. As was discussed in Section 3.4, the narrow and high-contrast distributions of the conversion in a polymerization spot are important for fluctuation-free structure fabrication.

In the stationary regime with the approximation of $Q_I \to 0$, the size of the inhibitor-free domain can be made arbitrarily small. In real writing conditions, the minimum size of the polymerization spot is limited by the scale on the order of L_I. The pure stationary regime can also be considered only as an ideal case. Thus, the real polymerization spot size is affected by the irradiation time/writing speed.

Use of the approach based on the quencher diffusion allowed the authors of [29, 39] to fabricate woodpile structures with an interlayer period of 400 nm, which is comparable to what has been achieved by the two-beam STED-inspired MPP technique currently regarded as state of the art.

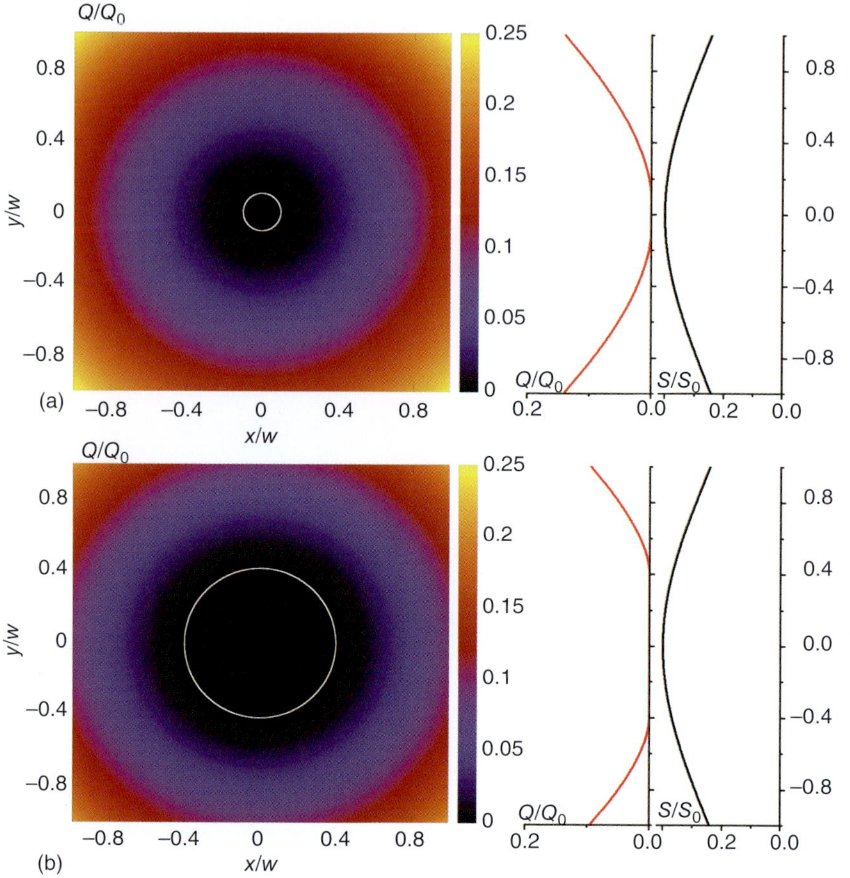

Figure 3.12 Calculated distribution of the quencher number density Q with respect to the initial quencher number density Q_0 (color maps and red line graphs). w is the width of the Gaussian radical source. The quencher-free domain is indicated with white circle. (Sakellari 2012 [29]. Reproduced with permission of American Chemical Society.)

3.6 Conclusion

In this chapter, we studied the physical phenomena that, in our opinion, have a major effect on the laser nanopolymerization process, concerning the limiting capabilities of MPP. They are the diffusion of molecules and fluctuations in polymer nanofeatures. By solving model problems, we demonstrated the approaches that can be employed for the calculations concerning the realistic polymerizable resins in writing experiment and technology.

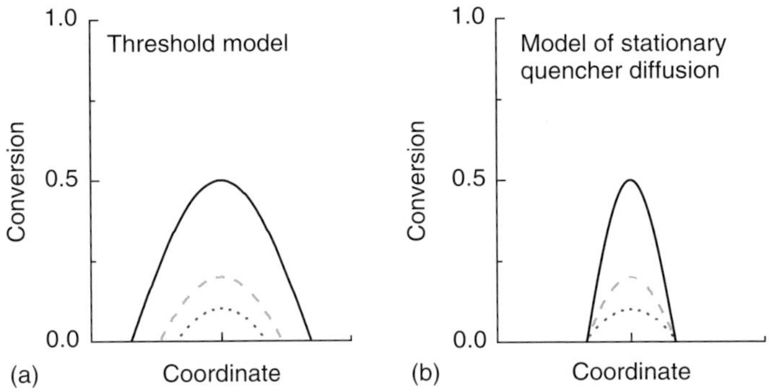

Figure 3.13 (a) Schematic of the conversion profiles in threshold polymerization regime for different exposures. (b) Schematic of the conversion profiles in the model of stationary quencher diffusion for different irradiation times and fixed irradiation intensity. (Sakellari 2012 [29]. Reproduced with permission of American Chemical Society.)

Acknowledgements

This work was funded by the Russian Federal Agency for Scientific Organizations as a part of the State Assignment for the Institute of Applied Physics RAS under Project No. 0035-2014-0019.

References

1. Sekkat, Z. and Kawata, S. (2014) Laser nanofabrication in photoresists and azopolymers. *Laser Photonics Rev.*, **8**, 1–26.
2. Malinauskas, M., Farsari, M., Piskarskas, A., and Juodkazis, S. (2013) Ultrafast laser nanostructuring of photopolymers: a decade of advances. *Phys. Rep.*, **533**, 1–31.
3. Kabouraki, E., Giakoumaki, A.N., Danilevicius, P., Gray, D., Vamvakaki, M., and Farsari, M. (2013) Redox multiphoton polymerization for 3D nanofabrication. *Nano Lett.*, **13**, 3831–3835.
4. Ovsianikov, A., Viertl, J., Chichkov, B., Oubaha, M., MacCraith, B., Sakellari, I., Giakoumaki, A., Gray, D., Vamvakaki, M., Farsari, M., and Fotakis, C. (2008) Ultra-low shrinkage hybrid photosensitive material for two-photon polymerization microfabrication. *ACS Nano*, **2**, 2257–2262.
5. Ovsianikov, A., Malinauskas, M., Schlie, S., Chichkov, B., Gittard, S., Narayan, R., Löbler, M., Sternberg, K., Schmitz, K.-P., and Haverich, A. (2011) Three-dimensional laser micro- and nano-structuring of acrylated poly(ethylene glycol) materials and evaluation of their cytoxicity for tissue engineering applications. *Acta Biomater.*, **7**, 967–974.
6. Nemani, K.V., Moodie, K.L., Brennick, J.B., Su, A., and Gimi, B. (2013) In vitro and in vivo evaluation of SU-8 biocompatibility. *Mater. Sci. Eng. C: Mater. Biol. Appl.*, **33**, 4453–4459.
7. Malinauskas, M., Žukauskas, A., Bičkauskaitė, G., Gadonas, R., and Juodkazis, S. (2010) Mechanisms of three-dimensional structuring of photo-polymers by tightly focussed femtosecond laser pulses. *Opt. Express*, **18**, 10209–10221.
8. Strickler, J.H. and Webb, W.W. (1991) Three-dimensional optical data storage

in refractive media by two-photon point excitation. *Opt. Lett.*, **16**, 1780–1782.
9. Maruo, S., Nakamura, O., and Kawata, S. (1997) Three-dimensional microfabrication with two-photon-absorbed photopolymerization. *Opt. Lett.*, **22**, 132–134.
10. Scott, T.F., Kowalski, B.A., Sullivan, A.C., Bowman, C.N., and McLeod, R.R. (2009) Two-color single-photon photoinitiation and photoinhibition for subdiffraction photolithography. *Science*, **324**, 913–917.
11. Li, L., Gattass, R.R., Gershgoren, E., Hwang, H., and Fourkas, J.T. (2009) Achieving lambda/20 resolution by one-color initiation and deactivation of polymerization. *Science*, **324**, 910–913.
12. Cao, Y., Gan, Z., Jia, B., Evans, R.A., and Gu, M. (2011) High-photosensitive resin for super-resolution direct-laser-writing based on photoinhibited polymerization. *Opt. Express*, **19**, 19486–19494.
13. Fischer, J. and Wegener, M. (2013) Three-dimensional optical laser lithography beyond the diffraction limit. *Laser Photonics Rev.*, **7**, 22–44.
14. Odian, G. (2004) *Principles of Polymerization*, 4th edn, John Wiley & Sons, Inc., Hoboken, NJ.
15. de Gennes, P.-G. (1979) *Scaling Concepts in Polymer Physics*, Cornell University Press.
16. Juodkazis, S., Mizeikis, V., Seet, K.K., Miwa, M., and Misawa, H. (2005) Two-photon lithography of nanorods in SU-8 photoresist. *Nanotechnology*, **16**, 846–849.
17. Bityurin, N.M., Genkin, V.N., Zubov, V.P., and Lachinov, M.B. (1981) On the gel-effect mechanism in radical polymerization. *Polym. Sci. U.S.S.R.*, **23**, 1873–1882.
18. Pikulin, A. and Bityurin, N. (2007) Spatial resolution in polymerization of sample features at nanoscale. *Phys. Rev. B*, **75**, 195430-11.
19. Stocker, M.P., Li, L., Gattass, R.R., and Fourkas, J.T. (2011) Multiphoton photoresists giving nanoscale resolution that is inversely dependent on exposure time. *Nat. Chem.*, **3**, 223–227.
20. Bityurin, N., Pikulin, A., and Alexandrov, A. (2003) Modeling of bleaching wave regime of UV laser polymerization of acrylates without initiators. *Appl. Surf. Sci.*, **208–209**, 481–485.
21. Jia, B., Kang, H., Li, J., and Gu, M. (2009) Use of radially polarized beams in three-dimensional photonic crystal fabrication with the two-photon polymerization method. *Opt. Lett.*, **34**, 1918.
22. Yang, L., El-Tamer, A., Hinze, U., Li, J., Hu, Y., Huang, W., Chu, J., and Chichkov, B.N. (2014) Two-photon polymerization of cylinder microstructures by femtosecond Bessel beams. *Appl. Phys. Lett.*, **105**, 041110.
23. Duocastella, M. and Arnold, C.B. (2012) Bessel and annular beams for materials processing. *Laser Photonics Rev.*, **6**, 607–621.
24. Born, M. and Wolf, E. (1999) *Principles of Optics*, 7th (expanded) edn, Cambridge University Press, Cambridge, pp. 484–492.
25. Serbin, J., Egbert, A., Ostendorf, A., Chichkov, B.N., Houbertz, R., Domann, G., Schulz, J., Cronauer, C., Fröhlich, L., and Popall, M. (2003) Femtosecond laser-induced two-photon polymerization of inorganic-organic hybrid materials for applications in photonics. *Opt. Lett.*, **28**, 301–303.
26. Pikulin, A.V. and Bityurin, N.M. (2012) Fluctuation limitations on the voxel minimal size at laser nanopolymerization. *Tech. Phys.*, **57**, 697–705.
27. Fischer, J. and Wegener, M. (2011) Three-dimensional direct laser writing inspired by stimulated-emission-depletion microscopy [Invited]. *Opt. Mater. Express*, **1**, 614.
28. Gan, Z., Cao, Y., Jia, B., and Gu, M. (2012) Dynamic modeling of super-resolution photoinduced-inhibition nanolithography. *Opt. Express*, **20**, 16871.
29. Sakellari, I., Kabouraki, E., Gray, D., Purlys, V., Fotakis, C., Pikulin, A., Bityurin, N., Vamvakaki, M., and Farsari, M. (2012) Diffusion-assisted high-resolution direct femtosecond laser writing. *ACS Nano*, **6**, 2302–2311.
30. Pikulin, A., Bityurin, N., and Sokolov, V.I. (2015) Model of diffusion-assisted

direct laser writing by means of nanopolymerization in the presence of radical quencher. *AIP Adv.*, **5**, 127215.

31. Mueller, J.B., Fischer, J., Mayer, F., Kadic, M., and Wegener, M. (2014) Polymerization kinetics in three-dimensional direct laser writing. *Adv. Mater.*, **26**, 6566–6571.
32. Sahimi, M. (1994) *Applications of Percolation Theory*, Taylor & Francis, London.
33. Stauffer, D. and Aharony, A. (1994) *Introduction to Percolation Theory*, 2nd rev. edn, Taylor & Francis, London.
34. Korolyov, G.V. and Mogilevich, M. (2009) *Three-Dimensional Free-Radical Polymerization*, Springer, New York.
35. Pikulin, A. and Bityurin, N. (2010) Spatial confinement of percolation: Monte Carlo modeling and nanoscale laser polymerization. *Phys. Rev. B*, **82**, 085406–9.
36. Lu, W.-E., Dong, X.-Z., Chen, W.-Q., Zhao, Z.-S., and Duan, X.-M. (2011) Novel photoinitiator with a radical quenching moiety for confining radical diffusion in two-photon induced photopolymerization. *J. Mater. Chem.*, **21**, 5650.
37. Takada, K., Sun, H.-B., and Kawata, S. (2005) Improved spatial resolution and surface roughness in photopolymerization-based laser nanowriting. *Appl. Phys. Lett.*, **86**, 071122.
38. Ligon, S.C., Husár, B., Wutzel, H., Holman, R., and Liska, R. (2014) Strategies to reduce oxygen inhibition in photoinduced polymerization. *Chem. Rev.*, **114**, 557–589.
39. Bickauskaite, G., Manousidaki, M., Terzaki, K., Kambouraki, E., Sakellari, I., Vasilantonakis, N., Gray, D., Soukoulis, C.M., Fotakis, C., Vamvakaki, M., Kafesaki, M., Farsari, M., Pikulin, A., and Bityurin, N. (2012) 3D photonic nanostructures via diffusion-assisted direct fs laser writing. *Adv. Optoelectron.*, **2012**, 927931.

Part II
Equipment and Techniques

Multiphoton Lithography: Techniques, Materials and Applications, First Edition.
Edited by Jürgen Stampfl, Robert Liska, and Aleksandr Ovsianikov.

4
Light Sources and Systems for Multiphoton Lithography

Ulf Hinze and Boris Chichkov

A setup for multi photon polymerization (MPP) comprises at least three key components:

1) light source
2) focusing optics
3) nano-positioning or beam-deflection unit.

The technical characteristics of these key components very much define the performance of an MPP setup. In addition to these components, an MPP setup usually includes other elements: some of them support the user in sample processing (e.g., various sample illumination techniques, real-time process observation, focus finder), and other elements allow selective process control or optimization (e.g., spatial light modulators (SLMs) or diffractive optical elements (DOEs) can be used for parallel processing).

In this chapter, we discuss laser light sources and systems for multiphoton lithography from the view points of the underlying concept, characteristics, and performance.

4.1
Laser Light Sources

The light source that is used for a MPP process strongly impacts the nonlinear process by its physical properties. Relevant properties include (but are not limited to)

- laser wavelength
- pulse energy
- pulse duration
- repetition rate
- spectral bandwidth
- beam quality and profile.

Multiphoton Lithography: Techniques, Materials and Applications, First Edition.
Edited by Jürgen Stampfl, Robert Liska, and Aleksandr Ovsianikov.

Such parameters as wavelength, pulse energy, pulse duration, spectral bandwidth, and energy distribution within the beam profile directly influence the nonlinear interaction in the photomaterial and, thereby, the result of MPP. Laser wavelength and spectral bandwidth have to be chosen with respect to the spectral sensitivity of the photo resist; the beam profile influences the shape and size of the polymerized volume pixels; the maximum available pulse energy and the repetition rate impose limits to the maximum processing speed.

The light source has to be chosen very carefully with respect to the research task in the field of MPP. However, two-photon processes (2PP) can often be operated quite stably, especially in materials that are specially designed for this process, providing a wide processing window.

4.2 Ultrashort-Pulse Lasers

Ultrashort-pulse lasers include pulsed lasers that emit laser light pulses with a pulse duration of 1 ps or less [1] (see Figure 4.1 for the basic characteristics of laser pulses),which can be generated with mode-locked laser systems. During the beginning of the 1990s, nobody would have expected that this type of lasers would play a significant role in material processing and additive manufacturing. However, things have changed significantly after 20 years of laser research and development efforts and the demonstration of precise laser–matter interaction with femtosecond laser pulses [2].

The first report that introduced two-photon absorption to the UV-photopolymerizing fabrication process to enhance longitudinal resolution was published in 1997 by Shoji Maruo, Osamu Nakamura, and Satoshi Kawata [3]. Their experimental work was based on the application of a Ti:sapphire ultrashort pulse laser, emitting at a wavelength of 790 nm with a pulse duration of 200 fs and a repetition rate of 76 MHz.

Subsequently, this type of laser became the workhorse of two-photon polymerization. We have surveyed more than 200 papers published between 2010 and 2015 and found that more than 70% of the published works were done with commercial or self-built Ti:sapphire lasers in the near-infrared spectral range of

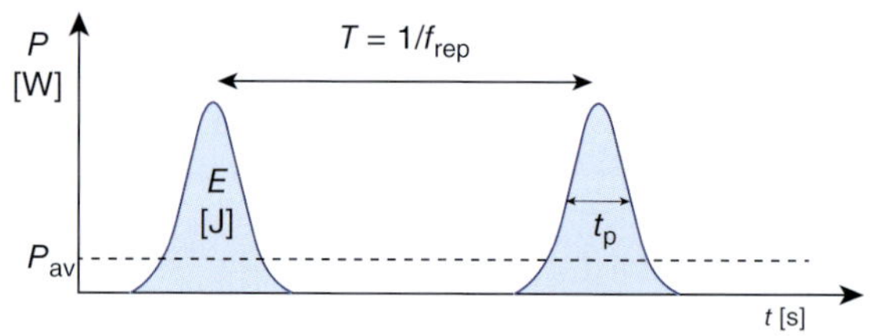

Figure 4.1 Basic characteristics of laser pulses: pulse power P, pulse energy E, average power $P_{av} = E*f_{rep}$, repetition rate f_{rep}, and pulse duration t_p. Note that the realistic time scale is very different for femtosecond laser pulses.

~800 nm. They are accompanied by some frequency-doubled Er-doped ultrafast fiber lasers, which emit at 780 nm. Most of the remaining experimental works were done in the green spectral range as discussed in the following.

During the last decade, Ti:sapphire lasers have much improved, for example, by semiconductor saturable absorber mirrors (SESAMs). This resulted in a wide choice of commercial turnkey lasers available from different manufacturers. For the MPP application discussed here, they are typically operated at a wavelength of ~780–820 nm with ~50–250 fs pulses at a repetition rate of ~80 MHz.

How much laser power is required to operate a two-photon polymerization setup very much depends on the concept of the optical setup and the photomaterial. Most authors report the use of an average laser power in the range 1–10 mW behind a tight focusing objective (100×, oil immersion, NA (numerical aperture) 1.4) at a repetition rate of 80 MHz. This corresponds to about 10–100 pJ in a focal spot, which is comparable to the wavelength of the applied laser. In addition, the laser power that is lost in the optical system of the MPP setup has to be considered. The optical system of an MPP setup typically involves several of the following elements: mirrors for adjustment and beam steering, a shutter or acoustooptic modulator, waveplate and polarizing optics for laser power control, lenses, diaphragms, and active optics for beam shaping, dichroic mirrors to adjust the laser, sample illumination, and process observation.

In total, pulse energy of 2 nJ (about 150 mW at 80 MHz, 100 fs) from an ultrafast Ti:sapphire laser oscillator is just sufficient to operate an optical system that allows polymerizing the frequently used photomaterials at a resolution in the 100 nm scale at a speed of about 100 μm s^{-1}. However, more laser power is preferable, as it opens access to higher writing speed and a wider range of materials. With more laser power, even objectives with lower magnification (e.g., 20×) can be used for faster structuring with micrometer resolution. Optical setups that include SLMs require even higher laser powers, as in such setups not only a single focus but several foci, larger areas, or an entire volume is illuminated. For such applications, femtosecond laser oscillators with pulse energies of 40 nJ or higher are available and used for MPP. These laser systems allow processing larger polymerization volumes at shorter processing times.

Amplified laser systems could provide much higher pulse energy, but they are not commonly used for nano and microstructuring with 2PP, as most amplifiers have to increase the pulse energy at the cost of a corresponding decrease of the repetition rate [3]. Nevertheless, these lasers are applied in industry for femtosecond laser ablation; consequently, they are taken into account for 2PP as robust and reliable alternatives. During the last years, even ultrashort-pulse lasers with high repetition rates in the gigahertz range at sufficient pulse energy have become available. However, these lasers are not commonly used for MPP, as the optical system requires increased design efforts, for example, to cope with the relatively low pulse energy, to maintain the characteristics of high-repetition-rate broadband pulses by broadband optics, or to compensate a significant divergence of the laser output.

Since the 1980s, much research effort has been put into the development of fiber lasers, and consequently a wide range of fiber laser concepts and models are available. Femtosecond fiber lasers are often based on transitions in rare-earth ions, for example, gain media doped with ytterbium, erbium, neodymium, or thulium. Erbium-doped fiber lasers at 1560 nm with subsequent second-harmonic generation at 780 nm have been reported to work in the same process window with pulse energy of up to 1.8 nJ, pulse duration of 100 fs, and repetition rate of 80 MHz. They are used in commercial systems [4]. The advantage of fiber lasers is their high efficiency, which results in a compact setup and relatively low price; however, the available laser pulse energy is limited.

About 20% of publications are based on laser light sources in the wavelength range 500–532 nm. A great variety of different lasers and parameters are reported for this spectral range. Most of them are picosecond microchip lasers (diode-pumped monolithic solid-state lasers with frequency doubling, e.g., based on Nd:YAG), which are relatively cheap and robust devices. However, they emit much longer laser pulses of about 500 ps, at a repetition rate of around 10 kHz only, and often come with a limited beam quality [5, 6].

Other green laser light sources reported for 2PP applications are based on a combination of diode-pumped laser crystals doped with ytterbium or neodymium in a mode-locked laser oscillator with frequency doubling. The authors report emission starting from a few nanojoules up to 10 μJ with repetition rates often of a few 100 kHz up to some 10 MHz range. Because of frequency doubling, the pulse duration is often longer compared to laser oscillators; typically values around 250 fs are reported [7–9]. However, frequency doubling can be even considered as an advantage, as it may help to suppress unwanted nanosecond pre and post pulses.

When choosing an ultrashort laser, also the expected service costs should be considered. Today's ultrashort laser systems are much more reliable than they used to be 10 years ago, as laser manufacturers have spent much effort to improve performance, stability, and lifetime of their products. Some of them even claim to sell industry-ready turnkey systems. However, in our experience, many of the commercially available lasers still incur significant service costs. For example, for low-power lasers, it should be kept in mind that already a slight degradation of the laser power might be critical: low laser power might be insufficient for polymerization or can significantly narrow the processing window and processing speed. For high-power laser systems, the service costs are often dominated by the pump diodes, which may cost 10% of the price of the laser with a guaranteed lifetime between 1000 and 10 000 h or 6 months and 3 years – whatever comes first.

4.3
Laboratory Systems and Processing Strategy

In the optical 2PP setup, the laser beam is expanded to overfill the back aperture of a high-quality objective or lens. Laser and objective together create a small focal spot inside the photomaterial where polymerization can be initiated. To trigger the polymerization reaction, the threshold of the photomaterial has to be overcome. The size of the polymerized volume pixel (voxel) is controlled by tuning the focal intensity with respect to the polymerization threshold, which allows the realization of voxels at a size well below the optical diffraction limit of the objective.

Complex three-dimensional (3D) structures are generated by moving this focus within the resin to the desired positions. The way to realize this is influenced by the positioning system or the beam deflection unit used. Most positioning systems for 2PP include three linear actuators to provide movement in all spatial directions. Frequently used actuators are linear motors and piezoelectric actuators; in addition, often galvanic laser scanners are used.

Each of these technologies has its own advantages and disadvantages: state-of-the-art piezoelectric actuators can provide sub-nanometer resolution in positioning; but at the same time they typically have a very short travel range of not more than a few hundred micrometers, limited speed of $\sim 100\,\mu m\,s^{-1}$, and may require sophisticated hysteresis compensation. Linear motors combine high speed (more than several centimeters per second) and acceleration with long travel ranges (even travel ranges of meters are available), but more work needs to be done to achieve nanometer precision. Galvanometer laser scanners can be even faster than linear motors or piezoelectric actuators, but the processing area is limited to the field of view of the objective (e.g., about $100 \times 100\,\mu m^2$ for a 100× NA 1.4 objective), and the effective speed of a laser scanner is reduced at the magnification factor of the objective. The combinations of the technologies mentioned earlier allows us to take advantage of their particular strengths, for example, combining a laser scanner with three linear motors results in a system that can be very fast (laser scanner) and at the same time cover large processing areas (linear motor).

It is important to consider how the positioning system is controlled. Some basic controllers do not support coordinated motion, which means that each axis of the positioning system travels at a different speed to the commanded position without taking into consideration the position and speed of the other axes. In particular, it is not even guaranteed that the movement is strictly monotonic, as some controllers use iterative algorithms to approach the set position. Nonetheless, such positioning systems can be used (and they are used) to build complex 3D structures realizing a dot-by-dot exposure of the photoresin. A system based on dot-by-dot exposure executes a sequence that commands the axes at the set position and waits until the set positions are reached and the axes are settled. Then, the laser is triggered to expose the resin for a defined time (Figure 4.2). After that, the generation of a volume dot is finished, and the sequence is repeated with the subsequent positions [10–12]. The advantage of this approach is that each voxel is polymerized at well-controlled parameters and it can be realized with nearly any type of positioning system. The obvious disadvantage of this approach is that a considerable amount of the processing time cannot be used for polymerization, as it is not possible to polymerize while the positioning system moves from one position to the next one.

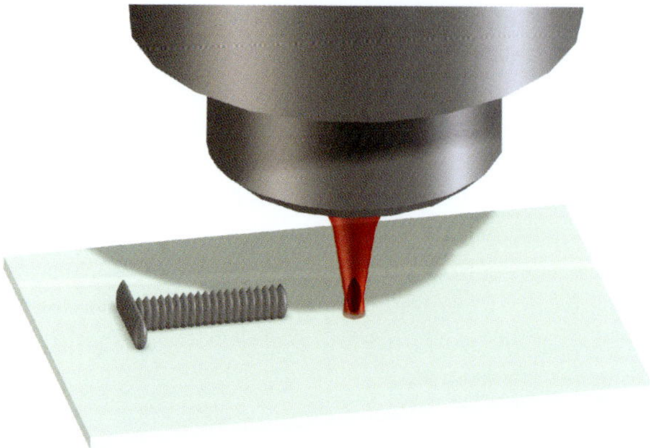

Figure 4.2 Dot-by-dot processing – each volume pixel is exposed separately.

This can be overcome with controllers and actuators that allow coordinated motion. Coordinated motion means that the controller is able to guide two or more axes of the positioning system synchronously so that the combined movement follows a well-defined 3D path. In general, a controller can achieve this either by relying on predefined axis parameters or by continuously comparing the set point and the actual value in real time during the movement. Some controllers, for example, piezo actuators, apply learning algorithms that use dry runs to measure the discrepancy between the set point and actual value (e.g., because of hysteresis or strong acceleration) and iteratively improve a certain movement vector.

In addition to coordinated motion, a position-synchronous switching of a trigger signal that can be used to control the exposure of the photoresist is required. If a positioning system allows coordinated motion and position-synchronous triggering, the dot-by-dot exposure can be replaced by a line-by-line exposure. A system based on line-by-line exposure accelerates the axis system to the desired polymerization speed and triggers the laser when the required start position is reached. Subsequently, it moves along a defined path at a defined speed to the end position, triggers the laser, and decelerates (Figure 4.3). This can result in a much more efficient use of the processing time, as the dead time between the exposure of two dots is eliminated; instead, the material is exposed continuously, which even results in smoother surfaces. However, with increasing processing speed v, longer acceleration paths are required, $s = v^2/(2a)$, which again creates dead times of $t = v/a$ before and after the processing of each line. Consequently, acceleration and speed should be well balanced during fast processing. Most currently used 2PP systems operate on the basis of this approach of coordinated motion.

It was demonstrated that with imaging devices such as DOEs, fly's-eye type lenses, digital mirror devices (DMDs), and liquid crystal (LC)-based SLMs, further optimization steps in 3D polymerization can be done. Instead of single line-by-line processing, a parallelization of the 2PP processing can be achieved with such devices. Fly's eye, microlens arrays, and DOEs provide static patterns of 10 or even of 100 spots, which can be used to modulate a single laser beam into a pattern

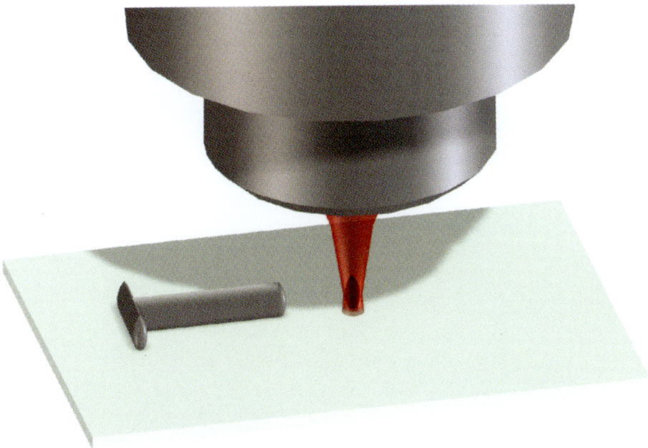

Figure 4.3 Line-by-line processing – synchronous axis movement and position-synchronous laser triggering allow generating linear structures in a single move.

of multiple beams, which allows the parallelization of the generation of periodic structures [13, 14]. DMDs [15, 16] are arrays of tiny micromechanical mirrors with a high fill factor (>90%), which can be switched between two tilt states at refresh rates of 10 kHz and more. The micromechanical mirrors provide high contrast (on/off); however, the gray scales have to be simulated by spatial or temporal averaging. LC-based SLMs allow the modulation of the intensity, phase, or polarization of a light beam spatially and in time by computer-generated patterns and holograms. They provide high fill factors (liquid crystal on silicon microdisplays provide 90% fill factor), and typically refresh at a rate of some 10 Hz, and each dot can be addressed in real gray scale. It was demonstrated that by means of a computer-generated hologram, the laser focus (behind the objective) can be modulated into several foci for parallelization of the 2PP processing [17, 18]. For example, with 10 foci instead of a single focus, the production of periodical structures can be speed up by a factor of 10 [11, 19, 20]. By this technique, arrays of hundreds and more foci can be generated, which has the potential of substantial reduction of the processing time of periodic structures. Even dynamic control of the foci is feasible, but it is limited to the refresh rate of the display. For a quite complex geometrical object, this was demonstrated in [21]: the authors polymerized a multi-dodecahedral microstructure moving the foci of five holographic beams in three dimensions at $9\,\mu m\,s^{-1}$ scanning speed.

SLMs allow the illumination of a complete resin layer in one step (Figure 4.4). A typical setup is shown in Figure 4.5. To preserve as much of the laser power as possible, this is usually done by calculating a phase hologram for the entire structure inside one layer. During exposure, the hologram is transformed by the objective into the image of a full layer, which is polymerized in a single step. However, it should be mentioned that some problems arise in application of this technology for a layer-by-layer exposition. Laser light does not only exist inside the desired layer, but also the light field evolves while passing through the focus. As a phase hologram on a display with finite resolution can carry only a finite amount

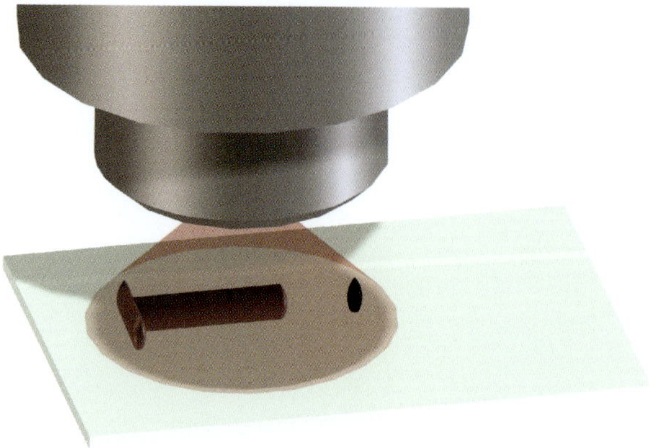

Figure 4.4 Spatial light modulator – the field of view of the objective can be exposed in a single shot.

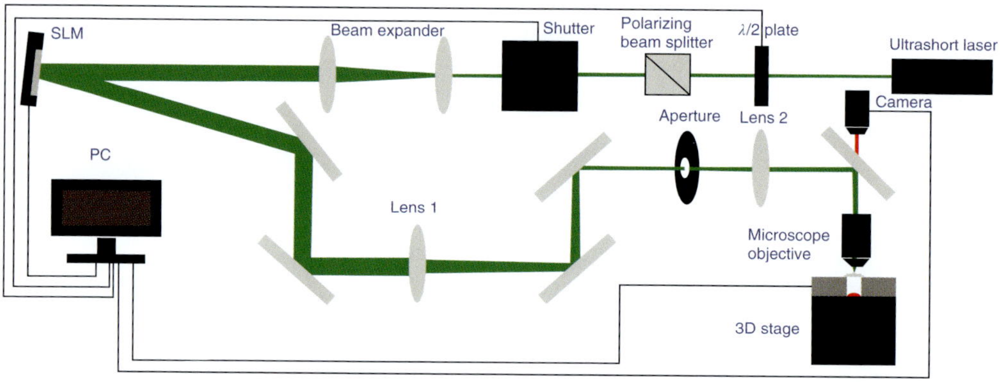

Figure 4.5 Typical setup with spatial light modulator. The ultrashort laser pulse passes through the polarizing optics for laser power control, the exposure time is controlled by a shutter, the beam is expanded to fill the SLM, an intermediate image is created between lens 1 and lens 2 for control, selection, and demagnification, and the microscope objective generates the final image for processing on a 3D stage.

of information, there is only limited control on the field distribution inside the predefined layer and – also worth mentioning – outside of this layer. As a result, structures generated with SLM typically show lower geometrical fidelity with less accurate edges and shape in comparison with line-by-line generated structures. For certain geometries this can be overcome, for example, if a photoresist layer is prepared that is thin in comparison with the length of the focal region by spin coating a sample. In this way, the SLM technique can be used to generate 2D structures with very well-defined height in a single step. For a layer-by-layer generation of complex 3D structures, careful control of the field distribution in the focal region is beneficial.

It is remarkable that an SLM can be even used to shape a laser beam three dimensionally. It would be, of course, the dream of technology developers to build an arbitrary 3D object in a single exposure. For certain geometries and structures, this has already been demonstrated; for example, cylindrical and helical microstructures could be designed and generated by SLM holographically shaped 3D beams [9, 22, 23]. However, there remain challenges in transferring this technique to the production of complex, arbitrary 3D objects: First and foremost, the calculation of the corresponding computer-generated holograms providing an appropriate electromagnetic field distribution inside and around the object in a nonlinear absorbing medium is a complex task. In addition, the amount of information that can be encoded into a hologram is limited by the SLM display, and, finally, a large polymerization volume requires a correspondingly high total laser power.

4.4 Further Processing Considerations

The production of a micro and nanostructure starts with a 3D sketch or model of the structure to be manufactured. This data has to be analyzed, converted, and prepared for the processing unit of the MPP setup. For stages and SLM, totally different approaches are required.

Linear stages, piezo systems, and galvanometer scanners usually require that the model is converted into movement trajectories and the corresponding movement commands. In most instances, the user starts with a 3D model from a computer-aided design (CAD) software that is exported to an STL file. The STL file format has become a standard in additive manufacturing, as it is an open standard, clear, and easy to implement. It is not only widely used in MPP but also in stereolithography and for 3D printers as well. In an STL file, only surfaces of an object are described, and this is done by segmenting the surface into many small triangles. To generate movement trajectories from STL files, algorithms are used that divide the model into slices and identify the edges of the model in each slice. To produce not only a surface but a solid 3D object, the algorithm has to identify the inside part of the object and add movement trajectories that fill the inner volume of the object, a method that is often referred to as *hatching*.

In addition, the algorithm has to consider requirements that are specific for the positioning system. For example, linear stages in the line-by-line mode usually require appropriate acceleration tracks; galvanometer scanners need special attention in synchronization, and some piezo systems can perform learning cycles before processing to achieve better fidelity. The result of the data analysis and processing is a program that can be executed by the system controller. Depending on the controller manufacturer, it might be in a standard programming language like computerized numerical control (CNC) code or in a device-specific, proprietary programming language.

Quite often, users of 2PP systems like to have very detailed control of the processing strategy to ensure essential properties of the fabricated structure. For instance, processing of micro lenses can be done with a standard rectangular grid hatching, but in many cases the surface quality of the resulting structure will be disappointing, as it still exhibits traces of the rectangular grid. A strategy based on a concentric writing algorithm can be much more successful in this case. As soon as freeform surfaces like aspheric lenses are targeted, it is worthwhile considering which processing approach is the most promising. In such a situation, it is essential to have full control of the hardware and to program the processing strategy in detail oneself, for example, in CNC code.

DMDs, SLMs, and related devices require images that are shown on their display. In the most basic case, these images are slices of the 3D model as described earlier, which are subsequently displayed on a DMD. With SLM, usually phase modulation with holographic images is used to minimize the loss of laser power. Different algorithms have been proposed and are used to calculate holographic images, of which the most popular is the Gerchberg–Saxton algorithm [24], which is used to iteratively create holograms for each image slice. However, similar to dedicated user programming of stages, calculation of holograms can be based on an analytic mathematical approach instead of an iterative numerical approach, particularly when single-shot 3D polymerization is intended.

Regardless of the positioning system or beam deflection system, an important aspect of an MPP setup is the precise alignment of the incident laser beam with respect to the sample substrate surface. For example, for a nanostructure of 100 μm width and an intended axial size voxel of 1 μm, an angular error of only 0.05° already corresponds to about 100 nm, which is 10% of the voxel size. Instead of aligning each substrate at the corresponding accuracy, this is often compensated by measuring three positions on the sample surface that are far enough apart from each other. Subsequently, the nanostructures are built onto a virtual plane that is defined by these positions.

Different technological approaches for two-photon polymerization described earlier result in a wide range of optical setups (Figure 4.6).

A very simple setup that allows dot-by-dot exposure can be built with a standard motorized *XYZ* positioning system. Most femtosecond or even picosecond lasers that might be already available in a laboratory can be used as a light source if they provide nanojoule pulse energy and good beam quality. In addition, an electromechanical shutter, a standard microscope objective, a camera module, and some optomechanical components for beam steering, laser power control, and sample illumination are required. This low-budget system will be of relatively low performance and require some development efforts and attention of the experimenter; however, it would work, and for certain tasks like material studies it can be sufficient.

An advanced setup typically includes a piezoelectric stage or a precision motorized *XYZ* stage operating considerably in the sub-micrometer range, a commercial ultrashort laser with <250 fs pulse duration, megahertz repetition rate and a pulse energy not less than 2 nJ, a high-quality high NA 100× objective, polarizing optics for power control, and a fast shutter.

A powerful 2PP system is based on a dedicated nano-positioning system with high dynamics and/or a laser scanner, an ultrashort laser with megahertz

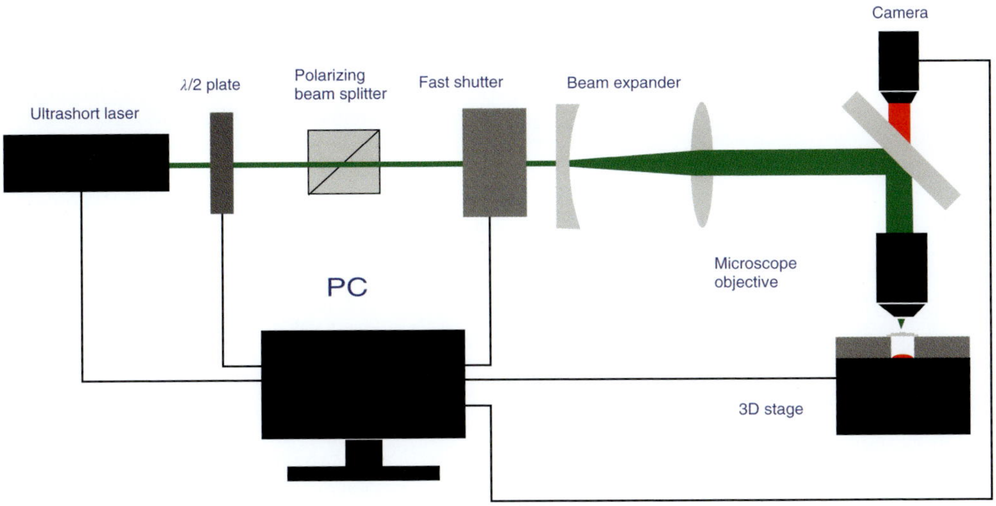

Figure 4.6 Typical universal arrangement for two-photon polymerization. The ultrashort laser pulse passes polarizing optics for laser power control, the exposure time is controlled by a fast shutter, the beam is expanded to fill the input aperture of the microscope objective, and the microscope objective generates the focus for processing on a 3D stage.

repetition rate, a high-quality, high-NA optical microscope objective, a fast optical shutter like an acoustooptic modulator, automated laser power control, and dedicated optics for beam shaping and sample illumination, and tools for identifying the resin interface [25].

Setups with an SLM require a very dedicated and extended optical system, as shown in Figure 4.4. For example, LC-based SLMs very much depend on the correct polarization and require a well-polarized light beam. The design and calculation of computer-generated holograms include assumptions on beam shape and wavefront properties, which require special attention. Finally, it is useful to be able to verify the exposure image on a reference camera to identify unexpected hologram design deviations.

In conclusion, a wide spectrum of ultrafast light sources and precise positioning systems is available in the literature and used for MPP. Low-budget laboratory systems can be self-built from standard laboratory equipment, but this requires much effort to characterize, develop, and optimize the performance of optomechanical components together with material, process, and software. Commercially available systems are much more than the sum of their optical and mechanical components: These systems would have been optimized with respect to their performance, and they are based on much experience in lasers, optical design, performance, and control of nanopositioning stages and laser scanners and laser–matter interaction. It is worthwhile considering whether the system should be extendable, as the progress in the field of MPP is very fast and the state of the art changes every year with new technologies coming up.

References

1. Diels, J.C. and Rudolph, W. (2006) *Ultrashort Laser Pulse Phenomena*, Elsevier. ISBN: 978-0-12-215493-5
2. Chicbkov, B.N., Momma, C., Nolte, S., von Alvensleben, F., and Tünnermann, A. (1996) Femtosecond, picosecond and nanosecond laser ablation of solids. *Appl. Phys. A*, **63** (2), 109–115.
3. Maruo, S., Nakamura, O., and Kawata, S. (1997) Three-dimensional microfabrication with two-photon-absorbed photopolymerization. *Opt. Lett.*, **22**, 132–134.
4. Thiel, M., Fischer, H., von Freymann, G., and Wegener, M. (2010) Three-dimensional chiral photonic superlattices. *Opt. Lett.*, **35** (2), 166–168.
5. A typical application of a picosecond laser is e.g. found in Turunen, S., Käpylä, E., Lähteenmäki, M., Ylä-Outinen, L., Narkilahti, S., and Kellomäki, M. (2014) Direct laser writing of microstructures for the growth guidance of human pluripotent stem cell derived neuronal cells. *Opt. Lasers Eng.*, **55**, 197–204.
6. Wang, I., Bouriau, M., Baldeck, P., Martineau, C., and Andraud, C. (2002) Three-dimensional microfabrication by two-photon-initiated polymerization with a low-cost microlaser. *Opt. Lett.*, **27**, 1348–1350.
7. Raimondi, M.T., Nava, M.M., Eaton, S.M., Bernasconi, A., Vishnubhatla, K.C., Cerullo, G., and Osellame, R. (2014) Optimization of femtosecond laser polymerized structural niches to control mesenchymal stromal cell fate in culture. *Micromachines*, **5**, 341–358.
8. Kurselis, K., Kiyan, R., Bagratashvili, V., Popov, V., and Chichkov, B. (2013) 3D fabrication of all-polymer conductive microstructures by two photon polymerization. *Opt. Express*, **21**, 31029–31035.
9. Burmeister, F., Steenhusen, S., Houbertz, R., Zeitner, U.D., Nolte, S., and Tünnermann, A. (2012) Materials and technologies for fabrication of three-dimensional microstructures with sub-100 nm feature sizes by two-photon polymerization. *J. Laser Appl.*, **24**, 042014.
10. Tian, Y., Lu, D., Jiang, H., and Lin, X. (2012) Preparation of a novel ferrofluidic photoresist for two-photon photopolymerization technique. *J. Magn. Magn. Mater.*, **324** (20), 3291–3294.
11. Gan, Z., Cao, Y., Jia, B., and Gu, M. (2012) Dynamic modeling of super-resolution photoinduced-inhibition nanolithography. *Opt. Express*, **20**, 16871.
12. Wu, L., Han, Z., Song, Y., Niu, S., and Ren, L. (2012) Replication of butterfly scales nano-structure with two-photon polymerization method and the optical effect analysis. *Appl. Mech. Mater.*, **101–102**, 1006–1009.
13. Dong, X.-Z., Zhao, Z.-S., and Duan, X.-M. (2007) Micronanofabrication of assembled three-dimensional microstructures by designable multiple beams multiphoton processing. *Appl. Phys. Lett.*, **91**, 124103.
14. Kato, J.-I., Takeyasu, N., Adachi, Y., Sun, H.-B., and Kawata, S. (2005) Multiple-spot parallel processing for laser micronanofabrication. *Appl. Phys. Lett.*, **86**, 044102.
15. Nielson, R., Kaehr, B., and Shear, J.B. (2009) Microreplication and design of biological architectures using dynamic-mask multiphoton lithography. *Small*, **5** (1), 120–125.
16. Mills, B., Grant-Jacob, J., Feinaeugle, M., and Eason, R. (2013) Single-pulse multiphoton polymerization of complex structures using a digital multimirror device. *Opt. Express*, **21**, 14853–14858.
17. Takahashi, H., Hasegawa, S., Takita, A., and Hayasaki, Y. (2008) Sparse-exposure technique in holographic two-photon polymerization. *Opt. Express*, **16** (21), 16592–16599.
18. Obata, K., Koch, J., Hinze, U., and Chichkov, B. (2010) Multi-focus two-photon polymerization technique based on individually controlled phase modulation. *Opt. Express*, **18**, 17193–17200.
19. Gittard, S.D., Nguyen, A., Obata, K., Koroleva, A., Narayan, R.J., and Chichkov, B.N. (2011) Fabrication of microscale medical devices by two-photon polymerization with multiple

foci via a spatial light modulator. *Biomed. Opt. Express*, **2**, 3167–3178.

20. Lin, H., Jia, B.H., and Gu, M. (2011) Dynamic generation of Debye diffraction-limited multifocal arrays for direct laser printing nanofabrication. *Opt. Lett.*, **36** (3), 406–408.
21. Vizsnyiczai, G., Kelemen, L., and Ormos, P. (2014) Holographic multi-focus 3D two-photon polymerization with real-time calculated holograms. *Opt. Express*, **22** (20), 24223.
22. Yang, L., El-Tamer, A., Hinze, U., Li, J., Hu, Y., Huang, W., Chu, J., and Chichkov, B.N. (2014) Two-photon polymerization of cylinder microstructures by femtosecond Bessel beams. *Appl. Phys. Lett.*, **105**, 041110, and further references therein.
23. Zhang, S.-J., Li, Y., Liu, Z.-P., Ren, J.-L., Xiao, Y.-F., Yang, H., and Gong, Q. (2014) Two-photon polymerization of a three dimensional structure using beams with orbital angular momentum. *Appl. Phys. Lett.*, **105**, 061101.
24. Gerchberg, R.W. and Saxton, W.O. (1971) Phase determination from image and diffraction plane pictures in electron-microscope. *Optik*, **34** (3), 275.
25. Farsari, M. and Chichkov, B. (2009) Materials processing: Two-photon fabrication. *Nat. Photonics*, **3**, 450–452.

5
STED-Inspired Approaches to Resolution Enhancement

John T. Fourkas

5.1
Introduction

Many potential applications of multiphoton lithography require extremely high resolution (which is defined as the minimum pitch of features). One important element in attaining high resolution is minimization of the feature size. The simplest means of reducing the size of voxels are (i) using an objective with the highest possible numerical aperture; (ii) using the shortest possible exposure wavelength; and (iii) using the smallest possible exposure dose at each point. The first two strategies arise directly from the Abbe criterion (i.e., that the resolution is approximately the exposure wavelength divided by twice the numerical aperture of the imaging system). However, the third strategy allows two-photon polymerization to achieve feature sizes and resolutions that are far smaller than the Abbe criterion predicts. For any negative-tone photoresist, there is an exposure threshold below which solubility is maintained. Two-photon polymerization is typically performed with a laser beam with a Gaussian spatial profile; and by operating under appropriate exposure conditions, the threshold dose is received only at the very top of this profile (Figure 5.1). This phenomenon allows for features with transverse dimensions on the order of 100 nm to be attained routinely using 800-nm exposure.

Each of the above strategies has been exploited to improve the feature size and resolution of multiphoton polymerization. For instance, schemes have been explored to increase the effective numerical aperture of the imaging system used for fabrication [1]. The influence of the refractive index of the photoresist on resolution has also been explored [2], which is important because a higher refractive index enables a higher numerical aperture to be used. The use of photoresists that can be exposed at shorter wavelengths has also proven to be an effective means of attaining higher resolution [3], although this strategy requires the use of ultrafast laser systems that are not as widely available as those producing light in the near-infrared. A number of groups have also explored the use of radical quenchers to create photoresists with a steeper exposure contrast

Multiphoton Lithography: Techniques, Materials and Applications, First Edition.
Edited by Jürgen Stampfl, Robert Liska, and Aleksandr Ovsianikov.

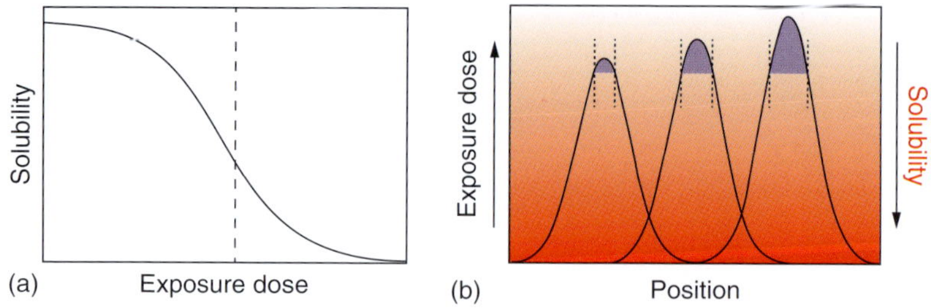

Figure 5.1 Exposure dose and voxel size. (a) Schematic depiction of the dependence of solubility on the exposure dose for a negative-tone photoresist. The dashed line indicates the threshold exposure dose to make the resist insoluble. (b) Illustration of how voxel dimensions can be considerably smaller than the Abbe limit. The red shading indicates the solubility as a function of exposure dose, and the curves are spatial Gaussian intensity profiles representing different laser intensities. Only the regions of the photoresist for which the Gaussian profile is above the exposure threshold remain after development. By operating at a dose such that just the very tip of the Gaussian profile is above the threshold, voxels as small as approximately 1/10 of the exposure wavelength can be created.

curve, which allows for finer control of the exposure threshold dose and thereby enables the creation of smaller voxels [4–6].

A facet of multiphoton lithography that can be inconvenient is that voxels are not spherical. This phenomenon arises from the fact that the multiphoton point-spread function (PSF) of a tightly focused laser beam is longer along the optical axis than it is perpendicular to this axis [7]. As a result, voxels are prolate spheroids that are typically a factor of 3–5 longer than their width [8]. This irregular shape presents significant challenges in the fabrication of high-resolution structures that require features that are of the minimum possible size in all dimensions. This issue can be addressed in part by using a different transverse and axial step size during fabrication; but for structures that require uniform, fine features, such a strategy limits the resolution to that in the axial direction. Furthermore, the elongated shape of voxels can cause the mechanical properties of structures that are fabricated parallel to the substrate to differ substantially from those of structures that are fabricated perpendicular to the substrate [9]; this phenomenon is all the more significant when different step sizes are used in different directions.

The approaches outlined above can provide significant enhancement of the resolution of two-photon polymerization. However, these methods cannot improve the resolution to the extent that is desired for many applications, and they have little effect on the voxel asymmetry problem. Thus, researchers have looked in new directions to address these issues. One of the most promising approaches is the creation of new photoresists that respond differently to two colors or two beams of light. In a typical implementation of such a photoresist, a beam of light of one color is used for two-photon exposure, and a beam of a second color of light is used to counteract the effects of the exposure. By appropriate spatial patterning of these two beams, it is possible to attain feature sizes and resolution that far surpass what can be achieved with two-photon exposure alone.

This chapter focuses on two-color approaches to resolution enhancement in two-photon polymerization [10–14]. We begin by discussing the optical technique that has inspired most of the approaches to two-color lithography, stimulated emission depletion (STED) fluorescence microscopy [15–17]. We then consider the various approaches that have been taken to apply some of the ideas of STED to two-color lithography, and discuss the state of the art for each method. We close the chapter by considering challenges that must be addressed to unlock the full potential of two-color techniques in multiphoton polymerization.

5.2 Stimulated Emission Depletion Fluorescence Microscopy

Traditional fluorescence microscopy faces some of the same challenges, as does photolithography. The Abbe criterion also dictates the minimum distance between resolvable features in imaging, for instance. However, in fluorescence microscopy, photodamage limits the shortest wavelength of radiation that can be employed, particularly in the case of live-cell imaging. The inability to use shorter wavelengths for biological imaging has been a driving force in the development of new, super-resolved techniques for optical microscopy, which include STED [15, 16] and others [18–20].

In a typical implementation of laser fluorescence microscopy, fluorescent molecules are excited by a tightly focused laser beam at wavelength λ_{ex}, and fluorescence is detected at wavelengths that are longer than λ_{ex}. In this scheme, fluorescence is collected from excited molecules everywhere along the beam path. By employing confocal detection, fluorescence generated outside of the center of the focal region can be rejected [21]. This effect can increase the resolution substantially, particularly in the axial direction. Confocal detection offers only limited resolution improvement in the transverse plane, however. The resolution in the transverse dimension is determined largely by how tightly the excitation beam can be focused (which depends upon λ_{ex}). Two features that are separated by a distance that is significantly less that λ_{ex} will be excited simultaneously, and therefore cannot be resolved.

Super-resolved techniques for optical microscopy were enabled by the realization that if one can use photochemical or photophysical means to prevent nearby fluorophores from being excited simultaneously, then there is no reason why the resolution of fluorescence microscopy cannot be far better than what is predicted by the Abbe criterion. Indeed, methods that achieve superresolution in this manner were the subject of the 2014 Nobel Prize in chemistry.

STED takes advantage of the fact that whereas fluorescence in a sample being imaged is usually emitted isotropically, stimulated emission always follows the direction of propagation of the stimulation source [22]. Thus, if one laser pulse is used to excite a molecule at wavelength λ_{ex} and another laser pulse stimulates emission from the molecule at wavelength λ_{SE} before the molecule has had time to fluoresce, then no emission will be detected from that molecule. We will here refer to the first laser as the "excitation laser" and the second laser as the "deactivation laser".

If the PSFs of the excitation and deactivation beams were to overlap completely, then no improvement in resolution would be realized. A key feature of STED microscopy is that a spatial phase mask in the path of the deactivation beam is used to engineer the PSF to contain a dark region in which stimulated emission will not occur. For instance, a spiral phase element [23] (Figure 5.2a) leads to a PSF with a dark region along its axis (Figure 5.2b), and a phase plate with a half-wave delay in its central region (Figure 5.2c) leads to a PSF with a dark region that is confined along the beam axis (Figure 5.2d). The former phase mask can be used to improve transverse resolution, whereas the latter can be used to improve axial resolution. It is also possible to employ two deactivation beams, each with its own phase mask, to improve the resolution in the transverse and axial dimensions simultaneously [24].

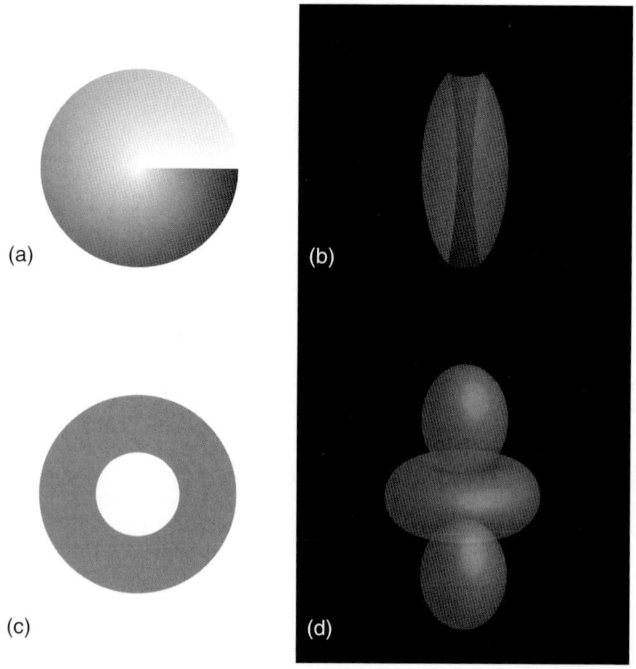

Figure 5.2 Phase plates used in STED microscopy. (a) The spiral phase element has a thickness that varies linearly with the angle about the circle away from the discontinuity. The phase varies from 2π of delay to zero in going around the circle. (b) Schematic of the point-spread function (PSF) generated by focusing of a beam that has passed through a spiral phase element. (c) Donut phase element, in which the phase delay varies by π between the white and gray regions. (d) Schematic of the PSF generated by focusing of a beam that has passed through a donut phase element with half of the beam intensity in the white area.

The dimensions of the dark region of the PSF of the deactivation beam are limited by diffraction. However, operating at an intensity that is in the saturation regime causes the effective dimensions of the bright region of a PSF to be enlarged. In the case of the excitation beam, this phenomenon is undesirable, but in the case of the deactivation beam the saturation effect can in principle be harnessed to make the effective dark region as small as desired [15]. In practice, there are limits to the ultimate resolution that can be attained through this technique, but resolutions that are far below the Abbe limit are readily achieved. In one recent, highly favorable case, 2.4-nm resolution was achieved using 532-nm excitation and 775-nm deactivation in a non-biological sample [25].

To understand how similar concepts can be applied to photolithography, it is instructive to consider first the molecular photophysics of STED (Figure 5.3). The excitation beam takes a fluorophore from the ground electronic state (S_0) to a vibrationally excited level of the first excited electronic state (S_1). Intramolecular

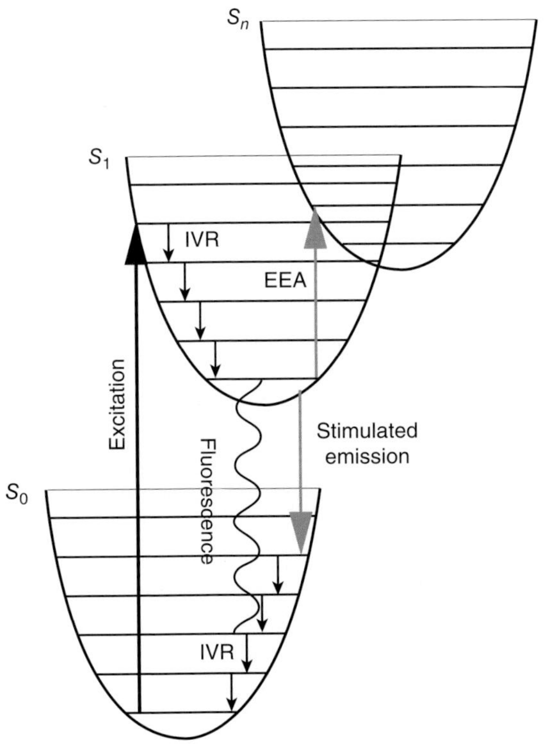

Figure 5.3 Schematic diagram of the photophysics of STED fluorescence microscopy. The excitation takes the molecule from the ground vibrational state of the ground electronic state S_0 to a vibrationally excited state of the first excited electronic state S_1. Following intramolecular vibrational redistribution, the molecule is in the ground vibrational state of the excited electronic state, from which it can fluoresce or undergo stimulated emission to a vibrationally excited state of S_0. Stimulated emission also competes with absorption from S_1 to higher electronic states.

vibrational redistribution (IVR) then causes the molecule to relax to the vibrationless level of S_1. IVR typically occurs on a timescale of a few picoseconds [26], which is much shorter than the fluorescence lifetime of a good fluorophore. Thus, the majority of fluorescence only occurs after IVR, and so can be considered to arise from the vibrationless level of S_1.

Excitation and emission transitions are typically vertical, which is to say that the nuclear positions to not change during the electronic transition. Because the minimum of the S_1 potential energy surface is usually shifted from that of the S_0 surface, the emission transition is generally red-shifted as compared to the excitation transition. This Stokes' shift can become substantially larger if the magnitude and/or direction of the molecular dipole moment differ in S_0 and S_1, in which case the surrounding medium will react to the excitation by reorganizing to stabilize the newly formed electronic excited state (which in turn destabilizes the ground electronic state) [27]. A large Stokes' shift is desirable in fluorescence microscopy, as it causes the excitation and emission to occur at substantially different wavelengths.

The stimulated emission pulse, which is timed to arrive after IVR is complete in S_1, drives molecules from the vibrationless excited electronic state to a vibrationally excited level of S_0. For a number of reasons, it is usually ideal if λ_{SE} can be considerably longer than the wavelength of the maximum of the fluorescence spectrum. First, this separation allows scattered deactivation light to be filtered out from the fluorescence signal. Second, having the stimulated emission wavelength far to the red of the origin of the absorption spectrum prevents the deactivation beam from exciting molecules, which is especially important if there are Franck–Condon-active, low-frequency vibrations that lead to the presence of hot bands. Third, the farther to the red that λ_{SE} lies, the smaller the probability of absorption to higher lying excited states occurring. For all of these reasons, it is desirable to perform STED microscopy using dyes that have large Stokes' shifts [17].

The potential for disruption of the STED process by higher lying excited states through excited-state/excited-state absorption (EEA) is worth considering in greater detail. A molecule that is excited to a higher excited state typically still fluoresces from S_1 [28], so in the best case EEA simply does not lead to deactivation. However, highly excited molecules also usually have a higher probability of undergoing intersystem crossing to a long-lived triplet state, or even of undergoing irreversible photochemistry.

For the above reasons, EEA should be avoided if possible in STED microscopy. However, all fluorophores have higher lying excited states, and the energy gap between S_1 and the lowest of these states is typically less than the energy gap between S_0 and S_1. It is therefore difficult to prevent EEA through control of λ_{SE} alone. Luckily, the quantum mechanical properties of fluorophores come to the rescue in this case. A sum rule dictates that the sum of the oscillator strengths out of a given electronic state must be a fixed quantity [29]. Typical fluorophores have large extinction coefficients so that they can be excited efficiently. This condition implies that the oscillator strength from S_0 to S_1 is large. In this case, the oscillator strength from S_1 to S_0 must also be large, which leads to a high fluorescence quantum yield. If the $S_1 - S_0$ oscillator strength is large, then the oscillator strength from S_1 to higher lying excited states must be small, which reduces the likelihood of EEA occurring.

In practice, there is a limit to the degree to which STED can enhance resolution. One important practical reason for this limit is that full saturation of the deactivation transition leads to degradation of the resolution. In the saturation limit, the populations of the lower and upper states involved in a transition become equal, and the ability to depopulate the upper state through stimulated emission is therefore lost. For this reason, it is optimal for the deactivation pulse in STED to have a duration that is significantly longer than the IVR time in the ground electronic state, which allows IVR to depopulate the lower state in the stimulated emission transition continuously. It is also desirable for the deactivation pulse to be much shorter than the fluorescence lifetime. Meeting both of these conditions allows the most efficient possible deactivation.

5.3 Stimulated Emission Depletion in Multiphoton Lithography

In principle, STED should also be able to enhance resolution in the multiphoton exposure of negative-tone photoresists. Here we will concentrate on the application of STED to radical photoresists, but similar principles apply to cationic photoresists [12, 30].

As a representative example, here we consider the aryl ketone family of radical photoinitiators, which is exemplified by benzophenone [31]. Upon excitation to S_1, these molecules can undergo intersystem crossing to the lowest triplet state T_1 (Figure 5.4). In this class of molecule, T_1 is usually a triplet diradical centered on the carbonyl, which can either abstract hydrogen atoms from other molecules or undergo homolytic bond cleavage [32]. In either case, two or more radicals are generated, and can initiate a polymerization chain reaction. However, returning the molecule to S_0 through stimulated emission before it has a chance to undergo intersystem crossing can turn off the initiation process. The timescale for intersystem crossing in these molecules is generally on the order of 100 ps [33]. Although this time is considerably less than the fluorescence lifetime of a good fluorophore, it is still long enough to provide plenty of time to intercept the intersystem crossing process using stimulated emission.

A number aspects of the photophysics and photochemistry of aryl ketones appear favorable for the use of STED to enhance resolution in multiphoton polymerization. In addition to there being sufficient time to deactivate excited photoinitiators using stimulated emission, the requirements for deactivation are considerably less stringent than in fluorescence microscopy. In imaging, the vast majority of excited molecules outside of the region of interest must be deactivated to create the greatest possible resolution enhancement. On the other hand, in

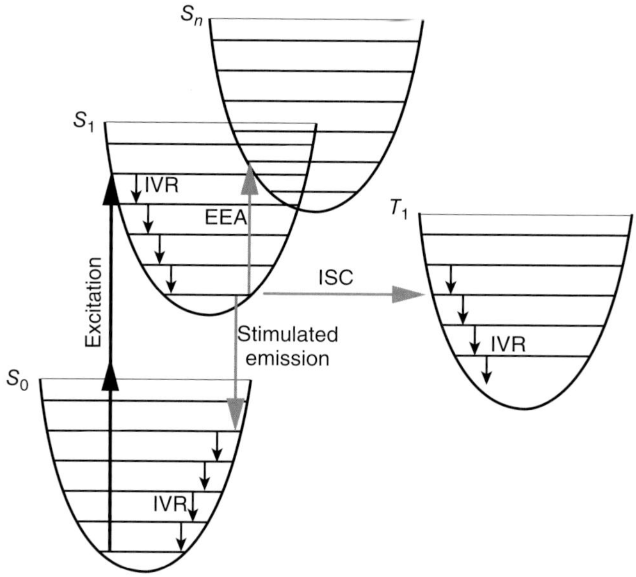

Figure 5.4 Schematic diagram of the photophysics of STED in multiphoton polymerization. Two-photon excitation takes the molecule from the ground vibrational state of the ground electronic state S_0 to a vibrationally excited state of the first excited electronic state S_1. Following intramolecular vibrational redistribution, the molecule is in the ground vibrational state of the excited electronic state, from which it can undergo intersystem crossing to the reactive triplet state T_1 or undergo stimulated emission to a vibrationally excited state of S_0. Stimulated emission also competes with absorption from S_1 to higher electronic states, which tends to lead to greater reactivity.

photolithography it is necessary only to decrease the concentration of excited photoinitiators outside of the region of interest to below the threshold exposure dose. This latter task could potentially be accomplished by deactivation of a small percentage of the excited photoinitiators. Furthermore, achieving a greater degree of crosslinking in an individual voxel generally entails making the voxel larger. If crosslinking can be inhibited effectively outside of the desired region of exposure, however, it is possible to crosslink the desired region to a higher degree than is typically possible without increasing the voxel size. This effect, which holds for any deactivation mechanism, offers the possibility of creating smaller structures with improved mechanical properties.

We developed an instrument to test whether STED could be applied to inhibit crosslinking following multiphoton excitation of a radical photoinitiator (Figure 5.5). In this apparatus, two independently tunable, mode-locked Ti:sapphire lasers (Coherent Mira 900-F) produce pulses of sub-200 fs duration. The two lasers are synchronized to one another with ~250 fs of timing jitter using commercial hardware (Coherent SynchroLock). One laser is used as the source for multiphoton excitation. The other laser is tuned either to a wavelength within the fluorescence spectrum of the photoinitiator or to twice the wavelength

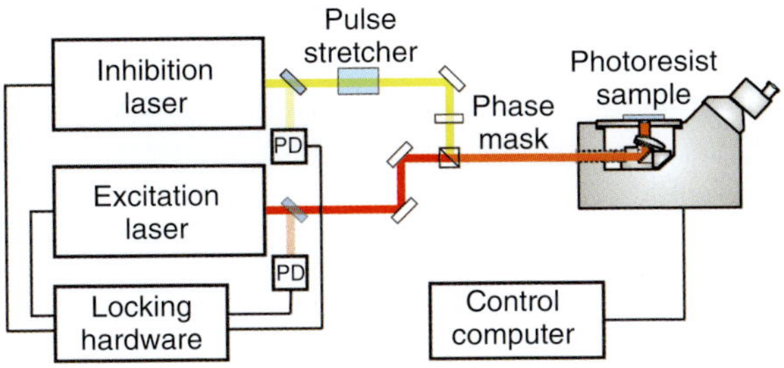

Figure 5.5 Schematic diagram of our experimental setup for using STED in multiphoton polymerization. Two independently tunable Ti:sapphire oscillators produce sub-200-fs pulses that are synchronized to within 250 fs of any desired delay up to the laser repetition time. The excitation laser is used to drive multiphoton absorption in the sample. The inhibition laser pulse is stretched to a duration of picoseconds (before which it can be frequency doubled, if desired) and then passes through a phase mask. The beams have orthogonal polarizations, and are recombined in a polarizing beam cube before being delivered to the sample through a high-numerical-aperture objective.

within the fluorescence spectrum (in which case the output is subsequently frequency-doubled). As described above, it is desirable for the duration of the deactivation pulse to be significantly longer than the IVR time in the ground electronic state. To meet this condition, the deactivation pulse is passed one or more times through a long-path-length cell containing a highly dispersive liquid. Depending on the deactivation wavelength, pulse durations of as much as 10 ps can be attained.

The two beams are of orthogonal polarizations, and are combined in a polarizing beam cube. If desired, a phase mask can also be placed in the path of the deactivation beam. The combined beams enter the reflected light port of an inverted microscope, and are focused into the sample through a high-numerical-aperture objective. Good overlap of the focal points of the two beams was ensured by using multiphoton-absorption-induced luminescence [34] from gold nanoparticles deposited on a cover slip.

Typical photoresist formulations contained equal weight percentages of ethoxylated(6) trimethylolpropane triacrylate and tris(2-hydroxyethyl)isocyanurate triacrylate. These trifunctional monomers create a highly crosslinked polymer with low shrinkage and good adhesion [35]. Photoinitiators were incorporated at 1–3 wt%.

We investigated a wide range of conventional radical photoinitiators that have been used widely for two-photon polymerization [35, 36], including 2-benzyl-2-(dimethylamino)-1-[4-(morpholinyl)phenyl)]-1-butanone (Irgacure 369), 2,4,6-trimethylbenzoylphenyl phosphinate (Lucirin TPO-L), 2-isopropylthioxanthone (ITX), and many others. For every photoinitiator tested, we found that the beam that was intended to inhibit polymerization instead increased polymerization, even though this beam did not lead to polymerization in the absence of the excitation beam.

Why did the deactivation beam have the opposite of the intended effect in these photoresists? As discussed above, stimulated emission always competes with EEA. Conventional radical photoinitiators differ in a number of important ways from the fluorophores used for STED microscopy. For instance, conventional photoinitiators generally have rather small Stokes' shifts. More importantly, these photoinitiators are typically designed to have relatively low extinction coefficients, so that it is possible to photocure thick films containing these materials. A small oscillator strength between S_0 and S_1 implies that S_1 must have large oscillator strength to higher excited states. Higher singlet states often have faster intersystem crossing rates than does S_1, and so can lead to even more efficient initiation of crosslinking.

Based on these issues, the use of STED to inhibit most conventional radical photoinitiators is likely to be challenging. The same is likely to hold true for the photoacid generators used in cationic polymerization as well. It should be noted, however, that Wegener and coworkers have successfully demonstrated the deactivation of ITX using continuous-wave (CW) radiation at 532 nm following pulse multiphoton excitation [37]. They suggested that the deactivation occurred through stimulated emission, as 532 nm is within the fluorescence band of ITX. They were able to achieve line widths of 65 nm, and the resolution enhancement was limited only by additional crosslinking caused at higher intensities of the deactivation light. Subsequent work has led to the alternative suggestion that some or all of the inhibition of polymerization in this system actually involves triplet states of ITX rather than stimulated emission [38, 39].

An alternative strategy for using STED to improve the resolution of multiphoton polymerization is to search for molecules that have high extinction coefficients and can also act as photoinitiators. Ideally, such molecules should also have high fluorescence quantum yields, so that STED can be effective. A molecule having a high fluorescence quantum yield would not typically be considered for use as a radical photoinitiator, as the quantum yield for radical formation would necessarily be low. However, such molecules may behave differently in a viscous photoresist than they do in a low-viscosity solvent. Additionally, multiphoton excitation often takes molecules to different excited states than does linear excitation, opening up new photochemical pathways. As an example, rhodamine 6G has a fluorescence quantum yield of 0.93 upon linear excitation in methanol [40], and its absorption spectrum peaks at about 528 nm [41]. However, its two-photon absorption spectrum has a strong peak at about 835 nm in the same solvent [42], which would correspond to a linear absorption peak at 417 nm. No such peak is seen in the linear absorption spectrum of rhodamine 6G in methanol [43], which suggests that the symmetry of this excited state is such that linear absorption from the ground state is forbidden. Thus, the photochemistry and photophysics upon multiphoton absorption may differ significantly from those observed for linear absorption. We have indeed observed that rhodamine 6G acts as an efficient multiphoton radical initiator in acrylic photoresists [44].

These sorts of ideas led Wegener and coworkers to discover a laser dye, 7-diethylamino-3-thenoylcoumarin (DETC), that can act as an effective multiphoton initiator and can be deactivated by CW 532 nm light [45, 46]. Although there is more than one mechanism for the deactivation of this dye [39, 46], at least one mechanism appears to be stimulated emission. This system was able to achieve significant axial and transverse resolution enhancement as compared to the same photoresist used without a deactivation beam. Wollhofen *et al.* subsequently demonstrated feature sizes of 55 nm at 120 nm pitch using a photoresist containing DETC with pulsed, 780-nm, multiphoton excitation and CW 532-nm deactivation [47].

It has been suggested by some authors that stimulated emission is the best choice for the photophysical deactivation of photopolymerization for resolution enhancement [13, 45]. It should be noted, however, that these sorts of claims have not yet been supported by any rigorous theoretical analysis or experimental comparisons. Indeed, the mechanism of initiation for the molecules that can be deactivated through STED is not yet known, and is clearly of importance for making comparisons with other techniques. In reality, there are inevitable tradeoffs for all deactivation schemes that have been demonstrated to date, and the optimal scheme is likely to be dependent upon the application.

STED has some undeniable advantages for resolution enhancement. Stimulated emission is a rapid process given a high-enough depletion intensity and oscillator strength. In the absence of other photophysical processes, stimulated emission returns the photoinitiator to its ground electronic state, which is the most desirable situation. STED may also generate less local heating that other deactivation schemes [45], as each excitation/deactivation cycle only leaves $2\hbar\omega_{ex} - \hbar\omega_{SE}$ of energy in the sample.

We have developed a detailed kinetic model of STED-based deactivation in two-photon lithography that reveals some of the potential disadvantages of this approach [12]. There are two key timescales for this process: the IVR time in the ground electronic state, and the apparent fluorescence lifetime (which is related directly to the timescale of initiation). Stimulated emission must take place on a timescale that is slower than the former process yet faster than the latter process. If either condition is not met, then initiation will not be prevented effectively. Because IVR must also occur in the excited electronic state before stimulated emission commences, the window for stimulated emission lasts from a minimum of a few picoseconds after excitation to a maximum of a few hundred picoseconds after excitation. The latter value depends on the acceptable amount of initiation in areas that are intended to be soluble after exposure. If deactivation is performed with CW radiation, then the vast majority of deactivation photons go waste, as the repetition time of the excitation laser is typically on a timescale of nanoseconds. In this case, the deactivation beam must be operated at high intensity so that deactivation is efficient within the useful time window. However, high intensity also leads to sample heating through Raman scattering and other effects. Thus, deactivation is best performed with pulses of a timescale on the order of picoseconds to tens of picoseconds. The maximum intensity that can be employed is limited by the potential for competition by EEA and by the requirement that the deactivation transition not be fully saturated. In addition, it is important to note that, if stimulated emission is efficient, then it is possible for a molecule to be excited and de-excited as often as once every excitation pulse. Thus, even if the amount of heat that is deposited during a single cycle is small, the ability to undergo up to one cycle per laser pulse can still allow a significant amount of heat to be deposited using STED.

Whether the disadvantages described above outweigh the advantages is strongly dependent on the desired application. With this in mind, we now proceed to discuss other reported methods of deactivation.

5.4 Photoinhibition

Photoinhibitors are molecules that form weakly reactive radicals upon photoexcitation. These radicals tend to be more reactive with other radicals than they are with closed-shell species, and so can act to terminate radical chain reactions. Indeed, these materials were originally developed for photoquenching radical polymerization so that detailed kinetic measurements of the polymerization process could be performed [48, 49].

McLeod and coworkers recognized that photoinhibitors could also provide a means of enhancing resolution in photopolymerization [50, 51]. They developed a methacrylic photoresist containing a photoinitiator that was excited at 473 nm and a photoinhibitor that was excited at 364 nm. Both the excitation and deactivation beams were CW, and the deactivation beam was phase-shaped as in STED microscopy. They were able to attain features as small as 64 nm in diameter with this technique, even though the excitation was based on linear absorption. Gu and coworkers developed a new photoresist formulation, which included the same photoinhibitor, and were able to achieve a 40-nm feature size with CW 488-nm excitation and 375-nm deactivation beams [52].

Gu and coworkers went on to apply the photoinhibition approach to resolution enhancement in multiphoton polymerization [53]. They achieved a minimum size for an isolated feature of 9 nm, and were able to create adjacent lines of 20 nm width on a pitch of 52 nm.

One of the attractive features of the photoinduced deactivation approach is that the initiation of polymerization is completely decoupled from the inhibition of polymerization. However, this property also means that both photoinitiator and photoinhibitor are used up every time the two beams are incident on a given area of the sample, instead of being regenerated, as in STED. Because of this effect, when fabricating close-packed features, the fabrication parameters must be adjusted based on the local density of features and the number of writing passes that have been performed already in that region [53]. The resolution enhancement declines substantially as a result of this effect. For instance, in creating 10 lines, Gu and coworkers found that the minimum pitch increased from 53 to 90 nm [53]. Additionally, although the photoinhibitor generates weakly reactive radicals, at high enough concentrations these radicals can still drive crosslinking reactions, which limits the maximum deactivation intensity that can be employed [50].

5.5 Inhibition Based on Photoinduced Electron Transfer

After our group recognized the challenges of performing STED on conventional radical photoinitiators, we pursued the strategy of searching for a molecule with a large extinction coefficient that could also act as a radical photoinitiator. Although we subsequently demonstrated that even highly fluorescent dyes could act as good radical photoinitiators [44], this fact was not appreciated at the time. We therefore focused our attention on molecules with high extinction coefficients and poor fluorescence quantum yields, with the hope that one of the nonradiative processes competing with fluorescence would lead to radical generation. One of the materials that we tested was malachite green carbinol base (MGCB), a common dye molecule. In low-viscosity solvents, MGCB undergoes a conformational change upon photoexcitation, preventing it from fluorescing efficiently [54]. However, in viscous solvents (such as a photoresist) this conformational change occurs more slowly, increasing the fluorescence quantum yield and fluorescence lifetime of the molecule [54]. Furthermore, the fluorescence spectrum peaks between 650 and 700 nm, but extends out well past 800 nm [55], making it ideal for stimulated emission at 800 nm using a Ti:sapphire laser.

As with many other dyes that we tested, MGCB was found to be an efficient radical photoinitiator with two-photon excitation at a wavelength of 800 nm. Initial tests with the deactivation laser of Figure 5.5 set to 800 nm (not using a phase mask) showed that crosslinking could be inhibited completely (Figure 5.6a) [56]. This was the first reported demonstration of a multiphoton photoresist that could be photodeactivated.

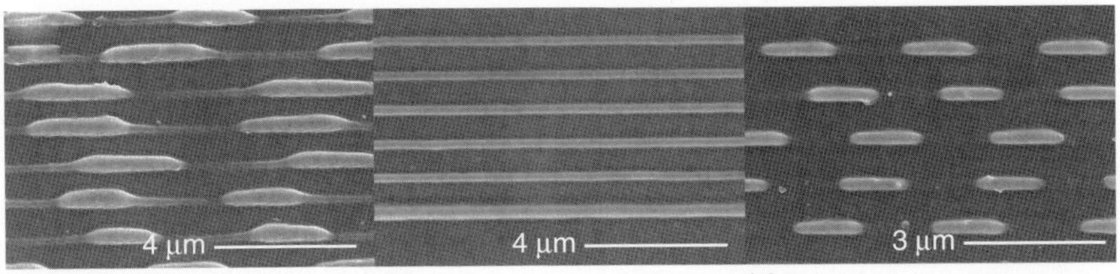

(a) (b) (c)

Figure 5.6 Representative experimental deactivation data from RAPID lithography. (a) Lines drawn with a femtosecond-pulsed excitation beam and a chopped, offset, picosecond-pulsed deactivation beam. Both beams were tuned to 800 nm. These data demonstrate that the photoresist can be deactivated efficiently, reducing the line width. (b) Lines drawn with the excitation beam only (bottom) and both excitation and deactivation beams at delay times ranging up to the laser repetition time of approximately 13 ns. The constant width of the lines indicates that the state that is deactivated has a lifetime that is considerably longer than the laser repetition time. (c) Lines drawn with a femtosecond-pulsed excitation beam and a chopped, collinear, continuous-wave (CW) deactivation beam. Both beams were tuned to 800 nm. These data demonstrate that the photoresist can be deactivated efficiently even with CW radiation, so that synchronization is not necessary. (Li 2009 [56]. Reproduced with permission of American Association for the Advancement of Science.)

Our expectation based on the reported photophysics of MGCB was that the observed deactivation was the result of stimulated emission. To determine the lifetime of the state that is deactivated, we set the intensity of the deactivation beam such that deactivation was only partial at short delay times between excitation and deactivation. We then fabricated lines at different delay times, ranging up to the ~13 ns repetition time of the Ti:sapphire lasers. We expected to be able to measure the excited state lifetime via the broadening of the lines with increasing delay time. Surprisingly, the efficiency of deactivation was completely independent of delay time over this window (Figure 5.6b) [56]. The lifetime of a strongly fluorescent excited state would be expected to be on the order of nanoseconds, so these results imply that MGCB is deactivated primarily through a mechanism that does not involve stimulated emission. The lifetime of the state that is deactivated must, instead, be considerably longer than the laser repetition time. This discovery led us to test whether deactivation could be accomplished using a CW laser. As shown in Figure 5.6c, CW radiation is indeed effective for deactivation, obviating the need for synchronized excitation and deactivation pulses.

Additional experiments revealed that at high enough excitation intensities, complete deactivation is not possible [56]. This observation indicates that there are at least two channels for polymerization and that only one of these channels can be deactivated with 800-nm light. We further found that the channel that cannot be deactivated grows in importance as the viscosity of the photoresist decreases.

The photochemistry and photophysics of triarylmethane dyes such as MGCB have been studied in great detail, and hundreds of papers have been published on the topic [57]. Many of the light-driven processes observed in these materials have had multiple explanations proposed for them, and it is likely the case that the dominant events depend strongly on factors such as the viscosity, polarity, and temperature of the medium. Based upon the above observations and evidence from the literature [57], we have suggested that initiation by MGCB is driven by photoinduced electron ejection [44, 56].

Our proposed mechanism for initiation and deactivation of MGCB is shown in Figure 5.7. An electron that is driven out of an MGCB molecule can be solvated in the surrounding photoresist, where it can lead to the relatively slow initiation of polymerization. If the electron remains in the vicinity of the MGCB cation that ejected it, subsequent irradiation can lead to photoinduced backtransfer of the electron, regenerating the photoinitiator. If the solvated electron diffuses away from the cation, then it will eventually initiate polymerization and cannot be deactivated. These ideas received additional support from experiments on other related species, which indicated that initiation does not occur directly from the state to which the molecule is excited [44]. Nevertheless, it has not yet been proven conclusively that this mechanism is responsible for the observed deactivation.

The use of CW light for deactivation allows considerably greater intensities to be employed than in the case of pulsed deactivation. Combining CW deactivation with the type of phase mask shown in Figure 5.2c, we were able to attain an axial resolution of 40 nm using 800-nm pulsed excitation and 800-nm, CW deactivation

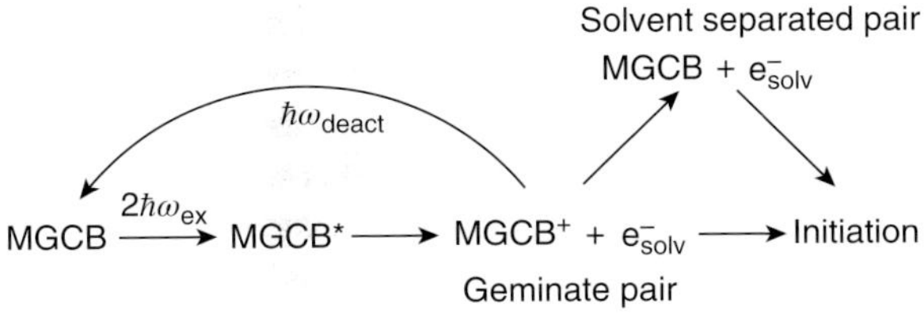

Figure 5.7 Proposed mechanism for deactivation in RAPID lithography. Multiphoton excitation of malachite green carbinol base leads to the ejection of an electron, which is solvated in the photoresist. Recombination of the electron with the cation can be achieved through photoinduced back transfer. The electron can initiate radical polymerization on a timescale much longer than the laser repetition time of approximately 13 ns. If the electron escapes the solvent shell that it shares with the MGCB cation, deactivation can no longer be achieved.

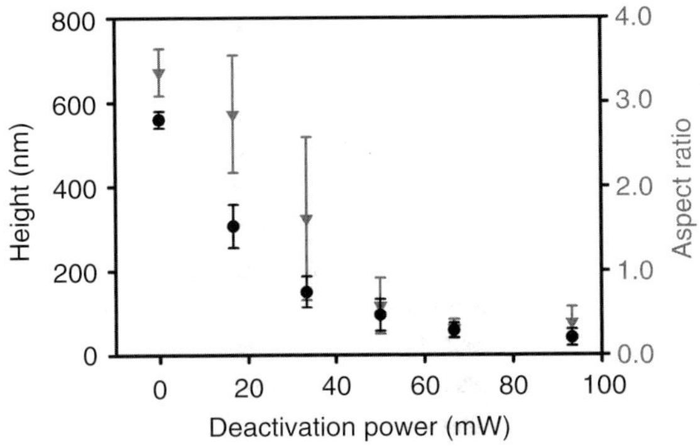

Figure 5.8 Height (black) and aspect ratio (gray) of voxels created with RAPID lithography as a function of the deactivation power. (Li 2009 [56]. Reproduced with permission of American Association for the Advancement of Science.)

(Figure 5.8). We call this scheme resolution enhancement through photo-induced deactivation (RAPID) lithography.

Our previous kinetic analysis revealed a number of criteria that are desirable in a material for resolution enhancement through photodeactivation [12]: (i) the state that is deactivated should not be the state that is excited, but it should be reached as quickly as possible after the first state is excited; (ii) the state that is deactivated should have as long a lifetime as possible so that initiation does not compete with deactivation; and (iii) the state that is deactivated should have a large extinction coefficient so that deactivation can occur efficiently. These conditions cannot all be met simultaneously in a STED scheme. Any system that meets conditions (i) and (iii) is capable of fast writing speeds under most conditions, although it may be necessary to use phase masks that produce dark lines rather than dark central regions. A system that can also meet condition (ii) allows higher contrast to be attained. At the same time, practical constraints require that the initiation time to be not overly long, which would place a limit on the minimum delay time between write cycles in a given region.

It is clear that RAPID with MGCB meets criterion (ii) above. Although the initiation time has not been measured yet, a lower limit would be on the order of 100 ns based on available data. The intensities that are needed for deactivation indicate that the extinction coefficient is comparable to that of STED systems as well. It will require further experimentation to ascertain whether the first criterion is met by this system. Also, as with STED, RAPID also has the advantage that the photoinitiator is regenerated by the deactivation step.

The implementations of RAPID so far also have some distinct disadvantages. First, the surface roughness with the MGCB photoresist is greater than for similar resists with conventional photoinitiators, even when deactivation is not employed. Second, the best resolution in a STED-inspired fabrication technique should arise when the deactivation wavelength is as short as possible. For the MGCB photoresist, the deactivation efficiency decays quickly as the wavelength used moves to smaller than 800 nm. Finally, as noted above, there is a polymerization channel that cannot be deactivated with the MGCB photoresist. This channel has a significant influence on the ultimate feature pitch that will be able to be attained. Further materials will need to be tested to determine whether these issues can be addressed.

5.6 Absorbance Modulation Lithography

A discussion of two-color schemes for resolution enhancement would not be complete without considering absorbance modulation lithography (AML) [58, 59], even though this method has yet to be used with multiphoton polymerization or any other 3D fabrication scheme. In AML, a thin photochromic overlayer is deposited on top of a thin layer of photoresist. Photochromics are bistable, photoswitchable molecules [60]. The photochromics used in AML can be switched between "open" and "closed" forms. The former absorbs light in the ultraviolet but is transparent to red light, whereas the latter absorbs red light but is relatively transparent in the ultraviolet. The key to AML is that the photoresist is exposed by ultraviolet light, but patterned red light is used to prevent the ultraviolet light from getting through the overlayer except at desired locations.

As with photoinhibitor schemes, AML decouples the excitation and inhibition steps. Indeed, the photoinitiators in AML are in an entirely different spatial region than the molecules that inhibit exposure. However, it is not clear so far whether such a scheme could be employed for 3D fabrication using multiphoton polymerization, because the overlayer effectively acts as a near-field mask. Perhaps a photochromic scheme that could use multiphoton absorption at two different wavelengths for excitation and inhibition could be used in a bulk material. Even if such a scheme could be realized, current photochromics have relatively long switching times and can only be cycled a limited number of times. These issues must be addressed for AML to have a major impact in 3D fabrication.

5.7
Challenges for Two-Color, Two-Photon Lithography

Two-color approaches hold great promise for enabling vast improvements in the resolution of multiphoton lithography. At the same time, there are many hurdles to be overcome before this promise can be fully realized. Every approach to two-color resolution enhancement would benefit from the development of optimized materials, and this development must go hand in hand with a deeper understanding of the photochemistry and photophysics that underlie the methods. Until such knowledge is available, it will be difficult to determine which approach is best for a given application.

All two-color approaches that have been developed to date suffer from the common shortcoming that the state that leads to initiation is the same state that must be deactivated. Thus, there is in inherent competition between initiation and deactivation, and this competition requires that deactivation occur on as fast a timescale as possible. Ultimately, not every molecule that is excited in a region in which crosslinking is desired can be deactivated before initiation occurs. Because a key goal of two-color approaches is to improve resolution, it is desirable to be able to fabricate structures on as tight a pitch as possible. However, every fabrication pass that is within the focal volume of the excitation beam contributes to a buildup of photoinitiation in areas in which it is not desired. As shown in Figure 5.9, this phenomenon can lead to a strong polymerization background, even when the degree of excitation in undesired regions is quite small. Figure 5.9 considers a two-dimensional case, and this problem is substantially worse when fabrication is performed in three dimensions. This problem does not arise in STED microscopy, in the sense that fluorescence regenerates ground-state molecules. In contrast, in lithography, once an initiation event has occurred, the

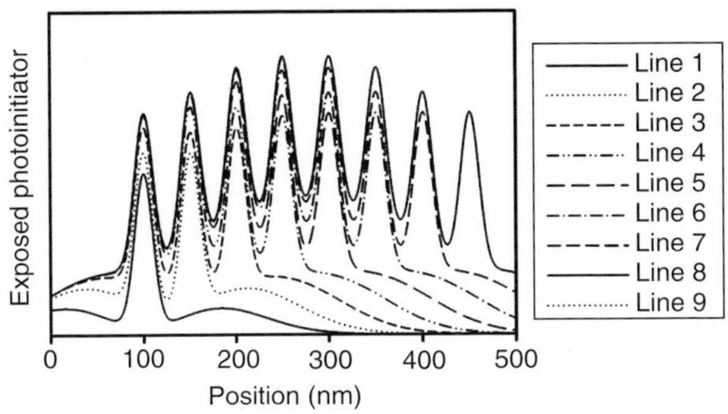

Figure 5.9 Simulation of the writing of many parallel lines at 50 nm pitch using two-color, two-photon polymerization. It is assumed that at the intensity maxima of the deactivation beam 95% of excited molecules are deactivated. The background of exposed photoinitiators that cause chemistry increases greatly with the fabrication of subsequent lines, raising the overall exposure in a given maximum well above the threshold for a single pass, thereby decreasing the resolution and dynamic range.

photoinitiator is used up. Undesired background polymerization is probably the greatest challenge to improving the resolution of two-color materials.

Another important challenge is the creation of materials that can be deactivated at the shortest possible wavelength, which will allow the dark region in the focus of the deactivation beam to be as small as possible. However, this advantage must be balanced against the fact that the deactivation PSF should envelop the entire excitation PSF to minimize background polymerization.

As voxels and structures created with two-photon polymerization become smaller, there is a nonlinear change in their mechanical properties [61–63]. Even though two-color approaches allow for a greater degree of crosslinking in a voxel of a given size, mechanical integrity may end up to be a major issue in pushing the resolution of 3D structures to its limit. Addressing this issue may require applying two-color approaches to other materials, such as hybrid organic/inorganics that have been used in two-photon polymerization [64].

Finally, as voxels become ever smaller, throughput becomes a major concern. Smaller voxels mean an increased number of scans to perform patterning over the same volume. It is therefore essential to develop highly sensitive materials that allow high writing speeds. Controllable phase masks that allow the voxel dimensions to be changed on the fly may also assist in the most rapid possible fabrication of 3D structures with features on a range of different size scales.

5.8 Conclusions

Two-color approaches have already enabled substantial resolution enhancement for two-photon lithography, and hold great promise for eventually pushing the resolution down to the 10–20 nm level. Success in this endeavor will require the pursuit of new materials and approaches that can address the challenges discussed above. With further advances, two-color techniques may lead to tabletop tools that can be used to perform 3D lithography with a resolution that rivals that of electron-beam lithography.

Acknowledgments

Our work in this area of research is funded by the U.S. National Science Foundations, Grants IIP-1318211 and CMMI-1449309. We appreciate the support of the Maryland NanoCenter and its NispLab. The NispLab is supported in part by the NSF as a MRSEC Shared Experimental Facility.

References

1. Burmeister, F., Zeitner, U.D., Nolte, S., and Tünnermann, A. (2012) High numerical aperture hybrid optics for two-photon polymerization. *Opt. Express*, **20** (7), 7994–8005.
2. Williams, H.E., Luo, Z., and Kuebler, S.M. (2012) Effect of refractive index mismatch on multi-photon direct laser writing. *Opt. Express*, **20** (22), 25030–25040.

3. Haske, W., Chen, V.W., Hales, J.M., Dong, W., Barlow, S., Marder, S.R., and Perry, J.W. (2007) 65 nm feature sizes using visible wavelength 3-D multiphoton lithography. *Opt. Express*, **15** (6), 3426–3436.
4. Takada, K., Sun, H.-B., and Kawata, S. (2005) Improved spatial resolution and surface roughness in photopolymerization-based laser nanowriting. *Appl. Phys. Lett.*, **86** (7), 071122.
5. Lu, W.-E., Dong, X.-Z., Chen, W.-Q., Zhao, Z.-S., and Duan, X.-M. (2011) Novel photoinitiator with a radical quenching moiety for confining radical diffusion in two-photon induced photopolymerization. *J. Mater. Chem.*, **21** (15), 5650–5659.
6. Sakellari, I., Kabouraki, E., Gray, D., Purlys, V., Fotakis, C., Pikulin, A., Bityurin, N., Vamvakaki, M., and Farsari, M. (2012) Diffusion-assisted high-resolution direct femtosecond laser writing. *ACS Nano*, **6** (3), 2302–2311.
7. So, P.T.C., Dong, C.Y., Masters, B.R., and Berland, K.M. (2000) Two-photon excitation fluorescence microscopy. *Annu. Rev. Biomed. Eng.*, **2**, 399–429.
8. Sun, H.B., Takada, K., Kim, M.S., Lee, K.S., and Kawata, S. (2003) Scaling laws of voxels in two-photon photopolymerization nanofabrication. *Appl. Phys. Lett.*, **83** (6), 1104–1106.
9. Dawood, F., Qin, S., Li, L., Lin, E.Y., and Fourkas, J.T. (2012) Simultaneous microscale optical manipulation, fabrication and immobilisation in aqueous media. *Chem. Sci.*, **3** (8), 2449–2456.
10. Fourkas, J.T. (2010) Nanoscale photolithography with visible light. *J. Phys. Chem. Lett.*, **1** (8), 1221–1227.
11. Fourkas, J.T. (2011) RAPID lithography: new photoresists achieve nanoscale resolution. *Opt. Photonics News*, **22** (6), 24–29.
12. Fourkas, J.T. and Petersen, J.S. (2014) 2-Colour photolithography. *Phys. Chem. Chem. Phys.*, **16** (19), 8731–8750.
13. Fischer, J. and Wegener, M. (2013) Three-dimensional optical laser lithography beyond the diffraction limit. *Laser Photonics Rev.*, **7** (1), 22–44.
14. Scott, T.F., Kloxin, C.J., Forman, D.L., McLeod, R.R., and Bowman, C.N. (2011) Principles of voxel refinement in optical direct write lithography. *J. Mater. Chem.*, **21** (37), 14150–14155.
15. Hell, S.W. and Wichmann, J. (1994) Breaking the diffraction resolution limit by stimulated emission: stimulated-emission-depletion fluorescence microscopy. *Opt. Lett.*, **19** (11), 780–782.
16. Klar, T.A. and Hell, S.W. (1999) Subdiffraction resolution in far-field fluorescence microscopy. *Opt. Lett.*, **24** (14), 954–956.
17. Müller, T., Schumann, C., and Kraegeloh, A. (2012) STED microscopy and its applications: new insights into cellular processes on the nanoscale. *ChemPhysChem*, **13** (8), 1986–2000.
18. Betzig, E., Patterson, G.H., Sougrat, R., Lindwasser, O.W., Olenych, S., Bonifacino, J.S., Davidson, M.W., Lippincott-Schwartz, J., and Hess, H.F. (2006) Imaging intracellular fluorescent proteins at nanometer resolution. *Science*, **313** (5793), 1642–1645.
19. Coltharp, C. and Xiao, J. (2012) Super-resolution microscopy for microbiology. *Cell. Microbiol.*, **14** (12), 1808–1818.
20. Huang, B., Wang, W.Q., Bates, M., and Zhuang, X.W. (2008) Three-dimensional super-resolution imaging by stochastic optical reconstruction microscopy. *Science*, **319** (5864), 810–813.
21. Gu, M. (1996) *Principles of Three-Dimensional Imaging in Confocal Microscopes*, World Scientific, Singapore.
22. Saleh, B.E.A. and Teich, M.C. (2007) *Fundamentals of Photonics*, John Wiley & Sons, Inc, Hoboken, NJ.
23. Beijersbergen, M.W., Coerwinkel, R.P.C., Kristensen, M., and Woerdman, J.P. (1994) Helical-wavefront laser beams produced with a spiral phase plate. *Opt. Commun.*, **112** (5–6), 321–327.
24. Wildanger, D., Medda, R., Kastrup, L., and Hell, S.W. (2009) A compact STED microscope providing 3D nanoscale resolution. *J. Microsc.*, **236** (1), 35–43.
25. Wildanger, D., Patton, B.R., Schill, H., Marseglia, L., Hadden, J.P., Knauer, S., Schönle, A., Rarity, J.G., O'Brien, J.L., Hell, S.W., and Smith, J.M. (2012)

Solid immersion facilitates fluorescence microscopy with nanometer resolution and sub-Ångström emitter localization. *Adv. Mater.*, **24** (44), OP309–OP313.

26. Gruebele, M. and Wolynes, P.G. (2004) Vibrational energy flow and chemical reactions. *Acc. Chem. Res.*, **37** (4), 261–267.
27. Fleming, G.R. and Cho, M.H. (1996) Chromophore-solvent dynamics. *Annu. Rev. Phys. Chem.*, **47**, 109–134.
28. Kasha, M. (1950) Characterization of electronic transitions in complex molecules. *Discuss. Faraday Soc.*, **9**, 14–19.
29. Hirschfelder, J.O., Curtiss, C.F., and Bird, R.B. (1954) *Molecular Theory of Gases and Liquids*, John Wiley & Sons, Inc., New York.
30. Saotome, M., Takano, S., Tokushima, A., Ito, S., Nakashima, S., Nagasawa, Y., Okada, T., and Miyasaka, H. (2005) Picosecond nanosecond laser photolysis studies of a photoacid generator in solutions: transient absorption spectroscopy and transient grating measurements. *Photochem. Photobiol. Sci.*, **4** (1), 83–88.
31. Kaur, M. and Srivastava, A.K. (2002) Photopolymerization: a review. *J. Macromol. Sci., C Polym. Rev.*, **42** (4), 481–512.
32. Crivello, J.V. and Dietliker, K. (1998) *Photoinitiators for Free Radical Cationic and Anionic Photopolymerisation*, 2nd edn, John Wiley & Sons, Ltd, Chichester.
33. Colley, C.S., Grills, D.C., Besley, N.A., Jockusch, S., Matousek, P., Parker, A.W., Towrie, M., Turro, N.J., Gill, P.M.W., and George, M.W. (2002) Probing the reactivity of photoinitiators for free radical polymerization: time-resolved infrared spectroscopic study of benzoyl radicals. *J. Am. Chem. Soc.*, **124** (50), 14952–14958.
34. Farrer, R.A., Butterfield, F.L., Chen, V.W., and Fourkas, J.T. (2005) Highly efficient multiphoton-absorption-induced luminescence from gold nanoparticles. *Nano Lett.*, **5** (6), 1139–1142.
35. Baldacchini, T., LaFratta, C.N., Farrer, R.A., Teich, M.C., Saleh, B.E.A., Naughton, M.J., and Fourkas, J.T. (2004) Acrylic-based resin with favorable properties for three-dimensional two-photon polymerization. *J. Appl. Phys.*, **95** (11), 6072–6076.
36. Schafer, K.J., Hales, J.M., Balu, M., Belfield, K.D., Van Stryland, E.W., and Hagan, D.J. (2004) Two-photon absorption cross-sections of common photoinitiators. *J. Photochem. Photobiol., A*, **162** (2–3), 497–502.
37. Fischer, J., von Freymann, G., and Wegener, M. (2010) The materials challenge in diffraction-unlimited direct-laser-writing optical lithography. *Adv. Mater.*, **22** (32), 3578–3582.
38. Harke, B., Bianchini, P., Brandi, F., and Diaspro, A. (2012) Photopolymerization inhibition dynamics for sub-diffraction direct laser writing lithography. *ChemPhysChem*, **13** (6), 1429–1434.
39. Harke, B., Dallari, W., Grancini, G., Fazzi, D., Brandi, F., Petrozza, A., and Diaspro, A. (2013) Polymerization inhibition by triplet state absorption for nanoscale lithography. *Adv. Mater.*, **25** (6), 904–909.
40. Magde, D., Wong, R., and Seybold, P.G. (2002) Fluorescence quantum yields and their relation to lifetimes of rhodamine 6G and fluorescein in nine solvents: Improved absolute standards for quantum yields. *Photochem. Photobiol.*, **75** (4), 327–334.
41. Lu, Y. and Penzkofer, A. (1986) Absorption behaviour of methanolic rhodamine 6G solutions at high concentration. *Chem. Phys.*, **107** (2–3), 175–184.
42. Albota, M.A., Xu, C., and Webb, W.W. (1998) Two-photon fluorescence excitation cross sections of biomolecular probes from 690 to 960 nm. *Appl. Opt.*, **37** (31), 7352–7356.
43. Noginov, M.A., Zhu, G., and Gavrilenko, V.I. (2007) Sensitized nonlinear emission of gold nanoparticles. *Opt. Express*, **15** (24), 15648–15655.
44. Stocker, M.P., Li, L., Gattass, R.R., and Fourkas, J.T. (2011) Multiphoton photoresists giving nanoscale resolution that is inversely dependent on exposure time. *Nat. Chem.*, **3** (3), 223–227.
45. Fischer, J. and Wegener, M. (2011) Three-dimensional direct laser writing

inspired by stimulated-emission-depletion microscopy. *Opt. Mater. Express*, **1** (4), 614–624.

46. Fischer, J. and Wegener, M. (2012) Ultrafast polymerization inhibition by stimulated emission depletion for three-dimensional nanolithography. *Adv. Mater.*, **24** (10), OP65–OP69.
47. Wollhofen, R., Katzmann, J., Hrelescu, C., Jacak, J., and Klar, T.A. (2013) 120 nm resolution and 55 nm structure size in STED-lithography. *Opt. Express*, **21** (9), 10831–10840.
48. Karatekin, E., Landis, M., Lem, G., O'Shaughnessy, B., and Turro, N.J. (2001) Photocopying living chains. 1. Steady-state. *Macromolecules*, **34** (23), 8187–8201.
49. Karatekin, E., Landis, M., Lem, G., O'Shaughnessy, B., and Turro, N.J. (2001) Photocopying living chains. 2. Time-dependent measurements. *Macromolecules*, **34** (23), 8202–8215.
50. Scott, T.F., Kowalski, B.A., Sullivan, A.C., Bowman, C.N., and McLeod, R.R. (2009) Two-color single-photon photoinitiation and photoinhibition for subdiffraction photolithography. *Science*, **324** (5929), 913–917.
51. Forman, D.L., Cole, M.C., and McLeod, R.R. (2013) Radical diffusion limits to photoinhibited superresolution lithography. *Phys. Chem. Chem. Phys.*, **15** (36), 14862–14867.
52. Cao, Y.Y., Gan, Z.S., Jia, B.H., Evans, R.A., and Gu, M. (2011) High-photosensitive resin for super-resolution direct-laser-writing based on photoinhibited polymerization. *Opt. Express*, **19** (20), 19486–19494.
53. Gan, Z., Cao, Y., Evans, R.A., and Gu, M. (2013) Three-dimensional deep sub-diffraction optical beam lithography with 9 nm feature size. *Nat. Commun.*, **4**, 2061.
54. Pellegrino, F., Dagen, A., and Alfano, R.R. (1982) Fluorescence polarization anisotropy and kinetics of malachite green measured as a function of solvent viscosity. *Chem. Phys.*, **67** (1), 111–117.
55. Nagasawa, Y., Ando, Y., and Okada, T. (1999) Solvent dependence of ultrafast ground state recovery of the triphenylmethane dyes, brilliant green and malachite green. *Chem. Phys. Lett.*, **312** (2–4), 161–168.
56. Li, L.J., Gattass, R.R., Gershgoren, E., Hwang, H., and Fourkas, J.T. (2009) Achieving $\lambda/20$ resolution by one-color initiation and deactivation of polymerization. *Science*, **324** (5929), 910–913.
57. Duxbury, D.F. (1993) The photochemistry and photophysics of triphenylmethane dyes in solid and liquid media. *Chem. Rev.*, **93** (1), 381–433.
58. Andrew, T.L., Tsai, H.Y., and Menon, R. (2009) Confining light to deep subwavelength dimensions to enable optical nanopatterning. *Science*, **324** (5929), 917–921.
59. Masid, F., Andrew, T.L., and Menon, R. (2013) Optical patterning of features with spacing below the far-field diffraction limit using absorbance modulation. *Opt. Express*, **21** (4), 5209–5214.
60. Kim, T., Zhu, L.Y., Al-Kaysi, R.O., and Bardeen, C.J. (2014) Organic photomechanical materials. *ChemPhysChem*, **15** (3), 400–414.
61. Sun, H.-B., Takada, K., and Kawata, S. (2001) Elastic force analysis of functional polymer submicron oscillators. *Appl. Phys. Lett.*, **79** (19), 3173–3175.
62. Nakanishi, S., Shoji, S., Kawata, S., and Sun, H.-B. (2007) Giant elasticity of photopolymer nanowires. *Appl. Phys. Lett.*, **91** (6), 063112.
63. Bayindir, Z., Sun, Y., Naughton, M.J., LaFratta, C.N., Baldacchini, T., Fourkas, J.T., Stewart, J., Saleh, B.E.A., and Teich, M.C. (2005) Polymer microcantilevers fabricated via multiphoton absorption polymerization. *Appl. Phys. Lett.*, **86** (6), 064105.
64. Serbin, J., Egbert, A., Ostendorf, A., Chichkov, B.N., Houbertz, R., Domann, G., Schulz, J., Cronauer, C., Fröhlich, L., and Popall, M. (2003) Femtosecond laser-induced two-photon polymerization of inorganic organic hybrid materials for applications in photonics. *Opt. Lett.*, **28** (5), 301–303.

Part III
Materials

Multiphoton Lithography: Techniques, Materials and Applications, First Edition.
Edited by Jürgen Stampfl, Robert Liska, and Aleksandr Ovsianikov.

6
Photoinitiators for Multiphoton Absorption Lithography

Mei-Ling Zheng and Xuan-Ming Duan

6.1
Introduction for Photoinitiators for Multiphoton Absorption Lithography

6.1.1
Multiphoton Absorption Lithography

In the past decade, three-dimensional (3D) multiphoton absorption lithography (MAL) technology based on two-photon photopolymerization (TPP) has been attracting much interest owing to its various applications to microelectromechanical systems (MEMS) [1], 3D optical devices [2, 3], lab on chip [4], and so on [5, 6]. TPP is considered one of the most important applications of two-photon absorption (TPA), which contributes to the intrinsic 3D fabrication capability and the high penetration depth. The commercial photoresist resins using conventional UV active initiators have commonly been utilized in this field. These conventional initiators usually have low TPA cross sections (δ), which results in low two-photon sensitivity. Therefore, large excitation power and long exposure time are needed to polymerize the resin, resulting in potential damage to the micro/nanostructures. From this perspective, the disadvantages of the conventional photoinitiators limit the versatility of 3D MAL and its scope for widespread application. Developing high-sensitivity initiators is essential to make the process more reliable and the structures to be patterned rapidly [7].

6.1.2
Photoinitiators for Multiphoton Absorption Lithography

6.1.2.1 History of the Design of Two-Photon Initiators

TPA has become easier to investigate with the development of subpicosecond pulsed lasers since the 1990s (particularly the Ti:sapphire laser). The invention of two-photon excited fluorescence microscopy (TPEF) by Webb and coworkers and the rapid assimilation of this technique by manufacturers of confocal microscopes have led to an explosion of interest in the investigation of various types of

Multiphoton Lithography: Techniques, Materials and Applications, First Edition.
Edited by Jürgen Stampfl, Robert Liska, and Aleksandr Ovsianikov.

two-photon chromophores [8, 9]. On the other hand, there is now a strong demand for efficient TPA dyes for a wide range of applications, including microscopy, microfabrication, 3D data storage, optical power limiting, upconverted lasing, and photodynamic therapy, as well as for the localized release of bioactive species [9]. Recently, this demand has been matched by rapid advances in the design and synthesis of TPA dyes.

Based on the understanding of TPA dyes, two-photon initiators (TPIs) has emerged, and two-photon polymerization has become a new research field [3]. Numerous studies on the design and synthesis of TPIs have been carried out. The classification and the application of TPIs are discussed in the following sections.

The TPIs investigated in the initial stage were conventional one-photon initiators, such as Irgacure 369 and Irgacure 184 [10, 11]. The cross section of these one-photon initiators is quite low – but their radical generation efficiency is high. They are applied in SCR500 and SU-8 photoresists [12].

6.1.2.2 **Property of Two-Photon Initiators**

The main difference between one-photon absorption (1PA) and two-photon absorption (2PA) is that 2PA involves the simultaneous interaction of two photons, and so it increases with the square of the light intensity, whereas 1PA depends linearly on the intensity. This is the reason why 2PA is observed only under intense laser beams, particularly focused pulsed lasers, which generate a very high instantaneous photon density. Most of the applications for 2PA result from this intensity dependence. Two-photon absorption is a way of accessing a given excited state by using photons of half the energy (or twice the wavelength) of the corresponding one-photon transition, thus leading to other applications.

One of the most important considerations for developing photoresists for TPP is high transparency. On the other hand, the prepolymer photoresist used for MAL is usually a viscous liquid, an amorphous solid, or a gel, to prevent the escape of the reactive components during TPP. Liquid samples are easy to fabricate and process, but will also bring disadvantages, such as the undesired deformation of structures during fabrication. Compared to liquid photoresists, solids and gels are more difficult and time consuming to prepare and process, but the complete restriction of motion in these media facilitates the fabrication of complex structures, such as those with free moving parts. There are two crucial components in a prepolymer photoresist for TPP: the photoinitiator and the monomers. Other components can include polymerization inhibitors to stabilize the photoresist and influence the feature size, photosensitizers, solvents to assist in casting, filler polymers to create gels or increase the viscosity of liquid resins, and other additives such as dye molecules that introduce new functions to the polymerized structures [13].

Generally, photoinitiators can be selected from any chemical compounds that can decompose into active species, such as free radicals and photoacid generators (PAGs) when exposed to light. Photoinitiators exist in nature and are widely used in the industry. In TPP, the photoinitiators play an important role as shown in Figure 6.1 [14]. Taking radical photopolymerization as an example, the photosensitizer in the photopolymerizable resin during the TPP process is excited by

Initiation: $S \xrightarrow{h\nu,\, h\nu} S^* \cdots \xrightarrow{I} I^* \rightarrow R$

Propagation: $R^{\cdot} + M \rightarrow RM^{\cdot} \rightarrow RMM \cdots \rightarrow RM_n$

Termination: $RM_n^{\cdot} + RM_m^{\cdot} \rightarrow RM_{n+m}R$

Figure 6.1 Schematic illustration of the TPP process. S stands for the photosensitizer, I denotes the photoinitiator, R• is the radical, and M is the monomer. S* and I* are the excited states of the photosensitizer and photoinitiator, respectively, after absorbing the photon energy.

absorbing two photons simultaneously, and then fluorescence is emitted in the UV–vis region. Consequently, the fluorescence will be absorbed by the photoinitiators with good chemical reactivity and give rise to radicals. Then, the radicals react with monomers or oligomers, resulting in the production of monomer radicals, which would expand in a chain reaction until two radicals meet and undergo termination. Finally, the low molecular weight monomers are polymerized into cross-linked, high molecular weight materials.

6.1.3 Characterization of Two-Photon Initiators

From both fundamental and application perspectives, it is thus necessary to monitor and quantify the strength of the 2PA process and to determine how material parameters and experimental conditions control it. Over the years, a number of different experimental techniques have been used to determine the 2PA spectra of materials and to measure their 2PA cross sections. The techniques most often used can be classified as direct methods, in which the attenuation of the excitation beam is monitored, and indirect methods, in which a secondary effect of the 2PA process, such as induced fluorescence, phosphorescence, or a temperature change in the sample, is measured [15]. Generally, direct methods of determining TPA are the nonlinear transmission (NLT) technique [16] and *Z*-scan measurement [17], while the indirect method is usually referred to as the TPEF method [18].

Direct measurement of TPA cross sections is usually difficult because only a small fraction of photons is absorbed in the two-photon process. The effect of 2PA is observed as a decrease in the intensity of the excitation beam itself [15]. TPA cross sections can be obtained by monitoring the transmittance changes of a sample when the incident intensity changes. NLT and *Z*-scan measurements are examples of techniques in this category.

In an NLT experiment, as shown in Figure 6.2, a sample is placed in the path of the excitation beam and the pulse energy is measured before and after the sample [16, 19, 20]. The measurement is repeated over a range of incident pulse energies. Because of the nonlinear absorption process, the transmittance depends on the incident energy. Knowing the optical configuration and pulse characteristics, it is possible to determine the TPA cross section of the material. It should also be pointed out that inherent in the NLT measurement is the assumption that TPA is the predominant process causing the observed intensity-dependent nonlinear absorption.

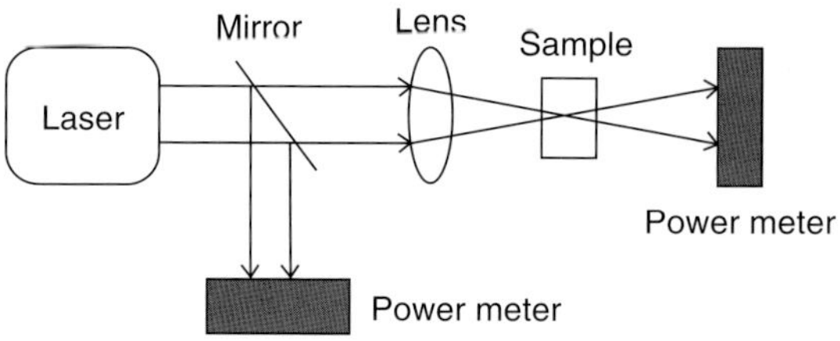

Figure 6.2 Schematic illustration of the NLT technique for the TPA cross-section measurement.

The deduced equations for a Gaussian beam in an ideal case are the following:

$$T_i = [\ln(1 + I_0 L\beta)]/(I_0 L\beta) \quad (6.1)$$

$$T_i = I_i/I_0 \quad (6.2)$$

$$\delta_{TPA} = (h\nu\beta/N_A) \times d \times 10^{-3} \quad (6.3)$$

where T_i is the transmittance, I_0 is the incident laser intensity, L is the thickness of the sample cuvette, β is the nonlinear absorption coefficient (centimeter per gigawatt), I_i is the laser intensity passing the sample cuvette (gigawatt per square centimeter), δ_{TPA} is two-photon absorption cross section (cm^4 s $photon^{-1}$), N_A is Avogadro constant (6.023×10^{23}), d is the concentration (M), h is the Planck constant (6.626×10^{-34} J s), and ν is the incident laser frequency.

In a Z-scan experiment, the pulse energy is kept constant, but change in intensity is achieved by moving the sample along the propagation direction z (hence the term *Z-scan*) of a focused beam and thus varying the beam cross-sectional area [17, 21–23]. As shown in Figure 6.3, the transmittance is measured at each position of the sample. For typical configurations (where the sample is thin, there is no 1PA, and all the transmitted light is collected), the transmittance is 1 when the sample is far from the focus of the beam, decreases as the sample is moved closer to the focus, and reaches a minimum at the focus (where the intensity is maximum). The trend is reversed as the sample is moved away from the focus in the opposite direction.

Indirect measurement of TPA cross sections is usually referred to as the TPEF method, in which the two-photon approach to determining the excitation (TPE)

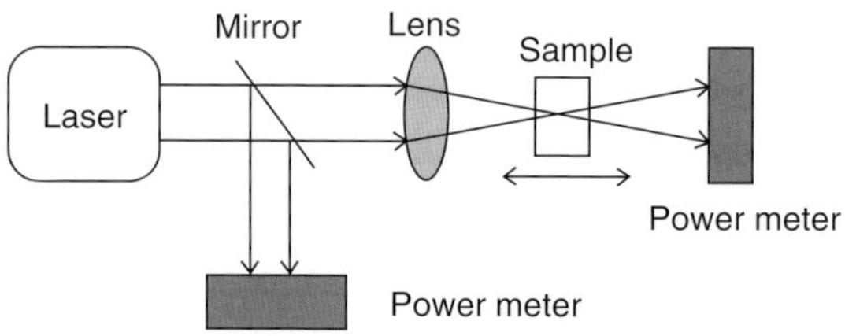

Figure 6.3 Schematic illustration of Z-scan measurement for characterizing TPA cross section.

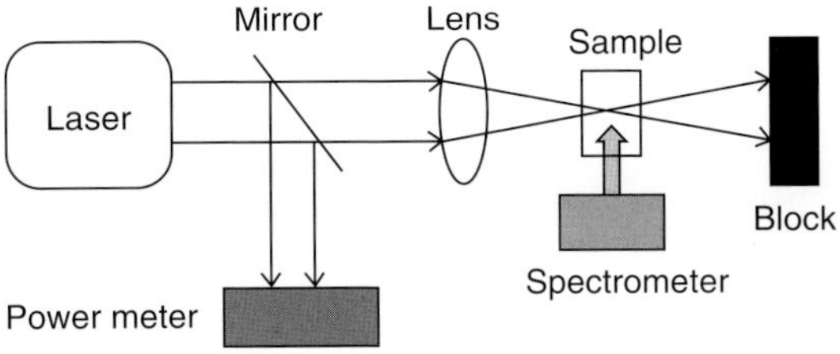

Figure 6.4 Schematic illustration of the TPA cross-section measurement by using TPEF method.

cross section is fluorescence [18, 24, 25]. The typical optics is shown in Figure 6.4. Pulsed lasers are the most commonly used excitation sources, which is due to the fact that TPA is a third-order nonlinear optical process requiring very high intensities. The advantages of using ultrafast pulses to drive TPA are illustrated by the Ti:sapphire laser. A typical Ti:sapphire laser gives out pulses with a duration of 100 fs or less at a repetition rate of approximately 80 MHz. The time between pulses is therefore on the order of 12 ns. Thus, a sufficient peak intensity to drive efficient TPA can be realized at a low average laser power. While the output of a Ti:sapphire laser is in the range of hundreds of milliwatts, two-photon excitation can generally be accomplished at average powers of a few milliwatts or less, depending on the focusing conditions and the species being excited. Most Ti:sapphire lasers operate at a center wavelength of about 800 nm, although with appropriate optics they can operate in the range 700–1100 nm. Phenomenologically, TPA is regarded as the third term in a conventional anharmonic oscillator model for depicting the vibrational behavior of molecules.

In order to characterize the TPA cross section, the fluorescence quantum yield (Φ) should be determined first. This is given by

$$\Phi_{\rm s} = \Phi_{\rm r}\left[\frac{A_{\rm r}(\lambda_{\rm r})}{A_{\rm s}(\lambda_{\rm s})}\right]\left[\frac{I_{\rm r}(\lambda_{\rm r})}{I_{\rm s}(\lambda_{\rm s})}\right]\left[\frac{n_{\rm s}}{n_{\rm r}}\right]^2\left[\frac{F_{\rm s}}{F_{\rm r}}\right] \tag{6.4}$$

where Φ is the quantum yield, n is the refractive index, $I(\lambda)$ is the relative intensity of the exciting light at wavelength λ, $A(\lambda)$ is the absorbance of the solution at the exciting wavelength λ, and F is the integrated area under the emission spectrum. The subscripts s and r refer to the sample and reference solution, respectively.

Then the TPA cross section is calculated as

$$\delta_{\rm s} = \delta_{\rm r}\frac{C_{\rm r}}{C_{\rm s}}\frac{n_{\rm r}}{n_{\rm s}}\frac{F_{\rm s}}{F_{\rm r}}\frac{\Phi_{\rm r}}{\Phi_{\rm s}} \tag{6.5}$$

where δ is the TPA cross section, C and n are the concentration and refractive index of the sample solution, and F is the integrated area obtained from the TPEF spectrum.

6.1.4 Molecular Design for Photoinitiators

The design and synthesis of highly active organic molecules have emerged with the development of MAL technology. Research on the design and synthesis of new molecules with enhanced TPA cross section δ has been widely carried out [9, 26]. Numerous molecules have been designed and synthesized for investigating the relationship between molecular structures and δ, in order to establish well-defined structure–property relationship with systematically varied molecular structural factors and reproducible characterization of two-photon properties. Detailed knowledge of molecular design parameters of new molecules with enhanced δ has been obtained.

In general, there are three types of molecular design to improve the sensitivity of TPI in TPP. Type I chromophores are symmetrical in nature, consisting of a π-electron-rich bridge flanked on either side by an electron-poor group or electron-rich group. It is to improve the δ of initiators by extending the conjugate system in the molecular structure design [27, 28], which is the most common way to improve TPI sensitivity. Some principles for designing TPA materials have been established. Type II chromophores are asymmetrical molecules consisting of a highly fluorescent π-electron-rich aromatic/olefinic bridge flanked on one side by a π-electron donor and on the other by a π-electron acceptor. This is to optimize two component systems by changing co-initiators with a known TPI [28, 29]. Type III is to fix a co-initiator unit, such as a sulfonium salt, into a large δ chromophore [27]. The targeted sulfonium groups covalently linked to bis(styryl)benzene dyes work as two-photon PAGs because (i) covalent attachment of the TPA donor and the sulfonium acceptor helps to ensure rapid forward electron transfer; (ii) triarylamine radical cations are known to be very stable and therefore should facilitate efficient bond cleavage; (iii) protonated triarylamino groups are strong acids (Ph_3N^+H, $pK_a = -5$), so the amine functionality should not inhibit reactions such as the ring-opening polymerization of epoxides; and (iv) non-nucleophilic anions such as PF_6^-, AsF_6^-, or SbF_6^- can be incorporated into these systems. Therefore, type III photoinitiators are strong two-photon absorbers and efficient initiators.

In principle, molecules with a π-conjugated system typically exhibit excellent TPA activity. The efforts of molecular design for enhancing TPA activity have been focused on the enhancement of the intramolecular electronic transfer by modifying the electronic structures of molecules by expanding the length of the π-conjugated system and using strong electronic donating and withdrawing groups at the ends of the π-conjugated system [30]. According to the molecular symmetry, the main strategy of molecular design for achieving large TPA activity can be summarized as shown in Figure 6.5. These can be classified into centrosymmetric ($A-\pi-A$, $D-\pi-D$) and noncentrosymmetric ($D-\pi-A$) molecules, in which D stands for an electron donor, A stands for an acceptor, and they are linked by a π-conjugate bridge. The centrosymmetric molecules are among the most commonly studied molecules. As shown in Figure 6.5, the extended $D-A-\pi-A-D$, $D-A-D-A-D$, and $D-D-A-D-D$ systems have

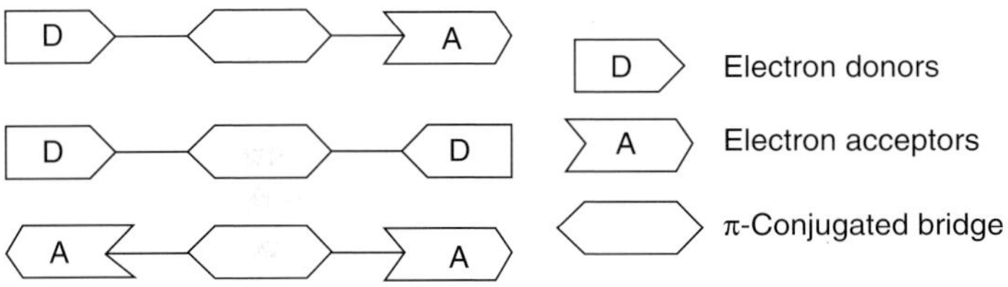

Figure 6.5 Schematic of molecular designs for enhancing TPA activity with different molecular symmetries: centrosymmetric (Type I) and noncentrosymmetric (Type II).

been proposed to increase TPA activity through constructing structures with improved intramolecular charge transfer (ICT) properties. We will introduce the research on the molecular design of photoinitiators briefly in this chapter.

6.2 Centrosymmetric Photoinitiators

Centrosymmetric photoinitiators are among the most widely studied types. There are many molecules based on carbazole derivatives, anthracene derivatives, and even those with multibranched structures and quenching effect. Marder *et al.* [31, 32] have demonstrated that symmetric stilbenes containing donor or acceptor groups linked by a π-conjugate bridge (D–π–D or A–π–A) exhibit large δ values. The δ values of molecules with D–π–A–π–D and A–π–D–π–A structures have been investigated. A series of bis dialkylamino- or diarylamino-substituted diphenylpolyenes and bis(styryl)benzenes are shown in Figure 6.6. Values of δ as large as 1420 GM (1 GM = 10^{-50} cm^4/photon molecule) have been reported for this kind of molecules with the bis-donor molecular structure. The effect of the type and length of the conjugated chain and of dialkylamino or diarylamino substitution on the position and magnitude of the peak two-photon absorptivity has been demonstrated. Furthermore, the structure–property relationships for 2PA chromophores have been confirmed, and the bis-donor substitution leads to an increase in the δ values. δ can also be increased by increasing the length of the conjugated linker in D–π–D molecules using diphenylpolyene or phenylene-vinylene-type bridges. It is pointed out that the increase of δ is essentially linear with the chain length and is related to the type of bridge and that those phenylene-vinylenes show a larger increase compared to diphenylpolyene.

The transition dipole moments for the transitions between the ground state and the first excited singlet state and between the first and second excited singlet states have been estimated using experimental data from the one- and two-photon spectra. It was found that increases in chain length mainly result in an increase in the transitions between the ground state and the first excited singlet state, whereas the addition of donor end groups or going from diphenylpolyene- to phenylene-vinylene-type bridges leads primarily to an increase in the transitions between the first and second excited singlet state. The trends in the energy of the lowest excited singlet states and in the transition moments for the diphenylpolyene series

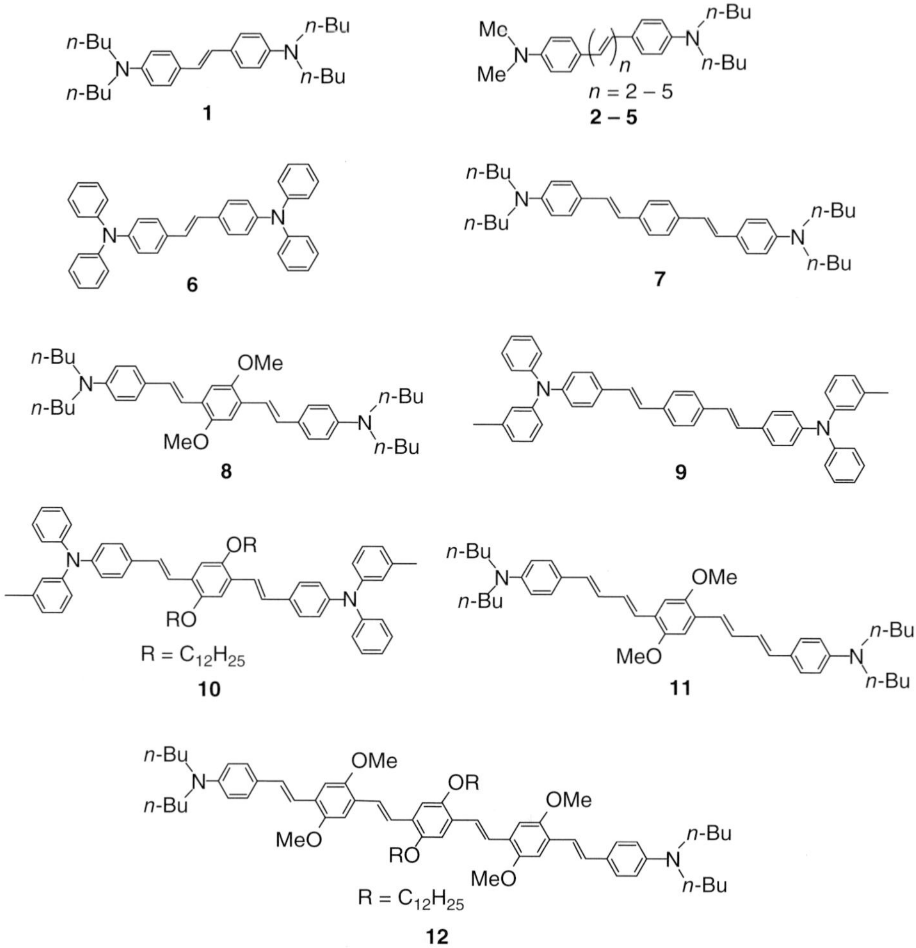

Figure 6.6 Molecular structure of the D−π−D chromophores. (Reference [31] © 2000 American Chemical Society. All Rights Reserved.)

as a function of chain length are in agreement with those calculated by quantum mechanical methods. These results suggest a link between structural features in these classes of molecules and the electronic dipole couplings and state energies that control the strength of the two-photon absorption. These studies of structure−property relationships can provide insight into the molecular mechanism of TPA, and further give general ideas about the role of the greater degree of intramolecular symmetric charge redistribution in molecules of D−π−A−π−D and A−π−D−π−A structural types. This study further improves our understanding of the molecular design of TPI in TPP.

Liska's group has presented a series of linear and cyclic benzylidene ketone-based TPIs containing double bonds and dialkylamino groups synthesized in one step via classical aldol condensation reactions [23]. Systematic investigations of the structure–activity relationships were conducted via quantum chemical calculations and experimental tests. The development of practical two-photon absorption photoinitiators has been slow due to their complicated syntheses often relying on expensive catalysts. These shortcomings have been a critical bottleneck for further advances in the promising field of two-photon-induced photopolymerization (TPIP) technology. The authors have designed and synthesized initiators with D–π–A–π–D structures and found that the size of the central ring significantly affected the excited state energetics and emission quantum yields as well as the two-photon initiation efficiency (Figure 6.7). In the TPIP tests, the 4-methylcyclohexanone-based initiator displayed much broader ideal processing windows than its counterparts with a central five-membered ring and the previously described highly active two-photon absorption photoinitiators. Surprisingly, a writing speed as high as 80 mm s^{-1} was obtained for the microfabrication of complex 3D structures employing acrylate-based formulations. These highly active two-photon absorption photoinitiators also exhibit excellent thermal stability and remain inert to one-photon excitation. Straightforward synthesis combined with high TPA initiation efficiency makes these novel initiators promising candidates for commercialization.

Carbazole-based compounds have been designed and used not only as two-photon fluorescence probes [33–35] but also as photoinitiators for TPP [7, 24]. Carbazole-based compounds usually possess C_v and C_{2v} symmetry and amphiphilic molecular structure. The property, for example, the affinity of molecular structure, can be tuned by changing both the molecular symmetry and the length of the hydrophobic chain in the carbazole ring. This provides further insight into the relationship between the molecular structure and its function in

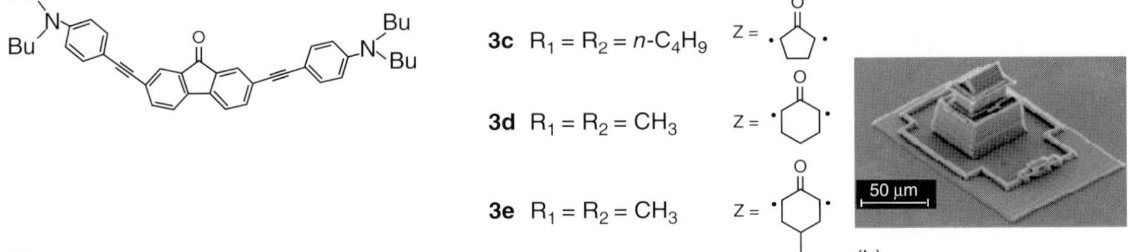

(a) (b)

Figure 6.7 (a) Molecular structure of the D–π–A–π–D chromophores **3a–3e**. (b) SEM image of an ancient tower in oriental style with **3e**. (Li 2013 [23]. Reproduced with permission of American Chemical Society.)

the selective staining [33]. Carbazole-based cyanine probes have an advantage for TPEF imaging: large δ values, which contribute to bright TPEF images even at low probe concentrations. A previous report has shown that related probes, C_{2v} symmetric molecules, exhibit δ values of over 1600 GM.

Carbazole is very easy to be modified at the 3- and 6-positions to form a C_{2v} symmetric molecule. Duan *et al.* [24] have reported C_{2v} symmetric carbazole-based 2D-ICT salts by introducing ionic groups at the 3- and 6-positions of the carbazole ring to elongate the π-conjugation system compared to its 1D analog. In this context, 1D and 2D carbazole-based hemicyanines, where methyl pyridinium, methyl indolium, and methyl benzothiazolium were used as the acceptor group, were synthesized by Knoevenagel condensation (as shown in Figure 6.8). The one-photon and two-photon photophysical properties of 1D-ICT organic salts (**1a–e**) and 2D-ICT organic salts (**2a–e**), where the pyridinium, indolium, and benzothiazolium cations were used as acceptor group, the *N*-alkyl-carbazolyl as donor group, and a carbon–carbon double bond as π-conjugated bridge, have been investigated. The experimental results indicated that the different ionic acceptors affected their one-photon and two-photon properties. Among them, 2D methyl pyridinium carbazole derivatives exhibited low quantum yields and large δ values (more than 1600 GM). The results showed that the 2D systems possessed larger δ than the 1D systems, although the 1D systems have higher fluorescence quantum yields.

Furthermore, the C_{2v} symmetric carbazole-based 2D stilbazolium compound 3,6-bis[2-(1-methylpyridinium)vinyl]-9-pentyl-carbazole ditosylate (**2c**) was used in TPP 3D lithography as a photoinitiator, and a 3D microstructured photonic crystal was successfully fabricated by TPP 3D lithography (Figure 6.8). The threshold of TPP for the V-shaped 2D-ICT salt **2c** was determined to be 3.0 mW, and it was successfully used as TPP photoinitiator for TPP microfabrication. This study indicates that the carbazole-based compounds could be utilized as effective photoinitiators of TPP.

Based on the understanding of the molecular structure and the corresponding two-photon optical property, Duan *et al.* [7] have further developed a type of novel A–π–D–π–A V-shaped 3,6-bis(phenylethynyl)carbazole-based

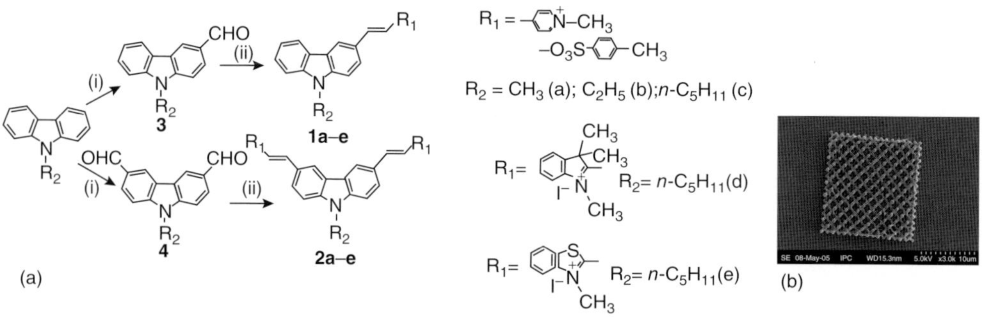

Figure 6.8 (a) Synthetic route and molecular structure of **1a–e** and **2a–e**. (b) SEM image of a 3D photonic crystal. (Gu 2007 [24]. Reproduced with permission of The Royal Society of Chemistry.)

chromophores as TPP initiators combining a large δ with facilitated radical formation. By introducing (i) a carbon–carbon triple bond to improve the δ, (ii) a carbazole moiety as a co-initiator unit, and (iii) an alkyl or benzyl group for radical stabilization into an A–π–D–π–A V-shaped molecule, a series of novel Type III initiators, rigid A–π–D–π–A V-shaped, 6-bis(4-nitro/carbonyl-phenylethynyl)carbazole derivatives (**4a–d**), have been designed, in which the carbazole moiety acts as a donor, phenylacetylenyl as a conjugate linker, and the nitro/carbonyl group as an acceptor. The synthetic route and molecular structure are shown in Figure 6.9. Chromophores **4a–d** were synthesized in a facile way by Sonogashira coupling reactions starting from 3,6-dibromocarbazole and 4-bromobenzaldehyde/4-bromonitrobenzene. 9-Benzyl-3,6-bis(4-nitrophenylethynyl)carbazole (**4d**) shows a strong TPA around 800 nm and exhibits very high TPP initiating sensitivity with a threshold power of 0.8 mW at the concentration of 0.18 mol%, which is much lower than the threshold power of 6.37 mW for benzil. The corresponding threshold of laser exposure intensity for TPP is 3.0×10^7 mJ cm^{-2}. The lowest loading of **4d** is up to 0.012 mol% with a threshold power of 3.2 mW. This high TPP-initiating sensitivity results from their large δ and low fluorescence quantum yield. The proposed concept of molecular design using a combination of radical stabilization and large δ within a molecular structure should provide a new direction for developing TPP initiators with high initiating efficiency.

As shown in Figure 6.10, two types of TPP initiators, 1-{(1E)-2-[4-(diphenylamino)phenyl]vinyl}-4-[4-N,N-di-methylamino]benzene (PVMB) and 1-[(1E)-2-(4-{[4-((1E)-2-{4-[4-N,N-di-methylamino]phenyl}vinyl)phenyl]phenylamino}phenyl)vinyl]-4-[4-N,N-di-methylamino] benzene (DPVMB) have been synthesized with solid-phase Wittig reaction [36]. The one-photon fluorescence, fluorescence quantum yields, lifetimes, and solvent effects of the initiators have been studied in detail, and both compounds exhibited solvent sensitivity. The δ values from 730 to 870 nm at intervals of 10 nm have been measured with the two-photon-induced fluorescence method, and the maximum values of δ

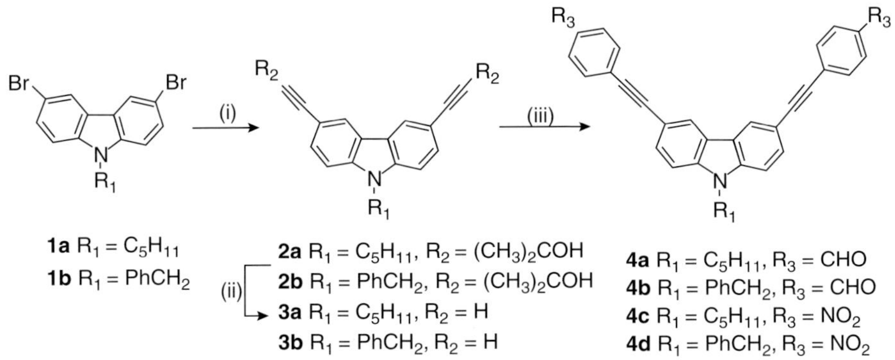

Figure 6.9 Synthetic route and molecular structure of **4a–d** [7]. (Reproduced from Ref. [7] with permission from The Royal Society of Chemistry.)

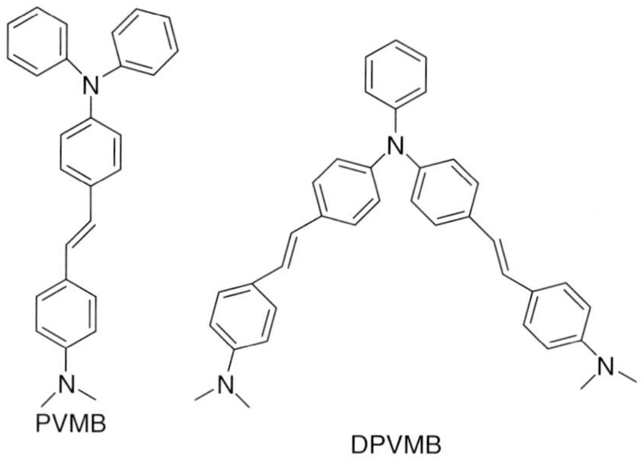

Figure 6.10 Synthetic route and molecular structure of PVMB and DPVMB [36]. (Reference [36]. Copyright © 2006 Elsevier B.V. All rights reserved.)

of PVMB and DPVMB are 1.6 and 19.9 GM at 800 nm, respectively. Finally, a microstructure has been successfully fabricated by using DPVMB as initiator. This proves that these two compounds are good two-photon absorbing chromophores and operative TPP initiators.

It is well known that the lateral spatial resolution (LSR) is mainly influenced by two factors: laser intensity and exposure time. Laser intensity can directly restrict the region of TPA, in which radicals are initially generated. Exposure time influences the number of radicals induced by TPA. On one hand, a low TPP threshold would reduce the size of the region where radicals are initially generated. On the other hand, a short exposure time should decrease the number and diffusion of radicals generated by TPA, resulting in localized polymerization in a tiny region. Therefore, it should be possible to improve the LSR of TPP by using a highly sensitive initiator that can contribute to a low threshold and short exposure time. Photoinitiators based on anthracene [37, 38] have attracted much attention as another kind of centrosymmetric initiators. LSR, which is an important factor for evaluating the capability of TPP, was improved to 80 nm by employing an

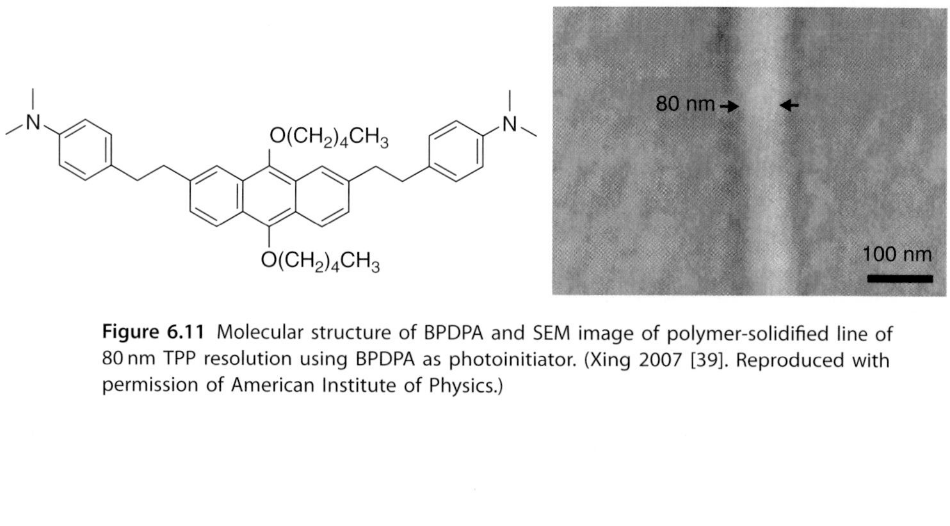

Figure 6.11 Molecular structure of BPDPA and SEM image of polymer-solidified line of 80 nm TPP resolution using BPDPA as photoinitiator. (Xing 2007 [39]. Reproduced with permission of American Institute of Physics.)

anthracene derivative 9,10-bis-pentyloxy-2,7-bis[2-(4-dimethylamino-phenyl)-vinyl]anthracene (BPDPA) as a highly sensitive and efficient photoinitiator (Figure 6.11). Photocurable resin containing 0.18 mol% BPDPA exhibited a low polymerization threshold of 0.64 mW at 800 nm. Theoretical calculations showed that LSR can be increased by reducing the laser power. This study indicates that LSR can be improved by using more sensitive initiators in future [39].

For better understanding the polymerization mechanism, a series of C_{2v} symmetrical two-photon absorption compounds with the anthracene core, 2,7-bis[2-(4-substituted phenyl)-vinyl]-9,10-dipentyloxyanthracenes designated as **I**, **II**, and **III** (the substituted groups at 4-position of phenyl of **I**, **II**, and **III** were dimethylamino, methyl and cyano, respectively) were designed and synthesized as initiators in TPP [37]. The anthracene ring was modified by linking vinylphenyl groups to the 2,7-positions to extend conjugation system length and two pentyloxy groups at 9,10-positions to serve as electronic donors. δ of **I** was around 300 GM, which was much larger than 10 GM of **II** and 29 GM of **III** at 800 nm. **I** of 0.18% molar ratio in a resin composed of methacrylic acid (MAA) and dipentaerythritol hexaacrylate exhibited a dramatically low threshold of 0.64 mW compared with commercial photoinitiator benzil at scanning speed $10\,\mu m\,s^{-1}$. Moreover, the threshold of photoinitiator **I** was increased to only 2.53 mW at the scanning speed $1000\,\mu m\,s^{-1}$. The dependence of the threshold on the concentration and exposure time was in accordance with theoretical calculation.

A reasonable mechanism of the two-photon initiating process was proposed, as shown in Figure 6.12. Anthracene derivatives absorb two photons to produce the activated species M1 when excited by a two-photon absorption process. The species M1 undergoes a photoxidation process to generate cationic radical species M2 in the 9,10-positions. Those with electron-donor groups could effectively increase the reduction potential and stabilize the cationic radical M2. Then, the cationic radical M2 undergoes a C–O cleavage to form a radical M3 at the 9,10-positions. Consequently, the free radical species M3 initiate the polymerization of methyl acrylate derivatives. The stability of the formed radicals is supposed to be in the order **I** > **II** > **III** since the electron-donating ability is in the order **I** > **II** > **III**. The larger δ at 800 nm and the stronger electron-donating ability of the dimethylamino group in **I** compared with the methyl group in **II** and the cyano group in **III** contribute to the higher sensitivity of **I**. Therefore, higher sensitivity and more rapid polymerization were observed when **I** was used as TPP initiator. This study provides good prospects for developing low-threshold photoinitiators in TPP.

Yang *et al.* [40] have reported a kind of triphenylamine-modified quinoxaline (TPAQ) derivatives and triphenylamine-modified benzoquinoxaline (TPABQ) as two-photon photoinitiators. Their results showed that TPAQ possesses a high fluorescence quantum yield and a high δ value of 4160 GM in the wavelength range 780–820 nm, and is superior to the most commonly studied benzil-type compounds as a photoinitiator in TPP. Perry's group [41] showed that quasi-linear quadrupolar D–π–D molecules exhibit peak δ values at least an order of magnitude larger than the corresponding unsubstituted counterparts and that

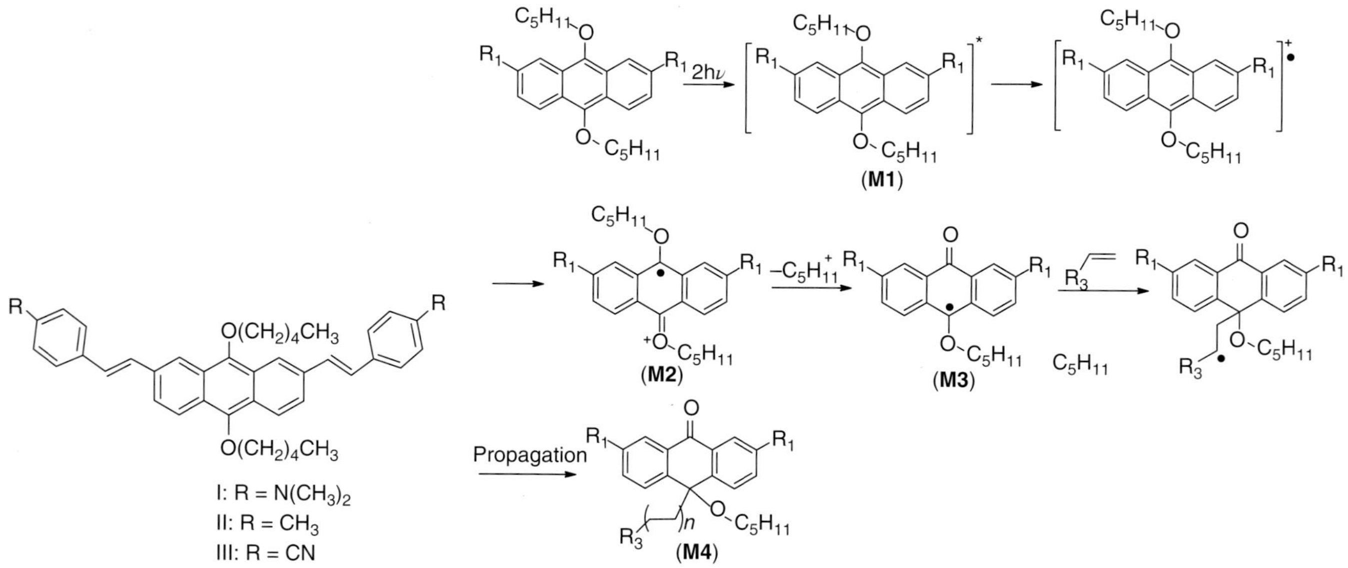

Figure 6.12 Molecular structure of the compounds with anthracene core, 2,7-bis[2-(4-substitutedphenyl)-vinyl]-9,10-dipentyloxyanthracenes designated as **I**, **II**, and **III** (the substituent groups at 4-position of phenyl of **I**, **II**, and **III** are dimethylamino, methyl, and cyano, respectively). The mechanism of two-photon photoinitiated polymerization at 800 nm [37]. (Reproduced from Ref. [37] with permission from the PCCP Owner Societies.)

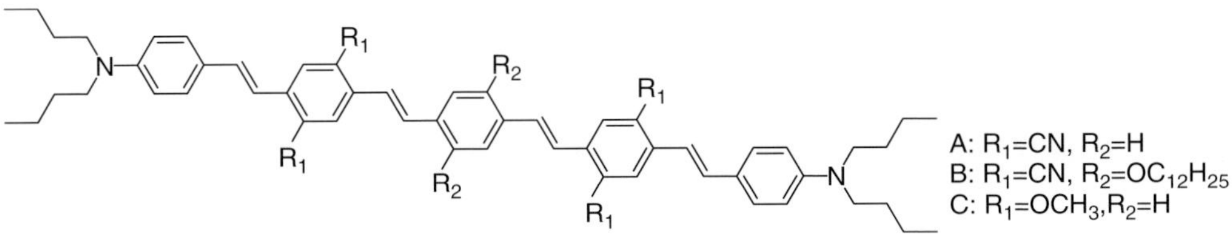

Figure 6.13 Molecular structure of extended D–A–D oligophenylenevinylenes A–C. (Reference [41] © 2005 American Chemical Society. All Rights Reserved.)

the additional presence of acceptors on the π-bridge (D–A–D) can augment the cross section further, as these substitutions can enhance the degree of ICT upon excitation. The three molecules examined are shown in Figure 6.13. Compounds **A–C** were synthesized using successive Horner–Emmons reactions. The substitution patterns can be described as D–A–π–A–D, D–A–D–A–D, and D–D–A–D–D, respectively, where the letters represent the type of substituent on each phenylene and π indicates that the ring is unsubstituted. The extended D–A–D oligophenylenevinylenes show large δ (>1000 GM) over a wide wavelength range and peak δ up to 5300 GM at almost twice the wavelength of the one-photon peak. These findings, which differ from those of similar molecules previously studied, indicate that relatively small changes in molecular structure can lead to large effects on the optical properties in this class of materials.

Novel water-soluble photoinitiators have been developed by Liska's group, which allow TPP of formulations with up to 80% cell culture medium [42, 43]. Highly efficient hydrophilic TPIs are necessary in order to produce structures from materials with high initial water content by TPP. Two benzylidene

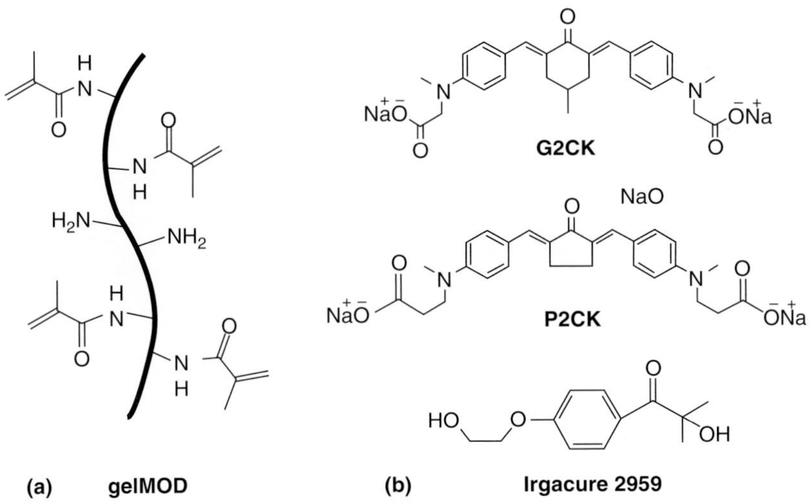

Figure 6.14 Chemical structures of (a) methacrylamide-modified gelatin (gelMOD) and (b) the water-soluble photoinitiators. (Reference [43] © 2013 American Chemical Society. All Rights Reserved.)

cycloketone-based TPIs G2CK and P2CK (Figure 6.14) were synthesized via classical Aldol condensation of appropriate aldehyde and cycloketones [42]. The TPA chromophores comprise typical D–π–A–π–D core structures with C_{2v} symmetry, where dialkylamino groups act as donors (D), vinyl as π-conjugated bridges, and carbonyl as an acceptor (A). Carboxylic sodium salts as hydrophilic functionalities were incorporated at the terminal amino group to ensure water solubility. Because of the long conjugation length and the presence of strong electron-donor and -acceptor groups, large TPA cross sections of 163 and 176 GM were obtained for aqueous solutions of G2CK and P2CK, respectively. These values surpass the TPA characteristics of commercially available initiators, most of which are not water-soluble, by a large margin [44]. Both initiators exhibited broad processing windows in TPP of hydrogel fabrication tests using formulations with water contents of up to 80%. Figure 6.15 shows the results of the live/dead staining of the sample containing four Yin–Yang structures produced by TPP in the presence of cells. This is the first report on the use of the two-photon polymerization technique for the fabrication of 3D hydrogel structures from a material containing living cells. The critical factor for this approach is the availability of efficient water-soluble initiators.

Recently, another kind of water-soluble initiator (WI) has been prepared through host–guest chemical interaction for microfabrication of 3D hydrogels via TPP by combining hydrophobic 2,7-bis(2-(4-pentaneoxy-phenyl)-vinyl)anthraquinone (N) with C_{2v} symmetrical structure and 2-hydroxypropyl-β-cyclodextrins [45]. This study has met the demand in TPP of 3D hydrogels, in which TPP initiators with high efficiency in an aqueous medium are very important. Both one- and two-photon optical properties of WI have been investigated. In aqueous medium, WI showed a δ of around 200 GM at the wavelength of 780 nm, which was much higher than that of commercial initiators. The threshold energy of TPP for the resin with WI as photoinitiator (the molar ratio of N in the resin is 0.03%) was 8.6 mW. As shown in Figure 6.16, 3D hydrogels with woodpile microstructure were further fabricated by using an average power of 9.7 mW and a scanning speed of 30 μm s^{-1}.

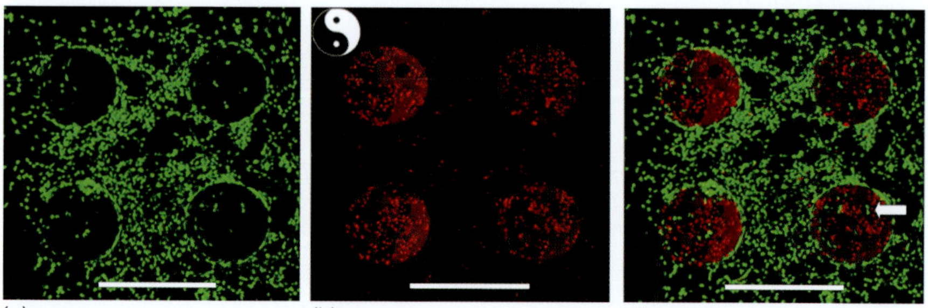

(a) (b) (c)

Figure 6.15 Viability of cells encapsulated in gelatine-based hydrogel structures produced by two-photon polymerization (2PP): (a) green-stained live cells; (b) red-stained dead cells; and (c) overlay, with an arrow indicating two live cells trapped within a cylindrical void. Scale bar represents 500 μm. (Ovsianikov 2013 [43]. Reproduced with permission of American Chemical Society.

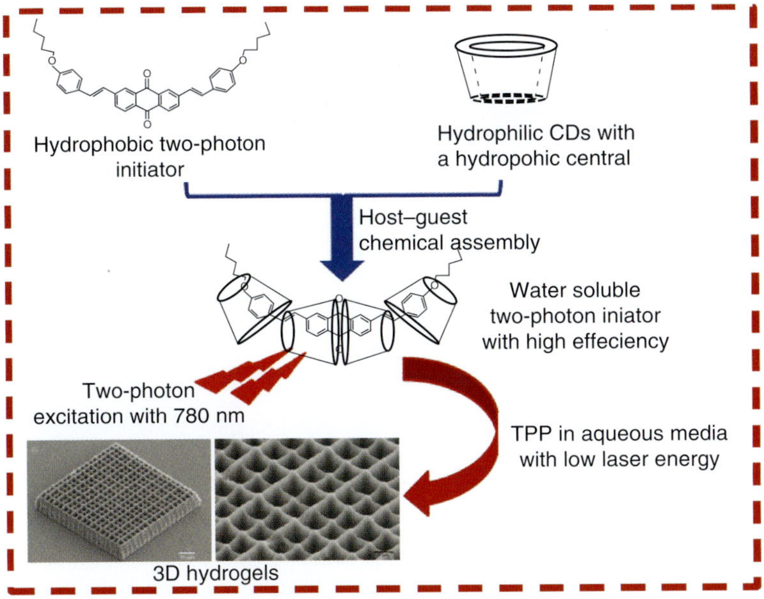

Figure 6.16 Design and synthesis of a water-soluble two-photon initiator with high efficiency and its application in TPP in aqueous media. (Xing 2014 [45]. Reproduced with permission of The Royal Society of Chemistry.)

Besides the photoinitiators mentioned above, multibranched structures have also been explored as TPP photoinitiators [28, 46, 47]. A new family of dendrimers with a naphthaline core flanked on both sides with triphenylamine branching was successfully synthesized and presented an increase of δ from 959 to 9575 GM with the generation number 1–3 (Figure 6.17). Wang's group found that these dendrimers could efficiently initiate the polymerization of acrylate resins to achieve regular diamond structures, and display higher TPP efficiency and sensitivity with the generation number. The overall TPP processes involved in TPA process, such as the ICT and intramolecular energy transfer as well as intermolecular electron transfer between the initiator and monomer, are illustrated in Figure 6.17. Steady-state fluorescence and time-resolved decay dynamics revealed that the light energy was absorbed by the peripheral triphenylamine unit and then transferred to the first generation, the energy funnel. Although strong interaction between dendritic initiator and monomer has been observed based on fluorescence quenching measurements, no intermolecular energy transfer but electron transfer was confirmed by the cyclic voltammograms and HOMO–LUMO measurements. That is, the dendritic initiator first produces the excited state via a TPA process, then transfers an electron to an acrylate monomer, and finally induces the latter to polymerize.

An alternative mechanism to radical polymerization is cationic polymerization [13]. A cationic photoinitiator works by generating a strong Bronsted acid that is capable of polymerizing epoxides and vinyl ethers. This photoacid is

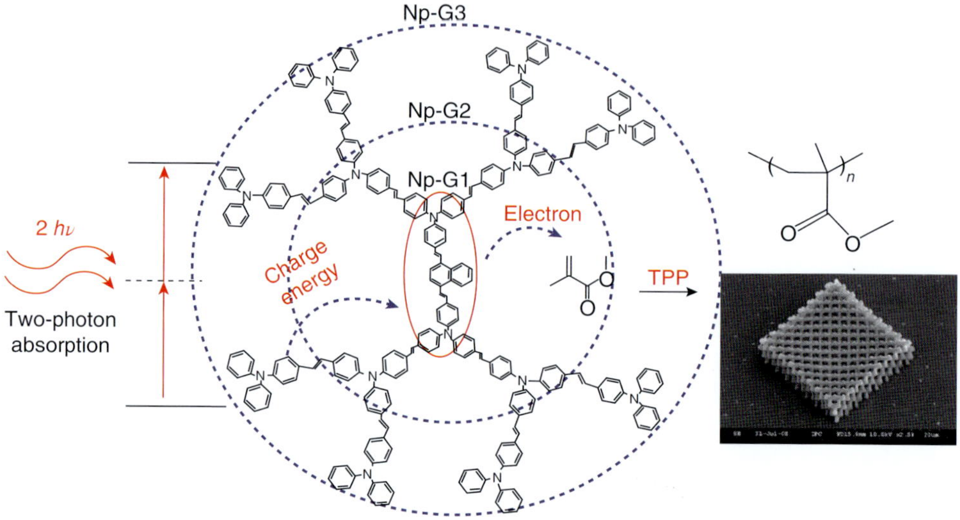

Figure 6.17 A new family of dendrimers with naphthaline core and triphenylamine branching as a TPA initiator. The overall TPP approach involves TPA, intramolecular charge transfer, intramolecular energy transfer, and intermolecular electron transfer. (Wang 2011 [47]. Reproduced with permission of American Chemical Society.)

catalytic, and so each PAG molecule can initiate multiple polymerizations [48]. Although singlet oxygen can quench radical polymerization, it does not affect cationic polymerization. Marder and Perry *et al.* [27] have developed a new PAG, BSB-S_2, containing a bis[(diarylamino)styryl]benzene core with covalently attached sulfonium moieties. It is reported that this PAGs offers as much as two orders of magnitude improved sensitivity relative to conventional PAGs. The bis[(diarylamino)styryl]benzene fragment is of the D–π–D structure. The chemical molecular structure is shown in Figure 6.18. BSB-S_2 exhibits a large δ, as high as 690 GM at 705 nm, and a high fluorescence quantum yield for the photochemical generation of acid ($\phi \approx 0.5$). This two-photon PAG has been applied in 3D microfabrication using a negative-tone cross-linkable epoxide

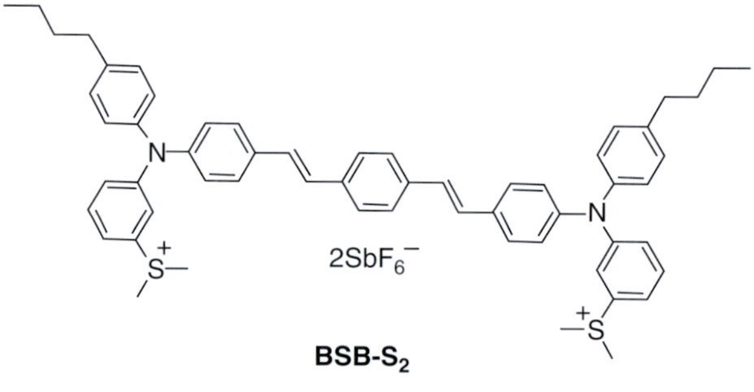

Figure 6.18 Chemical molecular structure of BSB-S_2 [27].

resin and a chemically amplified positive-tone resist. BSB-S2 has proven to be a high-sensitivity photoinitiator for cationic polymerization of various epoxide monomers under diverse TPA conditions utilizing nanosecond or femtosecond laser pulses. The advent of high-sensitivity initiators for two-photon patterning of radical- and acid-activated material should allow TPP to become more widely applicable.

Fourkas *et al.* [4] have demonstrated that by employing a novel PAG in an SU8 resist for TPP, high-AR (aspect ratio) masters with uniform, rectangular cross sections can be produced. The authors developed the fabrication of microfluidic channels with arbitrary, non-rectangular cross sections to combine channels with different ARs and cross sections in a single device. A specially formulated resist, FLUOR-SU8, was made by mixing dissolved epoxide oligomers with a fluorene-based PAG, 9,9-diethyl-[bis-(styrytriphenyl sulfonium hexafluoro phosphonate)]fluorine (FLUOR). This resist enables real-time monitoring of TPP. One can perform high-speed fabrication over centimeter-scale distances by employing writing speeds that are much faster than those typically used, enabling the fabrication of large-scale microfluidic masters on a practical time scale. The high-AR FLUOR-SU8 walls demonstrate the viability of using FLUOR-SU8 microstructures as masters for microfluidic devices. These walls have uniform cross sections and vertical side-wall profiles. Such master structures can assist diffusive mixing and allow for high packing densities in microfluidic devices. Channels of the molds that were 20 μm wide with ARs ranging from 5 to 10 were undamaged by the bonding process.

6.3 Noncentrosymmetric Photoinitiators

Prasad *et al.* [49–51] have pointed out that 1D D–π–A-type molecules containing fluorine or dithienothiophene (DTT) as the rigid π-conjugated backbone have large δ. The molecular structure–effective TPA relationships have been studied by careful characterization and systematic investigations of the varied changes in the organic structure of the dye molecules [49]. Such information can be useful in the design of more efficient two-photon dyes for imaging and power-limiting applications. The results indicate that dyes can be synthesized and have greatly increased effective δ with the incorporation of certain combinations of structural elements. It appears that several structural factors are critical in order to obtain large δ at 800 nm. In general, for the types of bridges studied, the structural elements are most important in increasing the effective conjugation between the donor and acceptor. In Type I chromophores, when a second thiophene ring is added to the bridge to increase the conjugation length, δ is increased but the upconverted emission is also red-shifted.

Tan *et al.* [50] have reported a series of linear, asymmetrical, diphenylaminofluorene-based chromophores with various strong π-electron acceptors, whose typical molecular structure is shown in Figure 6.19. These chromophores

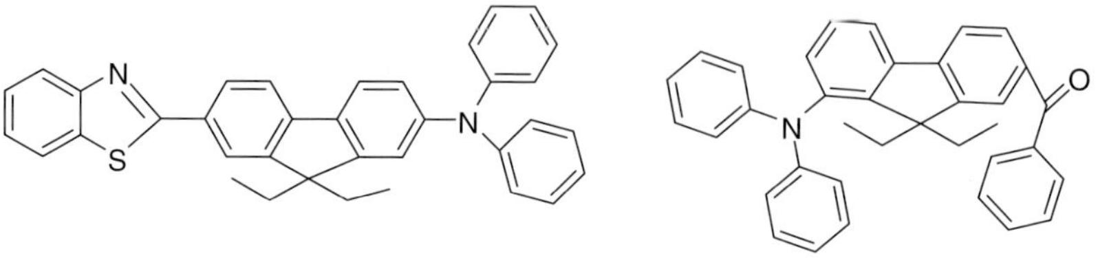

AF-240 **AF-370**

Figure 6.19 Chemical molecular structure of diphenylaminofluorene-based photoinitiators. (Reference [50] © 2001 American Chemical Society. All Rights Reserved.)

were studied to determine a suitable replacement for 2-(4-vinyl)pyridine; the nonlinear optical property was also studied. In addition to the benzoyl group (**AF-370**), these π-electron acceptors include 2-benzothiazolyl (**AF-240**), 2-benzoxazolyl, 2-(*N*-phenyl)benzimidazolyl, and 2-(3,4-diphenyl)imidazolyl moieties (five-membered heterocycles) and the 2-quinoxalinyl group (six-membered heterocycle). From nanosecond NLT measurements, these new chromophores were shown to have effective TPA at 800 nm, with $\delta = 9746$ GM for **AF-240** and 8432 GM for **AF-370**.

Kim *et al.* [51] reported a novel class of TPA chromophores bearing DTT as the π center, which comprises a three-component structure: D–π–D and D–π–A. Figure 6.20 shows two molecules, **13** and **14**, in which D is carbazole or an *N*,*N*-diphenylamine unit and A is 2-phenyl-(4-*tert*-butyl)-1,3,4-oxadiazole. The high δ values have been measured and calculated. This study indicates that the structural symmetry of the molecule and donor strength of the D component are important parameters for the enhanced TPA, but the crucial role is played by the π center, DTT.

Abbotto *et al.* [52, 53] designed and synthesized a series of heterocyclic push–pull TPA chromophores for frequency upconverted emission. Structural

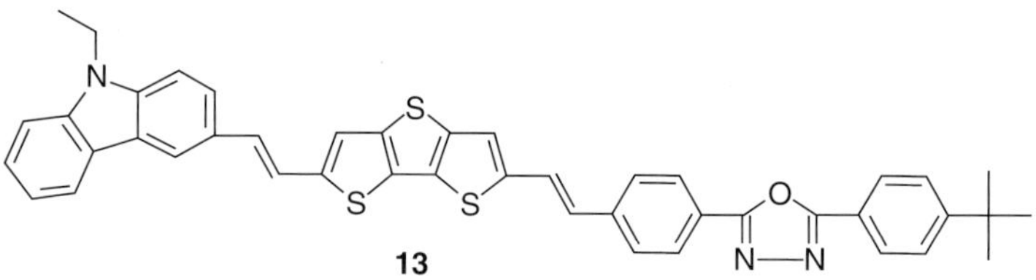

13

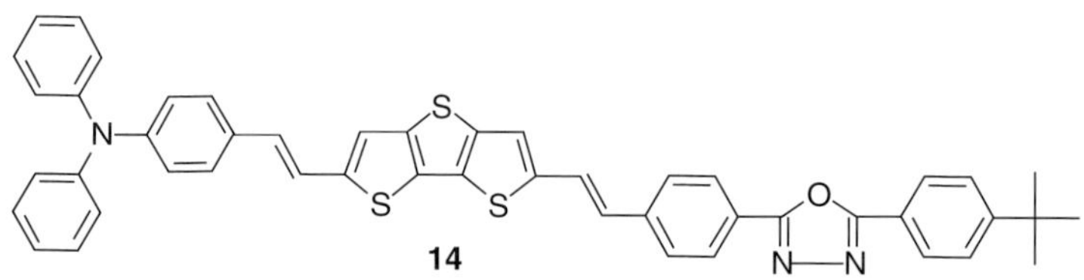

14

Figure 6.20 Molecular structure of the photoinitiator of type D–π–A. (Reference [51] © 2000 American Chemical Society. All Rights Reserved.)

modifications have been intentionally made to a minimal extent, since drastic changes in the nature of the donor and acceptor groups may lead to dramatic changes of nonlinear optical activity due to a combination of electronic and structural effects, which would be difficult to unravel. The molecules discussed are shown in Figure 6.21. Molecules **15** and **17** have one or two electron-donating substituents on the pyrrole ring, respectively. Dyes **16** and **18** carry an electron-withdrawing group on the electron-rich pyrrole and electron-poor pyridinium ring, respectively. By placing electron-withdrawing and electron-donating substituents on the π-rich and π-deficient heterocyclic rings, properties such as emission frequency and conversion efficiency can be tuned and directed to the desired target.

Ren *et al.* [54] have designed a two-photon, free-radical photopolymerization initiator, pyridine DBASVP, which possesses a large π-conjugated system and a strong donor and an acceptor group (Figure 6.22). Quantum chemistry calculations showed that the new initiator possesses a large change in dipole moment on transition to the excited state and a large transition moment. The calculated δ is as high as 881.34 GM. The single-photon and two-photon absorption and fluorescence properties in various solvents have been investigated. The experimental results indicate that the initiator DBASVP exhibits outstanding solvent sensitivity, which experimentally interprets the excellent electron delocalized and charge redistribution properties of the molecule upon excitation. A grating

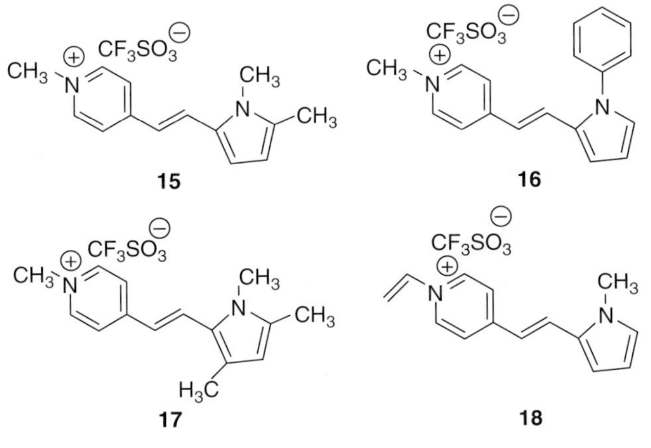

Figure 6.21 Heterocyclic push–pull TPA chromophores [52].

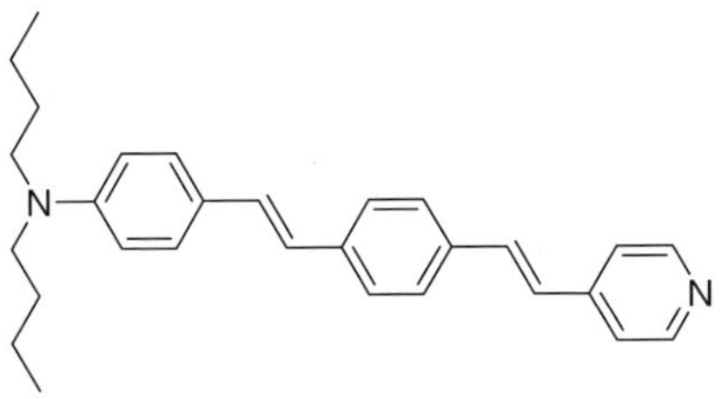

Figure 6.22 Chemical structure of DBASVP [54]. (Reproduced from Ref. [54] with permission from The Royal Society of Chemistry.)

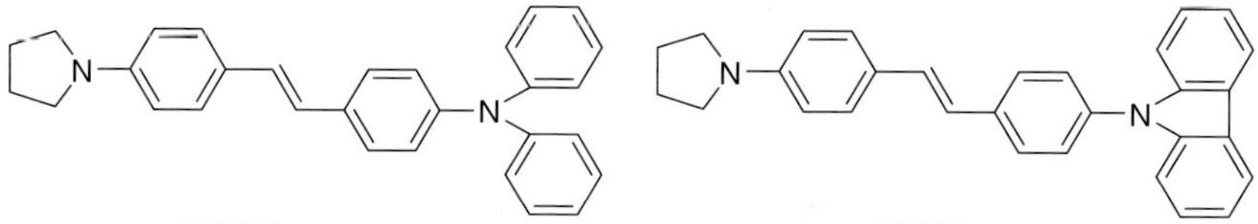

Figure 6.23 Chemical structure of DYPVA and PPVPA [55].

microstructure has been fabricated under irradiation at 800 nm using a 200-fs, 76-MHz Ti:sapphire femtosecond laser. This study proves that DBASVP is an effective TPP initiator.

Although Rumi *et al.* [31] have studied the structure–property relationship of some symmetrical TPA chromophores, in the case of asymmetrical compounds it is not well understood. To meet this requirement, Yan *et al.* [55] reported two new asymmetric, stilbene-type, free-radical TPP initiators, DYVPA and PPVPA, in which the pyrrolidino, diphenylamino, and carbazolyl groups are used as donors, which are found to be good electron donors (Figure 6.23). Linear absorption, one-photon fluorescence emission and their lifetimes, and two-photon fluorescence emission have been investigated. The calculated δ of DYVPA and PPVPA are 70.1 and 28.0 GM, respectively. Microfabrications via TPP were studied, and the possible photopolymerization mechanism was discussed in this study.

6.4 Application of Photoinitiators in Multiphoton Absorption Lithography

The most oftenly used TPP technique in the above-mentioned studies is radical polymerization, in which the radicals are first generated by TPA of photoinitiators, and then radical polymerization is initiated as normal. The photoinitiator plays a crucial role in the TPP process. Moreover, investigations have been made to improve the resolution. The resolution of TPP is influenced by the radicals produced at the focal spot of the laser as well as their diffusion in chain-growth radical polymerization. In the development of the fabrication technique, decreasing laser power and exposure time and using a large numerical aperture (NA) objective lens are very useful in confining the TPA region. By optimizing the TPP experimental conditions, nanostructures with high resolution have been achieved.

The influence of LSR and AR in TPP has been investigated [56]. Theoretical analysis based on the distribution of light intensity in this setup indicates that the LSR can be improved. An LSR of 50 nm and an AR of 1.38 were achieved for photocured polymer lines on the surface of a substrate by using the scanning mode. The asymmetric shrinkage of the voxel in the axial and lateral directions plays a significant role in obtaining features of low AR, which is a critical requirement for construction of micro/nanodevices by TPP. In TPP, the center of the focus was set on the surface of the substrate, which led to the cured portion of the resin fixed on the substrate surface to avoid shrinkage in the lateral direction. Figure 6.24a shows

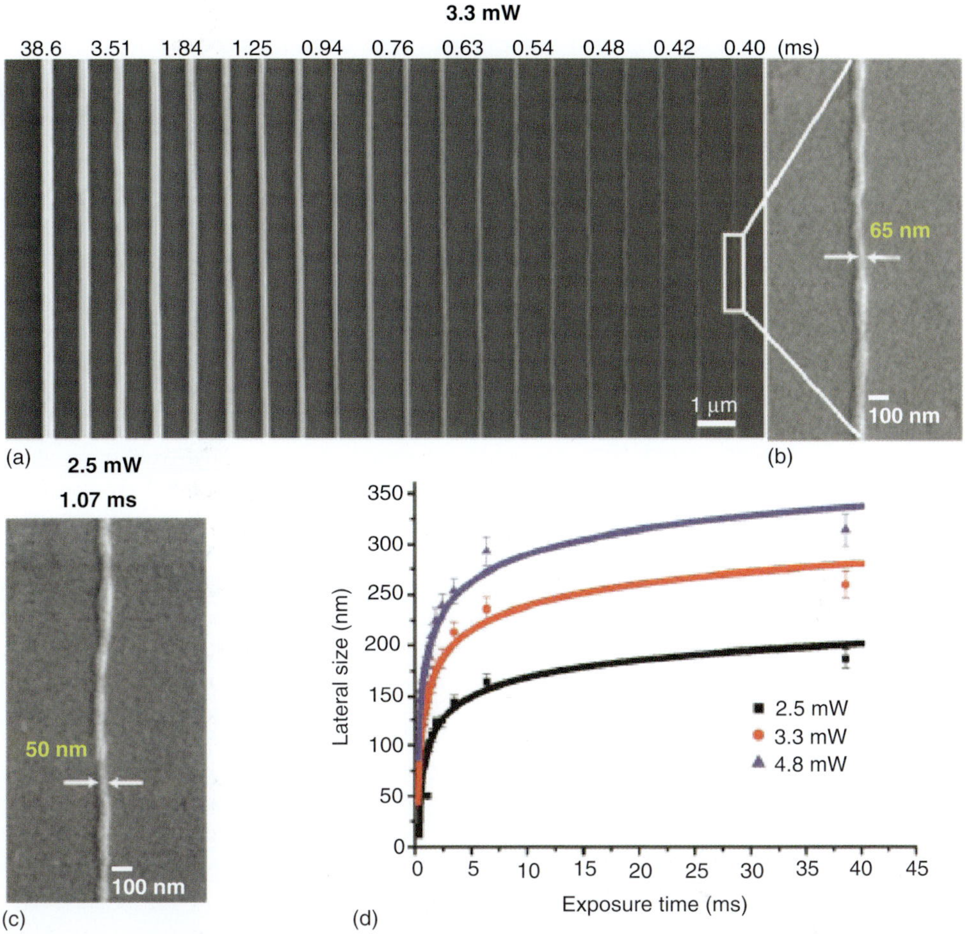

Figure 6.24 (a) SEM images of photocured polymer lines obtained using a laser power of 3.3 mW after various exposure times. (b) Enlarged image of a line fabricated with a laser power of 3.3 mW and an exposure time of 0.4 ms. (c) Polymer line of width 50 nm. (d) LSR versus exposure time under different laser powers (lines represent the calculated results). (Dong 2008 [56]. Reproduced with permission of American Institute of Physics.)

an example of a scanning electron microscopy (SEM) image of polymer lines cured by 3.3 mW laser power after various exposure times. The LSR is improved to 65 nm when the exposure time is reduced to 0.40 ms (Figure 6.24b). An LSR of 50 nm was achieved by using a laser power of 2.5 mW with an exposure time of 1.07 ms (Figure 6.24c). The results of the LSR as a function of exposure time under different laser powers are summarized in Figure 6.24d.

Recently, reducing the radical diffusion region in the polymerization process has been found to be a promising route to improving the resolution of TPP. The deactivation of radicals in the polymerization of photoresists mainly comes from radical termination by the combination of radicals and quenching molecules. TPP regions were well confined by the employment of the radical quenchers to a resin SCR500 [57, 58]. In that case, the radical quencher could deactivate the photoinduced radical around the focal spot, which resulted in chain reaction termination and prevented polymer growth. However, compared to the amount of photoinitiator used, in this reported approach a large amount of radical quencher was needed to obtain a significant confining effect.

A new method for increasing the resolution of direct femtosecond laser writing by TPP based on quencher diffusion has been proposed. Farsari *et al.* [59] have designed photomonomers by considering the induction of the quenching moiety in the molecular design. The material used for the fabrication of the 3D structures is an organic–inorganic hybrid, while the quencher is a photopolymerizable amine-based monomer that is bound to the polymer backbone upon fabrication of the structures [60]. It was produced by the addition of zirconium propoxide (Figure 6.25a, ZPO, 70% in propanol) to methacryloxypropyl trimethoxysilane (Figure 6.25b, MAPTMS). The monomer 2-(dimethylamino)ethyl methacrylate (Figure 6.25c, DMAEMA) was added as a quencher. MAPTMS, methacrylic acid (Figure 6.25d, MAA), and DMAEMA were used as the organic photopolymerizable monomers, while ZPO and the alkoxysilane groups of MAPTMS served as the inorganic network-forming moieties. Michler's ketone, 4,4-bis(diethylamino) benzophenone (Figure 6.25e, BIS), was used as the photoinitiator.

Correspondingly, the TPP technique was called "diffusion-assisted high-resolution direct femtosecond laser writing". This method relies on the

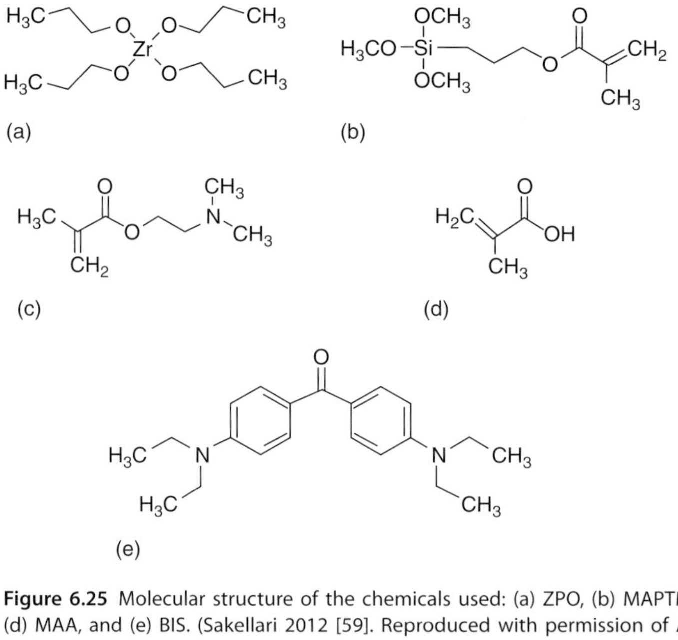

Figure 6.25 Molecular structure of the chemicals used: (a) ZPO, (b) MAPTMS, (c) DMAEMA, (d) MAA, and (e) BIS. (Sakellari 2012 [59]. Reproduced with permission of American Chemical Society.)

combination of a mobile quenching molecule with a low laser scanning speed, allowing the diffusion of the quencher in the scanned area and the depletion of the multiphoton generated radicals. By using diffusion-assisted high-resolution direct femtosecond laser writing technique, woodpile structures with a 400-nm intralayer period have been successfully achieved (Figure 6.26). This is comparable to the results produced by direct laser writing based on stimulated-emission–depletion microscopy, and the method is considered today as state of the art in 3D structure fabrication. These woodpile structures exhibited well-ordered diffraction patterns and stopgaps down to near-infrared wavelengths during optical characterization. Through simulations of the quencher

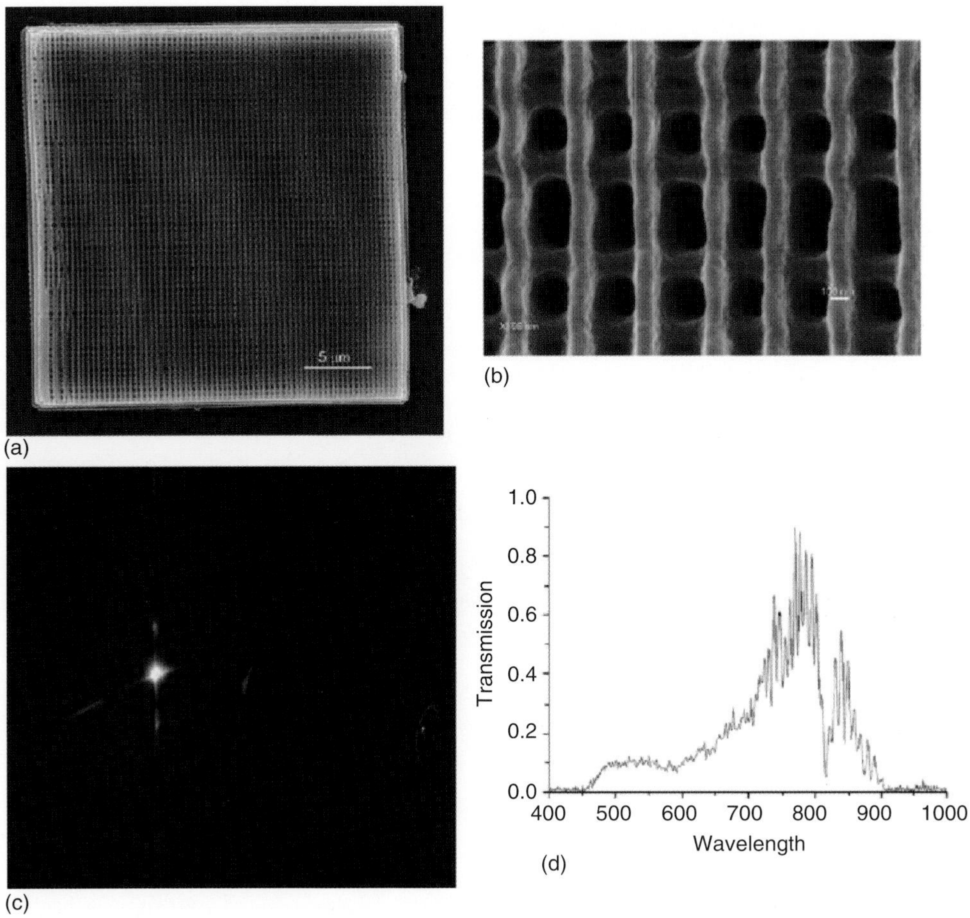

Figure 6.26 3D woodpile structure with 400 nm interlayer periodicity: (a) the whole structure, (b) magnified structure, (c) diffraction pattern generated when a white-light beam goes through the woodpile structure, and (d) measured bandgap. (Sakellari 2012 [59]. Reproduced with permission of American Chemical Society.)

diffusion, the authors showed that radical inhibition was responsible for the increased resolution.

Different from the diffusion-assisted high-resolution direct femtosecond laser writing mentioned above, a voxel tuning method using a radical quencher was proposed to increase the resolution by Yang *et al.* [58] Together with the photoinitiator P-Flu-TP2 (molecular structure shown in Figure 6.27), the radical quencher DBMP was added to the photoresist for TPP. The quenching effect as a function of the amount of radical quencher was experimentally investigated. Employing the proposed method, the lateral resolution of the line patterns was improved almost to 100 nm.

If the quenching moiety is connected in the photoinitiator molecule, the resolution for TPP would be further improved. Photoinitiators with the ability of radical diffusion control are desirable in TPP since the development of the TPP technique easily reaches its limit. Thus, the molecules combined with a high-efficiency TPP photoinitiator and radical quenching group can be expected to play an important role in TPP microfabrication with high resolution by confining radical diffusion by deactivating radicals in a tiny region. Duan *et al.* [61] have designed and synthesized a C_{2v} symmetric molecule, 3,6-bis[2-(4-nitrophenyl)-ethynyl]-9-(4-methoxy-benzyl)-carbazole (BNMBC, Figure 6.28) in which an

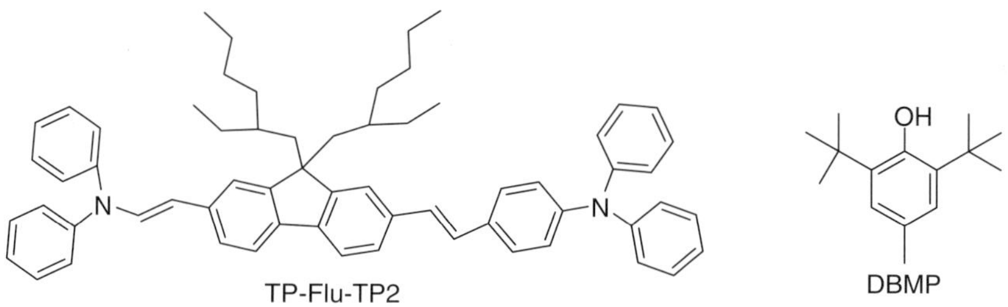

Figure 6.27 Molecular structure of the photoinitiator TP-Flu-TP2 and the radical quencher DBMP [58]. (With permission from Ref. [58].)

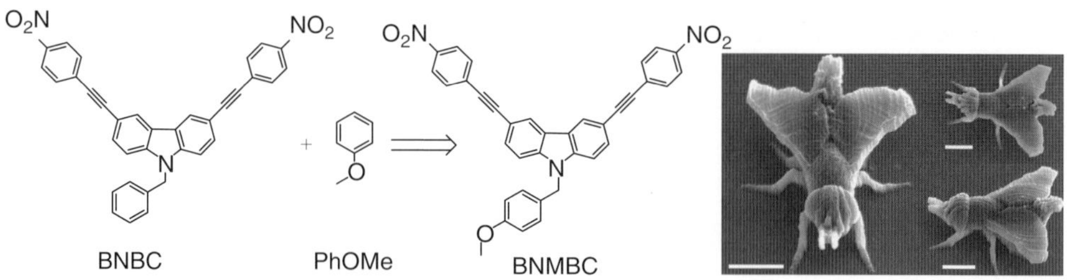

Figure 6.28 Chemical structure of photoinitiators BNBC and BNMBC, and the SEM images of a fly fabricated with a laser power of 7.70 mW and a scanning speed of 66 μm s^{-1} on the photoresist with BNMBC as photoinitiator. (Lu 2011 [61]. Reproduced with permission of The Royal Society of Chemistry.)

intramolecular charge transfer (ICT) conjugate system was introduced for high δ and a π-methoxybenzyl group for radical quenching. The authors investigated the optical properties and photoinitiating properties of BNMBC in comparison to its analogous 3,6-bis-[2-(4-nitrophenyl)-ethynyl]-9-benzyl-carbazole (BNBC, Figure 6.28), which was reported as a highly efficient TPP photoinitiator [7]. The evaluation with Z-scan measurement revealed that BNMBC had a large δ of 2367 GM, which was similar to that of BNBC.

TPP fabrication experiments showed that the volumes of fabricated suspended fibers were significantly decreased when BNMBC was used as photoinitiator compared to the comparable photoinitiating systems studied here. It was indicated that the introduction of the π-methoxybenzyl group into BNMBC led to a radical quenching effect in TPP. Meanwhile, a fine 3D microstructure with high TPP resolution was fabricated using the photoresist with BNMBC as the photoinitiator (Figure 6.28). The TPP fabrication exhibited that BNMBC possessed high TPP initiating efficiency and effective radical quenching effect. The volumes of polymer fibers fabricated by TPP of the photoresist using BNMBC as the photoinitiator were decreased to 20–30% of those using BNBC as the photoinitiator. The introduction of a radical quenching group into photoinitiator exhibited an effective confining effect of radical diffusion compared to the addition of same molar ratio of the radical quencher.

Furthermore, a mechanism of the confining effect on radical diffusion in TPP was proposed for a better understanding of the effect with the different photoinitiating systems. A schematic mechanism is illustrated in Figure 6.29. The density of two-photon-induced radicals was distributed depending on the intensity of the laser focused beam when the photoresist was exposed to light. Then, the radicals initiated the TPP to form the polymeric 3D networks via the cross-linking reaction between monomers and cross-linkers, which led to cured polymer formation. In the normal photoinitiating system such as that used in BNBC, the radicals were terminated by their combination during the chain-growth reactions after some average distance of radical diffusion. Once the radical quencher PhOMe was independently dispersed in the photoresist with BNBC, the radical diffusion distances would be limited as a result of the deactivation of radicals by quenchers, resulting in the resolution improvement of TPIP. Furthermore, when the photoinitiator BNMBC with the quenching moiety was introduced into photoresist, higher efficiencies of radicals quenching and chain termination could be obtained because of the higher chance of collision between radicals and the quenching group of BNMBC in shorter distances than when using BNBC/PhOMe, leading to higher TPP resolution. As a result, the smallest voxel size was obtained for the photoresist with BNMBC as photoinitiator out of the three photoinitiating systems studied here. However, in the photoinitiating system BNBC/PhOMe, the authors point out that this kind of photoinitiator with radical quencher should benefit the fabrication of precise structures.

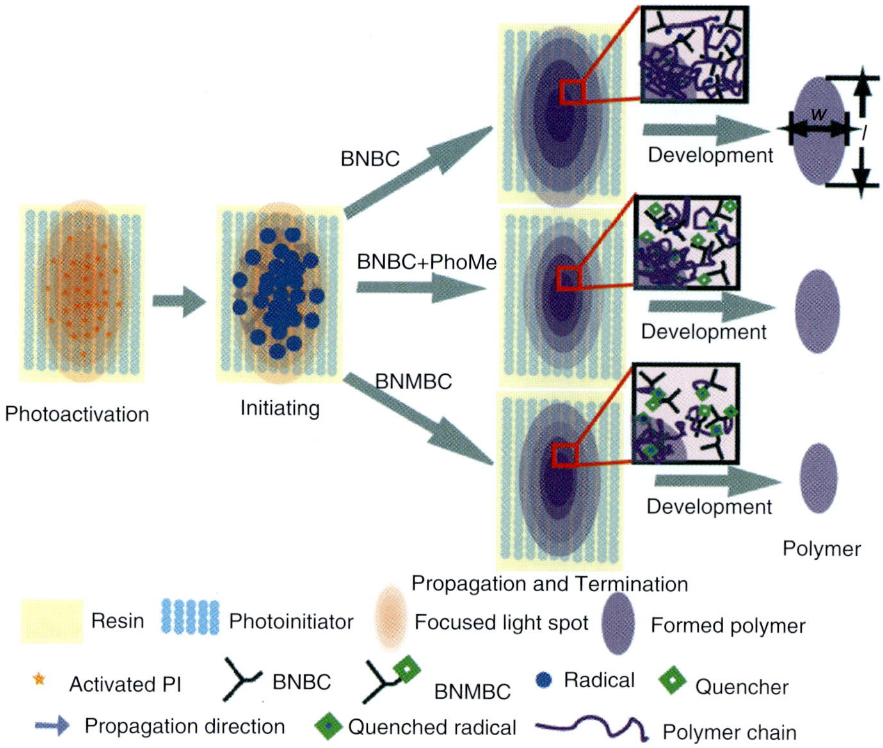

Figure 6.29 Schematic illustration of the mechanism on confining radical diffusion in TPP using different photoinitiators: BNBC, BNBC/PhOMe (molar ratio 1 : 1), and BNMBC [61]. (Lu 2011 [61]. Reproduced with permission of The Royal Society of Chemistry.)

6.5 Conclusion

In summary, the study of the photoinitiators for MAL has been briefly introduced in this section. The photoinitiators were introduced according to the classification of molecular symmetry. The two-photon absorption cross section is one of the most important factors determining the potential for the application as photoinitiators in multiphoton nanofabrication. Therefore, the molecular design, the structure–property dependence, the optical property, and the two-photon absorption property of some typical photoinitiators, focused on improving the two-photon absorption cross section, have been reviewed. Generally, photoinitiators with large two-photon absorption cross section, good solubility, and high initiating efficiency are preferred for MAL. Finally, the application of photoinitiators in MAL was discussed. The investigations of photoinitiators would provide good impetus for further development of MAL.

Acknowledgment

We would like to thank students and staff who provided us with the data shown in this chapter. We are grateful to the National Natural Science Foundation of China (Grant Nos. 91123032, 91323301, 61475164, and 61205194), the National Basic Research Program of China (2010CB934103), the International Cooperation Program MOST (2010DFA01180), and the CAS-JSPS Joint Research Project (GJHZ1411) for financial support.

References

1. Wang, W.-K., Sun, Z.-B., Zheng, M.-L., Dong, X.-Z., Zhao, Z.-S., and Duan, X.-M. (2011) *J. Phys. Chem. C*, **115** (22), 11275–11281.
2. Sun, Z.B., Dong, X.Z., Chen, W.Q., Nakanishi, S., Duan, X.M., and Kawata, S. (2008) *Adv. Mater.*, **20** (5), 914–919.
3. Kawata, S., Sun, H.-B., Tanaka, T., and Takada, K. (2001) *Nature*, **412** (6848), 697–698.
4. Kumi, G., Yanez, C.O., Belfield, K.D., and Fourkas, J.T. (2010) *Lab Chip*, **10** (8), 1057–1060.
5. Li, Z., Stankevičius, E., Ajami, A., Račiukaitis, G., Husinsky, W., Ovsianikov, A., Stampfl, J., and Liska, R. (2013) *Chem. Commun.*, **49** (69), 7635–7637.
6. Ovsianikov, A., Li, Z., Torgersen, J., Stampfl, J., and Liska, R. (2012) *Adv. Funct. Mater.*, **22** (16), 3429–3433.
7. Xing, J.-F., Chen, W.-Q., Gu, J., Dong, X.-Z., Takeyasu, N., Tanaka, T., Duan, X.-M., and Kawata, S. (2007) *J. Mater. Chem.*, **17** (14), 1433–1438.
8. Denk, W., Strickler, J.H., and Webb, W.W. (1990) *Science*, **248** (4951), 73–76.
9. Pawlicki, M., Collins, H.A., Denning, R.G., and Anderson, H.L. (2009) *Angew. Chem. Int. Ed.*, **48** (18), 3244–3266.
10. Sun, H.-B., Matsuo, S., and Misawa, H. (1999) *Appl. Phys. Lett.*, **74** (6), 786–788.
11. Nguyen, L.H., Straub, M., and Gu, M. (2005) *Adv. Funct. Mater.*, **15** (2), 209–216.
12. Wu, S., Serbin, J., and Gu, M. (2006) *J. Photochem. Photobiol A: Chem.*, **181**, 1–11.
13. Lafratta, C.N., Fourkas, J.T., Baldacchini, T., and Farrer, R.A. (2007) *Angew. Chem. Int. Ed.*, **46** (33), 6238–6258.
14. Lee, K.-S., Kim, R.H., Yang, D.-Y., and Park, S.H. (2008) *Prog. Polym. Sci.*, **33** (6), 631–681.
15. Rumi, M. and Perry, J.W. (2010) *Adv. Opt. Photonics*, **2**, 451–518.
16. Boggess, T.F., Bohnert, K.M., Moss, S.C., Boyd, I.W., and Smirl, A.L. (1986) *IEEE J. Quantum Electron.*, **22**, 360–368.
17. Sheik-Bahae, M., Said, A.A., Wei, T.-H., Hagan, D.J., and Stryland, E.W.V. (1990) *IEEE J. Quantum Electron.*, **26**, 760–769.
18. Xu, C. and Webb, W.W. (1996) *J. Opt. Soc. Am. B*, **13**, 481–491.
19. Tutt, L.W. and Bogges, T.F. (1993) *Prog. Quantum Electron.*, **17**, 299–338.
20. Lin, T.-C., He, G.S., Zheng, Q., and Prasad, P.N. (2006) *J. Mater. Chem.*, **16** (25), 2490–2498.
21. Kamada, K., Matsunaga, K., Yoshino, A., and Ohta, K. (2003) *J. Opt. Soc. Am. B*, **20**, 529–537.
22. Makarov, N.S., Campo, J., Hales, J.M., and Perry, J.W. (2011) *Opt. Mater. Express*, **1**, 551–563.
23. Li, Z., Pucher, N., Cicha, K., Torgersen, J., Ligon, S.C., Ajami, A., Husinsky, W., Rosspeintner, A., Vauthey, E., Naumov, S., Scherzer, T., Stampfl, J., and Liska, R. (2013) *Macromolecules*, **46**, 352–361.
24. Gu, J., Yulan, W., Chen, W.-Q., Dong, X.-Z., Duan, X.-M., and Kawata, S. (2007) *New J. Chem.*, **31** (1), 63–68.
25. Drobizhev, M., Stepanenko, Y., Dzenis, Y., Karotki, A., Rebane, A., Taylor, P.N., and Anderson, H.L. (2005) *J. Phys. Chem. B*, **109**, 7223–7236.

26. He, G.S., Tan, L.-S., Zheng, Q., and Prasad, P.N. (2008) *Chem. Rev.*, **108** (4), 1245–1330.
27. Kuebler, S.M., Braun, K.L., Zhou, W., Cammack, J.K., Yu, T., Ober, C.K., Marder, S.R., and Perry, J.W. (2003) *J. Photochem. Photobiol. A: Chem.*, **158** (2), 163–170.
28. Zhao, Y., Li, X., Wu, F., and Fang, X. (2006) *J. Photochem. Photobiol. A: Chem.*, **177** (1), 12–16.
29. Belfield, K.D., Ren, X., Van Stryland, E.W., Hagan, D.J., Dubikovsky, V., and Miesak, E.J. (2000) *J. Am. Chem. Soc.*, **122** (6), 1217–1218.
30. Cumpston, B.H., Ananthavel, S.P., Barlow, S., Dyer, D.L., Ehrlich, J.E., Erskine, L.L., Heikal, A.A., Kuebler, S.M., Lee, I.-Y.S., Mccord-Maughon, D., Qin, J., Röckel, H., Rumi, M., Wu, X.L., Marder, S.R., and Perry, J.W. (1999) *Nature*, **398** (6722), 51–54.
31. Rumi, M., Ehrlich, J.E., Heikal, A.A., Perry, J.W., Barlow, S., Hu, Z., Mccord-Maughon, D., Parker, T.C., Röckel, H., Thayumanavan, S., Marder, S.R., Beljoinne, D., and Bredas, J.-L. (2000) *J. Am. Chem. Soc.*, **122** (39), 9500–9510.
32. Albota, M., Beljonne, D., Brédas, J.-L., Ehrlich, J.E., Fu, J.-Y., Heikal, A.A., Hess, S.E., Kogej, T., Levin, M.D., and Marder, S.R. (1998) *Science*, **281** (5383), 1653–1656.
33. Zheng, M.-L., Fujita, K., Chen, W.-Q., Smith, N.I., Duan, X.-M., and Kawata, S. (2011) *ChemBioChem*, **12** (1), 52–55.
34. Zheng, Y.-C., Zheng, M.-L., Chen, S., Zhao, Z.-S., and Duan, X.-M. (2014) *J. Mater. Chem. B*, **2** (16), 2301–2310.
35. Kang, R.H., Zheng, M.L., Chen, W.Q., Zhao, Z.S., and Duan, X.M. (2012) *RSC Adv.*, **2** (28), 10478–10480.
36. Zhang, X., Yu, X., Sun, Y., He, W., Wu, Y., Feng, Y., Tao, X., and Jiang, M. (2006) *Opt. Mater.*, **28** (12), 1366–1371.
37. Xing, J.-F., Zheng, M.-L., Chen, W.-Q., Dong, X.-Z., Takeyasu, N., Tanaka, T., Zhao, Z.-S., Duan, X.-M., and Kawata, S. (2012) *Phys. Chem. Chem. Phys.*, **14** (45), 15785–15792.
38. Xing, J.-F., Chen, W.-Q., Dong, X.-Z., Tanaka, T., Fang, X.-Y., Duan, X.-M., and Kawata, S. (2007) *J. Photochem. Photobiol. A: Chem.*, **189** (2), 398–404.
39. Xing, J.-F., Dong, X.-Z., Chen, W.-Q., Duan, X.-M., Takeyasu, N., Tanaka, T., and Kawata, S. (2007) *Appl. Phys. Lett.*, **90** (13), 131106.
40. Cao, X., Jin, F., Li, Y.-F., Chen, W.-Q., Duan, X.-M., and Yang, L.-M. (2009) *New J. Chem.*, **33** (7), 1578–1582.
41. Chung, S.-J., Rumi, M., Alain, V., Barlow, S., Perry, J.W., and Marder, S.R. (2005) *J. Am. Chem. Soc.*, **127** (31), 10844–10845.
42. Li, Z., Torgersen, J., Ajami, A., Muhleder, S., Qin, X., Husinsky, W., Holnthoner, W., Ovsianikov, A., Stampfl, J., and Liska, R. (2013) *RSC Adv.*, **3** (36), 15939–15946.
43. Ovsianikov, A., MüHleder, S., Torgersen, J., Li, Z., Qin, X.-H., Van Vlierberghe, S., Dubruel, P., Holnthoner, W., Redl, H., and Liska, R. (2013) *Langmuir*, **30** (13), 3787–3794.
44. Schafer, K.J., Hales, J.M., Balu, M., Belfield, K.D., Van Stryland, E.W., and Hagan, D.J. (2004) *J. Photochem. Photobiol. A: Chem.*, **162** (2), 497–502.
45. Xing, J., Liu, J., Zhang, T., Zhang, L., Zheng, M.-L., and Duan, X.-M. (2014) *J. Mater. Chem. B*, **2**, 4318–4323.
46. Wang, X., Jin, F., Zhang, W., Tao, X., Duan, X.-M., and Jiang, M. (2011) *Dyes Pigm.*, **88** (1), 57–64.
47. Wang, X., Jin, F., Chen, Z., Liu, S., Wang, X., Duan, X., Tao, X., and Jiang, M. (2011) *J. Phys. Chem. C*, **115** (3), 776–784.
48. Zhou, W., Kuebler, S.M., Braun, K.L., Yu, T., Cammack, J.K., Ober, C.K., Perry, J.W., and Marder, S.R. (2002) *Science*, **296** (5570), 1106–1109.
49. Reinhardt, B.A., Brott, L.L., Clarson, S.J., Dillard, A.G., Bhatt, J.C., Kannan, R., Yuan, L., He, G.S., and Prasad, P.N. (1998) *Chem. Mater.*, **10** (7), 1863–1874.
50. Kannan, R., He, G.S., Yuan, L., Xu, F., Prasad, P.N., Dombroskie, A.G., Reinhardt, B.A., Baur, J.W., Vaia, R.A., and Tan, L.-S. (2001) *Chem. Mater.*, **13** (5), 1896–1904.
51. Kim, O.-K., Lee, K.-S., Woo, H.Y., Kim, K.-S., He, G.S., Swiatkiewicz, J., and Prasad, P.N. (2000) *Chem. Mater.*, **12** (2), 284–286.

52. Abbotto, A., Beverina, L., Bozio, R., Bradamante, S., Ferrante, C., Pagani, G.A., and Signorini, R. (2000) *Adv. Mater.*, **12** (24), 1963–1967.
53. Abbotto, A., Beverina, L., Bozio, R., Bradamante, S., Pagani, G., and Signorini, R. (2001) *Synth. Met.*, **121** (1-3), 1755–1756.
54. Ren, Y., Yu, X.-Q., Zhang, D.-J., Wang, D., Zhang, M.-L., Xu, G.-B., Zhao, X., Tian, Y.-P., Shao, Z.-S., and Jiang, M.-H. (2002) *J. Mater. Chem.*, **12** (12), 3431–3437.
55. Yan, Y.X., Tao, X.T., Sun, Y.H., Xu, G.B., Wang, C.K., Yang, J.X., Zhao, X., Wu, Y.Z., Ren, Y., and Jiang, M.H. (2005) *Opt. Mater.*, **27** (12), 1787–1792.
56. Dong, X.-Z., Zhao, Z.-S., and Duan, X.-M. (2008) *Appl. Phys. Lett.*, **92** (9), 091113.
57. Takada, K., Sun, H.-B., and Kawata, S. (2005) *Appl. Phys. Lett.*, **86** (7), 071122.
58. Park, S.H., Lim, T.W., Yang, D.-Y., Kim, R.H., and Lee, K.-S. (2006) *Macromol. Res.*, **14** (5), 559–564.
59. Sakellari, I., Kabouraki, E., Gray, D., Purlys, V., Fotakis, C., Pikulin, A., Bityurin, N., Vamvakaki, M., and Farsari, M. (2012) *ACS Nano*, **6** (3), 2302–2311.
60. Ovsianikov, A., Viertl, J., Chichkov, B., Oubaha, M., Maccraith, B., Sakellari, I., Giakoumaki, A., Gray, D., Vamvakaki, M., and Farsari, M. (2008) *ACS Nano*, **2** (11), 2257–2262.
61. Lu, W.-E., Dong, X.-Z., Chen, W.-Q., Zhao, Z.-S., and Duan, X.-M. (2011) *J. Mater. Chem.*, **21** (15), 5650–5659.

7
Hybrid Materials for Multiphoton Polymerization

Alexandros Selimis and Maria Farsari

7.1
Introduction

Multiphoton lithography (MPL) based on multiphoton polymerization (MPP) of photosensitive materials is a printing technique that enables the fabrication of three-dimensional (3D) structures with subwavelength resolution [1–5]. The polymerization is based on the phenomenon of multiphoton absorption. When an ultrafast laser beam is tightly focused into the volume of a transparent photosensitive material, the polymerization process can be initiated by nonlinear absorption within the focal volume. By moving the laser focus three dimensionally through the material, 3D structures can be fabricated. The technique has been implemented with a variety of acrylate and epoxy materials, and several components and devices have been fabricated such as photonic crystal templates [6], mechanical devices [7], and microscopic models [8].

A decisive parameter for the accurate fabrication of structures by MPL, besides the irradiation one, is the choice of the structuring material. Indeed, it is the material properties that ineluctably and crucially determine the success of the printing process, the resolution of the laser-fabricated structures, and the potentiality of their bulk and surface functionalization. The majority of the applications involve the use of commercially available photoresists. There have been examples of both negative and positive photoresists used [1, 3, 9], with the final fabricated structure telling them apart. In the case of negative photoresists, the two-photon exposure results in the cross-linking of the polymeric chains, making the exposed area insoluble in the solvent used to remove the nonpolymerized material. In this way, the structures to be fabricated in the sample are directly written. On the other hand, in the case of positive photoresists, the two-photon absorption leads to the cleavage of the polymeric chains, which are no longer diffused by the development solvent. That is to say, it is the reverse structure that is written on the sample. In any case, negative photopolymers are the most commonly used. However, besides the good structural results that these photosensitive ensure, they exhibit limited

Multiphoton Lithography: Techniques, Materials and Applications, First Edition.
Edited by Jürgen Stampfl, Robert Liska, and Aleksandr Ovsianikov.

flexibility concerning the bulk and surface functionalization of the laser-fabricated structures.

In order to overcome this obstacle, a great deal of research has recently focused on the development of organic–inorganic hybrid materials synthesized via the sol–gel route tailor-made for MPP [10]. Sol–gel technology provides a very powerful tool for the development of hybrid compounds [11] relying on the incorporation of inorganic compounds into organic molecules. These hybrid materials benefit from straightforward preparation, modification, and processing. In parallel, they are characterized by high optical quality, as well as post-processing chemical and electrochemical inertness. These features, along with their good mechanical and chemical stability, have set off the organic–inorganic hybrid materials' involvement in photonics, metamaterials, microelectromechanical systems (MEMS), microfluidics, and biomedical applications.

Here, we describe the recent developments in the technology of hybrid materials for MPP microfabrication, and the recent progress in the surface and bulk functionalization of the 3D fabricated structures; also, we discuss the applications of the technology.

7.2 Sol–Gel Preparation

The first materials employed in MPP were acrylic photopolymers and the negative photoresist SU8 [12]. Over the last few years, research on MPP materials has focused on photosensitive hybrid materials [13], such as the commercially available ORMOCER® (organically modified ceramic microresist) [14–16]. The sol–gel process is based on the phase transformation of a sol of metallic oxide or an alkoxide precursor to form a wet gel. A photosensitive sol–gel process usually involves the catalytic hydrolysis of the sol–gel precursors and the polycondensation of the hydrolyzed products and other sol–gel active components present in the reaction medium to form a macromolecular hybrid network structure. The formed gel is subsequently reacted through photopolymerization to give a product similar to glass.

Three-dimensional structure fabrication by MPP generally involves a four-step process, described in Figure 7.1. The first one is hydrolysis and condensation, in which precursors or monomers such as metal oxides or metal alkoxides are mixed

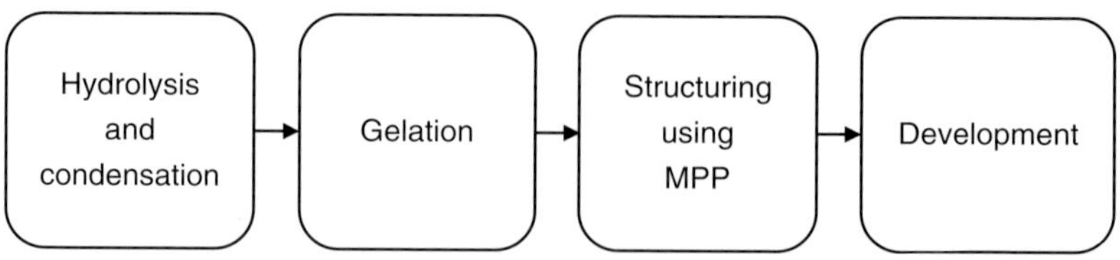

Figure 7.1 Process flow of a photosensitive sol–gel.

with water, which then undergo hydrolysis and condensation to form a porous interconnected cluster structure (Scheme 7.1). Either an acid such as HCl or a base like NH_3 can be employed as the catalyst.

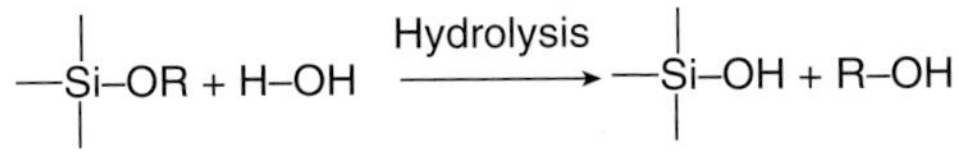

Scheme 7.1

The second step is gelation, where the solvent is removed and a gel is formed by heating at low temperature (Scheme 7.2), whereby a significant volume loss occurs.

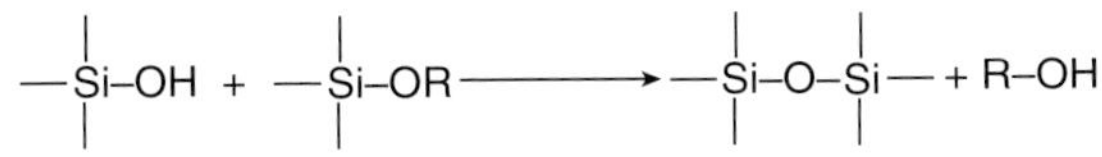

Scheme 7.2

The third step is photopolymerization. Provided a photoinitiator is present in the gel, the photoinduced radicals will cause the polymerization of the unsaturated moieties only in the area where the radicals are produced. At this step, there is no material removal and no volume loss.

The last step is the development, in which the sample is immersed in an appropriate solvent and the area of the sample that is not photopolymerized is removed.

7.3
Silicate Hybrid Materials

Silicate-based photopolymers have proven to be a very popular choice in microfabrication using MPP [14, 17–19] mainly due to the high reactivity they exhibit. They combine the properties of silicate glasses such as hardness, chemical and thermal stability, and optical transparency with the low-temperature laser processing of organic polymers – properties impossible to achieve with just inorganic or pure polymeric materials. The most widely used hybrid material is the commercially available photopolymer ORMOCER. This material comprises an inorganic (–Si–O–Si–) backbone which can be functionalized with a range of organic functionalities. ORMOCERs are often used as dental composites [20–23], as electrolytes for lithium batteries [24, 25], as membrane materials for fuel cell applications [26, 27], and in optical interconnect technology [28, 29]. The ORMOCER widely used in two-photon polymerization (2PP) was first developed by Ruth Houbertz [14, 15, 30, 31] and was used to fabricate photonic crystal woodpile structures, like the one shown in Figure 7.2 [32], proving its successful structuring ability.

The good mechanical, optical, and processing properties that ORMOCER possesses, combined with the fact that it is commercially available, make it a very

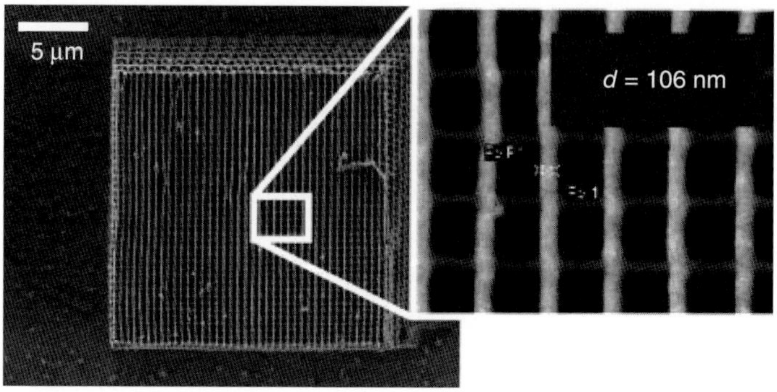

Figure 7.2 ORMOCER woodpile structures fabricated using MPP. (Burmeister 2012 [32].)

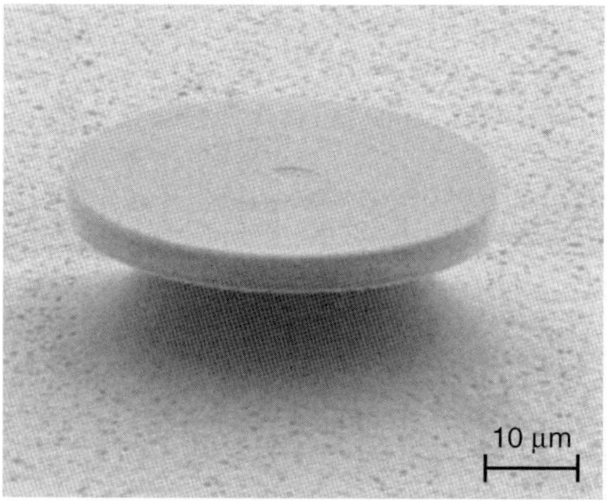

Figure 7.3 Microdisk fabricated using ORMOCER. (Grossmann 2011 [36].)

attractive option for the fabrication of photonic crystal structures and devices. In particular, it has been widely utilized for waveguide-coupled woodpile structures to investigate superprism phenomena [33, 34] and for the fabrication of diffractive optical elements [35] and optically active microdisks (Figure 7.3). It has also been used to fabricate woodpile structures and infiltrate them with quantum dots [37]. In addition, it has been shown that the maximum resolution of the laser-fabricated structures can be increased by post-treating ORMOCER structures with heat [38]. More recently, a low-threshold organic-semiconductor distributed-feedback laser was fabricated based on a surface grating structured in an ORMOCER material using MPP [39].

The commercially available ORMOCER is a viscous liquid that does not require any pretreatment before structuring, which makes its handling straightforward and easy. However, it also has certain drawbacks. As the liquid transforms into a solid during photopolymerization, shrinkage occurs [40], which results in the distortion of the photopolymerized structure and deviation from the original design. This drawback of laser-induced fabrication can be corrected using a compensation factor [41].

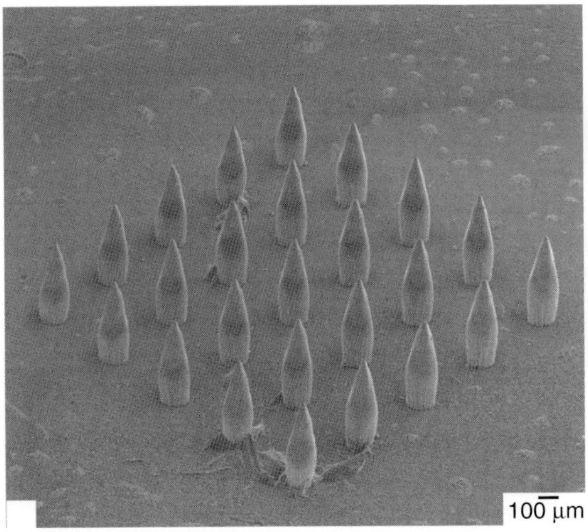

Figure 7.4 Micro-needles fabricated using 2PP. (Gittard 2009 [46]. Reproduced with permission of Institute of Physics.)

The biocompatibility and the chemical inertness that ORMOCER exhibits over and above its already mentioned properties have widened the range of ORMOCER 3D microstructural applications to biomedical microdevices and medical implants [41–48]. For instance, Figure 7.4 shows a micro-needle array that can be used in drug delivery devices, enabling the transdermal delivery of different pharmacological substances. MPP allows not only the miniaturization of such devices but also the adoption of more than one needle design in a single array, thereby optimizing the effect of geometry on mechanical and puncturing ability.

7.4
Composite Hybrid Materials

While ORMOCER and other silicate-only-based hybrid materials have provided the possibility to fabricate high-resolution 3D structures with good optical properties, they do not allow the optimization and "fine-tuning" of the materials properties for specific applications. By contrast, the versatile chemistry of sol–gel composites enables the co-polymerization of more than one metal alkoxide; this has been shown to enhance the material's mechanical stability and endows it with modified optical properties [49–53].

In addition to the above-mentioned materials employed in the laser-induced fabrication by MPP, there are several examples of composite photosensitive hybrid materials used in MPL applications [4, 54–56]. Ovsianikov *et al.* showed that under specific fluence conditions, specific material combinations can be structured into complex 3D structures without shrinkage [57, 58]. Psycharakis *et al.* showed that doped silicates are biocompatible and can enhance cell

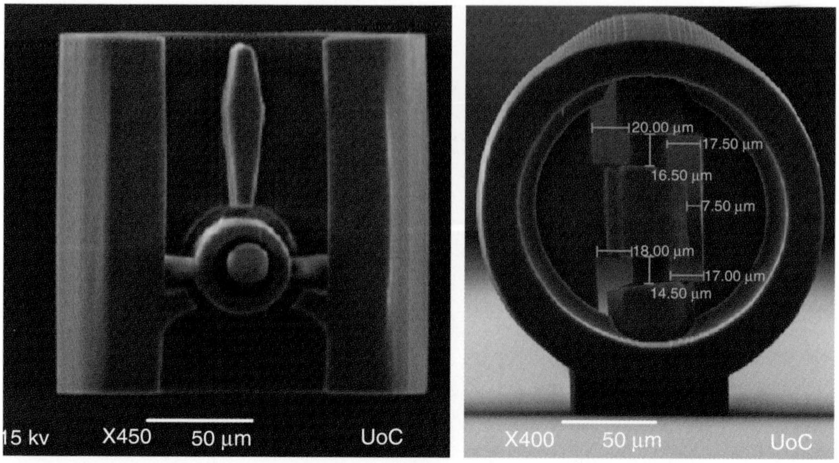

Figure 7.5 Microfluidic pump fabricated by MPL. (Galanopoulos 2014 [63].)

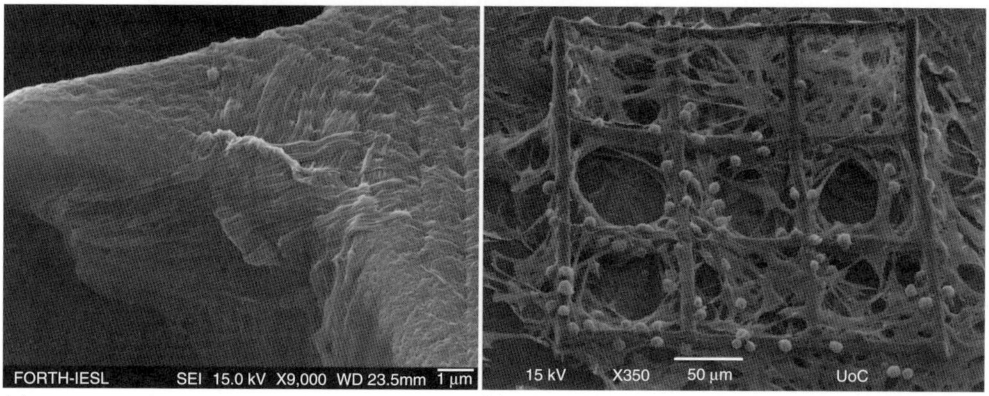

(a) (b)

Figure 7.6 (a) Mineralized 3D scaffold and (b) mineralized scaffold with osteoblastic cells. (Terzaki 2013 [66]. Reproduced with permission of Institute of Physics.)

growth [59]. A zirconium material was employed by Raimondi to fabricate niches for cell growth [60] and for the construction of microfluidic devices (Figure 7.5) [10, 61–63].

Different variants of zirconium–silicon composites have been investigated for their suitability for bone regeneration [64, 65]. Terzaki *et al.* decorated the surface of a zirconium hybrid with calcium-binding amyloid peptides, proposing a "scaffold-on-scaffold" hard tissue engineering strategy (Figure 7.6) [66].

In addition to metal alkoxides, hybrid material chemistry provides the possibility of the inclusion of functional groups, such as nonlinear optical molecules [47, 67], quantum dots [68], and metal-binding materials [69]. Recently, a zirconium silicate doped with a monomer containing amine moieties was used for the

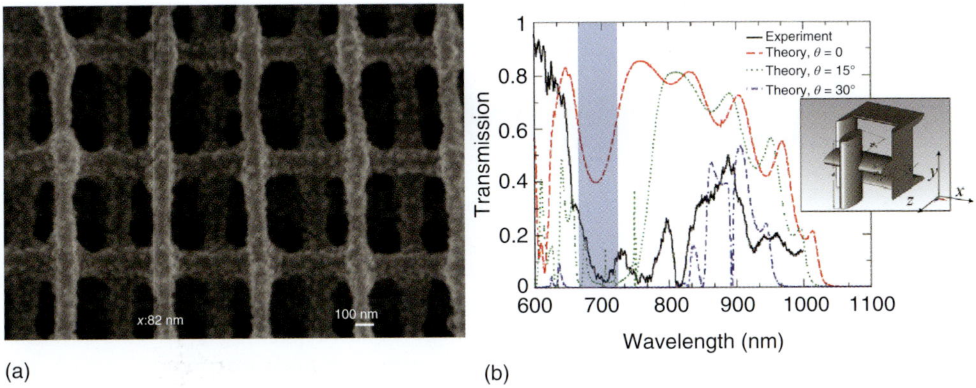

Figure 7.7 (a) Three-dimensional photonic crystal with sub-100 nm features. (b) Theoretical and experimental optical bandgaps depending on the incidence angle. (Vasilantonakis 2012 [70]. Reproduced with permission of Wiley.)

fabrication of 3D photonic crystals with optical bandgaps, for the first time, using MPL (Figure 7.7) [70].

Other material combinations investigated are silicon and titanium alkoxide [71] and silicon and vanadium alkoxide [72]. The latter one is particularly interesting, as it is the first and to date only example of a redox MPP. Here, one of the components, the vanadium organometallic complex, also acts as a photoinitiator. The composite employs multiphoton absorption to self-generate radicals by photoinduced reduction of the metal species from vanadium(V) to vanadium(IV).

7.5 Surface and Bulk Functionalization

As mentioned previously, the use of sol–gel chemistry for the production of hybrid organic–inorganic materials as photostructuring materials enables the bulk and surface functionalization of laser-fabricated 3D structures. In this section, we describe the potentiality of their selective metallization or mineralization. Indeed, recently, there has been an imperative request for the design and the fabrication of 3D metallic nanostructures for use in a great variety of applications extending from MEMS, plasmonics, and metamaterials to biosensors [2, 4, 73–75]. Regarding their fabrication, the traditional lithographic techniques involve the layer-by-layer structuring of metallic structures, which allows either the deposition of only a few layers or the creation of high-aspect-ratio two-dimensional structures. Therefore, the fabrication of complicated and fully 3D metallic structures of high resolution remains an open and challenging task. A potential concept that has been proposed toward this aim deals with the use of ultrafast lasers and employs the following three approaches: the multiphoton reduction of metal ions, MPL into a positive photoresist followed by the metal electrochemical deposition, and the MPL of a negative photoresist and the subsequent metallization of the laser-fabricated structures using electroless plating.

Among these approaches, the multiphoton reduction of metal ions is the only one that enables the fabrication of pure metallic and fully 3D nanostructures [76–79]. However, it is limited by the reduced transparency of the employed metal ion solutions at the commonly used laser wavelengths for fabrication (500–800 nm). A way to overcome this obstacle is to use diluted metal ion solutions, which leads to the fabrication of structures of, nevertheless, limited mechanical stability and integrity. On the other hand, MPL using a positive photoresist followed by electrochemical metal deposition is the most recent approach [9, 78]. It relies on the fabrication of hollow structures, which are subsequently filled by gold using the electrochemical deposition . The induced gold structures are characterized by excellent quality, uniform metal deposition, and high mechanical stability. Nonetheless, the gold deposition through the structuring holes is, at the same time, the Achilles heel of this approach since it is restricted to the fabrication of 2.5D structures. The third approach of laser-fabricated structures, namely metallization using electroless plating technique, subsequent to the MPL of negative photoresists, has proven to be the most attractive one to date [79–81]. The term *electroless plating* means that no electric current from an external source is employed for the metallic plating of the substrates. In general, this method is characterized by the reduction of metal ions selectively at the surface of a catalytic substrate that is immersed into an aqueous solution of metal ions with continued deposition on the substrate through the catalytic action of the deposit itself [82, 83]. In particular, either the use of negative photoresists is once more adopted, requiring an additional processing step for surface functionalization [84], or use is made of a metal-binding-site-doped photosensitive material [85]. The former method leads to very good resolution and structure integrity but of variable metallization quality since the density of the metal-binding sites on the structure cannot be controlled. On the other hand, the latter method ensures precise control of the density of the metal-binding sites and their distribution inside the photosensitive material and, thus, metallized structures of high quality. Recently, the selective metallization of a structure made of a zirconium-containing organic–inorganic photosensitive material doped with tertiary amine metal-binding moieties was reported [69].

In addition to the selective metallization of MPL-induced structures, the mineralization of laser-fabricated 3D scaffolds meets the fundamental needs of biological research [86–88]. Specifically, the mineralization of primary scaffolds made of an amine-containing photosensitive material has recently been reported [66]. These scaffolds are subsequently decorated with gold ions that immobilize self-assembled peptides, which specifically promote the deposition of calcium phosphate. The idea underlying this approach relies on the design of a bifunctional, self-assembling peptide that can be deposited on the gold-coated 3D scaffold acting as a secondary scaffold for calcium phosphate deposition. Yet, in the employment of surface-modified MPL-fabricated structures for biomedical applications, and besides the above-mentioned use of hybrid organic–inorganic materials, the commercially available ORMOCER cannot be excluded. For instance, ORMOCER was photo-structured following three-photon absorption upon irradiation at 1028 nm, and biotin was subsequently immobilized on the surface of the laser-fabricated structures by excimer laser photoactivation of photobiotin [48].

7.6 Replication

Nevertheless, it should be underlined that the typical writing speeds currently employed in laser structuring by MPP are limited to several millimeters per second. This results in low productivity of MPL and constitutes the main hindrance to the wider spread of MPL in cases concerning the realization of macroscopic structures such as, for example, those required for tissue engineering applications. An effective approach for improving the fabrication throughput relies on the combination of MPL with molding techniques [89, 90]. In particular, the time-consuming and expensive fabrication via MPL is required only once for the production of the master template, and a standard soft lithography replication method is used later on, which employs the thermopolymer PDMS (polydimethylsiloxane) [89, 91]. For instance, the improvement provided by this approach has been demonstrated in the case of the multiplication of identical 2.5D micro or nanostructured samples for cell growth experiments as well as for 2D optics for photonics and plasmonics. Indeed, the molding technique would speed up the production process dramatically.

In this way, a series of identical samples with a resolution of tens of nanometers can rapidly be produced as follows: The original laser-fabricated structure is used as a master for the production of the transparent PDMS elastomer mold, which is in turn used for stamping structures cured by ultraviolet polymerization [90, 92]. Most importantly, the stamping process can be replicated many times without any noticeable quality degradation. In addition, the stamp elasticity can be tuned by adjusting either the polymer/thermoinitiator or the curing time and temperature. Specifically, both a lower thermoinitiator concentration and temperature (the latter corresponding to longer curing times) make the mask softer and, thus, easier to peel off from the replicated structure; however, the resolution is affected. Moreover, it is worth mentioning that, in principle, PDMS can also be structured by MPL, enabling the direct fabrication of replica masks [92, 93]. Besides its structuring facility, PDMS is also known to be a biocompatible material that has been thoroughly studied for applications in micro-optics [94], microfluidics [95], and cell biology studies [96]. Finally, it is of great significance that by transferring the "inverted" structure, one can potentially use nontransparent materials containing electric or magnetic nanoparticles.

7.7
Conclusions

The potential of photostructurable hybrid materials for MPP fabrication has started to emerge. Their mechanical and chemical stability, combined with their superior MPP structuring ability and functionalization potential, makes them ideal candidates for a wide range of applications including photonics, metamaterials, and biomedical implants.

References

1. Sun, H.-B. and Kawata, S. (2004) in *NMR. 3D Analysis. Photopolymerization* (ed. N. Fatkullin), Springer, Berlin/Heidelberg, pp. 169–273.
2. LaFratta, C.N., Fourkas, J.T., Baldacchini, T., and Farrer, R.A. (2007) Multiphoton fabrication. *Angew. Chem. Int. Ed.*, **46**, 6238–6258.
3. Juodkazis, S., Mizeikis, V., and Misawa, H. (2009) Three-dimensional microfabrication of materials by femtosecond lasers for photonics applications. *J. Appl. Phys.*, **106**, 051101.
4. Farsari, M. and Chichkov, B.N. (2009) Two-photon fabrication. *Nat. Photonics*, **3**, 450–452.
5. Malinauskas, M., Farsari, M., Piskarskas, A., and Juodkazis, S. (2013) Ultrafast laser nanostructuring of photopolymers: a decade of advances. *Phys. Rep.*, **533**, 1–31.
6. Deubel, M., Wegener, M., Linden, S., von Freymann, G., and John, S. (2006) 3D-2D-3D photonic crystal heterostructures fabricated by direct laser writing. *Opt. Lett.*, **31**, 805–807.
7. Galajda, P. and Ormos, P. (2002) Rotation of microscopic propellers in laser tweezers. *J. Opt. B-Quantum Semiclassical Opt.*, **4**, S78–S81.
8. Deubel, M., Von Freymann, G., Wegener, M., Pereira, S., Busch, K., and Soukoulis, C.M. (2004) Direct laser writing of three-dimensional photonic-crystal templates for telecommunications. *Nat. Mater.*, **3**, 444–447.
9. Gansel, J.K., Thiel, M., Rill, M.S., Decker, M., Bade, K., Saile, V., von Freymann, G., Linden, S., and Wegener, M. (2009) Gold helix photonic metamaterial as broadband circular polarizer. *Science*, **325**, 1513–1515.
10. Schizas, C., Melissinaki, V., Gaidukeviciute, A., Reinhardt, C., Ohrt, C., Dedoussis, V., Chichkov, B.N., Fotakis, C., Farsari, M., and Karalekas, D. (2010) On the design and fabrication by two-photon polymerization of a readily assembled micro-valve. *Int. J. Adv. Manuf. Technol.*, **48**, 435–441.
11. Livage, J. and Sanchez, C. (1992) Sol-gel chemistry. *J. Non-Cryst. Solids*, **145**, 11–19.
12. Kuebler, S.M., Braun, K.L., Zhou, W.H., Cammack, J.K., Yu, T.Y., Ober, C.K., Marder, S.R., and Perry, J.W. (2003) Design and application of high-sensitivity two-photon initiators for three-dimensional microfabrication. *J. Photochem. Photobiol. A*, **158**, 163–170.
13. Penard, A.L., Gacoin, T., and Boilot, J.P. (2007) Functionalized sol-gel coatings for optical applications. *Acc. Chem. Res.*, **40**, 895–902.
14. Houbertz, R., Frohlich, L., Popall, M., Streppel, U., Dannberg, P., Brauer, A., Serbin, J., and Chichkov, B.N. (2003) Inorganic-organic hybrid polymers for information technology: from planar technology to 3D nanostructures. *Adv. Eng. Mater.*, **5**, 551–555.
15. Houbertz, R., Domann, G., Cronauer, C., Schmitt, A., Martin, H., Park, J.U., Frohlich, L., Buestrich, R., Popall, M., Streppel, U., Dannberg, P., Wachter, C., and Brauer, A. (2003) Inorganic-organic hybrid materials for application in optical devices. *Thin Solid Films*, **442**, 194–200.

16. Houbertz, R., Declerck, P., Passinger, S., Ovsianikov, A., Serbin, J., and Chichkov, B.N. (2007) Investigations on the generation of photonic crystals using two-photon polymerization (2PP) of inorganic – organic hybrid polymers with ultra-short laser pulses. *Phys. Status Solidi A: Appl. Mater. Sci.*, **204**, 3662–3675.
17. Straub, M., Nguyen, L.H., Fazlic, A., and Gu, M. (2004) Complex-shaped three-dimensional microstructures and photonic crystals generated in a polysiloxane polymer by two-photon microstereolithography. *Opt. Mater.*, **27**, 359–364.
18. Farsari, M., Filippidis, G., and Fotakis, C. (2005) Fabrication of three-dimensional structures by three-photon polymerization. *Opt. Lett.*, **30**, 3180–3182.
19. Jun, Y., Nagpal, P., and Norris, D.J. (2008) Thermally stable organic-inorganic hybrid photoresists for fabrication of photonic band gap structures with direct laser writing. *Adv. Mater.*, **20**, 606–610.
20. Manhart, J., Kunzelmann, K.H., Chen, H.Y., and Hickel, R. (2000) Mechanical properties of new composite restorative materials. *J. Biomed. Mater. Res.*, **53**, 353–361.
21. Rosin, M., Urban, A.D., Gartner, C., Bernhardt, O., Splieth, C., and Meyer, G. (2002) Polymerization shrinkage-strain and microleakage in dentin-bordered cavities of chemically and light-cured restorative materials. *Dent. Mater.*, **18**, 521–528.
22. Hurmuzlu, F., Kiremitci, A., Serper, A., Altundasar, E., and Siso, S.H. (2003) Fracture resistance of endodontically treated Premolars restored with ormocer and packable composite. *J. Endod.*, **29**, 838–840.
23. Al-Hiyasat, A.S., Darmani, H., and Milhem, M.M. (2005) Cytotoxicity evaluation of dental resin composites and their flowable derivatives. *Clin. Oral. Invest.*, **9**, 21–25.
24. Popall, M. and Durand, H. (1992) Inorganic organic copolymers as solid-state Li^+ electrolytes. *Electrochim. Acta*, **37**, 1593–1597.
25. Popall, M. and Du, X.M. (1995) Inorganic-organic copolymers as solid-state ionic conductors with grafted anions. *Electrochim. Acta*, **40**, 2305–2308.
26. Depre, L., Kappel, J., and Popall, M. (1998) Inorganic-organic proton conductors based on alkylsulfone functionalities and their patterning by photoinduced methods. *Electrochim. Acta*, **43**, 1301–1306.
27. Depre, L., Ingram, M., Poinsignon, C., and Popall, M. (2000) Proton conducting sulfon/sulfonamide functionalized materials based on inorganic-organic matrices. *Electrochim. Acta*, **45**, 1377–1383.
28. Buestrich, R., Kahlenberg, F., Popall, M., Dannberg, P., Muller-Fiedler, R., and Rosch, O. (2001) ORMOCER (R) s for optical interconnection technology. *J. Sol-Gel Sci. Technol.*, **20**, 181–186.
29. Popall, M., Dabek, A., Robertsson, M.E., Valizadeh, S., Hagel, O.J., Buestrich, R., Nagel, R., Cergel, L., Lambert, D., and Schaub, M. (2000) ORMOCER (R) S – inorganic-organic hybrid materials for e/o-interconnection-technology. *Mol. Cryst. Liq. Cryst.*, **354**, 711–730.
30. Popall, M., Houbertz, R., Frohlich, L., Streppel, U., Dannberg, P., Westenhofer, S., and Gale, M. (2003) ORMOCER (R) s – inorganic-organic hybrid materials for integrated, diffractive and refractive microoptics: synthesis, processing, and applications in optical components. *Glass Sci. Technol.*, **76**, 53–58.
31. Serbin, J., Egbert, A., Ostendorf, A., Chichkov, B.N., Houbertz, R., Domann, G., Schulz, J., Cronauer, C., Frohlich, L., and Popall, M. (2003) Femtosecond laser-induced two-photon polymerization of inorganic-organic hybrid materials for applications in photonics. *Opt. Lett.*, **28**, 301–303.
32. Burmeister, F., Steenhusen, S., Houbertz, R., Zeitner, U.D., Nolte, S., and Tünnermann, A. (2012) Materials and technologies for fabrication of three-dimensional microstructures with sub-100 nm feature sizes by two-photon polymerization. *J. Laser Appl.*, **24**, 042014.

33. Serbin, J. and Gu, M. (2006) Experimental evidence for superprism effects in three-dimensional polymer photonic crystals. *Adv. Mater.*, **18**, 221–224.
34. Serbin, J. and Gu, M. (2006) Superprism phenomena in waveguide-coupled woodpile structures fabricated by two-photon polymerization. *Opt. Express*, **14**, 3563–3568.
35. Jia, B.H., Serbin, J., Kim, H., Lee, B., Li, J.F., and Gu, M. (2007) Use of two-photon polymerization for continuous gray-level encoding of diffractive optical elements. *Appl. Phys. Lett.*, **90**, 073503.
36. Grossmann, T., Schleede, S., Hauser, M., Beck, T., Thiel, M., von Freymann, G., Mappes, T., and Kalt, H. (2011) Direct laser writing for active and passive high-Q polymer microdisks on silicon. *Opt. Express*, **19**, 11451–11456.
37. Li, J.F., Jia, B.H., Zhou, G.Y., Bullen, C., Serbin, J., and Gu, M. (2007) Spectral redistribution in spontaneous emission from quantum-dot-infiltrated 3D woodpile photonic crystals for telecommunications. *Adv. Mater.*, **19**, 3276–3280.
38. Li, J.F., Jia, B.H., and Gu, M. (2008) Engineering stop gaps of inorganic-organic polymeric 3D woodpile photonic crystals with post-thermal treatment. *Opt. Express*, **16**, 20073–20080.
39. Woggon, T., Kleiner, T., Punke, M., and Lemmer, U. (2009) Nanostructuring of organic-inorganic hybrid materials for distributed feedback laser resonators by two-photon polymerization. *Opt. Express*, **17**, 2500–2507.
40. Karalekas, D. and Schizas, C. (2009) Monitoring of solidification induced strains in two resins used in photofabrication. *Mater. Des.*, **30**, 3705–3712.
41. Ovsianikov, A., Schlie, S., Ngezahayo, A., Haverich, A., and Chichkov, B.N. (2007) Two-photon polymerization technique for microfabrication of CAD-designed 3D scaffolds from commercially available photosensitive materials. *J. Tissue Eng. Regener. Med.*, **1**, 443–449.
42. Ovsianikov, A., Chichkov, B., Adunka, O., Pillsbury, H., Doraiswamy, A., and Narayan, R.J. (2007) Rapid prototyping of ossicular replacement prostheses. *Appl. Surf. Sci.*, **253**, 6603–6607.
43. Narayan, R.J., Jin, C.M., Doraiswamy, A., Mihailescu, I.N., Jelinek, M., Ovsianikov, A., Chichkov, B., and Chrisey, D.B. (2005) Laser processing of advanced bioceramics. *Adv. Eng. Mater.*, **7**, 1083–1098.
44. Ovsianikov, A., Chichkov, B., Mente, P., Monteiro-Riviere, N.A., Doraiswamy, A., and Narayan, R.J. (2007) Two photon polymerization of polymer-ceramic hybrid materials for transdermal drug delivery. *Int. J. Appl. Ceram. Technol.*, **4**, 22–29.
45. Doraiswamy, A., Jin, C., Narayan, R.J., Mageswaran, P., Mente, P., Modi, R., Auyeung, R., Chrisey, D.B., Ovsianikov, A., and Chichkov, B. (2006) Two photon induced polymerization of organic-inorganic hybrid biomaterials for microstructured medical devices. *Acta Biomater.*, **2**, 267–275.
46. Gittard, S.D., Narayan, R.J., Jin, C., Ovsianikov, A., Chichkov, B.N., Monteiro-Riviere, N.A., Stafslien, S., and Chisholm, B. (2009) Pulsed laser deposition of antimicrobial silver coating on Ormocer (R) microneedles. *Biofabrication*, **1**, 041001.
47. Dinca, V., Kasotakis, E., Catherine, J., Mourka, A., Ranella, A., Ovsianikov, A., Chichkov, B.N., Farsari, M., Mitraki, A., and Fotakis, C. (2008) Directed three-dimensional patterning of self-assembled peptide fibrils. *Nano Lett.*, **8**, 538–543.
48. Drakakis, T.S., Papadakis, G., Sambani, K., Filippidis, G., Georgiou, S., Gizeli, E., Fotakis, C., and Farsari, M. (2006) Construction of three-dimensional biomolecule structures employing femtosecond lasers. *Appl. Phys. Lett.*, **89**, 144108.
49. Saravanamuttu, K., Blanford, C.F., Sharp, D.N., Dedman, E.R., Turberfield, A.J., and Denning, R.G. (2003) Sol-gel organic-inorganic composites for 3-D holographic lithography of photonic crystals with submicron periodicity. *Chem. Mater.*, **15**, 2301–2304.
50. Segawa, H., Yamaguchi, S., Yamazaki, Y., Yano, T., Shibata, S., and Misawa, H. (2006) Top-gathering pillar array of hybrid organic-inorganic material by means of self-organization. *Appl. Phys. A-Mater. Sci. Process.*, **83**, 447–451.

51. Que, W.X., Jia, C.Y., Sun, M., Sun, Z., Wang, L.L., and Zhang, Z.J. (2008) Photo-patternable GeO_2-contained organic-inorganic hybrid sol-gel films for photonic applications. *Opt. Express*, **16**, 3490–3495.
52. Di Maggio, R., Fambri, L., and Guerriero, A. (1998) Zirconium alkoxides as components of hybrid inorganic-organic macromolecular materials. *Chem. Mater.*, **10**, 1777–1784.
53. Schubert, U. (2003) Silica-based and transition metal-based inorganic-organic hybrid materials – a comparison. *J. Sol-Gel Sci. Technol.*, **26**, 47–55.
54. Winfield, R.J., Bhuian, B., O'Brien, S., and Crean, G.M. (2007) Refractive femtosecond laser beam shaping for two-photon polymerization. *Appl. Phys. Lett.*, **90**, 111115.
55. Winfield, R.J., Bhuian, B., O'Brien, S., and Crean, G.M. (2007) Fabrication of grating structures by simultaneous multi-spot fs laser writing. *Appl. Surf. Sci.*, **253**, 8086–8090.
56. Bhuian, B., Winfield, R.J., O'Brien, S., and Crean, G.M. (2006) Investigation of the two-photon polymerisation of a Zr-based inorganic–organic hybrid material system. *Appl. Surf. Sci.*, **252**, 4845–4849.
57. Ovsianikov, A., Viertl, J., Chichkov, B., Oubaha, M., MacCraith, B., Sakellari, L., Giakoumaki, A., Gray, D., Vamvakaki, M., Farsari, M., and Fotakis, C. (2008) Ultra-low shrinkage hybrid photosensitive material for two-photon polymerization microfabrication. *ACS Nano*, **2**, 2257–2262.
58. Ovsianikov, A., Xiao, S.Z., Farsari, M., Vamvakaki, M., Fotakis, C., and Chichkov, B.N. (2009) Shrinkage of microstructures produced by two-photon polymerization of Zr-based hybrid photosensitive materials. *Opt. Express*, **17**, 2143–2148.
59. Psycharakis, S., Tosca, A., Melissinaki, V., Giakoumaki, A., and Ranella, A. (2011) Tailor-made three-dimensional hybrid scaffolds for cell cultures. *Biomed. Mater.*, **6**, 045008.
60. Raimondi, M.T., Eaton, S.M., Lagana, M., Aprile, V., Nava, M.M., Cerullo, G., and Osellame, R. (2013) Three-dimensional structural niches engineered via two-photon laser polymerization promote stem cell homing. *Acta Biomater.*, **9**, 4579–4584.
61. Amato, L., Gu, Y., Bellini, N., Eaton, S.M., Cerullo, G., and Osellame, R. (2012) Integrated three-dimensional filter separates nanoscale from microscale elements in a microfluidic chip. *Lab Chip*, **12**, 1135–1142.
62. Eaton, S.M., De Marco, C., Martinez-Vazquez, R., Ramponi, R., Turri, S., Cerullo, G., and Osellame, R. (2012) Femtosecond laser microstructuring for polymeric lab-on-chips. *J. Biophotonics*, **5**, 687–702.
63. Galanopoulos, S., Chatzidai, N., Melissinaki, V., Selimis, A., Schizas, C., Farsari, M., and Karalekas, D. (2014) Design, fabrication and computational characterization of a 3D micro-valve built by multi-photon polymerization. *Micromachines*, **5**. doi: 10.3390/mi5030505
64. Skarmoutsou, A., Lolas, G., Charitidis, C.A., Chatzinikolaidou, M., Vamvakaki, M., and Farsari, M. (2013) Nanomechanical properties of hybrid coatings for bone tissue engineering. *J. Mech. Behav. Biomed. Mater.*, **25**, 48–62.
65. Terzaki, K., Kissamitaki, M., Skarmoutsou, A., Fotakis, C., Charitidis, C.A., Farsari, M., Vamvakaki, M., and Chatzinikolaidou, M. (2013) Pre-osteoblastic cell response on three-dimensional, organic-inorganic hybrid material scaffolds for bone tissue engineering. *J. Biomed. Mater. Res. A*, **101A**, 2283–2294.
66. Terzaki, K., Kalloudi, E., Mossou, E., Mitchell, E.P., Forsyth, V.T., Rosseeva, E., Simon, P., Vamvakaki, M., Chatzinikolaidou, M., Mitraki, A., and Farsari, M. (2013) Mineralized self-assembled peptides on 3D laser-made scaffolds: a new route toward 'scaffold on scaffold' hard tissue engineering. *Biofabrication*, **5**, 045002.
67. Zukauskas, A., Malinauskas, M., Kontenis, L., Purlys, V., Paipulas, D., Vengris, M., and Gadonas, R. (2010) Organic dye doped microstructures for

optically active functional devices fabricated via two-photon polymerization technique. *Lith. J. Phys.*, **50**, 55–61.
68. Jia, B., Buso, D., Embden, J.V., Li, J., and Gu, M. (2010) Highly Non-linear quantum dot doped nanocomposites for functional three-dimensional structures generated by two-photon polymerization. *Adv. Mater.*, **22**, 2463–2467.
69. Terzaki, K., Vasilantonakis, N., Gaidukeviciute, A., Reinhardt, C., Fotakis, C., Vamvakaki, M., and Farsari, M. (2011) 3D conducting nanostructures fabricated using direct laser writing. *Opt. Mater. Express*, **1**, 586–597.
70. Vasilantonakis, N., Terzaki, K., Sakellari, I., Purlys, V., Gray, D., Soukoulis, C.M., Vamvakaki, M., Kafesaki, M., and Farsari, M. (2012) Three-dimensional metallic photonic crystals with optical bandgaps. *Adv. Mater.*, **24**, 1101–1105.
71. Sakellari, I., Gaidukeviciute, A., Giakoumaki, A., Gray, D., Fotakis, C., Vamvakaki, M., Farsari, M., Reinhardt, C., Ovsianikov, A., and Chichkov, B.N. (2009) in *Metamaterials: Fundamentals and Applications II* (eds A.N. Mikhail, I.Z. Nikolay, D.B. Allan, and E. Nader), SPIE, p. 73920Y.
72. Kabouraki, E., Giakoumaki, A.N., Danilevicius, P., Gray, D., Vamvakaki, M., and Farsari, M. (2013) Redox multiphoton polymerization for 3D nanofabrication. *Nano Lett.*, **13**, 3831–3835.
73. Kabashin, A.V., Evans, P., Pastkovsky, S., Hendren, W., Wurtz, G.A., Atkinson, R., Pollard, R., Podolskiy, V.A., and Zayats, A.V. (2009) Plasmonic nanorod metamaterials for biosensing. *Nat. Mater.*, **8**, 867–871.
74. Gramotnev, D.K. and Bozhevolnyi, S.I. (2010) Plasmonics beyond the diffraction limit. *Nat. Photonics*, **4**, 83–91.
75. Rill, M.S., Plet, C., Thiel, M., Staude, I., Von Freymann, G., Linden, S., and Wegener, M. (2008) Photonic metamaterials by direct laser writing and silver chemical vapour deposition. *Nat. Mater.*, **7**, 543–546.
76. Cao, Y.Y., Dong, X.Z., Takeyasu, N., Tanaka, T., Zhao, Z.S., Duan, X.M., and Kawata, S. (2009) Morphology and size dependence of silver microstructures in fatty salts-assisted multiphoton photoreduction microfabrication. *Appl. Phys. A*, **96**, 453–458.
77. Cao, Y.Y., Takeyasu, N., Tanaka, T., Duan, X.M., and Kawata, S. (2009) 3D metallic nanostructure fabrication by surfactant-assisted multiphoton-induced reduction. *Small*, **5**, 1144–1148.
78. Gansel, J.K., Wegener, M., Burger, S., and Linden, S. (2010) Gold helix photonic metamaterials: a numerical parameter study. *Opt. Express*, **18**, 1059–1069.
79. Duan, X.M., Sun, H.B., Kaneko, K., and Kawata, S. (2004) Two-photon polymerization of metal ions doped acrylate monomers and oligomers for three-dimensional structure fabrication. *Thin Solid Films*, **453-54**, 518–521.
80. Kaneko, K., Yamamoto, K., Kawata, S., Xia, H., Song, J.F., and Sun, H.B. (2008) Metal-nanoshelled three-dimensional photonic lattices. *Opt. Lett.*, **33**, 1999–2001.
81. Sun, Z.B., Dong, X.Z., Chen, W.Q., Shoji, S., Duan, X.M., and Kawata, S. (2008) Two- and three-dimensional micro/nanostructure patterning of CdS-polymer nanocomposites with a laser interference technique and in situ synthesis. *Nanotechnology*, **19**, 035611.
82. Mallory, G.O. and Hajdu, J.B. (eds) (1990) *Electroless Plating: Fundamentals and Applications*, American Electroplaters and Surface Finishers Society, Orlando, FL.
83. Hrapovic, S., Liu, Y.L., Enright, G., Bensebaa, F., and Luong, J.H.T. (2003) New strategy for preparing thin gold films on modified glass surfaces by electroless deposition. *Langmuir*, **19**, 3958–3965.
84. Chen, Y.-S., Tal, A., Torrance, D.B., and Kuebler, S.M. (2006) Fabrication and characterization of three-dimensional silver-coated polymeric microstructures. *Adv. Funct. Mater.*, **16**, 1739–1744.
85. Takeyasu, N., Tanaka, T., and Kawata, S. (2008) Fabrication of 3D metal/polymer microstructures by site-selective metal coating. *Appl. Phys. A*, **90**, 205–209.
86. Hutmacher, D.W. (2000) Scaffolds in tissue engineering bone and cartilage. *Biomaterials*, **21**, 2529–2543.

87. Nasr, H.F., Aichelmann-Reidy, M.E., and Yukna, R.A. (1999) Bone and bone substitutes. *Periodontology*, **2000** (19), 74–86.
88. Palmer, L.C., Newcomb, C.J., Kaltz, S.R., Spoerke, E.D., and Stupp, S.I. (2008) Biomimetic systems for hydroxyapatite mineralization inspired by bone and enamel. *Chem. Rev.*, **108**, 4754–4783.
89. LaFratta, C.N., Li, L.J., and Fourkas, J.T. (2006) Soft-lithographic replication of 3D microstructures with closed loops. *Proc. Natl. Acad. Sci. U.S.A.*, **103**, 8589–8594.
90. LaFratta, C.N., Baldacchini, T., Farrer, R.A., Fourkas, J.T., Teich, M.C., Saleh, B.E.A., and Naughton, M.J. (2004) Replication of two-photon-polymerized structures with extremely high aspect ratios and large overhangs. *J. Phys. Chem. B*, **108**, 11256–11258.
91. Seidel, A., Ohrt, C., Passinger, S., Reinhardt, C., Kiyan, R., and Chichkov, B.N. (2009) Nanoimprinting of dielectric loaded surface-plasmon-polariton waveguides using masters fabricated by 2-photon polymerization technique. *J. Opt. Soc. Am. B-Opt. Phys.*, **26**, 810–812.
92. Maruo, S., Hasegawa, T., and Yoshimura, N. (2009) Replication of three-dimensional rotary micromechanism by membrane-assisted transfer molding. *Jpn. J. Appl. Phys.*, **48**. doi: 10.1143/jjap.1148.1106fh1105
93. Coenjarts, C.A. and Ober, C.K. (2004) Two-photon three-dimensional microfabrication of poly(dimethylsiloxane) elastomers. *Chem. Mater.*, **16**, 5556–5558.
94. Cadarso, V.J., Llobera, A., Villanueva, G., Plaza, J.A., Brugger, J., and Dominguez, C. (2010) Mechanically tuneable microoptical structure based on PDMS. *Sens. Actuators, A-Phys.*, **162**, 260–266.
95. Kumi, G., Yanez, C.O., Belfield, K.D., and Fourkas, J.T. (2010) High-speed multiphoton absorption polymerization: fabrication of microfluidic channels with arbitrary cross-sections and high aspect ratios. *Lab Chip*, **10**, 1057–1060.
96. Koufaki, N., Ranella, A., Aifantis, K.E., Barberoglou, M., Psycharakis, S., Fotakis, C., and Stratakis, E. (2011) Controlling cell adhesion via replication of laser micro/nano-textured surfaces on polymers. *Biofabrication*, **3**, 045004.

8
Photopolymers for Multiphoton Lithography in Biomaterials and Hydrogels

Mark W. Tibbitt, Jared A. Shadish, and Cole A. DeForest

8.1
Introduction

The advent of the polymer industry in the 1950s introduced a vast array of novel materials that have impacted all areas of life [1]. Today, polymers and polymerization reactions are utilized additionally in a range of biomedical applications, including as drug delivery systems [2], tissue engineering scaffolds [3], and as medical devices [4]. These biomedical applications have flourished as the available polymer chemistries and processing techniques have matured, enabling the creation of materials that can be tailored specifically for each use. In particular, advances in synthetic chemistry and photophysics have provided the tools needed to initiate and propagate polymerization reactions with light in a controlled manner [5, 6]. The use of light as the catalyst for polymerization provides the experimenter with an unprecedented level of spatial and temporal control over the reaction. That is, the reaction proceeds only where and when the user desires, allowing for facile production of geometrically defined polymeric materials on demand. Additionally, photopolymerization, which we define generally as the class of reactions that transforms solutions of low molecular weight monomers or prepolymers into high molecular weight species or crosslinked networks upon exposure to visible, ultraviolet (UV), or infrared (IR) light, affords high reaction rates under ambient conditions with relatively low energy input in both non-aqueous and aqueous solutions. To clarify, in this chapter photopolymerization will be used predominantly to describe the photocrosslinking of liquid precursors into solid or viscoelastic materials. Photopolymerization reactions have been leveraged within the biomaterials community to address niche design constraints, such as defined geometries (e.g., bone implants) or post-application sol–gel transition (e.g., *in situ* curing hydrogels) [7]. Concurrently, multiphoton lithography (MPL) techniques have been applied in photopolymerization, and the necessary tools are becoming more widely available to the biomedical community. In this chapter, we discuss, review, and provide perspective on the intersection between MPL and photopolymerization for biomaterials with a special emphasis

Multiphoton Lithography: Techniques, Materials and Applications, First Edition.
Edited by Jürgen Stampfl, Robert Liska, and Aleksandr Ovsianikov.

on the chemistries, platforms, and techniques that are relevant to the biological context.

Photopolymerizations have been instrumental in the advancement of several modern industries, including coatings and adhesives, integrated circuits, and optics, on account of their unique properties [8, 9]. Historically, the ancient Egyptians were the first to explore photopolymerization reactions as they utilized sunlight to crosslink oil-soaked linens to create environmental barriers during mummification [10]. In recent decades, photopolymerizations have been leveraged within modern medicine to fabricate biocompatible networks and hydrogels from monomers and end-functional polymers [11]. Here, networks refer to crosslinked polymeric materials that are utilized in the non-solvated state, whereas hydrogels refer to water-swollen crosslinked polymeric materials. The high swelling degree of hydrogels makes them attractive for biomedical applications, as they mimic the mechanical properties of tissues within the body and afford facile transport or nutrients, waste, and signaling molecules [12]. In either case, the spatial and temporal control afforded by photopolymerization enables the production of highly structured materials with predefined geometries and the ability to polymerize *in situ*. Photopolymerized biomaterials have been successfully implemented as barriers, cell delivery vehicles, tissue engineering scaffolds, and drug delivery systems [11].

For many applications, it is desirable to achieve increased feature resolution within 3D objects. Highly ordered or structured materials play an essential role in many modern technologies, including microprocessors, photonics, and optics [13]. Demands in these industries push for increased information content or feature density, which has fostered innovation in the materials science community to write micro- and nanoscale features into functional materials with high fidelity [14]. Photolithography and stereolithography remain the most accessible techniques for rapid and reliable fabrication at the micrometer scale for the fabrication of functional biomaterials. However, as our knowledge of biological systems is increasing, it is becoming clear that traditional monolithic photopolymerization reactions fail to provide materials with the information content needed for all applications. For example, the extracellular matrix (ECM), that is, the local environment that supports and instructs cell function, possesses a high degree of complexity with nanometer- and micrometer-scale architecture that varies with time [15]. Therefore, technologies used for *ex vivo* organ growth, three-dimensional (3D) cell culture, or as injectable cell carriers could all benefit by adopting strategies developed for high-content feature patterning in other industries.

Traditional photopolymerizations employed for biomaterial fabrication are often accessed through a photo-induced free-radical polymerization [16]. This approach minimally requires a precursor solution containing multifunctional monomers and a photoinitiator, as well as an appropriate light source (a comprehensive overview of these components is provided in Section 8.4). Photopolymerized materials are then fabricated by irradiating the precursor solution with light. Molds or masks constrain the geometry, while the dosage and intensity of the light control the extent and kinetics of reaction, respectively [17]. *In vivo* photopolymerization (or *in situ* photopolymerization) is achieved similarly by placing the precursor solution into the appropriate location within the body, and subsequently initiating the polymerization with light [18]. In this manner, biomaterials can be fabricated rapidly that conform to desired tissue geometries. Alternatively, interfacial photopolymerizations are achieved by adsorbing or grafting a photoinitiator to a biomaterial surface from which a photopolymerized surface or brush can be generated [19]. These approaches enable conformal coatings, molded materials, or *in vivo* implants; however, none of these approaches exploits the full 3D and spatiotemporal resolution that photoinitiation affords, as they are confined to planar patterning.

Increased spatial resolution has been achieved by leveraging photolithography techniques used extensively in the microelectronics industry for the fabrication of integrated circuits [20]. Here, a photomask is used to selectively constrain illumination of the underlying precursor solution. Regions of illumination are photopolymerized into insoluble blocks, while the other nonpolymerized regions can be washed away post fabrication. In this manner, the user can generate defined planar geometries with micrometer-scale resolution and access 3D structures with layer-by-layer fabrication [21]. However, photolithography requires high-resolution photomasks for each geometry, remains diffraction-limited, and can form only simple 3D structures. Soft lithography can also be adapted to photopolymerizations [22]. Here, a master mold is formed in an elastomeric material, such as polydimethylsiloxane (PDMS), with a desired geometry. The mold is filled with a precursor solution, which is photopolymerized to recover the desired features. This approach has been applied to microfabricate biomaterials for drug delivery [23], tissue engineering [24], and microfluidic biosensing [25]. Soft lithography is also limited in that it requires a master mold tailored for each application and involves a multistep process to achieve micrometer-scale materials. An alternative approach for rapid prototyping and micrometer-scale material fabrication is laser scanning lithography (LSL)-based photopolymerization, whereby the focal point of a laser is rastered or serially scanned through defined regions of a precursor solution [26]. Photoinitiation occurs only where the solution is illuminated by the focused laser light, which provides the necessary photons to drive polymerization. On account of scattering and attenuation of the focused laser light, LSL is typically confined to the surface of the precursor solution. In this manner, 3D objects can be generated adopting stereolithographic techniques, whereby an automated stage with *xyz* control moves the object through the focal point of the laser. Layer-by-layer photolithography and LSL combined with stereolithography are currently utilized in the 3D printing industry. In each of these approaches, material fabrication remains limited in spatial resolution to the micrometer scale and fails to access easily all arbitrary 3D geometries.

Specifically, developments in multiphoton technologies have made ultrafast lasers increasingly accessible to the materials science, biology, and biomaterials communities. As a result, researchers have begun to explore the use of multiphoton irradiation to facilitate nanometer- and micrometer-scale phenomena within the photopolymerization and biological context [27]. MPL – the use of ultrafast, multiphoton lasers to execute LSL – enables the experimenter to photopolymerize materials with feature content down to the nanoscale in three dimensions with high fidelity, owing to the narrow focal region of a multiphoton laser and minimal radical generation outside of the focal spot [28]. In this chapter, we will discuss, review, and provide perspective on the use of MPL as a technique for the fabrication and modification of biomaterials and hydrogels with special emphasis on the photopolymers and techniques that can be performed in or are amenable to the biological context.

8.2
Multiphoton Lithography (MPL) for Photopolymerization

The theory of multiphoton excitation was described initially by Maria Goeppert-Mayer in her doctorate dissertation; Goeppert-Mayer is also the recipient of the 1963 Nobel Prize for Physics for articulating the nuclear shell model of atomic nuclei [29]. The principle of multiphoton excitation, as proposed by Goeppert-Mayer, allows a photoactive molecule to enter an excited state by absorbing multiple photons of lesser energy (longer wavelength) than a single photon of greater energy (shorter wavelength) that would normally result in photoexcitation [29]. This phenomenon, now well tested and established, is traditionally achieved with two photons from the same laser possessing roughly half the energy of a single photon capable of excitation [29]. As both photons are needed to bump the molecule into the excited quantum state, they must both be absorbed within femtoseconds of each other. This is achieved most commonly with high-energy femtosecond pulsed laser sources that significantly increase the probability of absorbing multiple photons nearly simultaneously within the focal volume [30]. Owing to the nonlinear nature of the two photon process, the probability of excitation decays as a function of distance from the focal plane with quadratic dependence, which compresses the axial spread of the point spread function [30]. In practical terms, this means that nearly all of the excitation is confined to the focal plane, providing high resolution (<1 μm) in the axial or z-dimension (Figure 8.1a). Assuming reasonable optics, this translates to sub-femtoliter focal volumes. For additional reading, the concepts and physics behind multiphoton excitation are handled with sufficient rigor in Part I and Chapter 6.

Multiphoton techniques were first introduced into the biology and biomaterials communities through the development and proliferation of multiphoton microscopy (MPM) by Watt and colleagues [34, 35]. MPM is now widely available and an attractive tool for high-resolution fluorescent imaging in thick tissues as well as in live animals [30]. Beyond fluorescent imaging, multiphoton excitation is commonly employed for molecular uncaging of small molecules [36], fluorescence recovery after photobleaching (FRAP) [37], and biophysical investigations [38]. Importantly, multiphoton excitation has been incorporated directly into traditional laser scanning microscopes (LSMs) or LSL setups to image or fabricate materials with micrometer- and nanometer-scale features.

As multiphoton systems have become increasingly accessible, they have been adopted by the materials science and polymers community for high-resolution 3D photopolymerizations [28]. MPL has successfully generated a range of complex, 3D photopolymerized structures (Figure 8.1). The same principles above apply to photopolymerization reactions: if two photons with lesser energy are absorbed by the same photoinitiator nearly simultaneously, the molecule may undergo cleavage and generate initiating radicals within the sub-femtoliter focal volume [28]. Accordingly, rapid prototyping with MPL affords the fabrication of objects with sub-micrometer feature sizes. Initial work has explored the fabrication of open

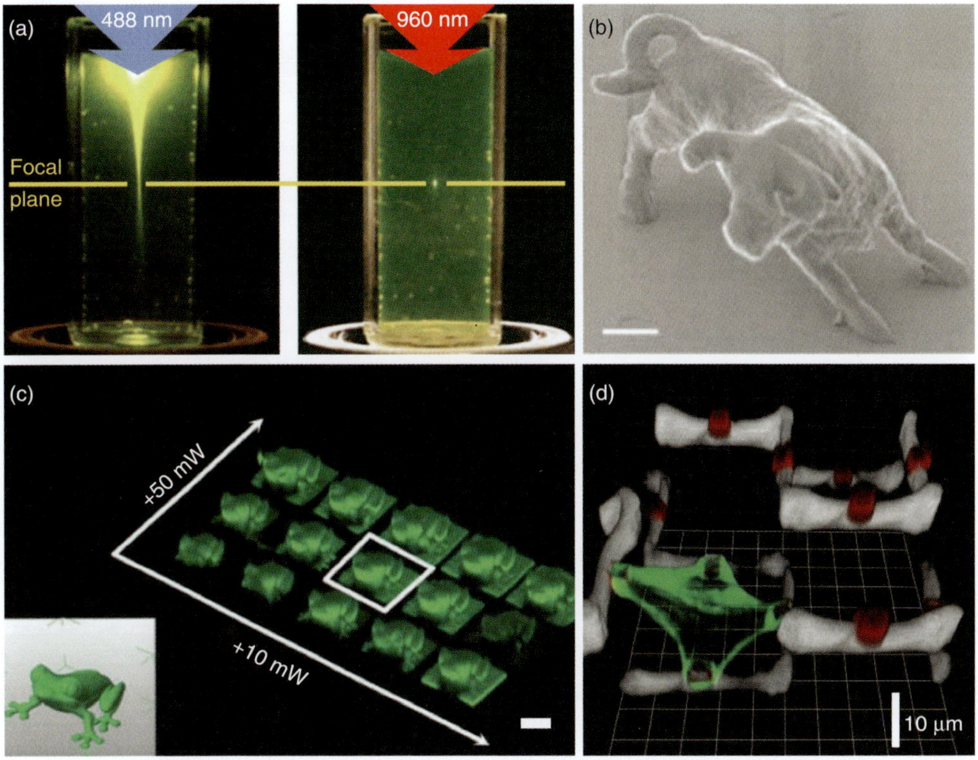

Figure 8.1 Materials fabricated by MPL photopolymerization. (a) The nonlinear nature of multiphoton excitation results in a decreased excitation volume (960 nm) compared to single photon excitation (488 nm). (Zipfel 2003 [30]. Reproduced with permission of Nature Publishing Group.) (b) A microbull fabricated by two-photon photopolymerization. Scale bar = 2 μm. (Kawata 2001 [31]. Reproduced with permission of Nature Publishing Group.) (c) Hydrogel frogs patterned with MPL. Scale bar = 60 μm. (Li 2013 [32]. Reproduced with permission of Royal Society of Chemistry.) (d) Primary chicken fibroblast interacting with composite microfabricated cell scaffold. Scale bar = 10 μm. (Klein 2011 [33]. Reproduced with permission of Wiley.)

cellular structures as biomaterials or photonic crystals [39, 40]. In principle, the technique is amenable to any rendering accessible through traditional computer-assisted drawing (CAD) software, and this has been demonstrated through the fabrication of many complex objects, such as a bull, frogs, and composite culture scaffolds (Figure 8.1) [31–33].

The main advantage of MPL over traditional LSL is the greatly increased z-resolution afforded because the axial dispersion of the point spread function is significantly decreased compared to single-photon excitation [30]. Additionally, the use of IR or near-IR light enables increased penetration depth, as these wavelengths are minimally absorbed and scatter less in most media, including biological samples [30]. These advantages afford true 3D patterning at size scales commensurate with the focal volume of the multiphoton laser (<1 fl or 1 μm^3), which conveniently accesses subcellular length scales pertinent to cell and tissue biology.

While MPL improves the voxel resolution for photopolymerized biomaterials and hydrogels, practical constraints of the technology still exist. MPL remains limited in total thickness without using advanced stages and by the throughput of fabrication, which remains low for high-feature-content, millimeter-sized objects. Additionally, the available photoinitiator toolbox, those photoinitiators that are amenable to multiphoton excitation, should be improved. Advances in computational quantum chemistry are beginning to elucidate novel photoinitiators that are susceptible to multiphoton excitation and cleavage to generate initiating radicals [41]. Excitingly, water-soluble initiators with favorable two-photon cross sections are under development for biological applications [32].

8.3
MPL Equipment for Biomaterial Fabrication

A complete discussion of the equipment and techniques for MPL can be found in Part II. For the polymer or biomaterial scientist, the user will likely employ either commercial LSMs or home-built setups. In fact, the main limitations are the requirement for a multiphoton laser source (e.g., a femtosecond pulsed Ti:sapphire laser), a suitable light path with focusing optics, a high-fidelity xyz stage, and software to raster the laser focal volume through the sample, or vice versa. All of these requirements are met with commercial MPMs (e.g., Zeiss 710 NLO, Nikon A1R MP+, Olympus Fluoview FVMPE series) designed primarily for imaging biological samples. Such systems are adapted for writing instead of reading by rastering the focal volume through a precursor solution within defined regions of interest (ROIs) that define the shape to be polymerized. Advances in CAD-supported ROI software are facilitating easy translation of arbitrary structures into 3D photopolymerized materials using commercial microscopes. Additionally, to achieve high-resolution patterning with high fidelity, MPL equipment requires an automated [42] stage with independent x-, y-, and z-dimension resolution at the sub-micrometer scale.

8.4 Chemistry for MPL Photopolymerizations

The major requirements for photopolymerizations are a precursor solution containing a photoinitiator and functional monomer or macromers as well as a light source for initiation. In this section, the focus will be on the chemistries and mechanisms for photopolymerization of biomaterials and hydrogels via MPL.

8.4.1 Photopolymerization

Photopolymerizations comprise a particularly attractive mechanism for the fabrication of biomaterials via MPL, as they rapidly convert a liquid precursor solution into a solid material only within the regions of irradiation. This allows easy removal of the unwanted material and recovery of the desired microstructured biomaterial. The initial step in a radical-mediated photopolymerization reaction is photoinitiation. Photoinitiation occurs when a photoinitiator molecule (PI) absorbs a photon of light – or in the case of MPL multiple photons nearly simultaneously – and enters an excited state. The excited-state photoinitiator (PI*) is capable of following one of two pathways: (i) dissociate directly into one or multiple primary radicals (Type I PI), or (ii) undergo a bimolecular reaction wherein the excited PI reacts with a second molecule to generate free radicals (Type II PI) [43]. The rate of initiation, R_{i}, for a Type I PI undergoing two photon photoinitiation from a typical pulsed laser is given by [44, 45]

$$R_{\mathrm{i}} = 2{\cdot}1.17\delta_{\mathrm{u}}\Phi_{\mathrm{u}}\frac{T}{\tau_{\mathrm{p}}}\left(\frac{\lambda}{\pi hcw_{xy}^{2}}\right)^{2}P_{\mathrm{avg}}^{2}\mathrm{VF[PI]} \tag{8.1}$$

where $\delta_{\mathrm{u}}\Phi_{\mathrm{u}}$ is the two-photon cross section (generally a function of λ), T is the period of the laser pulses, τ_{p} is the duration of the laser pulses, λ is the wavelength of the laser light, h is Planck's constant, c is the speed of light, w_{xy} is the lateral focal radius of the laser, P_{avg} is the average laser power, VF is a volume factor (taken as 0.63 for an axial cylinder), and [PI] is the local concentration of the photoinitiator. The prefactor of 2 assumes that two radicals are generated for each dissociated photoinitiator molecule. Similar rate expressions can be derived for other multiphoton initiation processes. Most single-photon photopolymerizations are hindered by light attenuation, which creates a spatial gradient in light intensity, as well as out-of-plane initiation. Owing to the limited scattering and absorption of IR light in most media, the light intensity or average power is assumed uniform over the spatial domain of the polymerization. Quantifying the rate of initiation is critical, as the rate of initiation affects the rate of polymerization and, thus, the final material properties. As shown in Eq. (8.1), the rate of initiation can be controlled by tailoring the PI concentration, selecting a PI with a favorable two-photon cross section, tuning the wavelength of light, or adjusting the laser power. As opposed to single-photon initiation, the rate of two-photon initiation varies quadratically with the input laser power.

Once radicals are generated within the precursor solution, they begin to propagate through the functional groups of the monomers or macromers, assuming the radicals do not immediately recombine within the solvent cage. In principle, polymerization reactions proceed through a chain or step-growth mechanism [43]. In a classic, radical-initiated, step-growth (or condensation) polymerization, a radical activates a functional group (such as a thiol) on one of the surrounding monomers, which then adds to another functional group (such as a vinyl) on a second monomer while the radical transfers to another functional group in solution. In this manner, dimers are formed. Dimers can then react with monomers or other dimers to form trimers or tetramers, which sequentially build up in mass in a stepwise manner. In contrast, a classic, chain-growth (or addition) radical-initiated polymerization proceeds when a radical activates a functional group (such as a (meth)acrylate) on one of the surrounding monomers, which then activates and adds to another monomer in solution. This proceeds iteratively through monomer addition to form long chains of monomers from each initiation event. In either case, polymerization results in high molecular weight species that form an entangled polymeric solid or, when multifunctional monomers/macromers are in solution, a crosslinked polymer network. During this process, radicals can also terminate as opposed to initiate or propagate the reaction, which decreases the concentration of radicals in solution and affects the kinetic rate of polymerization. Termination can occur through several mechanisms, including bimolecular combination between two active chain ends or an active chain end and an initiating radical; disproportionation, in which a hydrogen atom is abstracted from one active chain by another; interaction with inhibitors such as oxygen; and chain transfer to nonpropagating species [43]. The kinetic rate of polymerization, R_p, for chain growth polymerizations accounts for both propagation and termination as follows [43]:

$$R_\mathrm{p} = k_\mathrm{p}[M]\left(\frac{R_\mathrm{i}}{2k_\mathrm{t}}\right)^{1/2} \tag{8.2}$$

where k_p is the rate constant for propagation, $[M]$ is the macromer functional group concentration, and k_t is the rate constant for termination. This chapter is focused on the formation of crosslinked networks and hydrogels. In these systems, k_p and k_t vary as a function of macromer conversion or polymerization time on account of the changes in the diffusion constants of the reacting species caused by the reaction of the precursor solution [46]. At early macromer conversion, the viscosity of the precursor solution increases, limiting the radical diffusion rate and, thus, the rate constant for termination k_t. The viscosity-mediated decrease in k_t leads to an increase in the rate of polymerization termed *autoacceleration*. As the macromer conversion increases and a crosslinked network begins to form, the macromer functional groups also become diffusion-limited, causing a decrease in k_p and the an overall decrease in R_p. This phenomenon is referred to as *autodeceleration* and, combined with a consumption of functional groups, leads to a cessation of the polymerization.

The kinetics of photopolymerization reactions are more completely described elsewhere with extra considerations for polymer networks and hydrogels [6, 17, 47, 48]. In general, by accounting for multiphoton photoinitiation kinetics, these same principles can be applied to MPL photopolymerizations. Understanding the kinetics of the photopolymerization is critical in the fabrication of networks and hydrogels to be used as biomaterials, as the extent of reaction affects the final material properties and biological function. For instance, in hydrogels, the crosslinking density influences not only the mechanical properties of the gel but also the water swelling ratio, the transport of nutrients and waste, as well as the degradation kinetics [49].

8.4.2 Photoinitiator Selection

PIs are organic compounds capable of converting absorbed light into active chemical species that can initiate polymerization [50]. Depending on the chemical makeup of the parent PI, the active species can be cationic, anionic, or radical in nature, thereby enabling all types of chain-growth and step-growth polymerizations to be controlled with light. As both cationic and anionic polymerization methods are highly moisture-sensitive, polymerizations within a biological context are performed nearly exclusively by free-radical methodologies. Radical PI systems can contain one or more components. Type I initiators operate independently and are typically based on the heterolytic cleavage of aromatic ketones or disilanes. Type II PIs do not themselves cleave, but instead work in conjunction with a co-initiating species (usually non-photoactive) to induce polymerization, often through hydrogen abstraction. As greater energy is associated with bond photolysis than molecular excitation, type I PIs are largely confined to UV light sources ($\lambda < 400$ nm), while type II PIs can be initiated with light extending well into the visible range ($\lambda < 600$ nm).

A complete description of photoinitiators amenable to MPL is presented in Chapter 6. In general, this class of photoinitiators – those that can be activated by multiphoton excitation – can be applied to fabricate biomaterials; however, certain constraints apply. In all biomaterials, the constituent components as well as the associated degradation products or leachants must be minimally biocompatible. In this context, this means that cells, tissue, and organisms must tolerate the photoinitiators and their cleavage products post fabrication. Additional constraints apply for hydrogels, particularly those employed for cell encapsulation. Radical polymerization must operate under stringent physiological conditions if the interest is in performing the polymerization in the presence of live cells: body temperature (37 °C), pH (7.4), oxygen content (5%), and buffering salts must all be accommodated. Additionally, biological conditions further necessitate an aqueous environment, dramatically limiting the selection of available PIs. As most PIs absorb light through the presence of conjugated aromatic moieties, the significant hydrophobicity of these same functionalities leads to poor water solubility. If a candidate PI is able to meet these restrictive conditions, the small molecule need also be non-cytotoxic at the concentrations used to initiate photopolymerization, as well as in the range of light in which they operate. Typically, light with $\lambda \geq 365$ nm is considered appropriate for use in biological systems [51]. These wavelengths usually have enough energy to initiate polymerization, but not so much as to denature proteins and damage cellular DNA (known to occur rapidly at $\lambda = 254$ nm) [52, 53].

Despite the restrictive requirements posed in performing photopolymerization in the presence of live cells, several key PIs have been identified as invaluable for such tasks (Figure 8.2). Irgacure® 2959 [2-hydroxy-1-(4-(2-hydroxyethoxy)phenyl)-2-methylpropan-1-one] has perhaps the most widespread use, and decomposes under UV light ($\lambda = 365$ nm) to form both an acetone ketyl and a substituted benzoyl radical (Figure 8.2a) [51]. Though these active radicals initiate chain polymerization rapidly in aqueous systems, the parent PI suffers from poor water solubility (~1 wt%, ~50 mM). This, coupled with relatively low activity at cytocompatible wavelengths of light, dramatically limits the initiation rates [54]. More recently, lithium phenyl-2,4,6-trimethylbenzoylphosphinate (LAP) has been rediscovered as a water-soluble (~8.5 wt%, ~300 mM) type I PI that exhibits high molar absorptivity in the range $300 < \lambda < 400$ nm (Figure 8.2b) [55, 56]. Upon exposure to UV light, LAP dissociates into a 2,4,6-trimethylbenzoyl

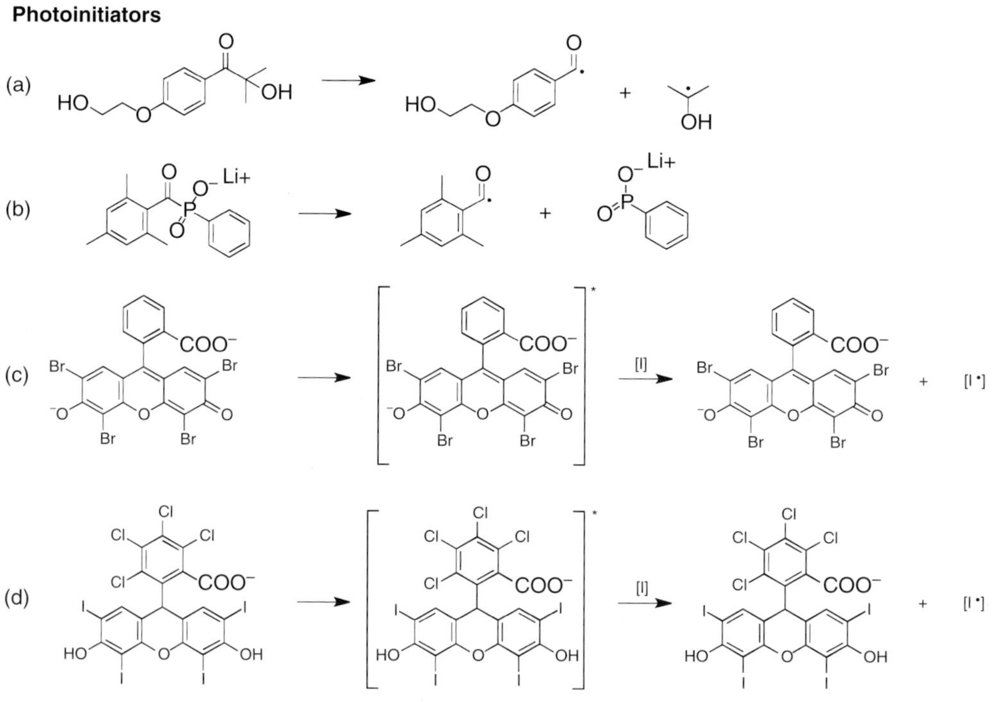

Figure 8.2 A selection of type I and type II photoinitiators amenable to MPL that have been employed in the presence of live cells. (a) Irgacure 2959. (b) Lithium phenyl-2,4,6-trimethylbenzoylphosphinate (LAP). (c) Eosin Y. (d) Rose Bengal.

and a phosphonyl radical, both of which can initiate polymerization efficiently in cellular systems. Though the LAP PI is not commercially available, its two-step synthesis is straightforward and nearly quantitative [56].

Several type II PIs have also been utilized for photopolymerization in the presence of cells. These most commonly used include rose Bengal, eosin Y (Figure 8.2c,d), and camphorquinone [57]. When exposed to light, these compounds get excited to a triplet state, which reacts with a hydrogen-donating co-initiator (e.g., an amine or thiol containing molecule) to produce an initiating free radical. Despite their wide usage, as well as their unique ability to perform effectively in the visible range, the photokinetics of type II systems are less understood than their type I counterparts [58].

8.4.3 Photopolymer Chemistries

8.4.3.1 Macromer Chemistries

The photopolymerization precursor solution also requires the selection of appropriate macromer chemistries for material fabrication. The macromer selection can be approached in two stages: (i) deciding on a backbone chemistry, and (ii) choosing a linking chemistry for network formation. Fortunately, single-photon photopolymerization of biomaterials and hydrogels has been employed extensively over the last few decades and has identified a broad range of backbone and linking chemistries for the formation of biocompatible polymer networks and hydrogels. In principle, the same toolbox can be adapted to MPL fabrication given that the photoinitiator and light source chosen provide a sufficient source of radicals to drive polymerization. This section serves to introduce a subset of the available toolboxes even though the specific chemistries have not yet been explicitly utilized with MPL. A brief overview of MPL biomaterials is presented in Section 8.5.

Typical non-hydrogel photopolymerized biomaterials employ a polymeric resin-like precursor solution that is composed of functional monomers, oligomers, or macromers and a suitable photoinitiator that can be polymerized rapidly into a solid. Common resins that have been applied to fabricate photopolymerized biomaterials include (meth)acrylates, epoxides, urethanes, styrenes, *n*-vinylpyrrolidone, and other vinylic small molecules [57]. In these systems, polymerization induces the formation of high molecular weight, insoluble, and entangled polymers that comprise a solid material. In parallel, functional oligomers or macromers are also used to make biomaterial based on non-hydrophilic backbone polymers. Such hydrophobic backbone chemistries include esters [59], carbamates [60], and carbonates [61] that are end-functionalized with polymerizable moieties such as vinyl groups or complementary thiol and ene functionalities [62]. Included in this class of materials are the biophotopolymers that have been used to fabricate microstructured biomaterials using MPL [63].

For the fabrication of hydrogels, the main constraint is that the backbone chemistry be hydrophilic. Decades of research on the use of hydrogels as biomaterials have provided a wealth of information on backbone chemistries for this application, all of which equally applies to MPL hydrogels [64]. In general, hydrophilic polymers can be segregated into two main classes: synthetic polymers and natural polymers. Synthetic polymers for hydrogel synthesis include poly(ethylene glycol) (PEG), poly(vinyl alcohol) (PVA), poly(acrylic acid) (PAA), and poly(2-hydroxyethylmethacrylate) (PHEMA) [65]. Of these, PEG or poly(ethylene oxide) (PEO) have found prolific use on account of their non-fouling behavior and ability to mimic the elasticity and transport of native tissue. Natural polymers for hydrogel fabrication include alginate, collagen/gelatin, hyaluronic acid, fibrin, chondroitin sulfate, and cellulose derivatives [66]. In either case, synthetic or natural, the polymers can be functionalized with a variety of polymerizable moieties (such as (meth)acrylates, (meth)acrylamides, or thiol–ene pairs) to render the backbone polymers amenable to photopolymerization. Once crosslinked via photopolymerization, the insoluble polymer network imbibes water and swells, resulting in a functional hydrogel [66].

In each material class – resins, networks, or hydrogels – the backbone chemistry often defines the bulk chemical properties of the material, which influence protein fouling, biocompatibility, cell adhesion, degradation, and mechanical integrity [12]. The linking chemistry defines the kinetics of formation, the microarchitecture, and crosslinking density [49]. Given the large toolbox, the orthogonal selection of backbone chemistry and linking chemistry spans a large parameter space of material properties. Additional information on the specific linking chemistries amenable to photopolymerization is provided in the following section.

8.4.3.2 Photochemical Polymerization and Degradation

Hundreds of photo-mediated reactions have been utilized for both the formation and degradation of polymers. In this section, we seek to highlight the diverse subset of these chemistries that have been performed within a biological context. Additive photoreactions are classified according to their utilization of PIs, photocaged reactive groups, or nonspecific radicals. Strategies that employ photolabile linkers to achieve subtractive degradation, as well as those offering reversible photofunctionalization, are also described. All discussed reactions are illustrated in Figures 8.3–8.7.

Initiator-Mediated Photopolymerization To date, the majority of photopolymerization reactions are initiator-mediated. As PI identity can be selected independently from that of the reactive species, reactions utilizing PIs offer increased tunability over initiation wavelengths, solvent composition, and reaction kinetics. Some common PI-mediated polymerization mechanisms that have been performed for biomaterial synthesis include vinyl chain polymerization, thiol–ene and thiol–yne reactions, and the photo-inducible copper-catalyzed azide–alkyne cycloaddition (pCuAAC) (shown in Figure 8.3).

Vinyl chain polymerization

Perhaps the most common photopolymerization reaction, namely radical chain polymerization of vinyl monomers (i.e., those that contain

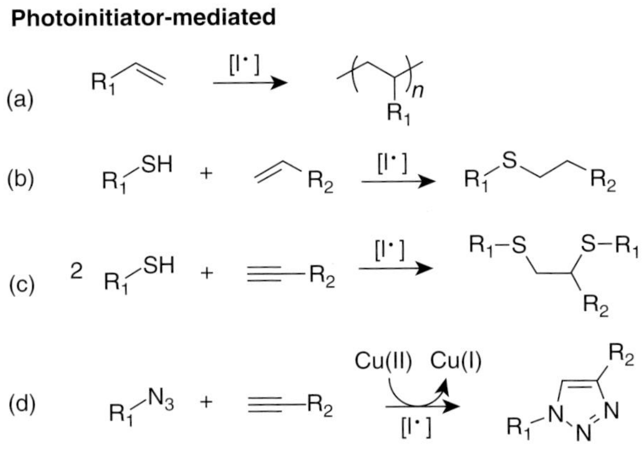

Figure 8.3 Photoinitiator-mediated polymerization reactions used for biomaterial synthesis and modification include (a) vinyl chain polymerization, (b) thiol–ene, (c) thiol–yne reactions, and (d) the photoinducible copper-catalyzed azide–alkyne cycloaddition.

carbon–carbon double bonds), occurs through a series of initiation, propagation, and termination steps [43]. During initiation, free-radical-containing initiator fragments attack the pair of π electrons on the vinyl monomer. This forms a bond between the initiator species and the monomer while simultaneously generating a new free radical at the chain end. The free radical on the growing polymer chains then reacts in a similar manner with additional monomer, iteratively increasing polymer length by one repeat unit during repeating steps of propagation. Polymer growth can be halted through a variety of different termination events, each of which destroys the active radical species. In a biological context, these can include reactions with oxygen, cellular proteins, DNA, and polysaccharides, making polymerizations in the presence of living cells more challenging [67]. Monomers are most typically substituted acrylates and methacrylates, where changes in substituent structure give rise to materials with different properties (e.g., temperature responsiveness, water content, compliance). These systems have been used extensively for the formation and modification of biomaterials and hydrogels [7, 26, 68].

Thiol–ene photoaddition

In this step-growth polymerization reaction between thiols and alkenes [62, 69], an active photoinitiator first abstracts a thiolic hydrogen to form a thiyl radical. The thiyl then propagates across the ene functional group to form a carbon-centered radical, which then abstracts a hydrogen from another thiol. Alternating steps of propagation and chain transfer give rise to a 1 : 1 addition of thiol- and ene-containing monomers, provided that the ene functionality exhibits no homopolymerization. Thiol–ene reactions proceed most quickly for electron-rich (e.g., vinyl ethers) or strained (e.g., norbornene) monomers. Notably, thiol–ene polymerizations work well with type II initiators, as the monomeric thiol can also serve as the hydrogen-donating co-initiator [70]. The reaction has been exploited for the formation of biomaterials derived from synthetic [71, 72] and natural polymers [73, 74], the formation of cyclic peptides [75], as well as in the biochemical decoration of ene-containing hydrogel networks [70, 76–78].

Thiol–yne photoaddition

Similar to the thiol–ene reaction, the thiol–yne reaction proceeds through the addition of an initiator-generated thiyl radical to an alkene [79]. The resulting carbon-centered radical abstracts a hydrogen from an additional thiol, thereby generating a vinyl sulfide intermediate and regenerating the sulfanyl radical. The monosubstituted vinyl sulfide can undergo further reaction with an additional thiyl radical via a thiol–ene reaction. The net result is the double addition of two thiols to a single alkyne to generate a dithioether. The uniqueness of this double reaction allows the facile creation of dendrimers [80, 81], the rapid generation of multivalent peptides [75], and crosslinked hydrogel materials to be formed from monomers containing just a single yne group [82, 83].

Photo-inducible copper-catalyzed azide-alkyne cycloaddition (pCuAAC)

The azide–alkyne Huisgen cycloaddition represents a 1,3-dipolar cycloaddition between an azide and a terminal or internal alkyne to give a triazole linkage. Though this reaction can be performed in the absence of a catalyst at high temperatures, pressures, or through the utilization of strained alkynes, the reaction kinetics can be increased dramatically through catalysis with Cu(I) (by a factor of $\sim 10^7$) [84–86]. In recent work by the Bowman group, photogenerated initiator fragments were shown to reduce Cu(II) to Cu(I) transiently, thereby providing photochemical control over the copper-catalyzed azide–alkyne cycloaddition (CuAAC) [87, 88]. This reaction has been utilized for photopolymerization/functionalization of hydrogel materials.

Photocage-Mediated Photoconjugation Photocages function as molecular gatekeepers for the regulation of chemical reactions. When intact, these compounds serve to block the reactivity of two species through physical masking. Upon photolysis, the reactive functional group is liberated, thereby permitting species reaction and photoconjugation to occur. To date, photocage-based regulation of the strain-promoted azide–alkyne cycloadditon (SPAAC), oxime ligation, Michael-type addition, and enzymatic crosslinking have proven the most promising for biological studies, and they are shown in Figure 8.4.

Strain-promoted azide–alkyne cycloadditon (SPAAC)

A strain-promoted reaction between cyclooctynes and azides has been established by the Bertozzi group as an effective means to perform traditional Huisgen cycloadditions rapidly without the need for a cytotoxic copper catalyst [89, 90]. The phototriggerability of the reaction has been demonstrated by the Popik group, who masked the triple bond of a dibenzocyclooctyne as cyclopropenone [91]. Upon exposure to UV light

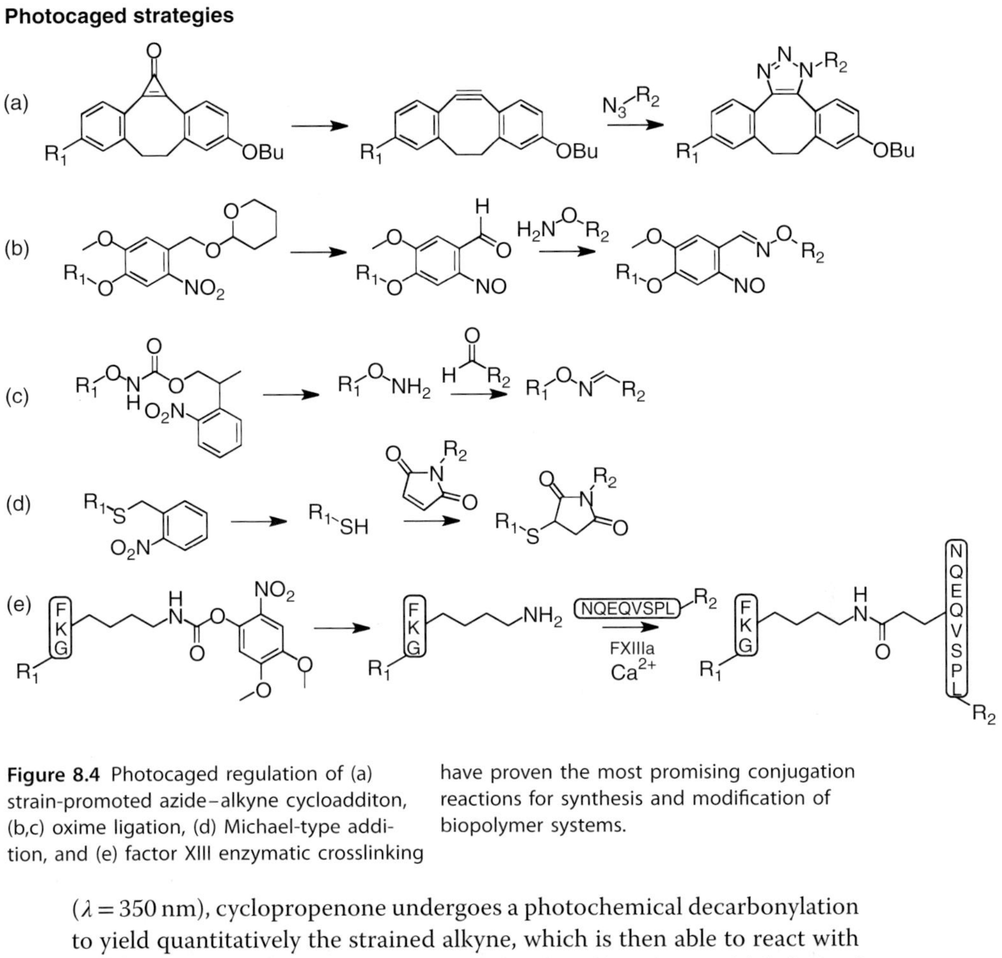

Figure 8.4 Photocaged regulation of (a) strain-promoted azide–alkyne cycloadditon, (b,c) oxime ligation, (d) Michael-type addition, and (e) factor XIII enzymatic crosslinking have proven the most promising conjugation reactions for synthesis and modification of biopolymer systems.

($\lambda = 350$ nm), cyclopropenone undergoes a photochemical decarbonylation to yield quantitatively the strained alkyne, which is then able to react with azides. This reaction has been utilized for the photochemical labeling of cellular glycoproteins and surfaces [92–94].

Oxime ligation

The condensation reaction of alkoxyamines with aldehydes or ketones yields oxime linkages that are both formed and stable under physiological conditions. Photocontrol over oxime ligation has been independently effected through strategies that cage each of the two reactive species. Barner-Kowollik and coworkers demonstrated that the photocleavage ($\lambda = 365$ nm) of *o*-nitrobenzyl derivatives, which typically leads to the generation of a nitrosobenzaldehyde, can be exploited to give spatiotemporal control over aldehyde formation [95]. The photogenerated aldehydes then react with hydroxylamines to form oxime linkages. This reaction sequence has been utilized to pattern fluorescent peptides onto reactive surfaces. In an alternative strategy developed by Dumy, the alkoxyamine has been photocaged with a 2-(2-nitrophenyl)propyloxycarbonyl (NPPOC) group [96]. Under brief irradiation with UV light ($\lambda = 365$ nm), NPPOC undergoes irreversible cleavage to yield CO_2, a styrene byproduct, and the reactive alkoxyamine. This reaction has been exploited for the conjugation of small molecules and peptides to glass surfaces [96–98].

Michael-type addition

The Michael-type addition of nucleophiles to electron-deficient alkenes represents one of the most common reactions in organic chemistry. Though a variety of nucleophilic Michael donors/acceptors exist, thiol–maleimide and amine–acrylate represent important variants, both of which have been controlled with light. The Shoichet group has demonstrated that thiols can be masked with both 2-nitrobenzyl- and coumarin-based photocages, and will subsequently react with maleimides upon irradiation [99–101]. Moreover, the Bowman group has demonstrated that NPPOC can serve as a photolabile protecting group for primary amines, which enables thiol-Michael additions with acrylates to be phototriggered [102]. These chemistries have been used to form hydrogel biomaterials [103] as well as to guide 3D cell function through immobilized biomolecules [99, 101].

Light-activated enzymatic crosslinking reaction

A strategy for light-activated enzymatic crosslinking utilizing factor XIII (FXIIIa) has been recently introduced by both the Lutolf and Segura groups [104, 105]. FXIIIa represents a transglutaminase enzyme that catalyzes covalent amide bond formation between the ε-amino group on a lysine (K) and the γ-carboxamide residue of glutamine (Q) of FKG and NQEQVSPL peptides. When the critical lysine residue is caged with an *o*-nitrobenzyl functional group, FXIIIa is unable to crosslink the two independent peptides. After cage photolysis, however, the two peptides are rapidly joined through FXIIIa-mediated chemoenzymatic ligation. These approaches have been utilized to pattern short peptides and synthetic proteins into hydrogels and used to guide cellular growth within a material.

Nonspecific Photoconjugation Several photoreactive groups have been developed that couple nonspecifically with target molecules upon activation with UV light exposure. The indiscriminant nature by which these moieties couple to local molecules enables rapid addition as well as covalent linkage between biomacromolecules that are in close proximity in 3D space. Aryl azides, diazirines, and benzophenones represent the most common functionalities for nonspecific photoconjugation and are highlighted in Figure 8.5.

Aryl azides

Perhaps the most popular of the nonspecific photoconjugation reactions, aryl azides form short-lived nitrenes upon exposure to UV light ($250 < \lambda < 350$ nm) [106]. These intermediate nitrenes can either insert nonspecifically into carbon–hydrogen bonds, or rearrange to form ring-expanded dehydroazepines prior to addition reaction with nucleophiles or carbon–hydrogen bonds. Photocrosslinking of aryl azides has

Nonspecific

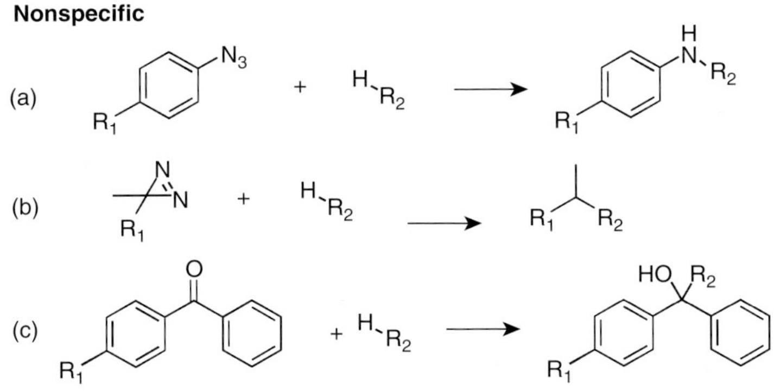

Figure 8.5 A variety of nonspecific chemical conjugation reactions have been utilized for photo-mediated biomaterial synthesis including those involving (a) aryl azides, (b) diazirines, and (c) benzophenones.

been extremely beneficial in capturing and probing protein interactions [107, 108], patterning cell-adhesive proteins to surfaces [109], and identifying peptide–receptor binding domains [110].

Diazirine

In the presence of UV light ($330 < \lambda < 370$ nm), diazirine compounds undergo photolysis to form a highly reactive carbine intermediate while simultaneously giving off nitrogen gas [111]. The resulting carbine yields efficient insertion into hydrogen–carbon and hydrogen–nitrogen bonds, as well as with unsaturated bonds within target molecules. These molecules have been useful in studying the effects of protein interactions in cells [112, 113], crosslinking protein-based hydrogels [114], and identifying targets of membrane receptors [115, 116].

Benzophenone

When exposed to UV light, benzophenone photolysis yields a highly reactive triplet-state ketone intermediate [117]. The energized electron of this activated benzophenone can insert into hydrogen–carbon bonds within the target molecules. Interestingly, the activated intermediate relaxes to yield the original benzophenone moiety when no covalent linkage is formed. The resulting potential for multiple activation gives rise to higher overall yields of photocrosslinking than with phenyl azide and diazirine crosslinkers. Though benzophenone as a photoaffinity tag permeates much of biochemistry [118], the molecule has also been used to attach peptides and polymer films to solid surfaces [119–121].

Light-Mediated Degradation Though the majority of photochemical reactions result in the formation of covalent bonds, there is growing interest in photoactive linkers that undergo cleavage in response to photonic irradiation. When incorporated into the backbone of polymeric species, these moieties provide a tunable handle for network depolymerization. Here, we highlight (Figure 8.6)

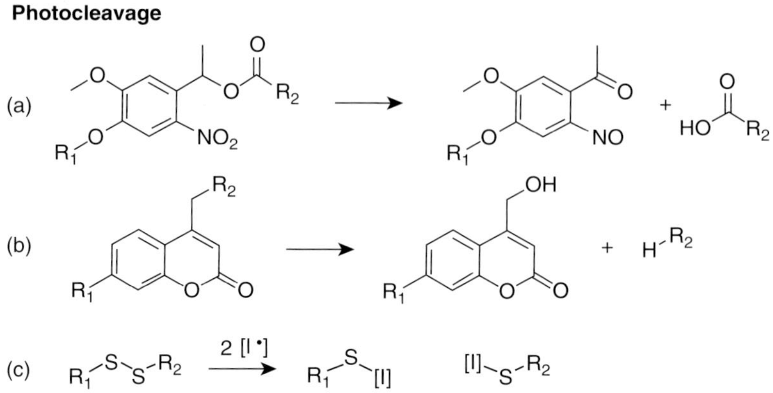

Figure 8.6 Photolabile linkages including (a) *o*-nitrobenzyl esters, (b) coumarins, and (c) disulfide-containing compounds can be photochemically cleaved under physiological conditions using MPL, enabling on-demand biopolymer depolymerization.

o-nitrobenzyl esters, coumarins, and disulfide-containing compounds as photocleavable species that can be controlled in a biological context.

***o*-Nitrobenzyl esters**
As one of the most heavily utilized photocleavable linkers, the use of *o*-nitrobenzyl esters for light-based molecular cleavage extends back to the 1970s [122]. Upon photoexcitation, the nitrobenzyl group forms an *aci*-nitro intermediate through either a singlet or triplet excited state. This intermediate further decays to generate both a nitroso and an acid/amide byproduct (depending on parent substitution) [123]. Cleavage rates and photoactive wavelengths ($250 < \lambda < 400$ nm) can be tuned by varying molecular substituents [124–126]. These reactions have been used for the uncaging of proteins [127, 128], controlling cell adhesions [129, 130], photodegrading polymers and biomaterials [70, 131–134], and probing the effects of physical material cues on cell function [135–137].

Coumarin methylester
When irradiated with UV light ($350 < \lambda < 450$ nm), coumarin methylesters undergo heterolytic ester photosolvolysis, leading to the corresponding coumarin methanol and a carboxylic acid [138–140]. Though some coumarins are known to exhibit reversible homodimerization upon exposure to light [141], the methylester variants preferentially undergo photocleavage when irradiated. Coumarins methylesters have been used in the synthesis of caged fluorophores for biological imaging [123], material degradation [142, 143], and drug delivery [144].

Disulfide displacement reactions
Disulfide linkages, formed through the oxidation of thiols, have been demonstrated to be susceptible to radical cleavage by carbon- and phosphorus-center radicals [145, 146]. As such, disulfide linkages can be

selectively displaced with radical species, including initiator fragments generated from exposure of PIs to light. Anseth *et al.* demonstrated that polymer-based hydrogels crosslinked with disulfide linkages could undergo complete and rapid photodegradation when exposed to sufficient amount of initiator radicals [147].

Reversible Photoconjugation Though the majority of the utilized chemistries have focused on either photomediated bond formation or cleavage, great interest lies in the development of photochemical processes that can be controlled in a fully reversible manner. Two of the most promising systems to date that can be used in a biological context include the reversible dimerization of anthracene and the allyl sulfide addition–fragmentation chain-transfer reaction. Both examples are shown in Figure 8.7.

Anthracene dimerization

Anthracene is capable of absorbing light over a broad spectrum ($250 < \lambda < 400$ nm), most frequently yielding a fluorescence emission ($390 < \lambda < 450$ nm). However, when anthracene is exposed to UV light ($\lambda = 360$ nm), a dimerization reaction can also occur [148, 149]. Interestingly, this reaction can be partially reversed when the dimers are illuminated with higher intensity light ($260 < \lambda < 280$ nm) [150]. These sequential reactions have been used for photocontrolled drug delivery [151], biomaterial formation and degradation [152, 153], and nanoparticle manipulation [154].

Allyl sulfide addition–fragmentation chain transfer

Allyl sulfides, typically employed as the transfer agent in reversible addition–fragmentation chain transfer (RAFT) polymerizations [155],

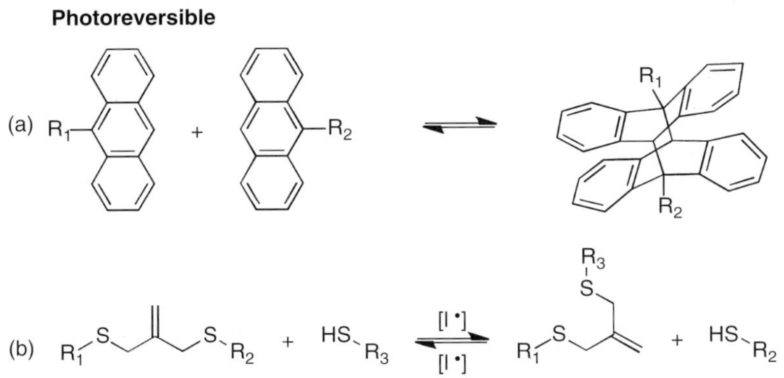

Figure 8.7 Reversible photochemistries open up the door to dynamic spatiotemporal control over biopolymer addition and degradation, as well as material formation and functionalization. The most promising examples to date include (a) the reversible dimerization of anthracene, and (b) the allyl sulfide addition–fragmentation chain transfer reaction.

can undergo reversible exchange with thiol-containing molecules [156, 157]. Here, a thiyl radical attacks the carbon–carbon π bond of the allyl sulfide, resulting in an unstable trifunctional intermediate. The instability of this intermediate leads to a β-cleavage, which results in the addition of the initial attacking species while regenerating a new double bond. As the regenerated double bond is susceptible to attack by other thiyl radicals, the linkage is said to be "living." Initial thiyl radicals are readily generated using PIs in the presence of light. Photoinitiated RAFT has been used extensively for the synthesis of biopolymers [158], to induce plasticity in crosslinked networks [159], to create self-healing materials [160], and to reversibly decorate hydrogel networks with synthetic peptides [161].

8.5 Biomaterial Fabrication

Initial work with MPL photopolymerization focused on the development on non-biological materials [162]. Polymerizable resins (or negative-tone photoresists) have been employed in MPL material fabrication, often starting as viscous liquids or amorphous solids. These systems can be easily transformed into complex structures such as interconnected microchannels [163], interlocking chains [164], or renditions of the Venus de Milo [165]. The field of photonics was an early adopter of the MPL paradigm for application-based materials. MPL polymerization has been employed to generate functional waveguides [166], interferometers [167], microlenses [168], and microlasers [169]. Based on the ability to generate features on the sub-micrometer scale, the community has adapted these approaches to generate photonic crystals with stop bands in the visible range. In other approaches, MPL polymerization has been exploited to generate active mechanical devices including springs, cantilevers, and nanorotors [162]. Based on the successful implementation of these techniques, researchers began exploring the use of polymer precursors for the fabrication of structured materials for biological applications.

Cells reside in the body within a complex, dynamic, and highly structured ECM, which choreographs cell fate through bidirectional cell–ECM interactions. Many spatiotemporal signals are integrated by the cell to inform output cellular functions including adhesion, proliferation, differentiation, migration, and tissue morphogenesis [170]. The specific mechanisms through which the ECM directs cell function is still not fully understood. Biomaterial strategies provide attractive platforms to dissect complex and dynamic cell–ECM communication as well as to engineer scaffolds to leverage these cues for regenerative medicine applications [15, 171]. Within this paradigm, photopolymerization reactions via MPL enable precise recapitulation of ECM properties (e.g., biophysical and biochemical signals) in 3D biomaterials with nanometer and micrometer resolution. When exploiting MPL to generate photopolymerized biomaterials, it is important to consider the essential functions and design constraints that apply to materials for cell culture applications. Minimally, the material should support cell colonization, migration, growth, and differentiation, as well as provide appropriate biophysical and biochemical cues for the specific cell type of interest [172]. In general, two classes of MPL photopolymerized materials have been introduced as biomaterials: biocompatible networks into which cells are introduced post fabrication and hydrogels into which cells can be directly encapsulated.

MPL polymerization affords the generation of highly ordered structures in three dimensions that are biocompatible, degradable, and easily fabricated – all important features for biological implementation. These techniques have been adapted to generate cytocompatible microneedles, biodegradable scaffolds, and ossicular replacement prostheses (Figure 8.8) [173, 174, 177]. To better recapitulate the context of the *in vivo* milieu, MPL has been extended to the fabrication of hydrated polymer networks in the form of hydrogels for cell culture applications [172]. Hydrogels are structurally similar to the native ECM and present an attractive class of materials for tissue engineering and regenerative medicine applications [171]. Specific examples include woodpile lattices, microfabricated porous structures, and vascular trees (Figure 8.8) [42, 175, 178, 179]. Additional work has focused on the manipulation of native proteins for biomedical applications (Figure 8.8) [176, 180–182].

8.6 Biomaterial Modulation

In addition to the synthesis of new biopolymers and biomaterials, photochemical reactions have been utilized to *modify* existing platforms at user-specified points in time and space. Through appropriately developed chemistries, both the physical and chemical material makeup can be altered on demand and with resolution below that of a single mammalian cell. Such high spatiotemporal control over biomaterial composition is critical in capturing the dynamic heterogeneity found in native tissue [15].

Spatiotemporally defined biochemical modulation of existing materials has been performed through both additive or subtractive light-based strategies. In perhaps the most common addition-based chemistry, radical chain polymerization of acrylate-functionalized peptides or proteins has yielded hydrogel materials that promote endothelial cell adhesion as well as enhanced vascular lumen formation through the patterned introduction of the arginylglycylaspartic acid (RGD) binding motif and growth factors such as vascular endothelial growth factor (VEGF) [183]. An alternative reaction between thiols and alkenes has been used to guide changes in cell function within a polymer hydrogel by photopatterning thiol-containing biomolecules into ene-functionalized networks (Figure 8.9a) [184, 185]. Similarly, simultaneous patterning of several proteins can be accomplished by covalently immobilizing protein physical-binding partners (e.g., biotin, barnase) via a photocaged thiol–maleimide reaction [101]. A light-activated enzymatic crosslinking employing factor XIII has also been used

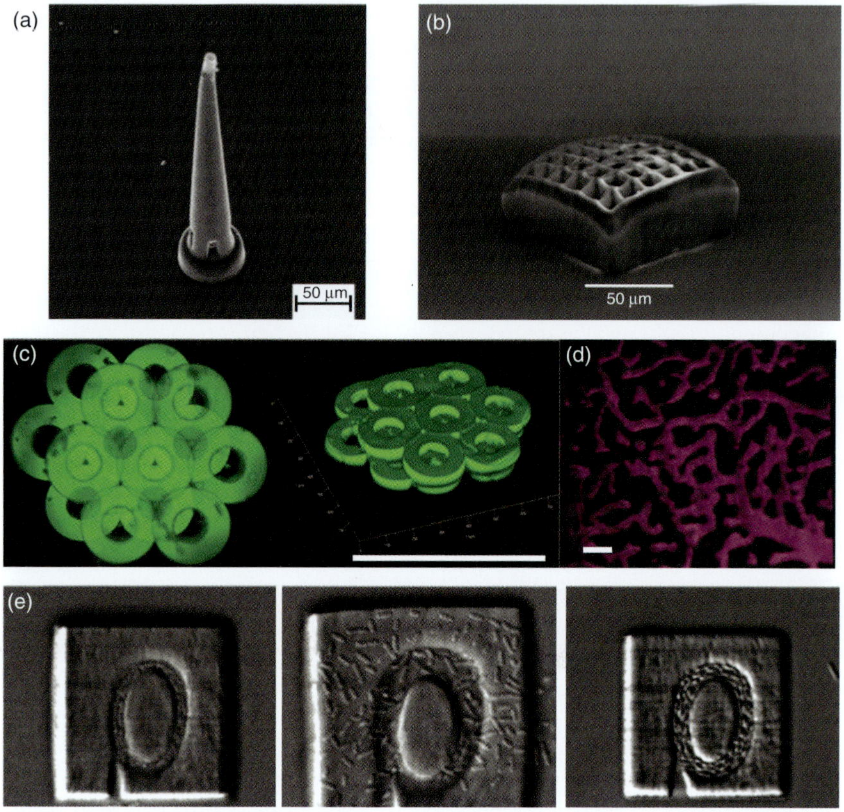

Figure 8.8 Biomaterials fabricated by MPL photopolymerization. (a) Biocompatible microneedles. Scale bar = 50 μm. (Doraiswamy 2006 [173]. Reproduced with permission of Elsevier.) (b) Structured cell culture scaffold. Scale bar = 50 μm (Claeyssens 2009 [174]. Reproduced with permission of American Chemical Society.). (c) Hydrogel cell culture scaffold. Scale bar = 500 μm. (Qin 2013 [175]. Reproduced with permission of Wiley.) (d) Patterned vascular tree. Scale bar = 50 μm. (Culver 2012 [42]. Reproduced with permission of Wiley.) (e) Biochemically responsive protein hydrogel. Scale bar = 5 μm. (Kaehr 2008 [176]. Reproduced with permission of PNAS.)

to photopattern peptides to direct mesenchymal stem cell outgrowth into a 3D material (Figure 8.9b) [104].

Photochemistries have also enabled material biochemical alteration through the selective removal of cues within a material. *o*-Nitrobenzyl (*o*NB) ester cleavage has been utilized for the release of RGD, thereby enhancing chondrogenesis of human mesenchymal stem cells [186]. Additionally, *o*NB cleavage has been used to create functional surfaces to study collective and patterned cell migration [187]. Photocleavage of coumarin derivative dimers has been utilized for a variety of drug delivery applications [141, 188, 189]. While the aforementioned techniques have been used to either introduce or remove biochemical signals, the combination of different photochemical reactions can enable the reversible immobilization of

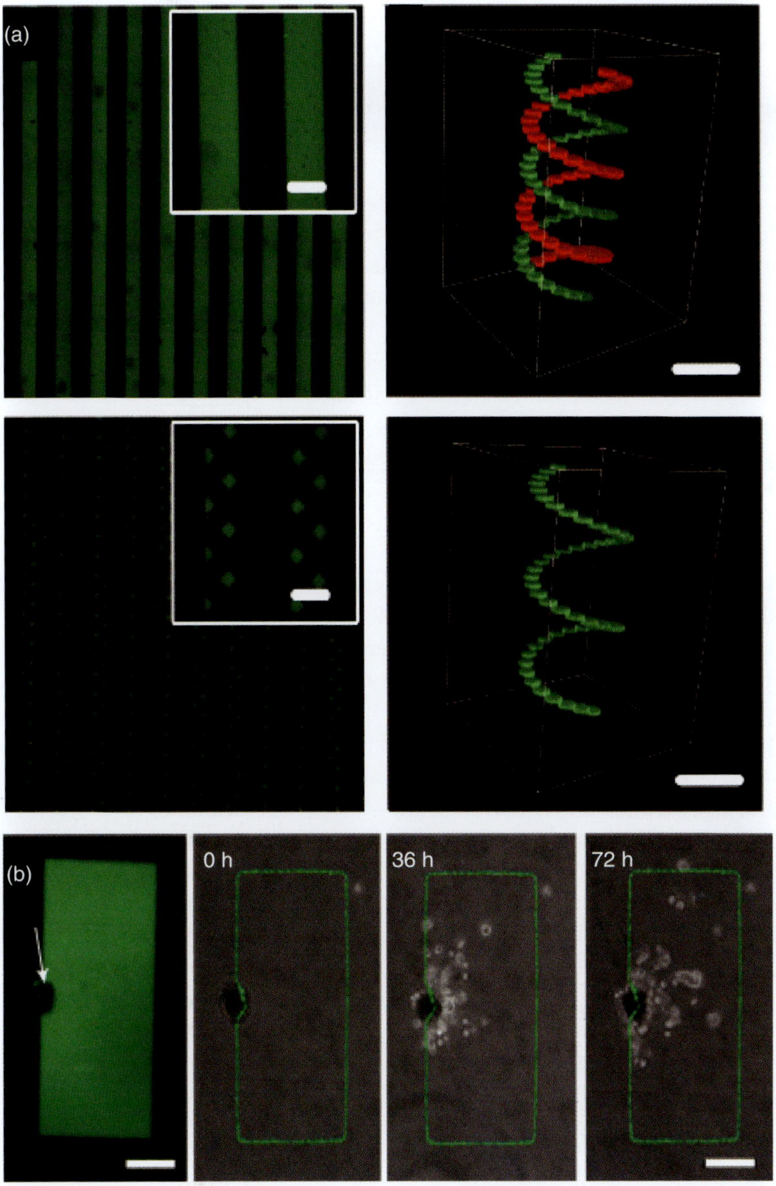

Figure 8.9 Photopatterned modulation of network biochemical properties has been achieved within 3D gels and in the presence of cells using MPL through a variety of chemistries. (a) Sequential thiol–ene coupling and *o*NB photocleavage enables engineered peptides to be reversibly immobilized within a gel network. Scale bars = 200 μm. (DeForest 2011 [184]. Reproduced with permission of Wiley.) (b) Factor XIII-mediated peptide patterning directs mesenchymal stem cell outgrowth into a 3D structure. Scale bars = 200 μm. (Mosiewicz 2013 [104]. Reproduced with permission of Nature Publishing Group.)

such cues. By combining a visible-light mediated thiol–ene addition reaction with a UV-mediated *o*NB cleavage, DeForest and Anseth demonstrated the reversible biochemical functionalization of a polymer-based material (Figure 8.9b) [184]. This approach allowed dynamic control over cell attachment to a material surface by introducing and, subsequently, removing adhesive ligands from the hydrogel in the presence of cells.

Photochemical reactions have also proven advantageous in controlling network biomechanics in the presence of cells. Strategies to increase culture platform compliance have focused on the photochemical increase in material crosslinking density. The radical chain photopolymerization of (meth)acrylates has been utilized to increase culture stiffness in the presence of cells [26, 190]. Through these same chemistries, photochemical alteration in network enzymatic degradability has enabled spatial control over mesenchymal stem cell differentiation (Figure 8.10a) [193, 194]. Thiol moieties have been photocaged in order to delay additional crosslinking with reactive vinyl groups upon exposure to light. These materials have been shown to influence the migration of human mesenchymal stem cells (hMSCs) [195].

Material stiffness can also be decreased through the use of photochemical reactions. The Anseth group has demonstrated extensively the bulk and local photodegradation of materials in the presence of live cells through the introduction of an *o*NB moiety in the backbone of hydrogel networks (Figure 8.10b) [70, 133, 134]. In a similar approach, coumarin moieties introduced into hydrogel backbones yielded visible-light-degradable materials [143]. The ability to control the mechanical properties of a gel dynamically has been used to guide cell migration and outgrowth [70, 133, 143], to probe cytoskeletal tension within cells (Figure 8.10c) [136], to assess biomechanical memory of hMSCs [137] and in a growing list of other applications.

Photochemical reactions have demonstrated precise control over many critical aspects of the cellular microenvironment, including those both chemical and physical in nature. Through the usage of MPL, micrometer-scale tunability in such parameters has been demonstrated. Given the size scale of a typical mammalian cell (~10 μm), we expect these strategies to prove useful in modulating material properties about individual cells and in better understanding biological response to dynamic extracellular signals.

8.7 Biological Design Constraints

In general, when fabricating materials with MPL photopolymerization, it is critical to balance the power needed to initiate polymerization against the power that will damage or cavitate the precursor solution. For a PEG-based hydrogel, it was found that pulse energies in excess of 2.3 nJ induced cavitation within the highly water swollen gel [136]. When material fabrication or modification is conducted in the presence of cells or organisms, it is also necessary to balance the photophysics of

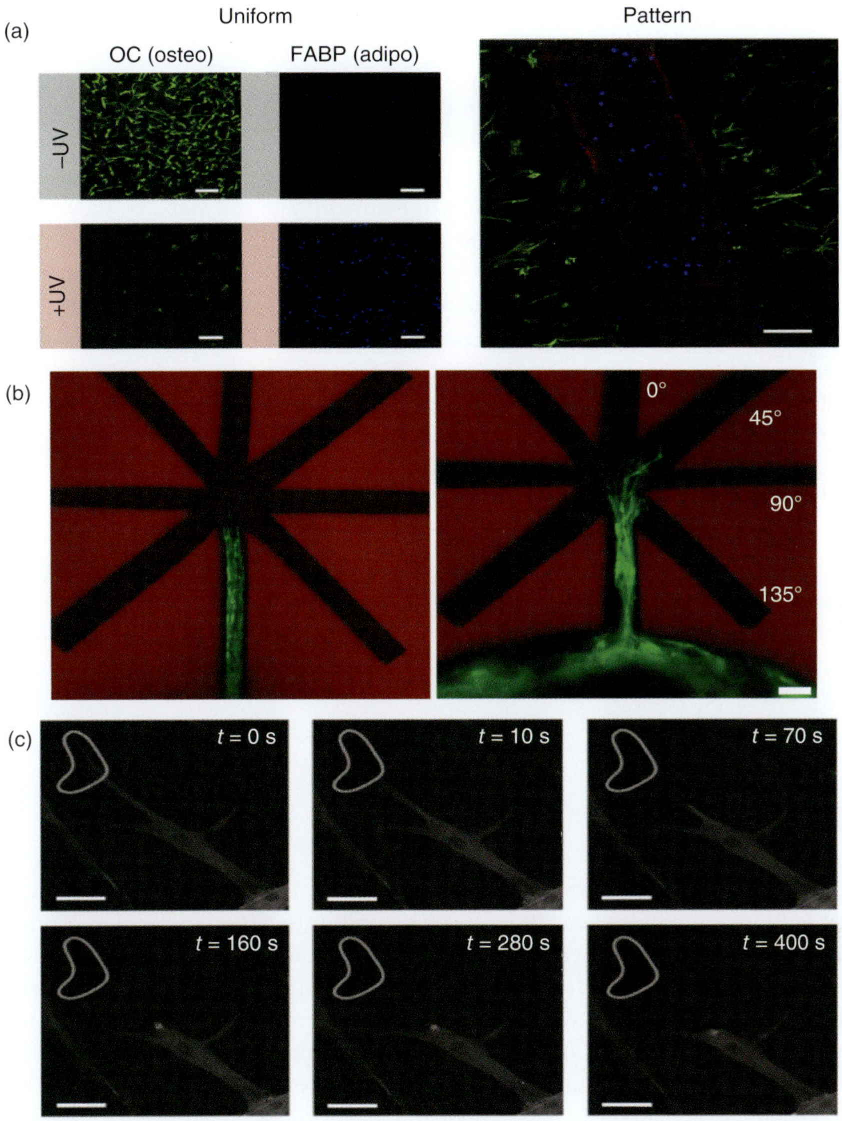

Figure 8.10 Photochemical alteration in network biomechanics has proven an indispensable tool in directing and understanding cell fate. (a) Patterned sequential network crosslinking has afforded spatial control over mesenchymal stem cell differentiation. Scale bars = 100 μm. (Khetan 2010 [191]. Reproduced with permission of Elsevier.) (b,c) MPL has been utilized to photodegrade complex channels within 3D gels [192] and to probe cytoskeletal tension within cells. Scale bars = 100 μm in (b) and 20 μm in (c). (Tibbitt 2010 [136]. Reproduced with permission of Royal Society of Chemistry.)

the reaction with the potential for photodamage to the biological specimen. For instance, it has been shown (and utilized for experimentation) that sufficient pulse energies can sever actin filaments within cells [38, 192, 196]. Additional work has shown that femtosecond pulsed two-photon lasers can be utilized in the presence of cells, but pulse energies above 1.5 nJ induced intracellular photoablation [196] and those above 4 nJ induced cell death [197]. In order to avoid nonspecific damage to intracellular structures and to maintain high cell viability, the experimenter should operate the laser with a pulse energy below 1.5 nJ. As a caveat, these guidelines were determined in non-radical-generating systems and there is the potential for additional cellular damage based on the type and concentration of radicals generated during polymerization. The constraints of biology may further confine the operating space for MPL photopolymerization in the presence of cells; however, the benefit of a small focal volume and little out-of-plane absorption confines potential damage to a narrow window through which the focal point is rastered.

Additional constraints exist in the use of MPL for biological applications. The materials need to be functional and integrate with biological systems, as well as be comprised of components that are compatible with living systems, including each of the constitutive components and potential leachants. While the majority of the chemistries in this chapter have been previously utilized for biomedical applications, novel chemistries should be assessed for compatibility with biological systems.

8.8 Biologic Questions

As advances in MPL fabrication of biocompatible materials are achieved, the class of biomaterials presents unique opportunities to address several questions in biology and regenerative medicine. Both fields are beginning to appreciate the utility of 3D scaffolds for cell culture, as opposed to 2D tissue-culture polystyrene (TCPS), that can better recapitulate critical aspects of the native ECM [198]. Three-dimensional systems facilitate improved tissue morphogenesis and more physiologic culture environments for many cell types of the body [171]. MPL-fabricated materials complement this need for 3D scaffolds, as they can be patterned with micrometer-scale resolution (sub-cellular length scales) that enable researchers to answer questions related to how 3D geometry, topography, and structure affect cell function and tissue organization [199–201]. Several reticulated materials are already being employed for 3D cell culture [172], and, uniquely, MPL allows rapid fabrication of geometrically defined scaffolds as tissue replacements for regenerative medicine. For example, patient-specific devices can be fabricated utilizing medical imaging to assess the geometric need, which can be translated into a digitally rastered structure and printed in a material of interest using MPL.

Excitingly, the ability to manipulate 3D hydrogel biomaterials with MPL (see Section 8.6) uncovers a unique ability to address questions related to how dynamic changes in the ECM influence cell function and tissue assembly. New directions in the field of 4D biology (3D scaffolds that are dynamic in time) rely on user-programmable and externally tunable chemistries to introduce or remove specific cues from the culture microenvironment in real time [15]. Combining these dynamic and user-tunable chemistries with MPL allows the user to control the biophysical and biochemical nature of the cellular environment on the micrometer scale. Similar approaches allow biologists to study how changes in material stiffness influence cell proliferation and differentiation [137] or to investigate how adhesive ligand presentation or removal influences tissue development and function on the cellular level [133]. Additionally, selective photochemistries now allow patterning and removal of bioactive ligands or full proteins to investigate how subcellular localization of specific biochemical moieties may direct intracellular signaling. Moreover, the ability to control these reactions in space and time opens the door to the creation of anisotropic materials and synthetic tissues to study heterogeneous tissue formation and function. Materials scientists and MPL experts must interface and interact with cell and developmental biologists to employ these burgeoning techniques to address critical questions in these fields.

8.9 Outlook

Combining traditional photopolymerization techniques with advances in multiphoton lithography through MPL enables the fabrication of a useful class of photopolymerized biomaterials with defined geometry, architecture, and micrometer-scale features. The broad scope of amenable chemistries for biomaterial and hydrogel synthesis (Section 8.4) and the increasing accessibility of multiphoton technologies will allow MPL-fabricated biomaterials to impact the fields of tissue engineering and regenerative medicine. Specifically, highly defined tissue replacements can be formed from spatially tailored MPL biomaterials for organ therapy. Furthermore, modulatable hydrogels with dynamic subcellular biophysical and biochemical patterning will unlock new understanding in stem cell biology.

As these approaches see continued use and broader applications, there remains a need for an extended toolbox for MPL photopolymerizations. Users will benefit from novel biocompatible photoinitiators with efficient multiphoton cross sections that respond to unique wavelengths. Water-soluble and biocompatible photoinitiators are particularly attractive for hydrogel manipulation via photopatterning in the presence of cells. Additionally, alternative chemistries for the crosslinking of polymer networks and hydrogels that are compatible with cell culture and especially those that facilitate dynamic or reversible crosslinking are desirable. Ideally, contributions will also be made to the toolbox of chemistries that enable multiphoton addition and cleavage reactions within fabricated biomaterials for advanced studies of how dynamic changes in the biophysical and biochemical nature of the ECM influence cell and tissue function. Expanding this toolbox will require interactions among polymer chemists, photochemists, materials scientists, and biologists to develop advanced techniques for MPL biomaterials.

Continued development on the access to multiphoton sources, which facilitate the necessary techniques for MPL and software that makes these approaches more user-friendly, will also increase use. As the 3D printing industry matures, many of the hardware and software approaches, in principle, can be adopted for advanced photopolymerization approaches via MPL. In total, the outlook for the use of MPL photopolymerizations for the fabrication and modification of biomaterials and hydrogels is bright. Continued interactions at the interface of photopolymerizations, materials science, and biology will allow MPL biomaterials find broad use in the fields of tissue engineering, regenerative medicine, and stem cell biology.

References

1. Lyman, D.J. (1974) Polymers in medicine. *Angew. Chem., Int. Ed. Engl.*, **13**, 108–112.
2. Langer, R. (1998) Drug delivery and targeting. *Nature*, **392**, 5–10.
3. Langer, R. and Vacanti, J.P. (1993) Tissue engineering. *Science*, **260**, 920–926.
4. Langer, R. and Peppas, N.A. (2003) Advances in biomaterials, drug delivery, and bionanotechnology. *AIChE J.*, **49**, 2990–3006.
5. Oster, G. and Yang, N.L. (1968) Photopolymerization of vinyl monomers. *Chem. Rev.*, **68**, 125–151.
6. Anseth, K.S., Bowman, C.N., and Peppas, N.A. (1994) Polymerization kinetics and volume relaxation behavior of photopolymerized multifunctional monomers producing highly cross-linked networks. *J. Polym. Sci. Part A - Polym. Chem.*, **32**, 139–147.
7. Burdick, J.A. and Anseth, K.S. (2002) Photoencapsulation of osteoblasts in injectable rgd-modified peg hydrogels for bone tissue engineering. *Biomaterials*, **23**, 4315–4323.
8. Dorkenoo, K.D., Klein, S., Bombenger, J.P., Barsella, A., Mager, L., and Fort, A. (2006) Functionalized photopolymers for integrated optical components. *Mol. Cryst. Liq. Cryst.*, **446**, 151–160.
9. Fouassier, J.P., Allonas, X., and Burget, D. (2003) Photopolymerization reactions under visible lights: principle, mechanisms and examples of applications. *Prog. Org. Coat.*, **47**, 16–36.
10. Decker, C. (1987) Uv-curing chemistry – past, present, and future. *J. Coat. Technol.*, **59**, 97–106.
11. Nguyen, K.T. and West, J.L. (2002) Photopolymerizable hydrogels for tissue engineering applications. *Biomaterials*, **23**, 4307–4314.
12. Peppas, N.A., Hilt, J.Z., Khademhosseini, A., and Langer, R. (2006) Hydrogels in biology and medicine: from molecular principles to bionanotechnology. *Adv. Mater.*, **18**, 1345–1360.
13. Juodkazis, S., Mizeikis, V., and Misawa, H. (2009) Three-dimensional microfabrication of materials by femtosecond lasers for photonics applications. *J. Appl. Phys.*, **106**, 051101.
14. Bates, C.M., Maher, M.J., Janes, D.W., Ellison, C.J., and Willson, C.G. (2014) Block copolymer lithography. *Macromolecules*, **47**, 2–12.
15. Tibbitt, M.W. and Anseth, K.S. (2012) Dynamic microenvironments: the fourth dimension. *Sci. Transl. Med.*, **4**, 160ps24.

16. Lovestead, T.M., O'Brien, A.K., and Bowman, C.N. (2003) Models of multivinyl free radical photopolymerization kinetics. *J. Photochem. Photobiol. A: Chem.*, **159**, 135–143.
17. Goodner, M.D. and Bowman, C.N. (2002) Development of a comprehensive free radical photopolymerization model incorporating heat and mass transfer effects in thick films. *Chem. Eng. Sci.*, **57**, 887–900.
18. Elisseeff, J., Anseth, K., Sims, D., McIntosh, W., Randolph, M., and Langer, R. (1999) Transdermal photopolymerization for minimally invasive implantation. *Proc. Natl. Acad. Sci. U.S.A.*, **96**, 3104–3107.
19. Sebra, R.P., Masters, K.S., Bowman, C.N., and Anseth, K.S. (2005) Surface grafted antibodies: controlled architecture permits enhanced antigen detection. *Langmuir*, **21**, 10907–10911.
20. Bratton, D., Yang, D., Dai, J.Y., and Ober, C.K. (2006) Recent progress in high resolution lithography. *Polym. Adv. Technol.*, **17**, 94–103.
21. Zhang, X., Jiang, X.N., and Sun, C. (1999) Micro-stereolithography of polymeric and ceramic microstructures. *Sens. Actuators, A: Phys.*, **77**, 149–156.
22. Xia, Y.N. and Whitesides, G.M. (1998) Soft lithography. *Annu. Rev. Mater. Sci.*, **28**, 153–184.
23. Guan, J.J., Ferrell, N., Lee, L.J., and Hansford, D.J. (2006) Fabrication of polymeric microparticles for drug delivery by soft lithography. *Biomaterials*, **27**, 4034–4041.
24. Khademhosseini, A., Langer, R., Borenstein, J., and Vacanti, J.P. (2006) Microscale technologies for tissue engineering and biology. *Proc. Natl. Acad. Sci. U.S.A.*, **103**, 2480–2487.
25. Kane, R.S., Takayama, S., Ostuni, E., Ingber, D.E., and Whitesides, G.M. (1999) Patterning proteins and cells using soft lithography. *Biomaterials*, **20**, 2363–2376.
26. Hahn, M.S., Miller, J.S., and West, J.L. (2006) Three-dimensional biochemical and biomechanical patterning of hydrogels for guiding cell behavior. *Adv. Mater.*, **18**, 2679–2684.
27. Marder, S.R., Bredas, J.-L., and Perry, J.W. (2007) Materials for multiphoton 3d microfabrication. *MRS Bull.*, **32**, 561–565.
28. Malinauskas, M., Farsari, M., Piskarskas, A., and Juodkazis, S. (2013) Ultrafast laser nanostructuring of photopolymers: a decade of advances. *Phys. Rep.: Rev. Sect. Phys. Lett.*, **533**, 1–31.
29. Pawlicki, M., Collins, H.A., Denning, R.G., and Anderson, H.L. (2009) Two-photon absorption and the design of two-photon dyes. *Angew. Chem. Int. Ed.*, **48**, 3244–3266.
30. Zipfel, W.R., Williams, R.M., and Webb, W.W. (2003) Nonlinear magic: multiphoton microscopy in the biosciences. *Nat. Biotechnol.*, **21**, 1369–1377.
31. Kawata, S., Sun, H.B., Tanaka, T., and Takada, K. (2001) Finer features for functional microdevices - micromachines can be created with higher resolution using two-photon absorption. *Nature*, **412**, 697–698.
32. Li, Z., Torgersen, J., Ajami, A., Muehleder, S., Qin, X., Husinsky, W., Holnthoner, W., Ovsianikov, A., Stampfl, J., and Liska, R. (2013) Initiation efficiency and cytotoxicity of novel water-soluble two-photon photoinitiators for direct 3d microfabrication of hydrogels. *RSC Adv.*, **3**, 15939–15946.
33. Klein, F., Richter, B., Striebel, T., Franz, C.M., von Freymann, G., Wegener, M., and Bastmeyer, M. (2011) Two-component polymer scaffolds for controlled three-dimensional cell culture. *Adv. Mater.*, **23**, 1341–1345.
34. Denk, W., Strickler, J.H., and Webb, W.W. (1990) Two-photon laser scanning fluorescence microscopy. *Science*, **248**, 73–76.
35. Xu, C., Zipfel, W., Shear, J.B., Williams, R.M., and Webb, W.W. (1996) Multiphoton fluorescence excitation: new spectral windows for biological nonlinear microscopy. *Proc. Natl. Acad. Sci. U.S.A.*, **93**, 10763–10768.
36. Denk, W. (1994) Two-photon scanning photochemical microscopy - mapping ligand gated ion-channel distributions. *Proc. Natl. Acad. Sci. U.S.A.*, **91**, 6629–6633.

37. Majewska, A., Tashiro, A., and Yuste, R. (2000) Regulation of spine calcium dynamics by rapid spine motility. *J. Neurosci.*, **20**, 8262–8268.
38. Kumar, S., Maxwell, I.Z., Heisterkamp, A., Polte, T.R., Lele, T.P., Salanga, M., Mazur, E., and Ingber, D.E. (2006) Viscoelastic retraction of single living stress fibers and its impact on cell shape, cytoskeletal organization, and extracellular matrix mechanics. *Biophys. J.*, **90**, 3762–3773.
39. Farsari, M., Vamvakaki, M., and Chichkov, B.N. (2010) Multiphoton polymerization of hybrid materials. *J. Opt.*, **12**, 124001.
40. Sun, H.B., Xu, Y., Juodkazis, S., Sun, K., Watanabe, M., Matsuo, S., Misawa, H., and Nishii, J. (2001) Arbitrary-lattice photonic crystals created by multiphoton microfabrication. *Opt. Lett.*, **26**, 325–327.
41. He, G.S., Tan, L.-S., Zheng, Q., and Prasad, P.N. (2008) Multiphoton absorbing materials: molecular designs, characterizations, and applications. *Chem. Rev.*, **108**, 1245–1330.
42. Culver, J.C., Hoffmann, J.C., Poche, R.A., Slater, J.H., West, J.L., and Dickinson, M.E. (2012) Three-dimensional biomimetic patterning in hydrogels to guide cellular organization. *Adv. Mater.*, **24**, 2344–2348.
43. Odian, G.G. (2004) *Principles of Polymerization*, 4th edn, John Wiley & Sons, Inc., Hoboken, NJ, p. xxiv, 812 pp.
44. Aujard, I., Benbrahim, C., Gouget, M., Ruel, O., Baudin, J.-B., Neveu, P., and Jullien, L. (2006) O-Nitrobenzyl photolabile protecting groups with red-shifted absorption: syntheses and uncaging cross-sections for one- and two-photon excitation. *Chem.-A Eur. J.*, **12**, 6865–6879.
45. Kiskin, N.I., Chillingworth, R., McCray, J.A., Piston, D., and Ogden, D. (2002) The efficiency of two-photon photolysis of a "caged" fluorophore, o-1-(2-nitrophenyl)ethylpyranine, in relation to photodamage of synaptic terminals. *Eur. Biophys. J. Biophys. Lett.*, **30**, 588–604.
46. Scranton, A.B., Bowman, C.N., Klier, J., and Peppas, N.A. (1992) Polymerization reaction dynamics of ethylene-glycol methacrylates and dimethacrylates by calorimetry. *Polymer*, **33**, 1683–1689.
47. Goodner, M.D. and Bowman, C.N. (1999) Modeling primary radical termination and its effects on autoacceleration in photopolymerization kinetics. *Macromolecules*, **32**, 6552–6559.
48. Andrzejewska, E. (2001) Photopolymerization kinetics of multifunctional monomers. *Prog. Polym. Sci.*, **26**, 605–665.
49. Kloxin, A.M., Kloxin, C.J., Bowman, C.N., and Anseth, K.S. (2010) Mechanical properties of cellularly responsive hydrogels and their experimental determination. *Adv. Mater.*, **22**, 3484–3494.
50. Fouassier, J.-P. and Lalevée, J. (2012) *Photoinitiators for Polymer Synthesis: Scope, Reactivity and Efficiency*, Wiley-VCH Verlag GmbH, Weinheim, p. xxviii, 476 pp.
51. Bryant, S.J., Nuttelman, C.R., and Anseth, K.S. (2000) Cytocompatibility of Uv and visible light photoinitiating systems on cultured Nih/3t3 fibroblasts in vitro. *J. Biomater. Sci. Polym. Ed.*, **11**, 439–457.
52. Cadet, J., Berger, M., Douki, T., Morin, B., Raoul, S., Ravanat, J.L., and Spinelli, S. (1997) Effects of Uv and visible radiation on DNA – final base damage. *Biol. Chem.*, **378**, 1275–1286.
53. Nakagawa, A., Kobayashi, N., Muramatsu, T., Yamashina, Y., Shirai, T., Hashimoto, M.W., Ikenaga, M., and Mori, T. (1998) Three-dimensional visualization of ultraviolet-induced DNA damage and its repair in human cell nuclei. *J. Invest. Dermatol.*, **110**, 143–148.
54. Fouassier, J.P., Burr, D., and Wieder, F. (1991) Water-soluble photoinitiators – primary processes in hydroxy alkyl phenyl ketones. *J. Polym. Sci. Part A-Polym. Chem.*, **29**, 1319–1327.
55. Majima, T., Schnabel, W., and Weber, W. (1991) Phenyl-2,4,6-trimethylbenzoylphosphinates as water-soluble photoinitiators – generation and

reactivity of O=P(C6h5)(O-) radical-anions. *Makromol. Chem.-Macromol. Chem. Phys.*, **192**, 2307–2315.
56. Fairbanks, B.D., Schwartz, M.P., Bowman, C.N., and Anseth, K.S. (2009) Photoinitiated polymerization of PEG-diacrylate with lithium phenyl-2,4,6-trimethylbenzoylphosphinate: polymerization rate and cytocompatibility. *Biomaterials*, **30**, 6702–6707.
57. Ifkovits, J.L. and Burdick, J.A. (2007) Review: photopolymerizable and degradable biomaterials for tissue engineering applications. *Tissue Eng.*, **13**, 2369–2385.
58. Jakubiak, J. and Rabek, J.F. (1999) Photoinitiators for visible light polymerization. *Polimery*, **44**, 447–461.
59. Shi, W.F. and Ranby, B. (1996) Photopolymerization of dendritic methacrylated polyesters.1. Synthesis and properties. *J. Appl. Polym. Sci.*, **59**, 1937–1944.
60. Dworak, C., Koch, T., Varga, F., and Liska, R. (2010) Photopolymerization of biocompatible phosphorus-containing vinyl esters and vinyl carbamates. *J. Polym. Sci. Part A: Polym. Chem.*, **48**, 2916–2924.
61. Heller, C., Schwentenwein, M., Russmueller, G., Koch, T., Moser, D., Schopper, C., Varga, F., Stampfl, J., and Liska, R. (2011) Vinylcarbonates and vinylcarbamates: biocompatible monomers for radical photopolymerization. *J. Polym. Sci. Part A: Polym. Chem.*, **49**, 650–661.
62. Hoyle, C.E. and Bowman, C.N. (2010) Thiol-ene click chemistry. *Angew. Chem. Int. Ed.*, **49**, 1540–1573.
63. Heller, C., Schwentenwein, M., Russmueller, G., Varga, F., Stampfl, J., and Liska, R. (2009) Vinyl esters: low cytotoxicity monomers for the fabrication of biocompatible 3D scaffolds by lithography based additive manufacturing. *J. Polym. Sci. Part A-Polym. Chem.*, **47**, 6941–6954.
64. Slaughter, B.V., Khurshid, S.S., Fisher, O.Z., Khademhosseini, A., and Peppas, N.A. (2009) Hydrogels in regenerative medicine. *Adv. Mater.*, **21**, 3307–3329.
65. Hoffman, A.S. (2002) Hydrogels for biomedical applications. *Adv. Drug Delivery Rev.*, **54**, 3–12.
66. Lee, K.Y. and Mooney, D.J. (2001) Hydrogels for tissue engineering. *Chem. Rev.*, **101**, 1869–1879.
67. von Sonntag, C. (1991) The chemistry of free-radical-mediated DNA damage. *Basic Life Sci.*, **58**, 287–317; discussion -21.
68. Sawhney, A.S., Pathak, C.P., and Hubbell, J.A. (1993) Bioerodible hydrogels based on photopolymerized poly(ethylene glycol)-co-poly(alpha-hydroxy acid) diacrylate macromers. *Macromolecules*, **26**, 581–587.
69. Hoyle, C.E., Lee, T.Y., and Roper, T. (2004) Thiol-enes: chemistry of the past with promise for the future. *J. Polym. Sci. Part A-Polym. Chem.*, **42**, 5301–5338.
70. DeForest, C.A. and Anseth, K.S. (2011) Cytocompatible click-based hydrogels with dynamically tunable properties through orthogonal photocoupling and photodegradation reactions. *Nat. Chem.*, **3**, 925–931.
71. Fairbanks, B.D., Schwartz, M.P., Halevi, A.E., Nuttelman, C.R., Bowman, C.N., and Anseth, K.S. (2009) A versatile synthetic extracellular matrix mimic via thiol-norbornene photopolymerization. *Adv. Mater.*, **21**, 5005–5010.
72. Aimetti, A.A., Machen, A.J., and Anseth, K.S. (2009) Poly(ethylene glycol) hydrogels formed by thiol-ene photopolymerization for enzyme-responsive protein delivery. *Biomaterials*, **30**, 6048–6054.
73. Munoz, Z., Shih, H., and Lin, C.C. (2014) Gelatin hydrogels formed by orthogonal thiol-norbornene photochemistry for cell encapsulation. *Biomater. Sci.*, **2**, 1063–1072.
74. Mergy, J., Fournier, A., Hachet, E., and Auzely-Velty, R. (2012) Modification of polysaccharides via thiol-ene chemistry: a versatile route to functional biomaterials. *J. Polym. Sci. Part A-Polym. Chem.*, **50**, 4019–4028.
75. Aimetti, A.A., Feaver, K.R., and Anseth, K.S. (2010) Synthesis of cyclic, multivalent arg-gly-asp using sequential

thiol-ene/thiol-yne photoreactions. *Chem. Commun.*, **46**, 5781–5783.

76. DeForest, C.A., Polizzotti, B.D., and Anseth, K.S. (2009) Sequential click reactions for synthesizing and patterning three-dimensional cell microenvironments. *Nat. Mater.*, **8**, 659–664.
77. DeForest, C.A., Sims, E.A., and Anseth, K.S. (2010) Peptide-functionalized click hydrogels with independently tunable mechanics and chemical functionality for 3D cell culture. *Chem. Mater.*, **22**, 4783–4790.
78. Malkoch, M., Vestberg, R., Gupta, N., Mespouille, L., Dubois, P., Mason, A.F., Hedrick, J.L., Liao, Q., Frank, C.W., Kingsbury, K., and Hawker, C.J. (2006) Synthesis of Well-defined hydrogel networks using click chemistry. *Chem. Commun.*, 2774–2776.
79. Hoogenboom, R. (2010) Thiol-yne chemistry: a powerful tool for creating highly functional materials. *Angew. Chem. Int. Ed.*, **49**, 3415–3417.
80. Chen, G.J., Kumar, J., Gregory, A., and Stenzel, M.H. (2009) Efficient synthesis of dendrimers via a thiol-yne and esterification process and their potential application in the delivery of platinum anti-cancer drugs. *Chem. Commun.*, 6291–6293.
81. Konkolewicz, D., Gray-Weale, A., and Perrier, S. (2009) Hyperbranched polymers by thiol-yne chemistry: from small molecules to functional polymers. *J. Am. Chem. Soc.*, **131**, 18075–18077.
82. Fairbanks, B.D., Scott, T.F., Kloxin, C.J., Anseth, K.S., and Bowman, C.N. (2009) Thiol-yne photopolymerizations: novel mechanism, kinetics, and step-growth formation of highly cross-linked networks. *Macromolecules*, **42**, 211–217.
83. Fairbanks, B.D., Sims, E.A., Anseth, K.S., and Bowman, C.N. (2010) Reaction rates and mechanisms for radical, photoinitated addition of thiols to alkynes, and implications for thiol-yne photopolymerizations and click reactions. *Macromolecules*, **43**, 4113–4119.
84. Meldal, M. (2008) Polymer "clicking" by cuAAC reactions. *Macromol. Rapid Commun.*, **29**, 1016–1051.
85. Rostovtsev, V.V., Green, L.G., Fokin, V.V., and Sharpless, K.B. (2002) A stepwise huisgen cycloaddition process: copper(I)-catalyzed regioselective "Ligation" of azides and terminal alkynes. *Angew. Chem. Int. Ed.*, **41**, 2596–2599.
86. Tornoe, C.W., Christensen, C., and Meldal, M. (2002) Peptidotriazoles on solid phase: [1,2,3]-triazoles by regiospecific copper(I)-catalyzed 1,3-dipolar cycloadditions of terminal alkynes to azides. *J. Org. Chem.*, **67**, 3057–3064.
87. Adzima, B.J., Tao, Y., Kloxin, C.J., DeForest, C.A., Anseth, K.S., and Bowman, C.N. (2011) Spatial and temporal control of the alkyne-azide cycloaddition by photoinitiated Cu(II) reduction. *Nat. Chem.*, **3**, 258–261.
88. Gong, T., Adzima, B.J., Baker, N.H., and Bowman, C.N. (2013) Photopolymerization reactions using the photoinitiated copper (I)-catalyzed azide-alkyne cycloaddition (CuAAC) reaction. *Adv. Mater.*, **25**, 2024–2028.
89. Baskin, J.M., Prescher, J.A., Laughlin, S.T., Agard, N.J., Chang, P.V., Miller, I.A., Lo, A., Codelli, J.A., and Bertozzi, C.R. (2007) Copper-free click chemistry for dynamic in vivo imaging. *Proc. Natl. Acad. Sci. U.S.A.*, **104**, 16793–16797.
90. Agard, N.J., Prescher, J.A., and Bertozzi, C.R. (2004) A strain-promoted [3 + 2] azide-alkyne cycloaddition for covalent modification of biomolecules in living systems. *J. Am. Chem. Soc.*, **126**, 15046–15047.
91. Poloukhtine, A.A., Mbua, N.E., Wolfert, M.A., Boons, G.J., and Popik, V.V. (2009) Selective labeling of living cells by a photo-triggered click reaction. *J. Am. Chem. Soc.*, **131**, 15769–15776.
92. Arumugam, S., Orski, S.V., Mbua, N.E., McNitt, C., Boons, G.J., Locklin, J., and Popik, V.V. (2013) Photo-click chemistry strategies for spatiotemporal control of metal-free ligation, labeling, and surface derivatization. *Pure Appl. Chem.*, **85**, 1499–1513.
93. Orski, S.V., Poloukhtine, A.A., Arumugam, S., Mao, L.D., Popik, V.V., and Locklin, J. (2010) High density orthogonal surface immobilization

via photoactivated copper-free click chemistry. *J. Am. Chem. Soc.*, **132**, 11024–11026.

94. McNitt, C.D. and Popik, V.V. (2012) Photochemical generation of oxa-dibenzocyclooctyne (ODIBO) for metal-free click ligations. *Org. Biomol. Chem.*, **10**, 8200–8202.

95. Pauloehrl, T., Delaittre, G., Bruns, M., Meissler, M., Borner, H.G., Bastmeyer, M., and Barner-Kowollik, C. (2012) (Bio)molecular surface patterning by phototriggered oxime ligation. *Angew. Chem. Int. Ed.*, **51**, 9181–9184.

96. Dendane, N., Hoang, A., Guillard, L., Defrancq, E., Vinet, F., and Dumy, P. (2007) Efficient surface patterning of oligonucleotides inside a glass capillary through oxime bond formation. *Bioconjugate Chem.*, **18**, 671–676.

97. Park, S. and Yousaf, M.N. (2008) An interfacial oxime reaction to immobilize ligands and cells in patterns and gradients to photoactive surfaces. *Langmuir*, **24**, 6201–6207.

98. Mancini, R.J., Li, R.C., Tolstyka, Z.P., and Maynard, H.D. (2009) Synthesis of a photo-caged aminooxy alkane thiol. *Org. Biomol. Chem.*, **7**, 4954–4959.

99. Luo, Y. and Shoichet, M.S. (2004) A photolabile hydrogel for guided three-dimensional cell growth and migration. *Nat. Mater.*, **3**, 249–253.

100. Wosnick, J.H. and Shoichet, M.S. (2008) Three-dimensional chemical patterning of transparent hydrogels. *Chem. Mater.*, **20**, 55–60.

101. Wylie, R.G., Ahsan, S., Aizawa, Y., Maxwell, K.L., Morshead, C.M., and Shoichet, M.S. (2011) Spatially controlled simultaneous patterning of multiple growth factors in three-dimensional hydrogels. *Nat. Mater.*, **10**, 799–806.

102. Xi, W.X., Krieger, M., Kloxin, C.J., and Bowman, C.N. (2013) A new photoclick reaction strategy: photo-induced catalysis of the thiol-michael addition via a caged primary amine. *Chem. Commun.*, **49**, 4504–4506.

103. Liu, Z.Z., Lin, Q.N., Sun, Y., Liu, T., Bao, C.Y., Li, F.Y., and Zhu, L.Y. (2014) Spatiotemporally controllable and cytocompatible approach builds 3D cell culture matrix by photo-uncaged-thiol michael addition reaction. *Adv. Mater.*, **26**, 3912–3917.

104. Mosiewicz, K.A., Kolb, L., Vlies, A.J.V.D., Martino, M.M., Lienemann, P.S., Hubbell, J.A., Ehrbar, M., and Lutolf, M.P. (2013) In situ cell manipulation through enzymatic hydrogel photopatterning. *Nat. Mater.*, **12**, 1072–1078.

105. Griffin, D.R., Borrajo, J., Soon, A., Acosta-Velez, G.F., Oshita, V., Darling, N., Mack, J., Barker, T., Iruela-Arispe, M.L., and Segura, T. (2014) Hybrid photopatterned enzymatic reaction (hyper) for in situ cell manipulation. *ChemBioChem*, **15**, 233–242.

106. Gilchrist, T.L. and Rees, C.W. (1969) *Carbenes Nitrenes and Arynes (Studies in Modern Chemistry)*, Appleton-Century-Crofts.

107. Kotzybahibert, F., Kapfer, I., and Goeldner, M. (1995) Recent trends in photoaffinity-labeling. *Angew. Chem., Int. Ed. Engl.*, **34**, 1296–1312.

108. Chin, J.W., Santoro, S.W., Martin, A.B., King, D.S., Wang, L., and Schultz, P.G. (2002) Addition of p-azido-L-phenylaianine to the genetic code of Escherichia coli. *J. Am. Chem. Soc.*, **124**, 9026–9027.

109. Carrico, I.S., Maskarinec, S.A., Heilshorn, S.C., Mock, M.L., Liu, J.C., Nowatzki, P.J., Franck, C., Ravichandran, G., and Tirrell, D.A. (2007) Lithographic patterning of photoreactive cell-adhesive proteins. *J. Am. Chem. Soc.*, **129**, 4874–4875.

110. Smith, J.W. and Cheresh, D.A. (1988) The Arg-Gly-Asp binding domain of the vitronectin receptor – photoaffinity cross-linking implicates amino-acid residues-61-203 of the beta-subunit. *J. Biol. Chem.*, **263**, 18726–18731.

111. Smith, R.A.G. and Knowles, J.R. (1973) Aryldiazirines – potential reagents for photolabeling of biological receptor sites. *J. Am. Chem. Soc.*, **95**, 5072–5073.

112. Hoppe, J., Brunner, J., and Jorgensen, B.B. (1984) Structure of the membrane-embedded F0 part of F1f0 ATP synthase from Escherichia-coli as inferred

from labeling with 3-(trifluoromethyl) 3-(m-[I-125]iodophenyl)diazirine. *Biochemistry*, **23**, 5610–5616.

113. Harter, C., James, P., Bachi, T., Semenza, G., and Brunner, J. (1989) Hydrophobic binding of the ectodomain of influenza hemagglutinin to membranes occurs through the fusion peptide. *J. Biol. Chem.*, **264**, 6459–6464.
114. Raphel, J., Parisi-Amon, A., and Heilshorn, S.C. (2012) Photoreactive elastin-like proteins for use as versatile bioactive materials and surface coatings. *J. Mater. Chem.*, **22**, 19429–19437.
115. Brunner, J. and Semenza, G. (1981) Selective labeling of the hydrophobic core of membranes with 3-(trifluoromethyl)-3-(m-[I-125]iodophenyl)diazirine, a carbene-generating reagent. *Biochemistry*, **20**, 7174–7182.
116. Vanderbend, R.L., Brunner, J., Jalink, K., Vancorven, E.J., Moolenaar, W.H., and Vanblitterswijk, W.J. (1992) Identification of a putative membrane-receptor for the bioactive phospholipid, lysophosphatidic acid. *EMBO J.*, **11**, 2495–2501.
117. Walling, C. and Gibian, M. (1965) Hydrogen abstraction reactions by the triplet states of ketones. *J. Am. Chem. Soc.*, **87**, 3361.
118. Dorman, G. and Prestwich, G.D. (1994) Benzophenone photophores in biochemistry. *Biochemistry*, **33**, 5661–5673.
119. Prucker, O., Naumann, C.A., Ruhe, J., Knoll, W., and Frank, C.W. (1999) Photochemical attachment of polymer films to solid surfaces via monolayers of benzophenone derivatives. *J. Am. Chem. Soc.*, **121**, 8766–8770.
120. Herbert, C.B., McLernon, T.L., Hypolite, C.L., Adams, D.N., Pikus, L., Huang, C.C., Fields, G.B., Letourneau, P.C., Distefano, M.D., and Hu, W.S. (1997) Micropatterning gradients and controlling surface densities of photoactivatable biomolecules on self-assembled monolayers of oligo(ethylene glycol) alkanethiolates. *Chem. Biol.*, **4**, 731–737.
121. Balakirev, M.Y., Porte, S., Vernaz-Gris, M., Berger, M., Arie, J.P., Fouque, B., and Chatelain, F. (2005) Photochemical patterning of biological molecules inside a glass capillary. *Anal. Chem.*, **77**, 5474–5479.
122. Rich, D.H. and Gurwara, S.K. (1973) Removal of protected peptides from an ortho-nitrobenzyl resin by photolysis. *J. Chem. Soc. Chem. Commun.*, 610–611.
123. Zhao, Y., Zheng, Q., Dakin, K., Xu, K., Martinez, M.L., and Li, W.H. (2004) New caged coumarin fluorophores with extraordinary uncaging cross sections suitable for biological imaging applications. *J. Am. Chem. Soc.*, **126**, 4653–4663.
124. Holmes, C.P. and Jones, D.G. (1995) Reagents for combinatorial organic-synthesis – development of a new O-nitrobenzyl photolabile linker for solid-phase synthesis. *J. Org. Chem.*, **60**, 2318–2319.
125. Holmes, C.P. (1997) Model studies for new O-nitrobenzyl photolabile linkers: substituent effects on the rates of photochemical cleavage. *J. Org. Chem.*, **62**, 2370–2380.
126. Griffin, D.R., Schlosser, J.L., Lam, S.F., Nguyen, T.H., Maynard, H.D., and Kasko, A.M. (2013) Synthesis of photodegradable macromers for conjugation and release of bioactive molecules. *Biomacromolecules*, **14**, 1199–1207.
127. Wu, Y.I., Frey, D., Lungu, O.I., Jaehrig, A., Schlichting, I., Kuhlman, B., and Hahn, K.M. (2009) A genetically encoded photoactivatable rac controls the motility of living cells. *Nature*, **461**, 104-U11.
128. Deiters, A. (2010) Principles and applications of the photochemical control of cellular processes. *ChemBioChem*, **11**, 47–53.
129. Ohmuro-Matsuyama, Y. and Tatsu, Y. (2008) Photocontrolled cell adhesion on a surface functionalized with a caged arginine-glycine-aspartate peptide. *Angew. Chem. Int. Ed.*, **47**, 7527–7529.
130. Kaneko, S., Nakayama, H., Yoshino, Y., Fushimi, D., Yamaguchi, K., Horiike, Y., and Nakanishi, J. (2011) Photocontrol of cell adhesion on amino-bearing

surfaces by reversible conjugation of poly(ethylene glycol) via a photocleavable linker. *Phys. Chem. Chem. Phys.*, **13**, 4051–4059.

131. Johnson, J.A., Finn, M.G., Koberstein, J.T., and Turro, N.J. (2007) Synthesis of photocleavable linear macromonomers by atrp and star macromonomers by a tandem atrp-click reaction: precursors to photodegradable model networks. *Macromolecules*, **40**, 3589–3598.

132. Johnson, J.A., Baskin, J.M., Bertozzi, C.R., Koberstein, J.T., and Turro, N.J. (2008) Copper-free click chemistry for the in situ crosslinking of photodegradable star polymers. *Chem. Commun.*, 3064–3066.

133. Kloxin, A.M., Kasko, A.M., Salinas, C.N., and Anseth, K.S. (2009) Photodegradable hydrogels for dynamic tuning of physical and chemical properties. *Science*, **324**, 59–63.

134. Kloxin, A.M., Tibbitt, M.W., and Anseth, K.S. (2010) Synthesis of photodegradable hydrogels as dynamically tunable cell culture platforms. *Nat. Protoc.*, **5**, 1867–1887.

135. Kloxin, A.M., Benton, J.A., and Anseth, K.S. (2010) In situ elasticity modulation with dynamic substrates to direct cell phenotype. *Biomaterials*, **31**, 1–8.

136. Tibbitt, M.W., Kloxin, A.M., Dyamenahalli, K.U., and Anseth, K.S. (2010) Controlled Two-photon photodegradation of PEG hydrogels to study and manipulate subcellular interactions on soft materials. *Soft Matter*, **6**, 5100–5108.

137. Yang, C., Tibbitt, M.W., Basta, L., and Anseth, K.S. (2014) Mechanical memory and dosing influence stem cell fate. *Nat. Mater.*, **13**, 645–652.

138. Schade, B., Hagen, V., Schmidt, R., Herbrich, R., Krause, E., Eckardt, T., and Bendig, J. (1999) Deactivation behavior and excited-state properties of (coumarin-4-yl)methyl derivatives. 1. Photocleavage of (7-methoxycoumarin-4-yl)methyl-caged acids with fluorescence enhancement. *J. Org. Chem.*, **64**, 9109–9117.

139. Geissler, D., Antonenko, Y.N., Schmidt, R., Keller, S., Krylova, O.O., Wiesner, B., Bendig, J., Pohl, P., and Hagen, V. (2005) (Coumarin-4-yl)methyl esters as highly efficient, ultrafast phototriggers for protons and their application to acidifying membrane surfaces. *Angew. Chem. Int. Ed.*, **44**, 1195–1198.

140. Schmidt, R., Geissler, D., Hagen, V., and Bendig, J. (2007) Mechanism of photocleavage of (coumarin-4-yl)methyl esters. *J. Phys. Chem. A*, **111**, 5768–5774.

141. Mal, N.K., Fujiwara, M., and Tanaka, Y. (2003) Photocontrolled reversible release of guest molecules from coumarin-modified mesoporous silica. *Nature*, **421**, 350–353.

142. Babin, J., Pelletier, M., Lepage, M., Allard, J.F., Morris, D., and Zhao, Y. (2009) A new two-photon-sensitive block copolymer nanocarrier. *Angew. Chem. Int. Ed.*, **48**, 3329–3332.

143. Azagarsamy, M.A., McKinnon, D.D., Age, D.L., and Anseth, K.S. (2014) Coumarin-based photodegradable hydrogel: design, synthesis, gelation, and degradation kinetics. *ACS Macro Lett.*, **3**, 515–519.

144. Azagarsamy, M.A. and Anseth, K.S. (2013) Wavelength-controlled photocleavage for the orthogonal and sequential release of multiple proteins. *Angew. Chem. Int. Ed.*, **52**, 13803–13807.

145. Bonifacic, M. and Asmus, K.D. (1984) Adduct formation and absolute rate constants in the displacement reaction of thiyl radicals with disulfides. *J. Phys. Chem.*, **88**, 6286–6290.

146. Tobolsky, A.V., Takahashi, M., and Macknight, W.J. (1964) Relaxation of disulfide and tetrasulfide polymers. *J. Phys. Chem.*, **68**, 787–790.

147. Fairbanks, B.D., Singh, S.P., Bowman, C.N., and Anseth, K.S. (2011) Photodegradable, photoadaptable hydrogels via radical-mediated disulfide fragmentation reaction. *Macromolecules*, **44**, 2444–2450.

148. Bowen, E.J. and Tanner, D.W. (1955) The photochemistry of anthracenes. 3. Inter-relations between fluorescence quenching, dimerization, and photo-oxidation. *Trans. Faraday Soc.*, **51**, 475–481.

149. Odonnell, M. (1968) Photodimerization of solid anthracene. *Nature*, **218**, 460–461.
150. Chandros, E.A. (1965) Photolytic dissociation of dianthracene. *J. Chem. Phys.*, **43**, 4175–4176.
151. Wells, L.A., Brook, M.A., and Sheardown, H. (2011) Generic, anthracene-based hydrogel crosslinkers for photo-controllable drug delivery. *Macromol. Biosci.*, **11**, 988–998.
152. Sako, Y. and Takaguchi, Y. (2008) A photo-responsive hydrogelator having gluconamides at its peripheral branches. *Org. Biomol. Chem.*, **6**, 3843–3847.
153. Zheng, Y.J., Mieie, M., Mello, S.V., Mabrouki, M., Andreopoulos, F.M., Konka, V., Pham, S.M., and Leblanc, R.M. (2002) Peg-based hydrogel synthesis via the photodimerization of anthracene groups. *Macromolecules*, **35**, 5228–5234.
154. Frank, P.G., Tuten, B.T., Prasher, A., Chao, D.M., and Berda, E.B. (2014) Intra- chain photodimerization of pendant anthracene units as an effi cient route to single- chain nanoparticle fabrication. *Macromol. Rapid Commun.*, **35**, 249–253.
155. Chiefari, J., Chong, Y.K., Ercole, F., Krstina, J., Jeffery, J., Le, T.P.T., Mayadunne, R.T.A., Meijs, G.F., Moad, C.L., Moad, G., Rizzardo, E., and Thang, S.H. (1998) Living free-radical polymerization by reversible addition-fragmentation chain transfer: the raft process. *Macromolecules*, **31**, 5559–5562.
156. Meijs, G.F., Rizzardo, E., and Thang, S.H. (1988) Preparation of controlled molecular-weight, olefin-terminated polymers by free-radical methods – chain transfer using allylic sulfides. *Macromolecules*, **21**, 3122–3124.
157. Meijs, G.F., Morton, T.C., Rizzardo, E., and Thang, S.H. (1991) Use of substituted allylic sulfides to prepare end-functional polymers of controlled molecular-weight by free-radical polymerization. *Macromolecules*, **24**, 3689–3695.
158. Moad, G., Rizzardo, E., and Thang, S.H. (2008) Radical addition-fragmentation chemistry in polymer synthesis. *Polymer*, **49**, 1079–1131.
159. Scott, T.F., Schneider, A.D., Cook, W.D., and Bowman, C.N. (2005) Photoinduced plasticity in cross-linked polymers. *Science*, **308**, 1615–1617.
160. Kloxin, C.J., Scott, T.F., Adzima, B.J., and Bowman, C.N. (2010) Covalent adaptable networks (cans): a unique paradigm in cross-linked polymers. *Macromolecules*, **43**, 2643–2653.
161. Gandavarapu, N.R., Azagarsamy, M.A., and Anseth, K.S. (2014) Photo-click living strategy for controlled, reversible exchange of biochemical ligands. *Adv. Mater.*, **26**, 2521–2526.
162. Li, L. and Fourkas, J.T. (2007) Multiphoton polymerization. *Mater. Today*, **10**, 30–37.
163. Coenjarts, C.A. and Ober, C.K. (2004) Two-photon three-dimensional microfabrication of poly(dimethylsiloxane) elastomers. *Chem. Mater.*, **16**, 5556–5558.
164. Adzima, B.J., Kloxin, C.J., DeForest, C.A., Anseth, K.S., and Bowman, C.N. (2012) 3D photofixation lithography in Diels-Alder networks. *Macromol. Rapid Commun.*, **33**, 2092–2096.
165. Serbin, J., Egbert, A., Ostendorf, A., Chichkov, B.N., Houbertz, R., Domann, G., Schulz, J., Cronauer, C., Frohlich, L., and Popall, M. (2003) Femtosecond laser-induced two-photon polymerization of inorganic-organic hybrid materials for applications in photonics. *Opt. Lett.*, **28**, 301–303.
166. Lee, W.M., Pruzinsky, S.A., and Braun, P.V. (2002) Multi-photon polymerization of waveguide structures within three-dimensional photonic crystals. *Adv. Mater.*, **14**, 271–274.
167. Klein, S., Barsella, A., Leblond, H., Bulou, H., Fort, A., Andraud, C., Lemercier, G., Mulatier, J.C., and Dorkenoo, K. (2005) One-step waveguide and optical circuit writing in photopolymerizable materials processed by two-photon absorption. *Appl. Phys. Lett.*, **86**, 211118-1.
168. Guo, R., Xiao, S.Z., Zhai, X.M., Li, J.W., Xia, A.D., and Huang, W.H. (2006) Micro lens fabrication by means of

femtosecond two photon photopolymerization. *Opt. Express*, **14**, 810–816.
169. Yokoyama, S., Nakahama, T., Miki, H., and Mashiko, S. (2003) Fabrication of three-dimensional microstructure in optical-gain medium using two-photon-induced photopolymerization technique. *Thin Solid Films*, **438**, 452–456.
170. Daley, W.P., Peters, S.B., and Larsen, M. (2008) Extracellular matrix dynamics in development and regenerative medicine. *J. Cell Sci.*, **121**, 255–264.
171. Tibbitt, M.W. and Anseth, K.S. (2009) Hydrogels as extracellular matrix mimics for 3D cell culture. *Biotechnol. Bioeng.*, **103**, 655–663.
172. Torgersen, J., Qin, X.-H., Li, Z., Ovsianikov, A., Liska, R., and Stampfl, J. (2013) Hydrogels for two-photon polymerization: a toolbox for mimicking the extracellular matrix. *Adv. Funct. Mater.*, **23**, 4542–4554.
173. Doraiswamy, A., Jin, C., Narayan, R.J., Mageswaran, P., Mente, P., Modi, R., Auyeung, R., Chrisey, D.B., Ovsianikov, A., and Chichkov, B. (2006) Two photon induced polymerization of organic-inorganic hybrid biomaterials for microstructured medical devices. *Acta Biomater.*, **2**, 267–275.
174. Claeyssens, F., Hasan, E.A., Gaidukeviciute, A., Achilleos, D.S., Ranella, A., Reinhardt, C., Ovsianikov, A., Xiao, S., Fotakis, C., Vamvakaki, M., Chichkov, B.N., and Farsari, M. (2009) Three-dimensional biodegradable structures fabricated by two-photon polymerization. *Langmuir*, **25**, 3219–3223.
175. Qin, X.-H., Torgersen, J., Saf, R., Muehleder, S., Pucher, N., Ligon, S.C., Holnthoner, W., Redl, H., Ovsianikov, A., Stampfl, J., and Liska, R. (2013) Three-dimensional microfabrication of protein hydrogels via two-photon-excited thiol-vinyl ester photopolymerization. *J. Polym. Sci. Part A-Polym. Chem.*, **51**, 4799–4810.
176. Kaehr, B. and Shear, J.B. (2008) Multiphoton fabrication of chemically responsive protein hydrogels for microactuation. *Proc. Natl. Acad. Sci. U.S.A.*, **105**, 8850–8854.
177. Ovsianikov, A., Chichkov, B., Adunka, O., Pillsbury, H., Doraiswamy, A., and Narayan, R.J. (2007) Rapid prototyping of ossicular replacement prostheses. *Appl. Surf. Sci.*, **253**, 6603–6607.
178. Torgersen, J., Ovsianikov, A., Mironov, V., Pucher, N., Qin, X., Li, Z., Cicha, K., Machacek, T., Liska, R., Jantsch, V., and Stampfl, J. (2012) Photo-sensitive hydrogels for three-dimensional laser microfabrication in the presence of whole organisms. *J. Biomed. Opt.*, **17**, 105008.
179. Jhaveri, S.J., McMullen, J.D., Sijbesma, R., Tan, L.-S., Zipfel, W., and Ober, C.K. (2009) Direct three-dimensional microfabrication of hydrogels via two-photon lithography in aqueous solution. *Chem. Mater.*, **21**, 2003–2006.
180. Pitts, J.D., Campagnola, P.J., Epling, G.A., and Goodman, S.L. (2000) Submicron multiphoton free-form fabrication of proteins and polymers: studies of reaction efficiencies and applications in sustained release. *Macromolecules*, **33**, 1514–1523.
181. Pitts, J.D., Howell, A.R., Taboada, R., Banerjee, I., Wang, J., Goodman, S.L., and Campagnola, P.J. (2002) New photoactivators for multiphoton excited three-dimensional submicron cross-linking of proteins: bovine serum albumin and type 1 collagen. *Photochem. Photobiol.*, **76**, 135–144.
182. Kaehr, B., Allen, R., Javier, D.J., Currie, J., and Shear, J.B. (2004) Guiding neuronal development with in situ microfabrication. *Proc. Natl. Acad. Sci. U.S.A.*, **101**, 16104–16108.
183. Leslie-Barbick, J.E., Shen, C., Chen, C., and West, J.L. (2011) Micron-scale spatially patterned, covalently immobilized vascular endothelial growth factor on hydrogels accelerates endothelial tubulogenesis and increases cellular angiogenic responses. *Tissue Eng. Part A*, **17**, 221–229.
184. DeForest, C.A. and Anseth, K.S. (2011) Photoreversible patterning of biomolecules within click-based hydrogels. *Angew. Chem. Int. Ed.*, **51**, 1816–1819.

185. McCall, J.D., Luoma, J.E., and Anseth, K.S. (2012) Covalently tethered transforming growth factor beta in peg hydrogels promotes chondrogenic differentiation of encapsulated human mesenchymal stem cells. *Drug Delivery Transl. Res.*, **2**, 305–312.

186. Salinas, C.N. and Anseth, K.S. (2008) The enhancement of chondrogenic differentiation of human mesenchymal stem cells by enzymatically regulated rgd functionalities. *Biomaterials*, **29**, 2370–2377.

187. Rolli, C.G., Nakayama, H., Yamaguchi, K., Spatz, J.P., Kemkemer, R., and Nakanishi, J. (2012) Switchable adhesive substrates: revealing geometry dependence in collective cell behavior. *Biomaterials*, **33**, 2409–2418.

188. Lin, H.M., Wang, W.K., Hsiung, P.A., and Shyu, S.G. (2010) Light-sensitive intelligent drug delivery systems of coumarin-modified mesoporous bioactive glass. *Acta Biomater.*, **6**, 3256–3263.

189. Hossion, A.M.L., Bio, M., Nkepang, G., Awuah, S.G., and You, Y. (2013) Visible light controlled release of anticancer drug through double activation of prodrug. *ACS Med. Chem. Lett.*, **4**, 124–127.

190. Khetan, S., Katz, J.S., and Burdick, J.A. (2009) Sequential crosslinking to control cellular spreading in 3-dimensional hydrogels. *Soft Matter*, **5**, 1601–1606.

191. Khetan, S. and Burdick, J.A. (2010) Patterning network structure to spatially control cellular remodeling and stem cell fate within 3-dimensional hydrogels. *Biomaterials*, **31**, 8228–8234.

192. McKinnon, D.D., Brown, T.E., Kyburz, K.A., Kiyotake, E., and Anseth, K.S. (2014) Design and characterization of a synthetically accessible photodegradable hydrogel for user-directed formation of neural networks. *Biomacromolecules*, **15**, 2808–2816.

193. Guvendiren, M. and Burdick, J.A. (2012) Stiffening hydrogels to probe short- and long-term cellular responses to dynamic mechanics. *Nat. Commun.*, **3**, 792.

194. Khetan, S., Guvendiren, M., Legant, W.R., Cohen, D.M., Chen, C.S., and Burdick, J.A. (2013) Degradation-mediated cellular traction directs stem cell fate in covalently crosslinked three-dimensional hydrogels. *Nat. Mater.*, **12**, 458–465.

195. Mosiewicz, K.A., Kolb, L., van der Vlies, A.J., and Lutolf, M.P. (2014) Microscale patterning of hydrogel stiffness through light-triggered uncaging of thiols. *Biomater. Sci.*, **2**, 1640–1651.

196. Heisterkamp, A., Maxwell, I.Z., Mazur, E., Underwood, J.M., Nickerson, J.A., Kumar, S., and Ingber, D.E. (2005) Pulse energy dependence of subcellular dissection by femtosecond laser pulses. *Opt. Express*, **13**, 3690–3696.

197. Shen, N., Datta, D., Schaffer, C.B., LeDuc, P., Ingber, D.E., and Mazur, E. (2005) Ablation of cytoskeletal filaments and mitochondria in live cells using a femtosecond laser nanoscissor. *Mech. Chem. Biosyst.*, **2**, 17–25.

198. Watanabe, W., Arakawa, N., Matsunaga, S., Higashi, T., Fukui, K., Isobe, K., and Itoh, K. (2004) Femtosecond laser disruption of subcellular organelles in a living cell. *Opt. Express*, **12**, 4203–4213.

199. Abbott, A. (2003) Cell culture: biology's new dimension. *Nature*, **424**, 870–872.

200. Dalby, M.J., Gadegaard, N., Tare, R., Andar, A., Riehle, M.O., Herzyk, P., Wilkinson, C.D.W., and Oreffo, R.O.C. (2007) The control of human mesenchymal cell differentiation using nanoscale symmetry and disorder. *Nat. Mater.*, **6**, 997–1003.

201. Xiong, F., Ma, W., Hiscock, T.W., Mosaliganti, K.R., Tentner, A.R., Brakke, K.A., Rannou, N., Gelas, A., Souhait, L., Swinburne, I.A., Obholzer, N.D., and Megason, S.G. (2014) Interplay of cell shape and division orientation promotes robust morphogenesis of developing epithelia. *Cell*, **159**, 415–427.

9
Multiphoton Processing of Composite Materials and Functionalization of 3D Structures

Casey M. Schwarz, Christopher N. Grabill, Jennefir L. Digaum, Henry E. Williams, and Stephen M. Kuebler

9.1
Overview

Three-dimensional (3D) multiphoton lithography (MPL) is highly appealing as a fabrication and manufacturing technology because it provides a means for creating previously inaccessible structures that can have transformative applications in optics, sensing, imaging, consumer electronics, microelectromechanical systems (MEMS), and medicine. Yet most applications require not only well-defined 3D forms but also functional properties that are manifest in controlled chemical, mechanical, optical, electronic, or thermal interactions with the environment. Indeed, for many micro-/nanodevices, the overall performance depends sensitively upon the synergistic combination of small-scale form and properties. A good example is meta-structured materials, such as photonic crystals (PCs), for which the optical function is derived from the combination of sub-wavelength-scale structure as well as the bulk optical properties of the constituent material itself [1]. Microfluidics provide another good example, as these can be engineered to have peculiar and useful flow and mixing characteristics that derive from the combination of the channel size, shape, and cross-section with the chemical nature at the interface formed by the microfluidic walls and the fluid [2, 3].

The laser-induced change most commonly leveraged for MPL is photo-induced polymerization or cross-linking of an organic monomer or oligomer. Creating 3D micro-/nanoscale polymeric structures by MPL continues to serve many purposes, but a growing number of current applications require a wider range of materials to achieve targeted functions. In a limited number of situations, it has been possible to identify alternative materials that can be patterned in much the same way as a photopolymer. For example, thermally deposited arsenic sulfide [4] and other chalcogenides [5] can be directly photocrosslinked and then chemically etched, behaving like an inorganic analog of a photoresist that can be used to directly create functional 3D semiconductor nanostructures. But such materials are by far the exception, and in most cases it has been necessary to engineer

Multiphoton Lithography: Techniques, Materials and Applications, First Edition.
Edited by Jürgen Stampfl, Robert Liska, and Aleksandr Ovsianikov.

new materials and processes for MPL that provide access to the wider range of properties and functions needed for a given application.

The imperative to create functional micro-/nanoscale structures has led many investigators to develop complementary approaches for multiphoton processing of composite materials and functionalization of 3D structures. The approaches that have been reported for integrating specific properties or function into 3D microstructures can be divided into three classes, as illustrated in Figure 9.1: (i) MPL using composites, (ii) post-MPL functionalization, and (iii) templated deposition or back-filling.

In the first approach (Figure 9.1b), a composite material [6] is formulated that is both suitable for direct fabrication of the targeted microscale form by MPL and contains one or more additives that endow the final microstructure with the targeted functionality. Such composites are typically formulated so that a large fraction of the total mass is comprised of a polymerizable monomer and/or oligomer and a multiphoton-activatable initiator that are known to be suitable for MPL. A wide range of functional structures can be accessed by this approach. For example, optically functional microstructures can readily be created by blending a molecular chromophore into a polymeric MPL medium [7]. Composites

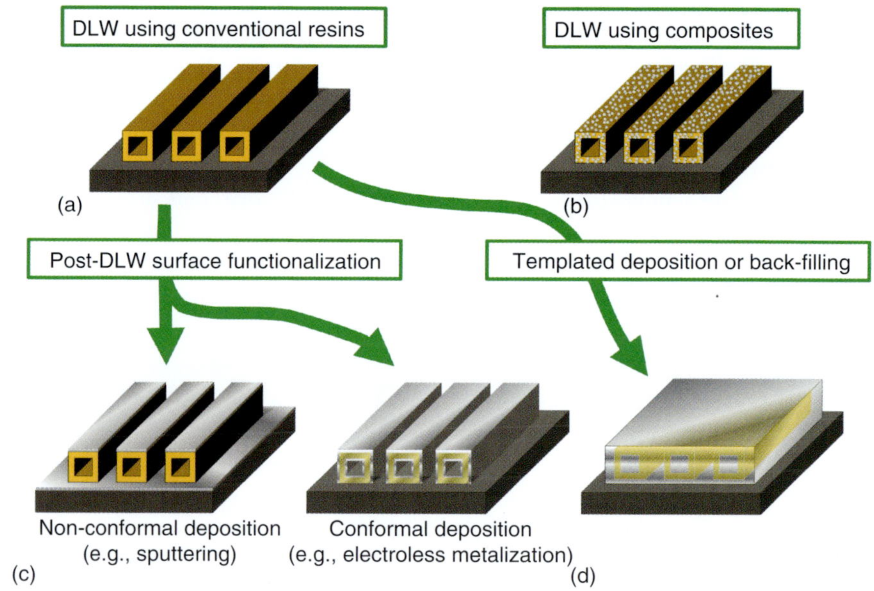

Figure 9.1 Schematic illustrating the general approaches that have been used to create functional microstructures by MPL in a wide range of materials. Rather than (a) fabricating 3D microstructures using a conventional photopolymer, (b) a composite can be engineered and used to fabricate structures with targeted properties or function. Alternatively, (c) a structure created by MPL using a conventional (or subtly modified) photopolymer can be functionalized after fabrication by chemical or physical deposition, or (d) used as a template for direct deposition of a second material onto the targeted form.

containing photo-reducible metal salts and/or metal nanoparticles have been formulated and used to create electrically conductive structures by MPL [8, 9]. The composite approach has the advantage that the functional structure can be fabricated in a single MPL step. But it has the disadvantage that the additives must be present at low weight loading so as to not interfere with the multiphoton excitation and polymerization processes.

In post-MPL functionalization (Figure 9.1c), a conventional MPL medium is used to create the targeted 3D form, and post-fabrication chemistry or physical deposition processes are used to deposit a second material onto the 3D form. Conformal and nonconformal deposition approaches can be selected as needed for the targeted application. Deposition approaches such as sputtering and physical vapor deposition tend to be nonconformal, depositing material primarily on those surfaces of the 3D structure that are exposed by "line of sight" to the source [10, 11]. Conformal deposition, in which all surfaces of the 3D microstructure are coated uniformly, can be achieved using a variety of approaches. Chemical-bath-based methods, such as electroless deposition [12, 13], can be used to deposit conductive metal conformally onto the 3D form. Other approaches, such as chemical vapor deposition (CVD) [14] and atomic layer deposition (ALD) [15] have also been used for conformal deposition. In some cases, the MPL medium may be modified slightly by incorporating an additive that makes the 3D structure active toward subsequent surface functionalization [16–19]. For polymeric media, unreacted functional groups at the surface can be sufficiently active so that surface functionalization can be carried out without the need to add additional components to the MPL medium [13, 20]. The post-MPL functionalization approach has the advantage that it decouples fabrication of the 3D structure from deposition of a targeted material, which eliminates potential complications that can arise in attempting to create a composite. A major disadvantage is that the functional material is usually present only on the outside of the structure, and not throughout the bulk of the 3D form, which can compromise performance. An example might be an electronic function, where the deposition of a thin skin of conductive metal would have much lower conductivity that an idealized bulk-metal structure, or one that could potentially be created by a composite approach. Bulk functionalization can occur when a surface functionalization agent is able to diffuse into the bulk of the crosslinked polymeric structure [20, 21].

With templated deposition, or backfilling (Figure 9.1d), MPL in a conventional photopolymer blend is used to create a free-standing 3D form upon which a second material is deposited to completely (or nearly completely) fill the void space between features of the 3D structure. This might be viewed as an extreme limit of post-MPL functionalization, where the deposition is simply continued until a maximally thick layer is deposited. The back-filled medium is the structural inverse of the original template. This has been achieved by ALD of silica [22] and by galvanic deposition of metal around a structure fabricated on a conductive surface [23]. The functional device could consist of the infiltrated template, or the final device could be obtained after using plasma etching or other approaches to remove the polymeric template, leaving the inverted structure in the second material. A second round of infiltration in another material can be used to re-invert the structure back to its original form in the final material. This approach was demonstrated by depositing silicon onto a silicate inverted structure as a means for preparing 3D silicon PCs [22]. Overall, templated deposition also offers the advantage that 3D patterning and targeted material deposition are decoupled, enabling both processes to be separately optimized, and it provides a means for creating bulk 3D structures from the second material. The primary disadvantage remains, again, that multiple fabrication steps are needed to produce the final structure, including alternate substrate bonding/removal steps.

Some methods for fabricating 3D micro/nanostructures in novel materials have employed processes that are essentially a combination of both the composite and surface functionalization approaches. For example, Kawata and coworkers reported a route to high-refractive-index PCs based on a polymeric "woodpile" lattice loaded with CdS nanoparticles [21]. The approach involved first patterning the woodpile lattice by MPL using a composite of a photopolymer doped with Cd^{2+} salts. After MPL, the woodpile structure was functionalized by exposing it to an atmosphere of H_2S. Molecules of H_2S diffused into the interstitial voids of the woodpile lattice as well as into the polymeric rods themselves, forming CdS nanocrystals throughout the structure. The resulting CdS-doped PC exhibited properties substantially different from those of a nonfunctionalized lattice.

Fabrication using composites, surface functionalization, or backfilling introduces a plethora of challenges. In addition to the general advantages and disadvantages listed above, the materials involved must be compatible with previous and subsequent steps, else the material function and 3D form can be compromised. Particular attention must be paid to chemical reactivity, solubility, swellability, and thermal tolerances associated with the materials and processing steps, as well as how these can disrupt the function or 3D form of the structure [20]. Adhesion and material compatibility are particularly important for interfaces created by surface functionalization or secondary deposition. Properties of the 3D form can also significantly affect those of the overall structure. For example, if the MPL patterning and processing creates features with high surface roughness, then a second material deposited onto the polymeric template could have desirable properties, such as high adhesion, as well as undesirable properties, such as high optical scattering. So invariably, fabrication involving multistep processing requires careful co-optimization of the processing conditions to achieve good performance from the 3D microstructure. It should also be noted that composite and surface functionalization can be used to create structures with either a passive or an active function. An actively functional device can be thought of as one that changes its form or physical properties in response to an environmental stimulus. These types of interactions often result from the combination of material, shape, or size of the device. One early example in the field of MPL involves a hydrogel composite that was formulated to respond to UV radiation, such that the structure underwent reversible mechanical deformation when exposed [24]. A wide range of applications can be effected by active-response devices, and their advancement depends heavily on the progress in composite development and microstructure functionalization.

Over the past decade, many examples of the fabrication strategies described above have been reported. Each of these has substantially extended the technological impact of MPL by opening new routes to functional 3D micro/nanoscale structures that can have tailored function for a given application. Several reviews of MPL and extensions of the method have been published over the past decade [25–40], providing a wealth of information concerning methods for fabricating functional micro/nanoscale devices using a range of materials. The remainder of this chapter reviews methods for fabricating and functionalizing composite structures in terms of the material classes, rather than processing method, as applications are typically dependent upon the class of material needed for a targeted function, and the means by which it is deposited are a secondary concern.

9.2 Polymer–Organic Composites

Most commonly, MPL is performed using an organic composite, which is a homogeneous mixture of two molecular species: (i) a photoinitiator, responsible for multiphoton excitation, and (ii) a monomer, responsible for forming a polymer chain or crosslinked polymeric matrix, where the polymerization reaction is activated by the excited photoinitiator. Combinations of monomers have been used to improve specific properties, such as increasing the structure hardness or stability, or reducing shrinkage that can occur during polymerization or post-MPL processing [41]. Using higher performance materials avoids the need to add supports or solid frames that can compensate for anticipated shrinkage or distortion, but alters the overall structural form [42]. Using composites also provides a means for enhancing different material, optical, and chemical properties. Monomers such as acrylates [43] and epoxides [44] have been used interchangeably or in combination to fabricate different components and devices. However, most of these compositional changes yield passive devices that are limited in properties and use. To enhance the properties of fabricated structures or to create active devices, additional components have been incorporated, such as fluorescent dyes, optically active chromophores, and electrically active species.

9.2.1 Fluorescent-Dye-Doped Organic Microstructures

Fluorescent 3D organic microstructures have been realized via MPL using acrylate composites doped with fluorescent dyes and through post-fabrication chemical functionalization [18, 41]. Doping with a fluorescent dye yields active structures that can be used for sensing, photonics, or biomedical functions [45, 46]. A microstructure created with a polymeric composite containing rhodamine 610 at 0.5 wt% exhibited red fluorescence upon one- and two-photon excitation at wavelengths of $\lambda = 480$ and 780 nm, respectively [47]. Incorporating rhodamine did not adversely affect the fabrication process, as the structure displayed good structural integrity and resolution. Optically active microstructures containing the conductive and luminescent polymer poly[2-methoxy-5-(2-ethylhexyloxy)-1,4-phenylenevinylene] (MEH-PPV) were observed to retain the fluorescence spectra of the dye after structure fabrication [41]. Complex 3D active microstructures were fabricated in acrylate resins mixed with the laser dye LD490 [48]. Two-photon confocal fluorescence showed that the microstructures were uniformly stained with the dye.

Microstructures that have a broad and tunable fluorescence emission from approximately $\lambda = 350$ to 700 nm have been prepared (Figure 9.2) [45]. The polymeric resin used was doped with a specific fluorescent dye, namely stilbene 420, disodium fluorescein, or rhodamine B, which emit at center wavelengths 420, 535, and 580 nm, respectively. The microfabrication was done in stages. First, the polymeric resin doped with stilbene was used to create a portion of the microstructure by MPL. After development, the polymeric resin doped with disodium fluoride was deposited onto the same substrate. A second part of the structure was fabricated by MPL and then developed again. Finally, the third part of the microstructure was fabricated using the polymeric resin doped with rhodamine. Fluorescence imaging showed that the three parts of the microstructure fluoresced at the different wavelengths, with no apparent evidence of the fluorophores having mixed between the parts. In this process, the order of the stages is interchangeable depending on which part of the microstructure is required to emit at a given wavelength. A fluorescing artificial scaffold with promising applications in cell growth and tracking was fabricated in the hybrid sol–gel photopolymer organically modified silicate (ORMOSIL) doped with rhodamine 6G, fluorescein, LC6500, and coumarin 152 [46]. Biofunctionalization was demonstrated by first immobilizing biotin onto the surface of a microstructure made using an organically modified ceramics (ORMOCER) followed by exposure to fluorescent-dye-labeled avidin [18] or streptavidin [19].

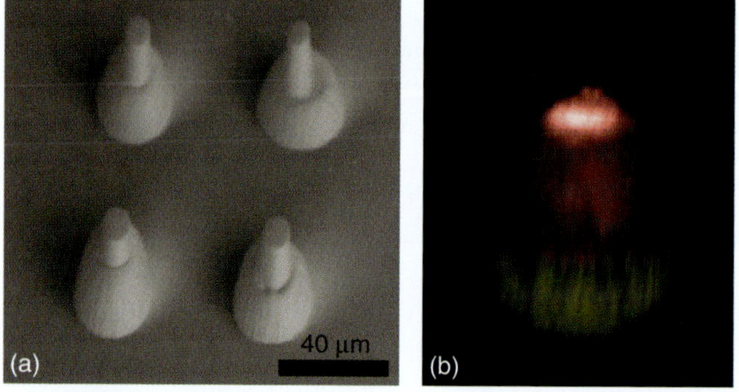

Figure 9.2 (a) Scanning electron microscopy (SEM) image and (b) confocal microscope image of double-doped microstructure. (Otuka 2012 [45]. Reproduced with permission of Optical Society of America.)

9.2.2
Organic Composites for Lasing Microstructures

Organic microlasers (OMLs) have attracted much attention because of their potential for enabling functional devices, particularly lab-on-a-chip systems. The advantages of having a microlaser that uses organic and polymer materials as a gain medium include low cost, compact size, low power consumption, versatility, tunability, and easy integration [46, 49]. However, when using organic components as an active medium, the photochemical processes can become complex because of the potential interplay of absorption and energy transfer between the photoinitiator and chromophore [50]. Additional challenges may include poor solubility of the chromophores in the resin, clustering, self-aggregation, and fluorescence self-quenching at high concentrations [49, 50]. To overcome these problems, and to increase the concentration of the gain medium, a solid-state microcavity was fabricated using a 4-(dicyanomethylene)-2-methyl-6-(4-dimethlyaminostyril)-4*H*-pyrane (DCM)-doped dendrimer incorporated into a pre-polymer material [50]. The dendrimer was made using repeated ester-branching groups and 64 end groups reacted with acrylic L-phenylalanine and FMOC-L-phenylalanine. The photopolymerized dendrimer encapsulates the laser dye, providing site isolation which limits intermolecular energy transfer and fluorescence quenching. The DCM-doped dendrimer gain medium concentration reached a maximum of 4.2 wt% compared to only 0.5 wt% without the dendrimer. The microcavity was optically characterized and found to lase at $\lambda = 612$ nm with a spectral bandwidth of 0.15 nm above threshold under an optical excitation intensity of 0.05 μJ per pulse [50]. Using a carbosilane dendrimer and fluorescein as the gain medium, the laser dye doping concentration was increased to 4.95 wt% in the photocured resin [49]. Using MPL, a distributed feedback resonator composed of cuboid arrays with suspended fibers was created. The resonator was observed to have an amplified spontaneous emission at $\lambda = 556$ nm with a threshold of 0.30 μJ per pulse.

A cuboid-shaped OML was fabricated using the 1,3,5,7,8-pentamethyl-2,6-di-*tert*-butylpyrromethen-difluoroborate complex (PM-597) as the gain medium, doped in a SR9020 triacrylate monomer mixed with a photoinitiator without the aid of dendrimer to increase the concentration [51]. The concentration of the OML gain medium was 0.8 wt%. The OMLs were fabricated directly on the glass substrate or on top of a body-centered tetragonal (BCT) woodpile PC pedestal (Figure 9.3). Fabricating the OML atop a PC decoupled the laser from the substrate and reduced the lasing threshold by a factor of 30.

9.2.3
Organic Composites for Electrically Conductive Microstructures

Electrically conductive 3D microstructures have many applications, including ion-exchange media, chemical sensors, organic photovoltaics, MEMS, micro-actuators, and lab-on-a-chip devices. A microstructure was fabricated in a

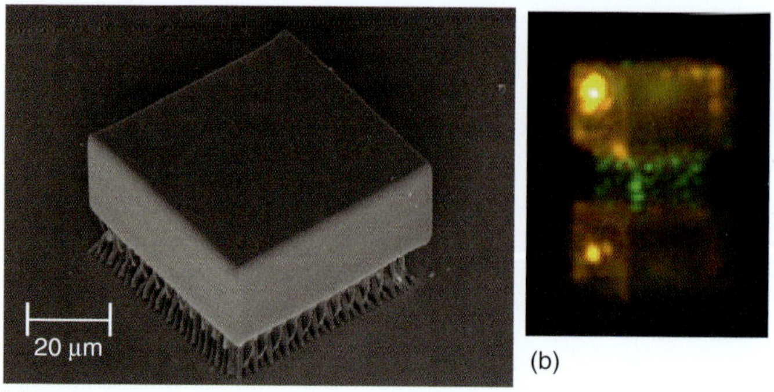

(a) (b)

Figure 9.3 (a) SEM image and (b) optical image of an organic micro-laser fabricated on top of a body-centered tetragonal woodpile PC. (Chen 2014 [51]. Reproduced with permission of Optical Society of America.)

polymer composite containing single-wall carbon nanotubes (SWCNTs) at 0.01 wt% via MPL at $\lambda = 800$ nm [52]. Aggregated SWCNTs linearly absorb at 800 nm and compete with the photoinitiator for the absorption of photons. This competition affects the MPL threshold, causing portions of the fabricated structure to shrink and appear distorted. To limit excess absorption, avoid damage to the SWCNTs, and improve structural integrity, the wavelength for MPL was shifted to $\lambda = 780$ nm, which corresponds to a valley between absorption peaks of the SWCNTs while still being suitable for two-photon polymerization. With this adjustment, robust 3D structures, including nanowires with lateral spatial resolutions as low as 200 nm, were fabricated in the composite, and the SWCNTs were observed to be uniformly embedded throughout the structure [52]. The SWCNTs were also found to be partially aligned within the nanowires, and it was suggested that this alignment could be exploited to create devices with improved mechanical and electrical properties.

An electrically conductive all-polymer 3D microstructure was fabricated by MPL and shown to have a conductivity of $0.04\,\mathrm{S\,cm^{-1}}$ (Figure 9.4) [53]. A blend of different amounts of 3,4-ethylenedioxythiophene (EDOT) dissolved in polyethylene glycol diacrylate (PEG-DA) was stirred for 30 min and degassed in a 1×10^{-2} mbar vacuum chamber for 1 h. The free-radical photoinitiator 2-hydroxy-4-(2′-hydroxyethoxy)-2-methyl-propiophenone (5 wt%) was added to different concentrations of the PEG-DA/EDOT blend, syringed through a 0.2-μm pore filter, and degassed for 5 min at 1×10^{-2} mbar. The PEG-DA/EDOT blend was then used to fabricate microstructures by MPL. The free-standing microstructures were subjected to oxidative polymerization in a bath of iron(III)chloride hexahydrate melt, which converted EDOT to poly-3,4-ethylenedioxythiophene (PEDOT) and increased electrical conductivity. The conductivity was found to be

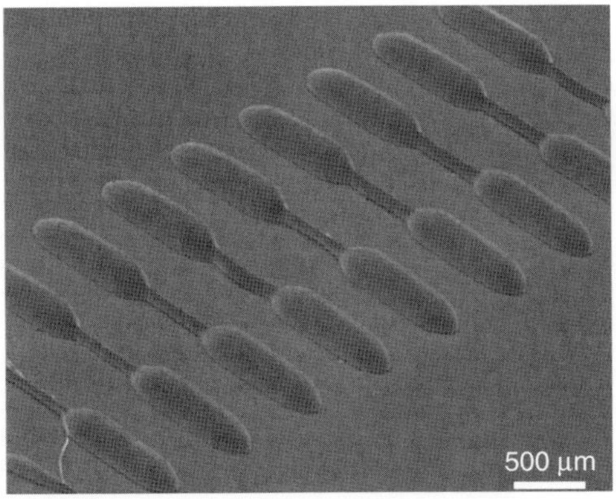

Figure 9.4 SEM image of the all-polymer conductive microstructure fabricated using two-photon polymerization. (Kurselis 2013 [53]. Reproduced with permission of Optical Society of America.)

directly limited by the concentration of the EDOT and reached a maximum of $0.04\,S\,cm^{-1}$ at 17 vol% EDOT.

9.2.4 Other Optically Active Microstructures

Optically birefringent microstructures with properties that can be altered by external means have direct application in optical data storage, optical switches, and waveguides. A reversibly birefringent microstructure was fabricated by MPL using a composite consisting of a triacrylate resin doped with the azo-aromatic chromophore disperse red 13 (DR13) [54]. A birefringence of 5×10^{-5} was achieved using 1 wt% DR13. The birefringence arises as a result of optical ratcheting of DR13 molecules that occurs under continuous exposure using linearly polarized excitation at $\lambda = 514.5$ nm. The birefringence could be induced and later erased by either heating the sample or by exposing it to circularly polarized laser light. As azo-aromatic chromophores have low residual birefringence, microstructures were fabricated using an azo-polymer HEMA-DR13, which was shown to increase the residual birefringence by ~35% [55].

A nonlinear 3D PC was fabricated by MPL using the nonlinear chromophore disperse red 1 (DR1) (Figure 9.5) [7]. DR1 was reacted with (3-isocyanatopropyl) triethoxysilane to form the functionalized silicon alkoxide precursor SGDR1, which was then mixed with methacryloxypropyl trimethosilane (MAPTMS) and the photoinitiator Irgacure 369. This composite was used to fabricate a woodpile PC. The structure did not exhibit a complete photonic bandgap due to the low refractive index contrast of SGDR1 and air, which is roughly 0.5. Fourier transform infrared spectroscopy (FTIR) indicated that a stopband was present in certain directions at infrared wavelengths.

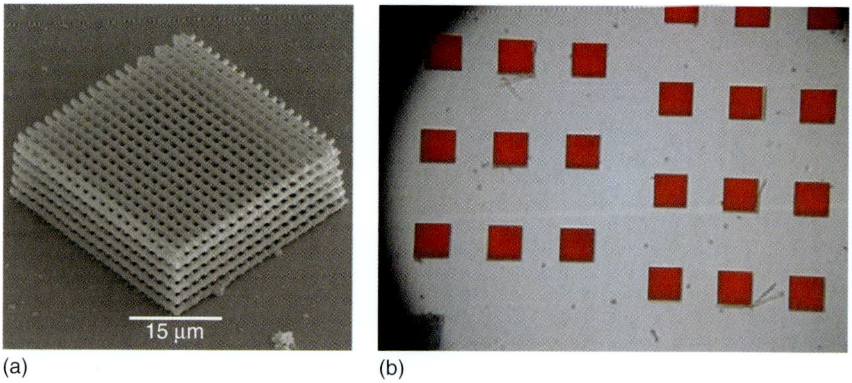

Figure 9.5 (a) SEM image and (b) optical image of the PC made by an active nonlinear chromophore SGDR1:MAPTMS composite. (Farsari 2008 [7]. Reproduced with permission of Springer.)

9.3
Multiphoton Processing of Oxide-Based Materials

Bulk metal oxides can have unique properties that depend sensitively upon their chemical composition and structure. These include large mechanical hardness, large and often tunable refractive index, wide optical bandgap, either high or low electrical resistivity depending upon structure and composition, photemissivity, and ferromagnetism. Reports have shown that some of these properties can be obtained by blending oxides into organic materials, thereby forming composites. Metal oxide nanoparticles are known for their promising electronic properties, which can be exploited for cancer treatment, chemical and biosensing, and photovoltaics [56].

Several approaches have been demonstrated for creating metal oxide microstructures by MPL. Metal oxide nanoparticles have been homogenously dispersed in polymers, forming photo-patternable composites for MPL [57, 58]. Inorganic–organic hybrid polymers have also been employed, where the inorganic backbone contains the species needed for the targeted metal oxide, and post-MPL processing is used to transform the patterned hybrid into a metal oxide 3D form [59–61]. Metal oxides have been introduced through sol–gel processing [62]. The simplest approach involves MPL using polymer composites containing metal oxide nanoparticles. Ideally, the weight loading of the metal oxide should be as high as possible; however, it is limited because high weight loading can interfere with the MPL process. Additionally, high weight loading increases the viscosity of the resin, making it difficult to mix and process, and possibly also resulting in phase separation [8, 63]. Surface functionalization of metal oxide nanoparticles has been used to suppress phase separation and achieve high weight loading [64]. Creating a metal oxide backbone via sol–gel processing can provide access to structures with specific properties [65–67]. Sol–gel processing is relatively simple and gives the added benefit of mechanical and chemical stability, although structures made from this process also suffer from shrinkage and distortion. Sol–gel processing has been used extensively for Si–O inorganic–organic hybrids, and is reviewed in detail elsewhere in this volume.

9.3.1
Titanium Dioxide

Titanium dioxide (TiO_2, titania) has both high refractive index and high transparency at visible wavelengths, which is not available with many other materials. Along with somewhat enhanced robustness, these properties have made composites of TiO_2 very worthwhile to explore. Structures of titania composites have been fabricated by processing complexes of titanium(IV) acrylate or titanium(IV) ethoxide blended with a photopolymer [8]. After fabrication, the nanoparticle composite contained titanium hydroxide as a result of air exposure. The structure was then converted to TiO_2 by annealing to drive off the organic components.

Titanium(IV) *n*-butoxide complexed with benzoylacetone undergoes bond cleavage when irradiated with UV light or by MPL, rendering it insoluble in organic solvents like acetone. The resulting composite contains TiO_2, which can be annealed, changing the titania morphology and eliminating organic components. This process yields a structure with high refractive index and was used to create the first titania PC having a complete bandgap at visible wavelengths [68].

Three-dimensional structures have been fabricated by MPL using composites based on titania nanoparticles or silicon/titanium sol–gel hybrids. Sol–gel hybrids were created by combining various amounts of titanium(IV) isopropoxide (TIP) and MAPTMS. This approach was used to tune the refractive index by varying the ratio of TIP and MAPTMS in the composite (Figure 9.6) [69, 70]. These and related approaches for creating 3D structures of nearly pure TiO_2 [71, 72] have shown how effective the methods are for creating microscale photonic devices with very high or very specific refractive index.

9.3.2
Zinc Oxide

Zinc oxide (ZnO) is a semiconductor that can give enhanced conductivity, photoemission, or hardness when added to composite structures. In one account, ZnO nanoparticles having an average diameter of 50 nm were added to 1 wt% cinnamate acrylate monomers [73]. Nanoparticles of shapes other than spheres were also used. ZnO nanowires with a length of 1–3 μm were prepared using a hydrothermal technique and directly added to acrylate monomers up to 5 wt% [56, 57]. The composites containing ZnO fluoresced at 570 nm. Other emission/absorption bands were obtained with rod-shaped nanoparticles. Three-diensional structures based on these composites exhibited enhanced fluorescence at ~450 nm [57].

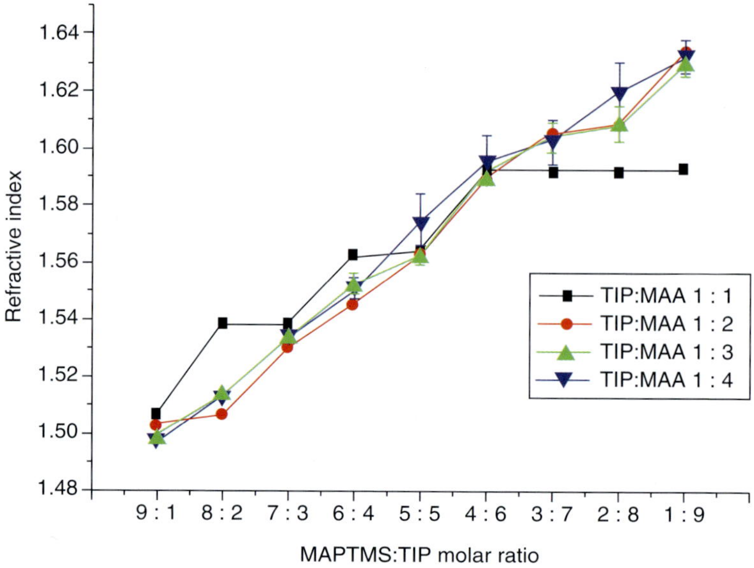

Figure 9.6 Dependence of the refractive index of the content of titanium(IV) isopropoxide (TIP) within sol–gel composites used for MPL. (Sakellari 2010 [69]. Reproduced with permission of Springer.)

9.3.3 Zirconium Dioxide

Zirconium dioxide (ZrO_2) is a hard material with high refractive index, which makes it well suited for composite microstructures. Chichkov and coworkers explored the use of the Zr-containing polymer sol–gel composite SZ2080 for MPL (Figure 9.7) [65, 74–76]. The zirconium oxide increases the hardness and reduces the shrinkage of the material upon processing, which are two common problems associated with fabricating high-resolution 3D structures by MPL. The resolution of this system was improved by formulating composites containing 2-dipentaerythritol pentamethacrylate [61]. This molecule features a large number of reactive acrylate groups, which substantially increases crosslinking and serves to reinforce the material. Minimum feature widths of ~100 nm were achieved with the composite [61].

9.3.4 Iron Oxide

Powered MEMS are needed for many applications, such as microfluidics, lab-on-a-chip devices, and clinical diagnostic equipment. However, such devices are difficult to actuate because their small size increases the complexity of power transfer. One viable and cost-effective method for making powered MEMS is to fabricate parts containing the ferromagnetic material iron oxide (Fe_3O_4) and then

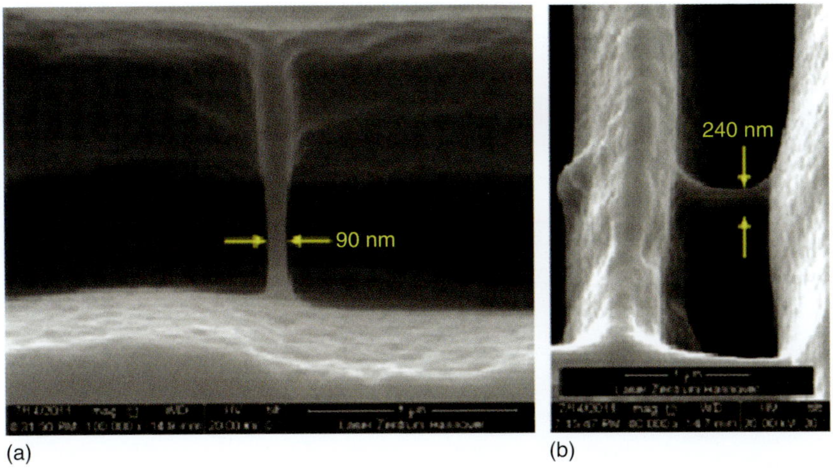

Figure 9.7 (a) SEM image of a feature having an axial dimension of 90 nm created by MPL in a ZrO composite. (b) The same structure observed at an angle of 20°. The minimum height of the structure is 240 nm. (Paz 2012 [61]. Reproduced with permission of Laser Institute of America.)

using magnets to actuate the structures remotely. The magnetic field intensity needed to drive a device can be reduced by using a high weight loading of Fe_3O_4. Although direct doping of ferric oxide nanoparticles into an organic photoresist has given little success, much higher weight loading has been achieved by modifying the surface of the nanoparticles using organic functionalizing agents. In early reports, the iron oxide nanoparticles were capped with oleic acid to disperse them into an acrylate photoresist up to 2.4 wt% [58]. A similar method was used to coat the surface of cantilevers fabricated in the crosslinkable epoxide SU-8 (MicroChem), which enhanced their sensitivity for detecting volatile molecules [77]. The performance was further improved by chemically modifying the surface of the nanoparticles using 3-(trimethoxysilyl)propyl methacrylate or 6-(methacryloyloxy) hexanoic acid, which stabilized the nanoparticle dispersion toward crosslinking during MPL [78, 79]. The nanoparticle concentration was further increased to 4–5 wt% by functionalizing with 6-(methacryloyloxy)hexanoic acid (Figure 9.8) [64].

9.3.5 Tin Dioxide

Structures of tin dioxide (SnO_2) have properties that are sensitive to the environment and can therefore be used as gas sensors. In one report, a unique metal oxide–organic composite was prepared by mixing and chelating tin(II) dichloride with acetylacetone [80]. The complex is soluble in organic solvents, and forms a gel at high weight loading, but decomposes to an organic-insoluble chelate when exposed to UV light or multiphoton absorption. This material and structures

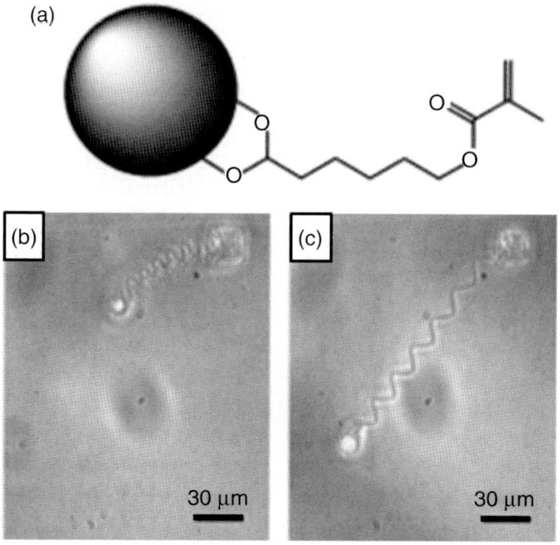

Figure 9.8 (a) Schematic illustration of a Fe_3O_4 magnetic nanoparticle coated with 6-(methacryloyloxy)-hexanoic acid. (b) Optical micrographs of a ferromagnetic microspring fabricated by MPL using an iron oxide nanoparticle composite. (c) The micro-spring was extended while immersed in ethanol by applying a magnetic field. (Tian 2012 [64]. Reproduced with permission of Elsevier.)

created with it by MPL can be calcined to form crystalline SnO_2. Nanowires of SnO_2 fabricated with this composite exhibited electric resistance that was sensitive to the relative humidity of air (Figure 9.9) [80].

9.3.6
Germanium Dioxide

Like titania, germanium dioxide (GeO_2) is widely used for optics because of its high transparency at near-infrared (NIR) and optical wavelengths. As a result, there is great interest in processes that can be used to create germanium dioxide structures by MPL. This has been achieved by sol processing silicon alkoxides mixed with up to 40 wt% of germanium alkoxide [60]. The refractive index of these composites can be tuned by changing the germanium content, but it was noted that high germanium content compromises the achievable resolution [59].

9.3.7
Silicon Dioxide

There are numerous examples of microstructures fabricated by MPL using composites based on silicon dioxide (SiO_2) nanoparticles and hybrid polymers containing a silicon–oxygen backbone [81, 82]. Silicon oxide nanoparticles blended into composites without any organic modification improves hardness

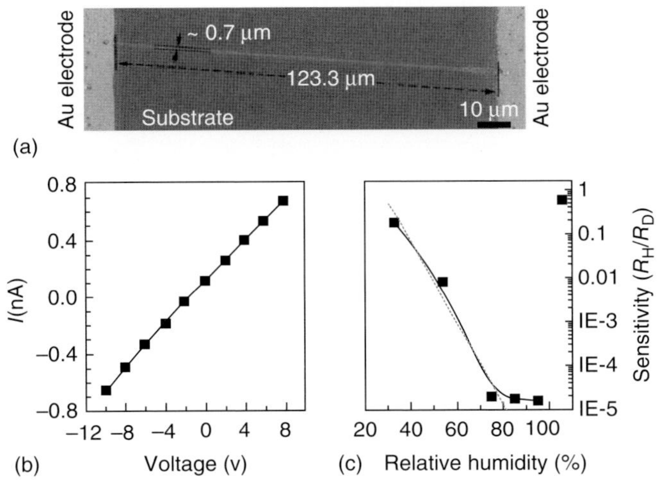

Figure 9.9 (a) SEM image of a SnO_2 nanowire between two gold electrodes. (b) *I–V* curve of the SnO_2 nanowire. (c) Dependence of resistance of the SnO_2 nanowire on relative humidity in air. (Guo 2010 [80]. Reproduced with permission of Optical Society of America.)

and reduces shrinkage. A wide range of Si–O hybrid polymers can be created by leveraging the large library of available silane precursors. The precursors are sol–gel-processed, patterned by MPL, and then developed to form silica-containing microstructures. High-refractive-index composites have been created by using phenyl-substituted polymers [66]. High-index siloxane waveguides have been patterned within a cladding of lower index material [83]. A composite based on silanol-terminated polysiloxane and methyl-substituted acryloxysilane has been developed for fabricating optical waveguides by MPL [67]. The commonly used silicon oxide hybrid polymers known as ORMOCERs and ORMOSILs are discussed in detail elsewhere in this volume.

9.4 Multiphoton Processing of Metallic Composites and Materials

Metals can be used to create structures having high electrical and thermal conductivity, high mechanical strength, good optical reflectivity, and other novel properties. Several approaches have been developed for creating metallized structures in conjunction with MPL. These include physical and chemical deposition of metal after MPL, coating functionalized surfaces with metal NPs after fabrication, and direct fabrication of metal-containing structures using composites that include components of the targeted metal. The simplest way to create a metallized 3D structure is to fabricate it by MPL using a conventional polymeric resin and then deposit the targeted metal physically or chemically. This can be achieved through CVD or by physical metal vapor deposition, where the metal is vaporized by resistive heating, electron beam (e-beam) evaporation, or magnetron sputtering (MS).

Electrochemical methods can also be used to deposit metal onto the surface after fabrication. Metal structures can be written directly using photo-patternable composites that contain a source of the target metal, present as metal salts, stabilized nanoparticles, or combinations of the two. These approaches are discussed in greater detail below.

9.4.1
Thermal Evaporation

In thermal evaporation (TE), a metal source is resistively heated under high vacuum and the metal vapor is deposited from a single direction because of the long mean free path of the particles in the deposition chamber [84]. TE is attractive because of its simplicity, but it is limited in that a fabricated structure cannot be conformally coated. Additionally, the heating associated with the process can deform the polymeric microstructure. However, TE can be effective if the fabricated device requires that only the metal is deposited onto a single surface and if suitable conditions can be found that preserve the integrity of the structures. Jia and coworkers fabricated U-shaped split-ring resonators (USRRs) by MPL in the photopolymer AR N 4340, and then metallized the USRRs with a 50-nm gold layer [11]. Any unpolymerized resin, as well as the gold deposited on top of that resin, was then removed using acetone, leaving freestanding structures with a 50-nm gold coating on top. These USRRs were shown to function as optical filters/polarizers in the 2–7 μm wavelength range.

9.4.2
e-Beam Evaporation

e-Beam evaporation is a sputtering technique in which an electron beam is used to deposit metal onto a target under high vacuum [85]. This technique is more complicated than TE, but the issue of high temperatures of the sputtered metal is avoided. Abaddi *et al.* used e-beam evaporation in conjunction with MPL to fabricate a conductive circuit [10]. The structural form of the circuit was created by MPL using the commercial positive-tone photoresist AZ 5214. e-Beam evaporation was then used to deposit a 10-nm-thick layer of titanium, which acted as an adhesion layer, followed by e-beam evaporation of a 100-nm layer of gold, which served as the conductive element of the circuit.

9.4.3
Magnetron Sputtering

MS can generate high-quality films that can outperform other deposition techniques in terms of smoothness and metal grain uniformity [86]. Additionally, as MS involves low vacuum, some conformal deposition can be achieved because the metal vapor collides with other gaseous atoms and molecules during deposition. Previously discussed techniques require high vacuum, so these collisions are much more infrequent. MS was used by Xiong and coworkers to coat fabricated structures with a 35-nm layer of gold [87, 88]. In this work, various optical functional devices were fabricated using the commercially available MPL photoresist IP-L (Nanoscribe). Polarization-dependent [88] and wavelength-dependent [87] absorbers were fabricated and coated on metals to increase the structures' refractive index contrast, increasing the efficiency of the device (Figure 9.10).

9.4.4 Chemical Vapor Deposition

CVD uses gaseous organometallic precursors to deposit metal onto a structure via a chemical reaction that occurs at the structure's surface [89]. The structure is heated to an elevated temperature, which depends on the organometallic precursor. When the gaseous precursor comes into contact with the heated structure, it decomposes, generating organic fragments that largely diffuse away from the structure leaving behind the residual metal. Because CVD is performed using a comparatively low vacuum, molecular collisions cause metal to deposit conformally on the target surface. However, CVD is slow compared to most other metal coating techniques, sometimes requiring as much as 1 h to deposit 2 nm of the target metal [90]. Additionally, it is not necessarily selective for deposition onto a microstructure versus a supporting substrate or other features on the surface. Wegener's team created functional split-ring resonators (SRRs) by fabricating the forms by MPL in SU-8, coating with a protecting semiconducting oxide layer, and then depositing silver onto the protected structures by CVD using the precursor Ag(hfac)(cod) [90]. Careful control over chamber pressure and temperature, precursor evaporation temperature, and structure temperature yielded a smooth coating of silver metal.

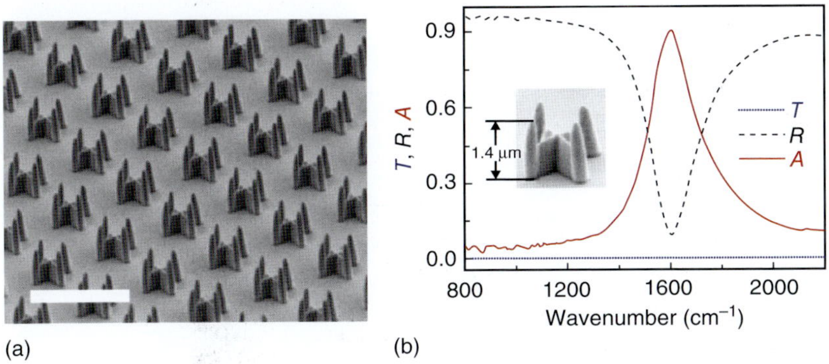

Figure 9.10 (a) SEM image of metal-coated four-tined fish-spear-like resonators (FFRs). The scale bars represent 5 μm. (b) The measured transmission (*T*), reflection (*R*), and absorbance (*A*) for a tine-height of 1.4 μm. (Xiong 2013 [87]. Reproduced with permission of Wiley.)

9.4.5
Functionalization by Attachment of Nanoparticles

Metal NPs synthesized *ex situ* can be chemically bound to a polymeric 3D microstructure created by MPL. Although this technique does not provide complete coverage of the structure, the process can deposit NPs onto all sides of a microstructure, including overhangs. The major advantage of producing NPs separately lies in being able to control the identity of the metal, the NP shape, or the chemical protecting group surrounding the NP. This can allow NPs with specific photonic or binding properties, which can be used in sensing, photonics, or any number of other applications.

Aekbote *et al.* [91] used the Turkevich synthesis [92] to produce citrate-capped Au NPs which were bound to a 3D microstructure to increase their ability to be captured in an optical trap (Figure 9.11). MPL was used to fabricate the 3D structures in SU-8. These were then treated with 3-(aminopropyl)-triethoxysilane to functionalize the microstructure surfaces with amino groups. Immersing the amine-functionalized structures in a low-pH solution protonated the surface and attracted negatively charged citrate-capped Au NPs onto the structures.

Laser-induced transfer (LIT) is another approach for depositing metal NPs onto a surface. In LIT, an amplified laser is focused onto a noble metal surface, causing localized melting and pressure gradients at the focal point. This results in a jet-like structure that ejects molten metal from the area in which the laser is focused. The surface tension of the small droplets causes the metal to form NPs. Reinhardt *et al.* used LIT to coat an organic–inorganic hybrid polymer with Au NPs [93]. First a woodpile PC was created by MPL in an organic–inorganic hybrid polymer. Next, a 50-nm gold film was produced on a separate substrate by MS. The PC was then placed near the gold film, and LIT was used to transfer gold NPs onto the surface. This technique produced Au NPs with a uniform size across the PC surface, although the coating was not conformal.

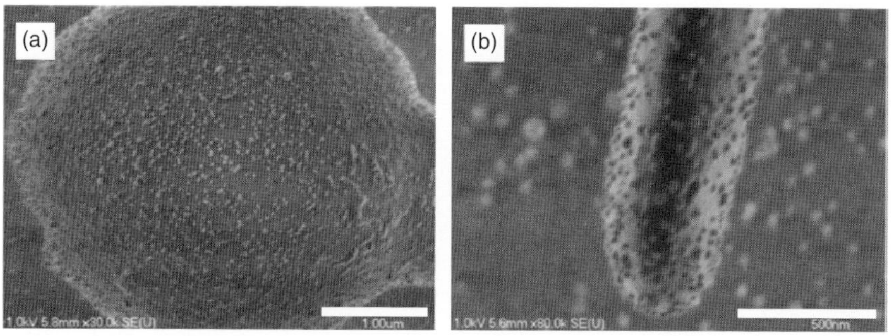

Figure 9.11 Two-photon-polymerized SU-8 microstructures for optical trapping applications. (a) SEM top view image of the spherical and (b) tip areas of the microstructures coated with 24 nm Au NPs. (Aekbote 2012 [91]. Reproduced with permission of Elsevier.)

9.4.6
Electroless Metallization from Solution

Microstructures can be conformally and rapidly coated with metal by electroplating or electroless deposition in solution. Electroplating of polymeric surfaces is inherently difficult because most are not electrically conductive. With electroless metallization (EM), a conductive surface is not required. In EM, a metal ion precursor and a reducing agent are co-present in the solution, along with a stabilizing agent that prevents them from reacting spontaneously. A surface to be metallized is then chemically sensitized and immersed into the EM bath. The sensitizer enables rapid reaction of the metal salt and the reducing agent, forming a conformal coating of metal on the surface. Polymeric surfaces can be sensitized by reaction with $SnCl_2$, binding metal NPs, or other means. Both EM and electroless deposition produce metal with a nanoscale morphology that is sensitive to the metal type, the deposition conditions, and additives that can be co-present to control the kinetics of the reaction [12, 94, 95].

The works by Kawata *et al.* [17, 96] and Fourkas *et al.* [13, 97] are among the earliest reports of using EM to deposit metal conformally onto 3D structures created by MPL. A widely used approach for depositing silver onto microstructures involves using a modified Tollens' reagent [98]. This EM bath consists of silver cations dissolved in basic aqueous media and an aldehyde reducing agent. The reducing agent can be a sugar, such as glucose, which exists in solution in an equilibrium between the cyclic form and the aldehyde-terminated open form. Kawata and coworkers fabricated a wide range of microstructures by MPL using a commercial photopolymer, activated their surfaces using $SnCl_2$, then deposited silver using a modified Tollens' reagent (Figure 9.12) [96]. They also demonstrated that large-area metallized structures could be created by patterning structures in parallel using a microlens array and then metallizing the structures. Tollens'-like

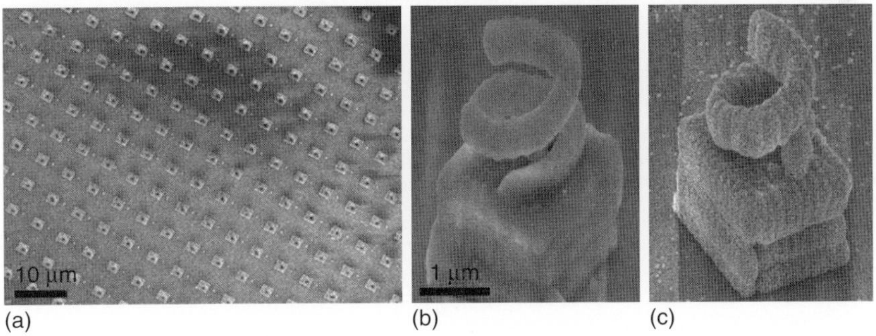

Figure 9.12 (a) SEM image of 3D periodic silver-coated microstructures fabricated on a hydrophobic coated glass surface. (b) Tilted magnified view of an individual uncoated polymer structure composed of a cube (2 μm in size) holding up a spring (height = 2.2 μm, inner diameter = 1 μm). (c) SEM image of an individual silver-coated microstructure after electroless metallization. (Formanek 2006 [96]. Reproduced with permission of Optical Society of America.)

approaches have been used to create a wide range of metallized 3D microstructures [95, 99–101].

One of the appeals of EM is that the approach can be selective toward a given surface. Several groups have shown that EM can be used to deposit metal on a polymeric microstructure selectively, leaving the supporting substrate largely metal-free [12, 17, 94, 102]. The selective functionalization was achieved by either copious cleaning or plasma-etching the substrate prior to processing [12, 94, 102], or through chemical modification of the substrate, such as application of a hydrophobic coating prior to casting the photosensitive resin [17].

Selective metallization can be achieved through a two-step process in which a microstructure is fabricated using two different resins, only one of which is active toward EM. Kawata and coworkers fabricated part of a microstructure by MPL using the methacrylate resin KC1102 (JSR), developed to remove unpolymerized material, and then continued fabrication using a composite of KC1102 blended with a methacrylamide, and developed again. The surface of the finished structure was treated with a Tollens'-like reagent. Silver deposited selectively onto those portions of the structure that were fabricated using the methacrylamide-doped resin [103]. Fourkas *et al.* used a similar approach in which parts of a microstructure were fabricated in an acrylate-based resin and others were fabricated using a methacrylate-based resin [13]. After fabrication, the sample was treated with a diamino alkane, which selectively adds to acrylates by Michael addition, and then exposed to a copper EM medium [104]. The metal was deposited selectively at the amine-functionalized surfaces, leaving portions of the structure created from methacrylates un-metallized. This technique was used to fabricate a functional micro-inductor that was electrically characterized (Figure 9.13).

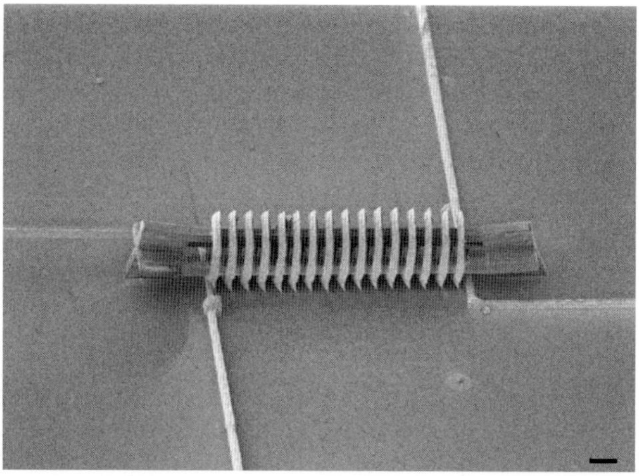

Figure 9.13 SEM image of a copper-coated micro-inductor with a methacrylate support (scale bar corresponds to 10 μm). (Farrer 2006 [13]. Reproduced with permission of American Chemical Society.)

Copper EM has also been used to create movable metallic 3D microstructures [105, 106]. Three-dimensional microstructures were fabricated using an acrylic resin consisting of Sartomer SR-368 and Sartomer SR-499 and functionalized with ethylene diamine. The surface of the amine-functionalized microstructure was sensitized using $PdCl_2$. Copper metal was then deposited by EM using 0.6 M Rochelle salt, 0.14 M NaOH, 0.28 M $CuSO_4$, and 0.3 M EDTA. The parts of the microstructure attached to the substrate were then laser-ablated, which freed the microstructure from the surface, allowing the microstructure to move. Some microstructures remained attached to the substrate, such as a cantilever [105], while other microstructures were not attached to the substrate but held in place by interconnecting parts, such as a micro-rotor [106].

Kuebler *et al.* used a modification of an EM process first reported by Danscher [107, 108] to deposit electrically conductive and highly reflective silver conformally onto 3D microstructures and woodpile PCs fabricated by MPL in SU-8. The surfaces of polymeric microstructures were activated toward EM by attachment of Au NPs [109, 110]. Silver deposited in this way was shown to have a conductivity of $1 \times 10^4\ \Omega^{-1}\ cm^{-1}$, which is 1.5% of that of bulk silver metal [12, 94]. Terzaki and coworkers have compared several approaches for silver deposition and used them to create metallized microstructures created with organic–inorganic hybrid polymers [95].

Kuebler *et al.* fabricated functional metallo-dielectric PCs by using EM to deposit copper onto polymeric lattices created by MPL in SU-8 (Figure 9.14) [102]. Infrared reflectance spectra showed that EM significantly transformed the optical properties of these PCs. Whereas the polymeric PCs exhibited the usual narrow stopband, metallization created a broad "plasmonic" stopband, characterized by high reflectance at all wavelengths below the stopband and low reflectance at wavelengths above the stopband. These metallo-dielectric PCs behaved like "hot mirrors" in which the structure transitions from reflecting to nonreflecting at a wavelength that can be tuned by simply changing the period of the lattice.

Misawa *et al.* fabricated woodpile PCs in SU-8 and coated these structures in nickel [111, 112]. This process involves fabricating a microstructure by MPL in SU-8, sensitizing the surface using aqueous $SnCl_2$, activating the surface for EM with aqueous $PdCl_2$, and then finally immersing the structure in a nickel EM bath consisting of $NiCl_2$, NH_4Cl, NaH_2PO_2, and NH_4OH at pH = 9.2 at 40 °C. The microstructures fabricated using this process were shown to be partially reflective in the infrared. Ren *et al.* used a similar process to create nickel-plated PCs and showed that they behaved like thermal emitters when heated (Figure 9.15) [113]. Wang *et al.* also used a nickel EM process to create ferromagnetic microstructures that could be manipulated using a magnetic field [114]. The metal coating also increased the hardness and modulus of the microstructure by 10- and 9-fold, respectively.

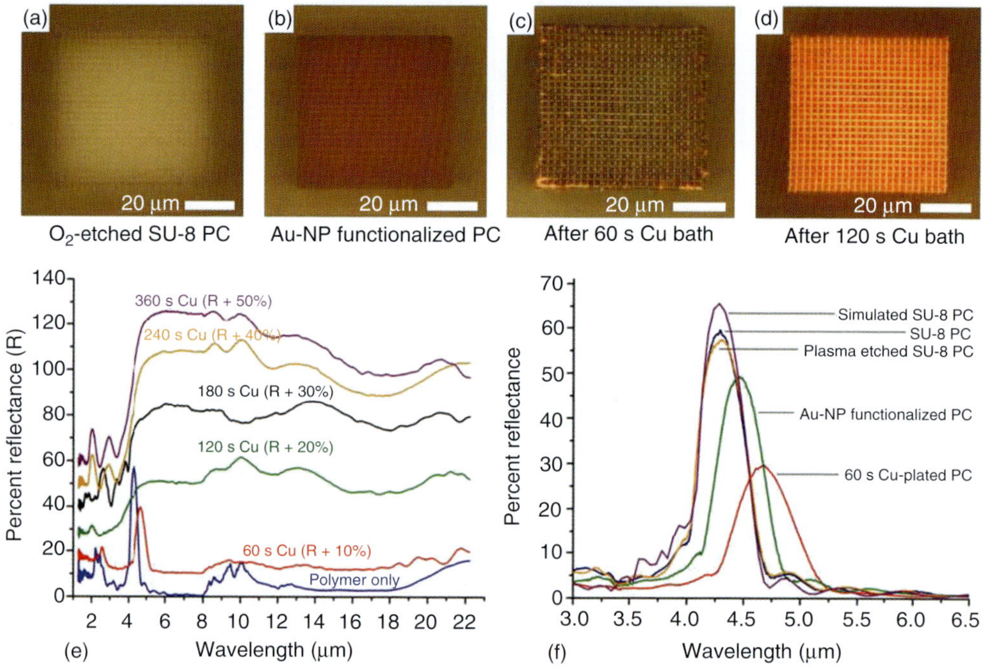

Figure 9.14 (a–d) Optical reflection image of SU-8 PCs at various stages of the copper deposition process. (e,f) FTIR reflection spectra of PCs at various stages of copper deposition. Each spectrum in (e) is offset vertically by 10% for clarity. (Tal 2007 [102]. Reproduced with permission of Optical Society of America.)

9.4.7 Multiphoton Lithography of Nanoparticles Supported in a Polymer Matrix

Metallized 3D microstructures can also be prepared by incorporating metal ions in the photopolymer used to create the structure. This approach can yield structures with metal distributed throughout the interior of the 3D microstructure, rather than merely on the surface. This can be advantageous for certain applications and, in principle, could simplify the fabrication process, insofar as the metallized structure can be produced in a single MPL step.

Shukla *et al.* introduced a method for creating Au NP-containing SU-8 microstructures for various photonic applications [115, 116]. In their approach, $HAuCl_4$ and the two-photon-absorbing dye AF 380 were blended with SU-8. AF 380 was included as a photo-activatable reducing agent that could transfer electrons to Au^{3+} cations upon photoexcitation, leading to the formation of Au NPs. After MPL, the microstructures were developed as usual. Figure 9.16a shows lines fabricated on a glass substrate under different exposure conditions. The line widths did not change with varying scan speed, but the number density of Au NPs was found to increase with exposure. Microstructures fabricated in the

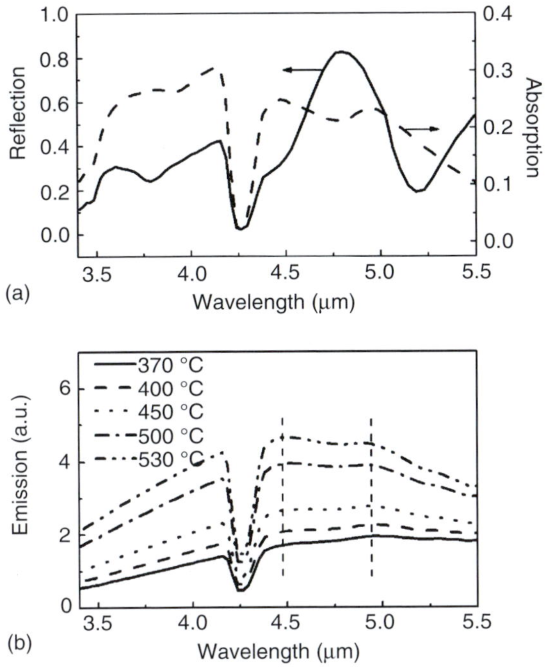

Figure 9.15 (a) Infrared reflection (solid line) and absorption (dashed line) of metallized PCs. (b) Thermal emission spectra of the PCs as a function of temperature. (Ren 2013 [113]. Reproduced with permission of Royal Society of Chemistry.)

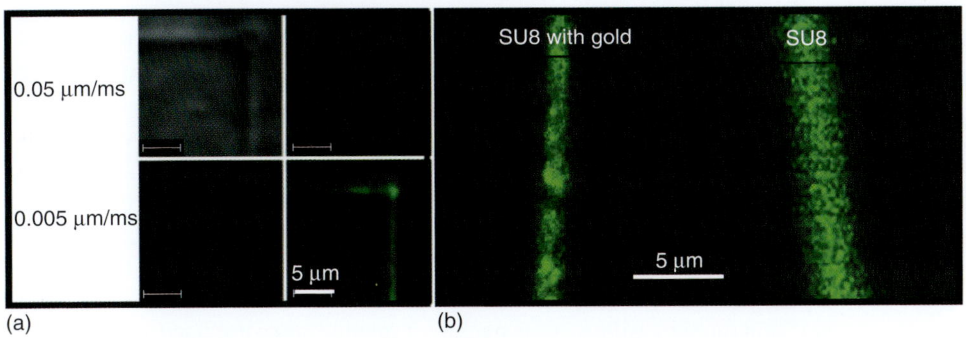

Figure 9.16 (a) Confocal transmission (left) and fluorescence (right) images of lines produced at scanning speeds of 50 μm s^{-1} (top) or 5 μm s^{-1} (bottom) at 10 mW. (b) Confocal images of gold precursor-doped SU-8 line and undoped polymeric feature at the same power (40 mW) and scan speed (50 μm s^{-1}). (Shukla 2011 [116]. Reproduced with permission of American Chemical Society.)

presence of the gold precursor had smaller line widths than those fabricated with neat SU-8 (Figure 9.16b). The fabricated lines were conductive to within an order of magnitude of bulk gold due to the high loading fraction of the gold precursor. Microstructures with metal NPs have been fabricated in several other materials, including Ag NPs in poly(vinylpyrrolidone) [117], Ag NPs in SU-8 [118], Ag NPs in a porous SiO_2 framework [62], Au NPs in an acrylate matrix [119], and Au, Cu, and Ag NPs in borosilicate glass [56].

Tanaka *et al.* took advantage of the fact that Au^{3+} quenches the fluorescence of rhodamine B dye, whereas Au NPs do not, to develop a direct-write medium for 3D optical data storage [120]. The investigators doped poly(methylmethacrylate) matrix (PMMA) with rhodamine B and $HAuCl_4$. Exposing the composite to focused femtosecond laser pulses locally excited the rhodamine B, leading to photo-induced reduction of Au^{3+} and formation of Au NPs. In regions where Au NPs were created, fluorescence was restored. Data could be stored digitally by writing Au NP-rich pixels within the volume of the material and reading out the data by laser-induced fluorescence. This system is useful for write-once–read-many (WORM) applications, like long-term data storage.

9.4.8 Direct Writing of Continuous-Metal Microstructures

When the goal is to create microstructures of *continuous* metal, then directly writing metallic features would appear to be the preferred route. However, this approach is not straightforward, because most chemical systems for forming metal in solution or viscous media tend to generate poly-nucleated and/or discontinuous NPs, rather than the bulk metal. The first report of direct fabrication of metallized structures was presented by Stellacci *et al.* [9]. In their approach, a composite was formulated of poly(vinylcarbazole) (PVK), Ag NPs, $AgBF_4$, and a two-photon-excitable dye that could be spin-coated onto surfaces. Laser excitation of the dye activated electron transfer to the Ag^+ ions, which caused Ag NPs to grow further and fuse into continuous metal within the PVK matrix. The host polymer could then be dissolved using dichloromethane, leaving a freestanding metallic nanostructure. Maruo *et al.* also used a host matrix as a support to directly write silver lines and pyramids in a PVP matrix containing silver nitrate [121]. Metal–organic frameworks (MOFs) have also been used as a host matrix by Ameloot *et al.* to write micro-QR codes and microcircuits [122].

Although it is generally advantageous to use a host matrix, this may not always be possible. Directly preparing metal microstructures from a liquid solution provides easier processing, but also pose new challenges in that NPs or other metal generated by MPL within a solution medium may not have a firm point of attachment on the substrate or other portions of the microstructure. Ishikawa *et al.* addressed this issue by scanning the focused laser at a slower rate and focusing at the surface of previously deposited metal [123]. This approach caused newly created NPs to aggregate, sinter, and become fused with the underlying layer (Figure 9.17). Lu *et al.* developed a medium for MPL of continuous gold

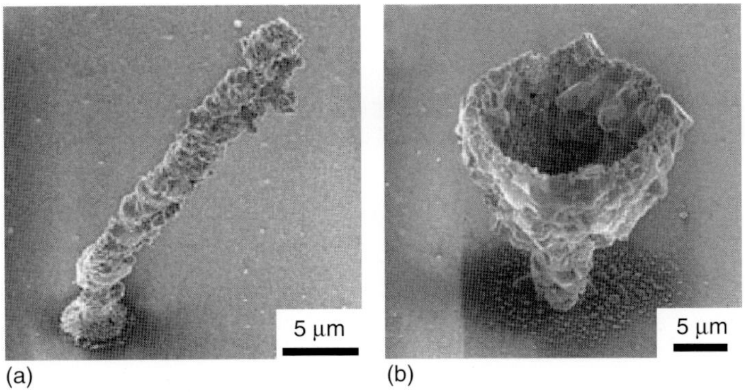

Figure 9.17 SEM images of a freestanding (a) silver tilted rod and (b) silver cup on a substrate. The length of the rod and the angle relative to the substrate are 35 μm and 60°, respectively. The height and the top and bottom diameters of the cup are 26, 20, and 5 μm, respectively. (Ishikawa 2006 [123]. Reproduced with permission of American Institute of Physics.)

metal based on Au^{3+} and an ionic liquid. The ionic liquid helped the formation of gold NPs upon laser irradiation and encouraged their aggregation into continuous gold metal features [124].

Fourkas *et al.* reported a hybrid two-step approach for creating continuous metal features [97, 125]. First, silver nitrate/PVP solutions were spin-coated onto substrates forming films. Then MPL was used to photo-reduce Ag^+ in the PVP films, forming silver lines [125]. This process could also be carried out on the surface of microstructures fabricated earlier by MPL using a metal-free photopolymer [97]. The silver lines were then enhanced with by copper EM, forming a microstructure with highly selective and conductive metallization.

9.4.9 Metal Backfilling by Electroplating

As mentioned previously, it is inherently difficult to electroplate polymeric microstructures because of their lack of conductivity. However, Gansel *et al.* have found a way to use electroplating to backfill a micro-patterned polymeric film with gold [23]. The positive-tone resist AZ 9260 (MicroChemicals GmbH) was used to create helical voids on top of an indium-tin oxide (ITO) substrate, which served as the cathode. After MPL and development, a platinized tin mesh was placed on top of the polymer block, to serve as an anode. The two-electrode sample was placed in a solution of sodium disulfitoaurate(I), sodium sulfite, ethylenediamine for stabilization, and disodium ethylenediamine tetraacetic acid. The solution was held at constant temperature while a fixed potential was applied at constant current to reduce Au^+ to gold metal at the ITO electrode. This resulted in the growth of continuous bulk metal within the helical holes of the patterned polymer film. Plasma etching was then used to remove the polymer matrix, leaving behind a structure of freely standing individual gold helices. These structures were shown to selectively reflect circularly polarized light of one handedness and transmit the other.

9.5 Multiphoton Processing of Semiconductor Composites and Materials

Semiconductor devices are important for electronic, optical, and sensing applications. Significant strides have been achieved in extending MPL for patterning semiconductors through the development of specialized photo-patternable chalcogenides, polymer composites that contain semiconductor NPs or their chemical precursors, and post-patterning processes, such as ALD. Understandably, many investigations have focused on demonstrating new material processes by creating functional 3D PCs, waveguides, or other optical devices, for which semiconductor materials are often best suited.

9.5.1 Structures Functionalized with Nanoparticles

The most commonly employed NPs consist of lead sulfide, lead selenide, or cadmium sulfide. Core/shell compositions are commonly used because the radiative properties of core-only NPs can be adversely affected by oxidation, contamination, and photobleaching [126]. By having a shell of a wider bandgap semiconductor, the quantum yield of the emitter is increased and stability is improved. Semiconductor NPs can have high quantum efficiency and narrow, tunable emission lines, which makes them well suited for micro and nanophotonic devices [127]. NPs can also be used to engineer linear or nonlinear optical properties and to increase quantum efficiency in certain wavelength ranges. The refractive index, absorption and emission wavelengths, and stop gaps of the hybrid material structure can be tuned via the composition and structure of the NPs. Fabrication parameters of the PCs and size distribution of the NPs can be altered to obtain targeted embedded NP emission wavelengths at or near the bandedge of the PCs [128]. NP-based microstructures can be formed by depositing the NPs onto a polymeric structure created by MPL, or performing MPL using a composite that contains the NPs or an NP precursor.

9.5.2 Structures Functionalized using NP–Polymer Composites

Free-standing 3D microstructures containing semiconductor NPs have been fabricated by MPL using polymeric photoresists that are preloaded with semiconductor NPs or by dispersing the NPs into a resin using the sol–gel method [127–130]. PbS and PbSe NPs have been used because of their NIR emission band, large Bohr radius, and high third-order nonlinearity [127]. After photo-patterning, the unpolymerized material could be washed away, leaving the 3D nanocomposite structure [131].

Li *et al.* fabricated 3D woodpile PCs with a photonic bandgap in the NIR using a commercial PbSe NP-doped composite material (1550 nm PbSe Evicomposite); however, aggregation during the fabrication process was observed to contribute to structure defects [128, 131]. Three-dimensional woodpile PCs with stopbands of 80% suppression in transmission in the telecommunications wavelength region were created by Jia *et al.* using the hybrid organic–inorganic resin ORMOCER containing PbS NPs [127, 129, 130]. The NPs were found to be evenly dispersed in the resin, and Z-scan measurements confirmed high third-order nonlinearity of the polymerized nanocomposite film [127, 129, 130]. The stopbands red-shifted with increasing NP concentration as a result of the increase of effective refractive index. At both low and high concentrations, the nanocomposites were transparent at visible wavelengths [127, 129, 130]. Venture *et al.* synthesized PbS NPs and had their surface chemistry modified in order to reduce aggregation before dispersing into the commercial photopolymer NOA63 [132]. The investigators were able to control the size of the NPs by varying either Pb:S ratio or the reaction time. PbS NPs were treated with 3-mercaptopropyl-trimethoxysilane followed by ultrasonication in order to overcome NP aggregation before dispersal in NOA63, and the resulting NP nanocomposites showed an increased refractive index. PCs made in the composite material containing modified PbS NPs in NOA63 were observed to have stopbands at telecommunication wavelengths.

9.5.3
Structures Functionalized by In Situ NP Formation

NP-containing microstructures can be created by MPL in a liquid composite that contains chemical precursors of the NPs [21, 133, 134]. As the focused laser beam is scanned within the composite, it both forms the polymeric matrix of the microstructure and induces chemical formation of NPs. After laser patterning, the unpolymerized material can be washed away, leaving the 3D nanocomposite polymer structure.

Sun *et al.* fabricated 3D log-pile PCs using a composite that formed CdS NPs *in situ* during MPL (Figure 9.18) [21]. The composite consisted of acrylate monomers, oligomers, a photoinitiator, and cadmium methacrylate [21, 134]. After MPL, the microfabricated structures were treated with hydrogen sulfide gas to form the CdS NPs [21, 134]. Nanocomposites with tunable fluorescence maxima were created by controlling the amount of crosslinker in the resin [134].

Gan *et al.* used PbSe–NP composites to create narrow-bandgap PCs [133]. The PbSe precursors lead oleate (PbOA) and TOPSe (Se dissolved in trioctylphosphine) were dissolved in 1-octadecene (ODE) and then blended with a photopolymer. The high solubility of the chemical precursors in the organic solution allowed laser beam-induced formation of structures containing a high loading of sufficient amounts of Pb and Se elements.Three-dimensional PbSe structures consisting of overlaying lines with sub-micrometer widths of 500 nm were fabricated using this

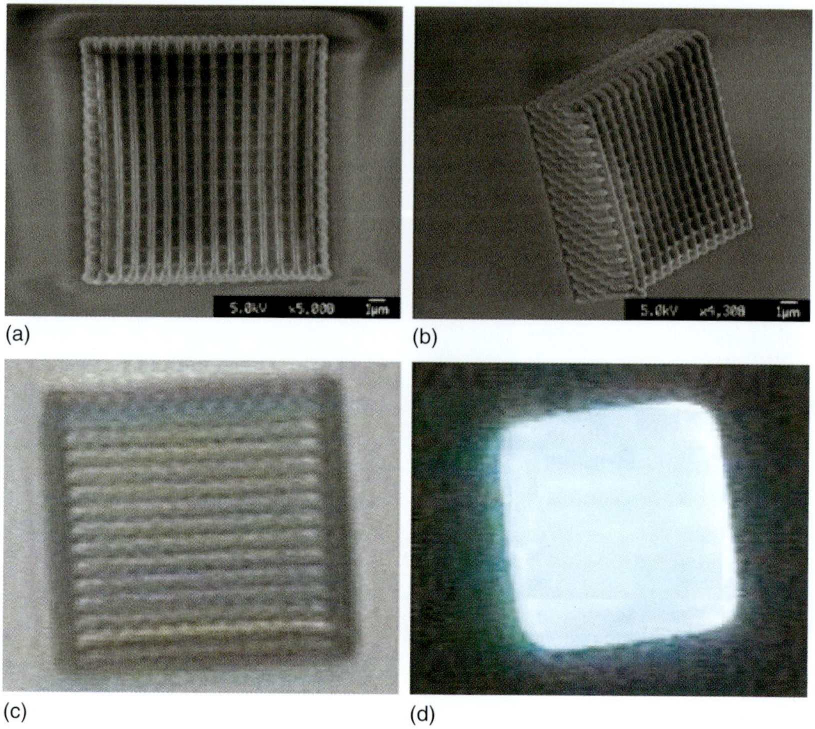

Figure 9.18 Top (a) and side (b) view SEM images of 20-μm-wide 3D PCs formed in a CdS NP–polymer nanocomposite. (c) Optical and (d) fluorescence microscopy images of the PCs. (Sun 2007 [21]. Reproduced with permission of Springer.)

composite. Energy-dispersive spectrum (EDS) measurements of the fabricated structure confirmed the formation of the PbSe [133].

9.5.4
Structures Functionalized by NP Coating

Semiconductor NPs can also be integrated into microstructures by first fabricating them by MPL in a photopolymer and then chemically functionalizing the microstructure surface with the NPs [12, 135]. Three-dimensional woodpile PCs with a bandgap centered at a wavelength of 2.04 μm were fabricated using an ORMOCER and then infiltrated with PbSe NPs for the first time by Li *et al.* [128, 135]. PbSe NPs were prepared by the method described by Yu *et al.* [136] and had a quantum efficiency of more than 85% [135]. The NPs were dissolved in a chloroform or toluene solution, and then drop-casted onto the PC, which dispersed them evenly throughout the structure. The stopband of the PC was observed to increase with concentration of NPs in the solution because of the increased fill factor and refractive index. The emission spectrum of the structure exhibited redistributed spontaneous emission [135]. Three-dimensional woodpile PCs with planar defects were fabricated in an ORMOCER and coated with PbSe/CdSe core–shell NPs [137]. A molecular linking method was used to ensure that the NPs were attached to the PC surface (Figure 9.19). The polymerized ORMOCER structure was functionalized with an amino group prior to immersing in the NP solution to obtain multiple layers. The emission peak overlapped with the stop gap of the PC, which enhanced the spontaneous emission.

Buso *et al.* demonstrated an effective method for tuning the refractive index of ORMOCER 3D woodpile PCs coated with CdS NPs [138]. The PC surface was first pretreated with 3-aminopropyl-triethoxysilane solution in toluene to ensure NP attachment [138]. Then, CdS NPs were deposited using a modified chemical bath deposition (CBD) approach. As the thickness of CdS layers deposited onto the PC lattice increased, the stopband increased in amplitude and shifted to longer wavelength by as much as 375 nm. The shift in the stopband was attributed to an

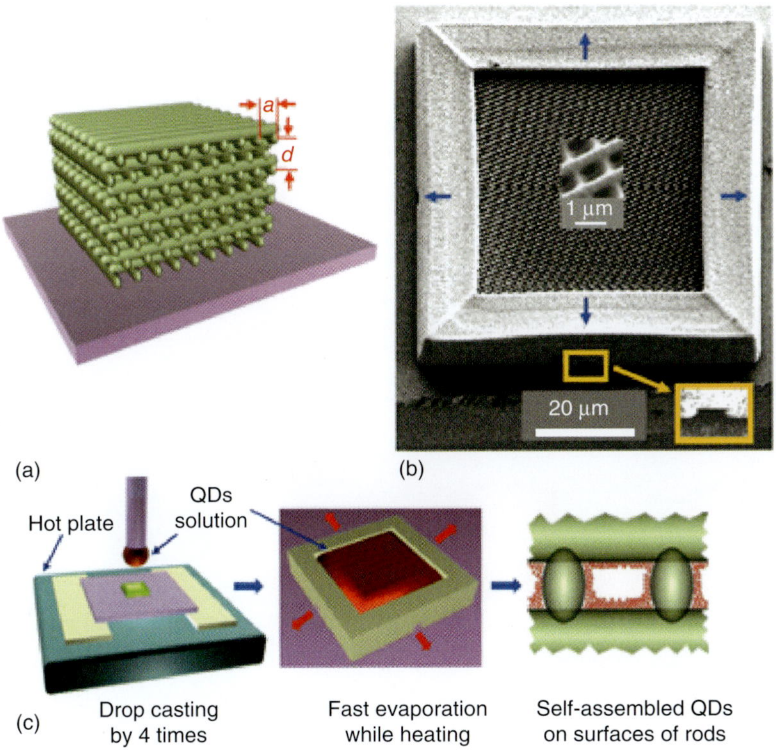

Figure 9.19 (a) Schematic of a 3D woodpile PC fabricated on a cover glass. For the face-centered cubic (fcc) symmetry, the lattice period *a* and the height of every four layers *d*, satisfy $d/a = 2$. (b) SEM image of a 3D woodpile PC with a 10-μm-wide frame. An enlarged view is superposed at the center of the image. A square through-hole was fabricated within each side of the frame (noted by arrows). (c) Schematic of the infiltration process. (Li 2007 [135]. Reproduced with permission of Wiley.)

increase in both the rod thickness and the effective refractive index, whereas the increase in amplitude was attributed to the increase in effective refractive index.

9.5.5
Structures Functionalized by Silicon Inversion

Silicon has a high refractive index ($n = 3.45$) and is optically transparent in the infrared, which makes it a good material for creating PCs. But techniques for microfabrication using Si are time consuming and expensive, requiring costly photolithography masks, and multistep procedures [139]. Some of these limitations were overcome by the introduction of the Si double-inversion (SDI) technique, which was reported in 2006 by Tétreault *et al.* and used to create a Si replica of a polymeric PC (Figure 9.20) [15]. PC templates were first fabricated by MPL using a photo-polymer, and then infiltrated with silica (SiO_2) via ALD. The polymeric lattice was removed by plasma etching or calcination in air to reveal the SiO_2 inverse lattice. If needed, the SiO_2 inverse structure was re-infiltrated with additional SiO_2 by ALD to fine-tune the size of the rods comprising the PC. Next, low-pressure CVD was used to deposit Si within the SiO_2 lattice. Hydrofluoric acid was then used to remove the SiO_2, leaving behind a doubly inverted lattice of Si that was a replica of the original polymeric PC, but now in a material with much higher refractive index. Using this approach, a Si PC was created having a complete photonic bandgap of 8.6%. A similar method was followed by Gratson *et al.* to create hollow woodpile PCs (Figure 9.21) [139].

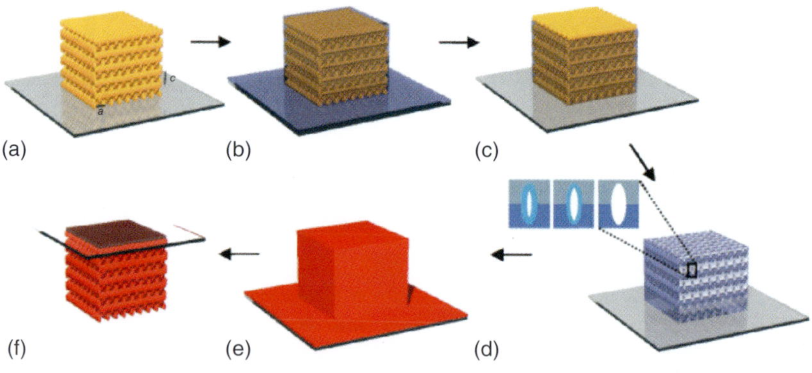

Figure 9.20 Schematic representation of the silicon double inversion (SDI) method. (a) An SU-8 PC template is fabricated by MPL using a photopolymer. (b) The structure is then fully infiltrated with SiO_2 through ALD. (c) Anisotropic reactive ion etching is used to remove the top SiO_2 layer, exposing the SU-8 interior. (d) O_2 plasma etching or calcination in air is then used to remove the SU-8 layer, leaving a SiO_2 inverse woodpile; inset: re-infiltration of the SiO_2 inverse woodpile by SiO_2 ALD to fine-tune the rod-filling fraction. (e) Low-pressure CVD was then used to re-infiltrate the SiO_2 lattice with Si. (f) The structure was attached to a HF-resistant substrate with a polymer adhesive and the SiO_2 inverse lattice was removed by chemical etching in aqueous hydrofluoric acid (HF), leaving the double-inverted Si woodpile replica. (Tétreault 2006 [15]. Reproduced with permission of Wiley.)

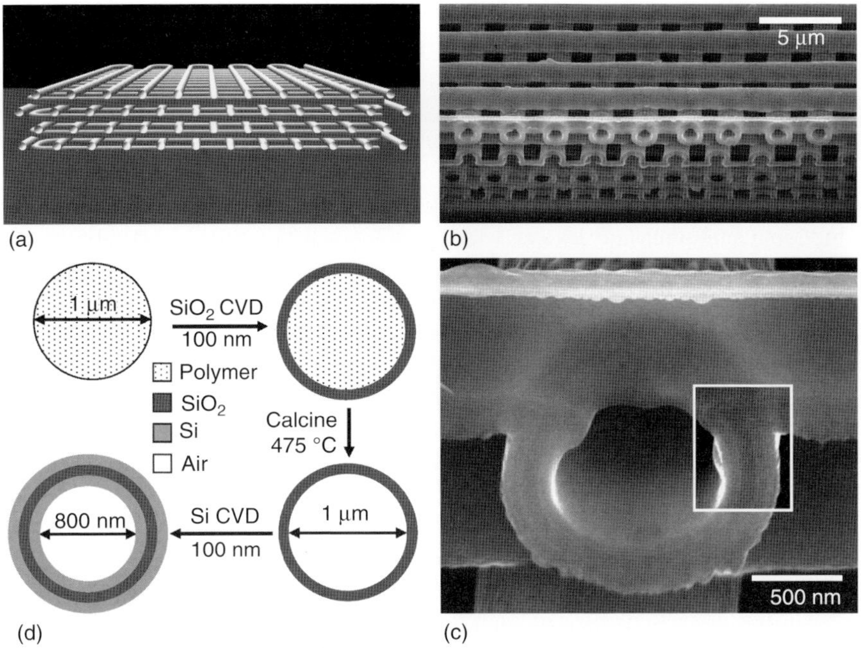

Figure 9.21 (a) Schematic illustration of a hollow-woodpile structure. (b) Low and (c) high magnification SEM images of a $Si/SiO_2/Si$ hollow woodpile structure after focused ion beam milling (eight-layer structure, $d = 2.8\,\mu m$). The contrast-enhanced inset in (c) reveals the internal structures of the trilayer tube wall. (d) Process sequence for templated assembly of such structures. (Gratson 2006 [139]. Reproduced with permission of Wiley.)

Hermatschweiler *et al.* introduced a modification of the SDI procedure in which a single inversion was used to create a Si inverse PC lattice [22]. In this method, a polymeric PC template was created by MPL and then conformally coated with a thin layer of SiO_2 by ALD. Next CVD was used to completely infiltrate the lattice forming a PC in Si, which was an inverse of the original polymeric template. By applying a thin coating of silica, the aspect ratios of the elliptical woodpile rods and volume fill fraction of the final Si structure were improved and the thin coating provided sufficient stability for the high-temperature Si CVD process. Using this approach, the team fabricated high-quality woodpile Si structures with gap/midgap ratios of 14.2% centered at a wavelength of 2.5 μm.

Staude *et al.* created 3D Si woodpile PCs with a complete photonic bandgap (PBG) in the telecommunication window by the SDI method using the photoresist IP-L (Nanoscribe) [140]. An additional step of tempering the Si woodpiles at 600 °C was introduced. This transformed the amorphous Si generated by CVD into polycrystalline Si, which possessed a lower refractive index (shifting the bandgap to shorter wavelengths) and lower absorption coefficient. Si woodpiles with embedded acceptor-type waveguide arrays were optically characterized, and reflectance measurements showed the presence of a broad stopband.

9.5.6
Functional Structures Fabricated in Bulk Chalcogenide Glasses

MPL using chalcogenide glasses (ChGs) has been explored as a means for creating functional nanophotonic devices, without the need for multistep post-fabrication chemical processing. ChGs have a high nonlinearity and a wide IR transmission window with direct application in sensors, photonic waveguides, acoustooptics, and IR optics [141]. Arsenic trisulfide (As_2S_3) is a common ChG chosen because of its transparency in the IR (620 nm to 11 μm) and potential for optical applications [142]. Waveguide structures [143] and gratings [142] have been written in bulk- and fiber-form As_2S_3. Raman spectroscopy revealed that light exposure breaks As–S bonds and forms As–As homopolar bonds, increasing in the disorder of the glass [143].

McMillen *et al.* and Ródenas *et al.* created waveguides by MPL in bulk GaLaS (GLS) and $75GeS_2-15Ga_2S_3-4CsI-2Sb_2S_3-4SnS$ (GCIS) [141, 144]. GLS and GCIS were chosen because of their high transmission in the mid-infrared. GLS was chosen in particular for its high nonlinear figure of merit (FOM) and superior environmental stability [141]. A single-mode buried waveguide was written in GLS and GCIS for use at $\lambda \leq 11$ μm [144]. Spectral beam shaping and temporal pulse-width tuning were employed to optimize the optical properties of the waveguides [141].

9.5.7
Structures Fabricated in ChG Film

Certain ChGs can be thermally deposited as thin glassy films (200 nm to 20 μm) of photosensitive molecular clusters. As_2S_3 is an attractive material because it has high transparency in the IR, high refractive index ($n = 2.45-2.53$), and a low-energy band edge (530 nm), which makes it a good candidate for applications in sensing, waveguiding, and acoustooptics [4, 5]. Raman spectroscopy has been used to show that thermally deposited As_2S_3 is highly photosensitive when it contains a high fraction of As_4S_6 and clusters [4]. Light exposure crosslinks the material, transforming the As_4S_6 cages and other clusters back to As_2S_3 network glass, red-shifting the band edge, and increasing the local refractive index [145]. MPL can be used to photo-pattern thin films of thermally deposited As_2S_3, and then a chemical etchant can be used to remove the unexposed, non-cross-linked regions. Thermally deposited As_2S_3 can be thought of as an inorganic analog of a negative-tone photopolymer. ChGs have the advantage over photopolymers that they do not exhibit swelling or shrinking during the etching process [146]. The shape, size, and optical properties of features produced by MPL in deposited As_2S_3 films have been shown to be sensitive to the conditions used for thermal deposition, photo-exposure, and etching [4, 5, 145, 147–153].

Single-channel waveguides were fabricated in As_2S_3 thin films by Zoubir *et al.* [145]. Von Freymann *et al.* fabricated functional 3D PCs in As_2S_3 [146]. Three-dimensional woodpile PCs with a complete PBG were fabricated in As_2S_3 and etched using increasing concentrations of diisopentylamine (DIPA) in dimethylsulfoxide (DMSO) [4]. Higher etch selectivity was obtained by using etchants based on novel hindered amines, such as *N*-(4-methoxybenzyl)-(pyren-1-yl)amine. This system produced features as small as 180 nm (Figure 9.22) [149]. Nicoletti *et al.* developed high-quality 3D PhCs with large multi-order NIR stopgaps fabricated in thick (20 μm) As_2S_3 films using multistep DIPA in DMSO etching solutions [154].

Thermal deposition conditions were observed to influence the morphology, size, and appearance of nanostructures in thin As_2S_3 films. Schwarz *et al.* demonstrated that fabricated structures with low deposition rates yielded high etch selectivity, whereas high deposition rates yielded low etch selectivity [5]. Nicoletti *et al.* used a multiple-line scanning method to help reduce the effects of focal spot elongation that result from refractive index mismatch when writing in As_2S_3 [150]. This approach was used to create high-quality PCs with wide-angle stopbands. ChGs were also found to be sensitive to the repetition rate of the laser beam [151]. By investigating the heat accumulation at the laser focal spot and the material's response, an optimal condition for high-resolution fabrication in As_2S_3 was found [151]. Using these MPL conditions, nanorods with dimensions of 68 nm

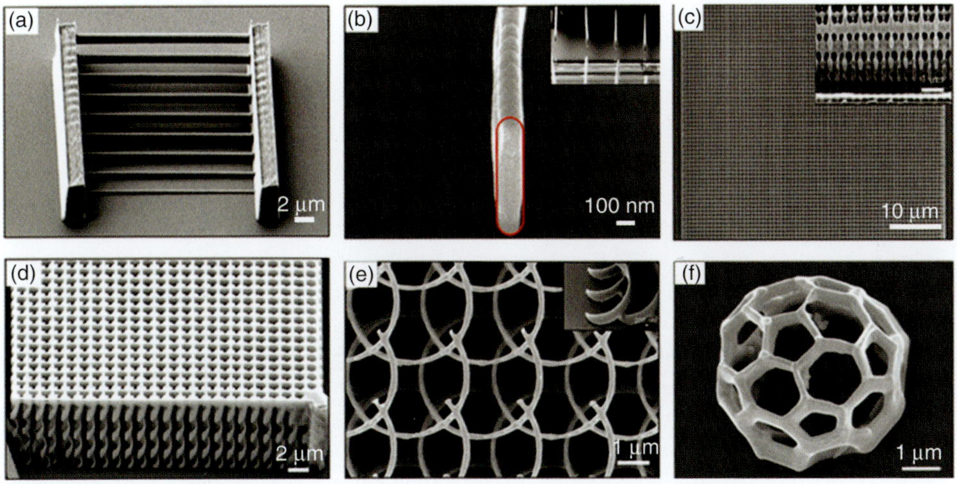

Figure 9.22 Three-dimensional As_2S_3 microstructures created by MPL and chemical etching. (a) SEM image of a "stepladder" test structure. (b) Close-up view of the cross section of (a), demonstrating that the smallest feature size that is definable using this system is on the order of 180 nm. (c) 3D woodpile PC with a lattice constant of 700 nm. The inset shows a cross section created by ion milling. (d) 3D array of spirals. (e) Top view of a 3D array of spirals with individual feature sizes on the order of 180 nm. (f) Close-up of a micro-bucky ball. (Wong 2007 [149]. Reproduced with permission of American Chemical Society.)

($\lambda/12$) and good mechanical properties could be created as well as 3D woodpile PCs with a stopgap in the NIR [151]. Cumming *et al.* used a spatial light modulator to remove spherical aberration, which was found to decrease the size of written features with fabrication depth. This system was used to create high-quality PCs and chiral gyroid PCs in lithium niobate and As_2S_3 films [155, 156].

9.5.8
Structures Fabricated in ChG–NP Composites

As_2S_3 was doped with erbium (Er) to fabricate microstructures that possess both high refractive index and room-temperature photoluminescence (PL). Er provides PL at 1.5 μm. Utilizing the gas-phase direct doping (GPDD) technique, Er can be co-evaporated with As_2S_3 forming Er-doped As_2S_3 thin films. Wong *et al.* fabricated high-quality 3D PCs via MPL in the Er/As_2S_3 photoresist with minimum features sizes of 180 nm and PL at 1.5 μm [147, 148]. Preliminary studies by Nicoletti *et al.* suggest that PCs fabricated in As_2S_3 can be functionalized by coating with PbSe/CdSe core–shell NPs [153]. Diaminoheptane was used to create a self-assembled monolayer of functional ligands, which act as selective linkers so that NPs can be added to the ChG PC in a continuous surface layer. The lattice constant of the PC was chosen to overlap with the NIR emission of the NPs to control their spontaneous emission [153].

9.6
Conclusion

The growing need for functional microstructured materials and devices has driven the rapid development of new processes for MPL. Three-dimensional microfabrication using photopolymers has led to exciting developments and applications. Yet conventional photopolymers alone cannot provide many of the physical and chemical properties that are essential for existing applications, let alone future needs. Other materials are required to create microstructures having low electrical resistivity, high mechanical toughness, high refractive index, optoelectronic function, or resilience under extreme temperature, to give but a few examples. This has led investigators to develop innovative photopolymer composites or nonpolymeric photo-responsive media that can be used to directly create structures with the required properties. Creative and complementary methods of post-MPL processing have also been reported – including surface functionalization, templated deposition, and backfilling – which can be used to direct one or more materials with the desired properties into a targeted microscale form defined by MPL using typical photopolymers.

More such developments can be expected as the technology of MPL matures and as interest in its use grows, particularly for industrial manufacturing. Areas for fruitful exploration include developing methods for patterning multifunctional materials, "smart" materials that actively respond to external stimuli, and active-form materials that cause the shape of the microstructure itself to change in response to environmental conditions. The existing and emerging material systems for MPL can be expected to form the foundation for dramatic advances in the fields of micro and nanophotonics, opto-fluidics, MEMS, optomechanical systems, sensors, and biomedical devices.

Acknowledgments

This work was partially supported by NSF CAREER Award DMR/CHE-0748712; the Space Research Initiative Program, through the Florida Space Institute hosted at the University of Central Florida; and the National Aeronautics and Space Administration through the University of Central Florida's NASA–Florida Space Grant Consortium.

References

1. Soukoulis, C.M. and Wegener, M. (2011) Past achievements and future challenges in the development of three-dimensional photonic metamaterials. *Nat. Photonics*, **5**, 523–530.
2. Yakhshi-Tafti, E., Cho, H.J., and Kumar, R. (2011) Diffusive mixing through velocity profile variation in microchannels. *Exp. Fluids*, **50**, 535–545.
3. Nguyen, N.-T. and Wu, Z. (2005) Micromixers – A review. *J. Micromech. Microeng.*, **15**, R1–R16.
4. Wong, S., Deubel, M., Perez-Willard, F., John, S., Ozin, G.A., Wegener, M., and von Freymann, G. (2006) Direct laser writing of three-dimensional photonic crystals with a complete photonic bandgap in chalcogenide glasses. *Adv. Mater.*, **18**, 265–269.
5. Schwarz, C.M., Williams, H.E., Grabill, C.N., Lewis, A.M., Kuebler, S.M., Gleason, B., Richardson, K.A., Pogrebnyakov, A., Mayer, T.S., Drake, C., and Rivero-Baleine, C. (2014) Processing and properties of arsenic trisulfide chalcogenide glasses for direct laser writing of 3D micro structures. Proceedings of SPIE: Advanced Fabrication Technologies for Micro/Nano Optics and Photonics VII, San Francisco, CA, February 1–6, 2014, SPIE, Vol. 8974, pp. 89740P-1–89740P-11.
6. Allen, S.M. and Thomas, E.L. (1999) *The Structure of Materials*, John Wiley & Sons, Inc., New York.
7. Farsari, M., Ovsianikov, A., Vamvakaki, M., Sakellari, I., Gray, D., Chichkov, B.N., and Fotakis, C. (2008) Fabrication of three-dimensional photonic crystal structures containing an active nonlinear optical chromophore. *Appl. Phys. A Mater. Sci. Process.*, **93**, 11–15.
8. Duan, X.M., Sun, H.B., Kaneko, K., and Kawata, S. (2004) Two-photon polymerization of metal ions doped acrylate monomers and oligomers for three-dimensional structure fabrication. *Thin Solid Films*, **453-454**, 518–521.
9. Stellacci, F., Bauer, C.A., Meyer-Friedrichsen, T., Wenseleers, W., Alain, V., Kuebler, S.M., Pond, S.J.K., Zhang, Y., Marder, S.R., and Perry, J.W. (2002) Laser and electron-beam induced growth of nanoparticles for 2D and 3D metal patterning. *Adv. Mater.*, **14**, 194–198.
10. Abaddi, M.A., Sasso, L., Dimaki, M., and Svendsen, W.E. (2012) Fabrication of 3D nano/microelectrodes via two-photon-polymerization. *Microelectron. Eng.*, **98**, 378–381.
11. Jia, Y.-P., Zhang, Y.-L., Dong, X.-Z., Zheng, M.-L., Li, J., Liu, J., Zhao, Z.-S., and Duan, X.-M. (2014) Complementary chiral metasurface with

strong broadband optical activity and enhanced transmission. *Appl. Phys. Lett.*, **104**, 011108-1–011108-5.
12. Chen, Y.S., Tal, A., Torrance, D.B., and Kuebler, S.M. (2006) Fabrication and characterization of three-dimensional silver-coated polymeric microstructures. *Adv. Funct. Mater.*, **16**, 1739–1744.
13. Farrer, R.A., LaFratta, C.N., Li, L., Praino, J., Naughton, M.J., Saleh, B.E.A., Teich, M.C., and Fourkas, J.T. (2006) Selective functionalization of 3-D polymer microstructures. *J. Am. Chem. Soc.*, **128**, 1796–1797.
14. Rill, M.S., Plet, C., Thiel, M., Staude, I., von Freymann, G., Linden, S., and Wegener, M. (2008) Photonic metamaterials by direct laser writing and silver chemical vapour deposition. *Nat. Mater.*, **7**, 543–546.
15. Tétreault, N., von Freymann, G., Deubel, M., Hermatschweiler, M., Pérez-Willard, F., John, S., Wegener, M., and Ozin, G.A. (2006) New route to three-dimensional photonic bandgap materials: silicon double inversion of polymer templates. *Adv. Mater.*, **18**, 457–460.
16. Ovsianikov, A., Li, Z., Torgersen, J., Stampfl, J., and Liska, R. (2012) Selective functionalization of 3D matrices via multiphoton grafting and subsequent click chemistry. *Adv. Funct. Mater.*, **22**, 3429–3433.
17. Formanek, F., Takeyasu, N., Tanaka, T., Chiyoda, K., Ishikawa, A., and Kawata, S. (2006) Selective electroless plating to fabricate complex three-dimensional metallic micro/nanostructures. *Appl. Phys. Lett.*, **88**, 083110-1–083110-3.
18. Drakakis, T.S., Papadakis, G., Sambani, K., Filippidis, G., Georgiou, S., Gizeli, E., Fotakis, C., and Farsari, M. (2006) Construction of three-dimensional biomolecule structures employing femtosecond lasers. *Appl. Phys. Lett.*, **89**, 144108-1–144108-3.
19. Farsari, M., Filippidis, G., Drakakis, T.S., Sambani, K., Georgiou, S., Papadakis, G., Gizeli, E., and Fotakis, C. (2007) Three-dimensional biomolecule patterning. *Appl. Surf. Sci.*, **253**, 8115–8118.
20. Kuebler, S.M., Narayanan, D.A., Karas, D.E., and Wilburn, K.M. (2014) Low-distortion surface functionalization of polymeric microstructures. *Macromol. Chem. Phys.*, **215**, 1533–1542.
21. Sun, H.-B., Dong, X.-Z., Nakanishi, S., Chen, W.-Q., Duan, X.-M., and Kawata, S. (2007) Log-pile photonic crystal of CdS–polymer nanocomposites fabricated by combination of two-photon polymerization and in situ synthesis. *Appl. Phys. A*, **86**, 427–431.
22. Hermatschweiler, M., Ledermann, A., Ozin, G.A., Wegener, M., and von Freymann, G. (2007) Fabrication of silicon inverse woodpile photonic crystals. *Adv. Funct. Mater.*, **17**, 2273–2277.
23. Gansel, J.K., Thiel, M., Rill, M.S., Decker, M., Bade, K., Saile, V., von Freymann, G., Linden, S., and Wegener, M. (2009) Gold helix photonic metamaterial as broadband circular polarizer. *Science*, **325**, 1513–1515.
24. Watanabe, T., Akiyama, M., Totani, K., Kuebler, S.M., Stellacci, F., Wenseleers, W., Braun, K., Marder, S.R., and Perry, J.W. (2002) Photoresponsive hydrogel microstructure fabricated by two-photon initiated polymerization. *Adv. Funct. Mater.*, **12**, 611–614.
25. Farsari, M., Vamvakaki, M., and Chichkov, B.N. (2010) Multiphoton polymerization of hybrid materials. *J. Opt.*, **12**, 124001-1–124001-16.
26. Fischer, J. and Wegener, M. (2013) Three-dimensional optical laser lithography beyond the diffraction limit. *Laser Photonics Rev.*, **7**, 22–44.
27. Gu, M., Jia, B., Li, J., and Ventura, M.J. (2010) Fabrication of three-dimensional photonic crystals in quantum-dot-based materials. *Laser Photonics Rev.*, **4**, 414–431.
28. Juodkazis, S., Mizeikis, V., and Misawa, H. (2008) in *Photoresponsive Polymers I*, vol. 213 (eds S.R. Marder and K.-S. Lee), Springer, Berlin, pp. 157–206.
29. Kuebler, S.M. and Rumi, M. (2004) in *Encyclopedia of Modern Optics* (eds R.D. Guenther, D.G. Steel, and L. Bayvel), Elsevier, Oxford, pp. 189–206.
30. LaFratta, C.N., Fourkas, J.T., Baldacchini, T., and Farrer, R.A. (2007)

Multiphoton fabrication. *Angew. Chem. Int. Ed.*, **46**, 6238–6258.
31. Lee, K.-S., Yang, D.-Y., Sang, H.-P., and Kim, R.-H. (2006) Recent developments in the use of two-photon polymerization in precise 2D and 3D microfabrications. *Polym. Adv. Technol.*, **17**, 7.
32. Lee, K.-S., Kim, R.H., Yang, D.-Y., and Park, S.H. (2008) Advances in 3D nano/microfabrication using two-photon initiated polymerization. *Prog. Polym. Sci.*, **33**, 631–681.
33. Li, L. and Fourkas, J.T. (2007) Multiphoton polymerization. *Mater. Today*, **10**, 30–37.
34. Malinauskas, M., Farsari, M., Piskarskas, A., and Juodkazis, S. (2013) Ultrafast laser nanostructuring of photopolymers: a decade of advances. *Phys. Rep.*, **533**, 1–31.
35. Maruo, S. and Fourkas, J. (2008) Recent progress in multiphoton microfabrication. *Laser Photonics Rev.*, **2**, 100–111.
36. Ovsianikov, A., Passinger, S., Houbertz, R., and Chichkov, B.N. (2007) in *Laser Ablation and Its Applications*, vol. 129 (ed. C. Phipps), Springer, New York, pp. 121–157.
37. Sekkat, Z. and Kawata, S. (2014) Laser nanofabrication in photoresists and azopolymers. *Laser Photonics Rev.*, **8**, 1–26.
38. Sun, H.-B. and Kawata, S. (2004) *NMR • 3D Analysis • Photopolymerization*, vol. 170, Springer, Berlin, pp. 169–273.
39. Wu, S., Serbin, J., and Gu, M. (2006) Two-photon polymerisation for three-dimensional micro-fabrication. *J. Photochem. Photobiol. A-Chem.*, **181**, 1–11.
40. Zhang, Y.-L., Chen, Q.-D., Xia, H., and Sun, H.-B. (2010) Designable 3D nanofabrication by femtosecond laser direct writing. *Nano Today*, **5**, 435–448.
41. Mendonca, C.R., Correa, D.S., Marlow, F., Voss, T., Tayalia, P., and Mazur, E. (2009) Three-dimensional fabrication of optically active microstructures containing an electroluminescent polymer. *Appl. Phys. Lett.*, **95**, 113309-1–113309-3.
42. Deubel, M., von Freymann, G., Wegener, M., Pereira, S., Busch, K., and Soukoulis, C.M. (2004) Direct laser writing of three-dimensional photonic-crystal templates for telecommunications. *Nat. Mater.*, **3**, 444–447.
43. Fitilis, I., Fakis, M., Polyzos, I., Giannetas, V., and Persephonis, P. (2010) Two-photon polymerization of a diacrylate using fluorene photoinitiators-sensitizers. *J. Photochem. Photobiol., A*, **215**, 25–30.
44. Digaum, J.L., Pazos, J.J., Chiles, J., D'Archangel, J., Padilla, G., Tatulian, A., Rumpf, R.C., Fathpour, S., Boreman, G.D., and Kuebler, S.M. (2014) Tight control of light beams in photonic crystals with spatially-variant lattice orientation. *Opt. Express*, **22**, 25788–25804.
45. Otuka, A.J.G., Tribuzi, V., Correa, D.S., and Mendonca, C.R. (2012) Emission features of microstructures fabricated by two-photon polymerization containing three organic dyes. *Opt. Mater. Express*, **2**, 1803–1808.
46. Zukauskas, A., Malinauskas, M., Kontenis, L., Purlys, V., Paipulas, D., Vengris, M., and Gadonas, R. (2010) Organic dye doped microstructures for optically active functional devices fabricated via two-photon polymerization technique. *Lith. J. Phys.*, **50**, 55–61.
47. Correa, D.S., Tribuzi, V., Cardoso, M.R., Misoguti, L., and Mendonca, C.R. (2011) Selective excitation through tapered silica fibers of fluorescent two-photon polymerized structures. *Appl. Phys. A-Mater.*, **102**, 435–439.
48. Sun, H.B., Tanaka, T., Takada, K., and Kawata, S. (2001) Two-photon photopolymerization and diagnosis of three-dimensional microstructures containing fluorescent dyes. *Appl. Phys. Lett.*, **79**, 1411–1413.
49. Li, C.F., Dong, X.Z., Jin, F., Jin, W., Chen, W.Q., Zhao, Z.S., and Duan, X.M. (2007) Polymeric distributed-feedback resonator with sub-micrometer fibers fabricated by two-photon induced photopolymerization. *Appl. Phys. A-Mater.*, **89**, 145–148.

50. Yokoyama, S., Nakahama, T., Miki, H., and Mashiko, S. (2003) Fabrication of three-dimensional microstructure in optical-gain medium using two-photon-induced photopolymerization technique. *Thin Solid Films*, **438**, 452–456.
51. Chen, V.W., Sobeshchuk, N., Lafargue, C., Mansfield, E.S., Yom, J., Johnstone, L.R., Hales, J.M., Bittner, S., Charpignon, S., Ulbricht, D., Lautru, J., Denisyuk, I., Zyss, J., Perry, J.W., and Lebental, M. (2014) Three-dimensional organic microlasers with low lasing thresholds fabricated by multiphoton and UV lithography. *Opt. Express*, **22**, 12316–12326.
52. Ushiba, S., Shoji, S., Masui, K., Kuray, P., Kono, J., and Kawata, S. (2013) 3D microfabrication of single-wall carbon nanotube/polymer composites by two-photon polymerization lithography. *Carbon*, **59**, 283–288.
53. Kurselis, K., Kiyan, R., Bagratashvili, V.N., Popov, V.K., and Chichkov, B.N. (2013) 3D fabrication of all-polymer conductive microstructures by two photon polymerization. *Opt. Express*, **21**, 31029–31035.
54. Mendonca, C.R., Baldacchini, T., Tayalia, P., and Mazur, E. (2007) Reversible birefringence in microstructures fabricated by two-photon absorption polymerization. *J. Appl. Phys.*, **102**, 013109-1–013109-4.
55. Tribuzi, V., Fonseca, R.D., Correa, D.S., and Mendonca, C.R. (2013) Birefringent microstructures fabricated by two-photon polymerization containing an azopolymer. *Opt. Mater. Express*, **3**, 21–26.
56. Almeida, J.M.P., Tribuzi, V., Fonseca, R.D., Otuka, A.J.G., Ferreira, P.H.D., Mastelaro, V.R., Brajato, P., Hernandes, A.C., Dev, A., Voss, T., Correa, D.S., and Mendonca, C.R. (2013) Femtosecond laser processing of glassy and polymeric matrices containing metals and semiconductor nanostructures. *Opt. Mater.*, **35**, 2643–2648.
57. Fonseca, R.D., Correa, D.S., Paris, E.C., Tribuzi, V., Dev, A., Voss, T., Aoki, P.H.B., Constantino, C.J.L., and Mendonca, C.R. (2014) Fabrication of zinc oxide nanowires/polymer composites by two-photon polymerization. *J. Polym. Sci. B Polym. Phys.*, **52**, 333–337.
58. Wang, J., Xia, H., Xu, B.-B., Niu, L.-G., Wu, D., Chen, Q.-D., and Sun, H.-B. (2009) Remote manipulation of micronanomachines containing magnetic nanoparticles. *Opt. Lett.*, **34**, 581–583.
59. Malinauskas, M., Žukauskas, A., Purlys, V., Gaidukeviciute, A., Balevicius, Z., Piskarskas, A., Fotakis, C., Pissadakis, S., Gray, D., Gadonas, R., Vamvakaki, M., and Farsari, M. (2012) 3D microoptical elements formed in a photostructurable germanium silicate by direct laser writing. *Opt. Laser Eng.*, **50**, 1785–1788.
60. Malinauskas, M., Gaidukevičiūtė, A., Purlys, V., Žukauskas, A., Sakellari, I., Kabouraki, E., Candiani, A., Gray, D., Pissadakis, S., Gadonas, R., Piskarskas, A., Fotakis, C., Vamvakaki, M., and Farsari, M. (2011) Direct laser writing of microoptical structures using a Ge-containing hybrid material. *Metamaterials*, **5**, 135–140.
61. Paz, V.F., Emons, M., Obata, K., Ovsianikov, A., Peterhänsel, S., Frenner, K., Reinhartd, C., Chichkov, B., Morgner, U., and Osten, W. (2012) Development of functional sub-100 nm structures with 3D two-photon polymerization technique and optical mthods for characterization. *J. Laser Appl.*, **24**, 042004-1–042004-7.
62. Wu, P.W., Cheng, W., Martini, I.B., Dunn, B., Schwartz, B.J., and Yablonovitch, E. (2000) Two-photon photographic production of three-dimensional metallic structures within a dielectric matrix. *Adv. Mater.*, **12**, 1438–1441.
63. Acharya, S., Hill, J.P., and Ariga, K. (2009) Soft Langmuir-Blodgett technique for hard nanomaterials. *Adv. Mater.*, **21**, 2959–2981.
64. Tian, Y., Lu, D., Jiang, H., and Lin, X. (2012) Preparation of a novel ferrofluidic photoresist for two-photon photopolymerization technique. *J. Magn. Magn. Mater.*, **324**, 3291–3294.
65. Ovsianikov, A., Viertl, J., Chichkov, B., Oubaha, M., MacCraith, B., Sakellari, I.,

Giakoumaki, A., Gray, D., Vamvakaki, M., Farsari, M., and Fotakis, C. (2008) Ultra-low shrinkage hybrid photosensitive material for two-photon polymerization microfabrication. *ACS Nano*, **2**, 2257–2262.

66. Bichler, S., Feldbacher, S., Woods, R., Satzinger, V., Schmidt, V., Jakopic, G., Lander, G., and Kern, W. (2012) Functional flexible organic-inorganic hybrid polymer for two photon patterning of optical waveguides. *Opt. Mater.*, **34**, 772–780.

67. Woods, R., Feldbacher, S., Zidar, D., Langer, G., Valentin, S., Schmidt, V., Pucher, N., Liska, R., and Kern, W. (2014) 3D optical waveguides produced by two photon photopolymerisation of a flexible silanol terminated polysiloxane containing acrylate functional groups. *Opt. Mater. Express*, **4**, 486–498.

68. Frölich, A., Fischer, J., Zebrowski, T., Busch, K., and Wegener, M. (2013) Titania woodpiles with complete three-dimensional photonic bandgaps in the visible. *Adv. Mater.*, **25**, 3588–3592.

69. Sakellari, I., Gaidukevičiūtė, A., Giakoumaki, A., Gray, D., Fotakis, C., Farsari, M., Vamvakaki, M., Reinhartd, C., Ovsianikov, A., and Chichkov, B. (2010) Two-photon polymerization of titanium-containing solgel composites for thre-dimensional structure fabrication. *Appl. Phys. A*, **100**, 359–364.

70. Guo, Q., Ghadiri, R., Xiao, S., Esen, C., Medenbach, O., and Ostendorf, A. (2012) in *Laser Applications in Microelectronic and Optoelectronic Manufacturing XVII*, vol. 8243 (eds G. Henning, X. Xu, B. Gu, and Y. Nakata), SPIE, San Francisco, CA, pp. 824304-1–824304-7, 21-26 January 2012.

71. Jia, B., Li, J., and Gu, M. (2007) Two-photon polymerization for three-dimensional photonic devices in polymers and nanocomposites. *Aust. J. Chem.*, **60**, 484–495.

72. Heinroth, F., Münzer, S., Feldhoff, A., Passinger, S., Cheng, W., Reinhartd, C., Chichkov, B., and Behrens, P. (2009) Three-dimensional titania pore structures produced by using a femtosecond laser pulse technique and a dip coating procedure. *J. Mater. Sci.*, **44**, 6490–6497.

73. Buruiana, E.C., Jitaru, F., Matei, A., Dinescu, M., and Buruiana, T. (2012) Influence of UV irradiation and two photon processing on the cinnamate monomers polymerization and formation of hybrid composites with nanosized ZnO. *Eur. Polym. J.*, **48**, 1976–1987.

74. Ovsianikov, A., Gaidukevičiūtė, A., Chichkov, B.N., Oubaha, M., MacCraith, B., Sakellari, I., Giakoumaki, A., Gray, D., Vamvakaki, M., Farsari, M., and Fotakis, C. (2008) Two-photon polymerization of hybrid sol-gel materials for photonics applications. *Laser Chem.*, **2008**, 493059-1–493059-7.

75. Malinauskas, M., Žukauskas, A., Purlys, V., Belazaras, K., Momot, A., Paipulas, D., Gadonas, R., Piskarskas, A., Gilbergs, H., Gaidukevičiūtė, A., Sakellari, I., Farsari, M., and Juodkazis, S. (2010) Femtosecond laser polymerization of hybrid/integrated micro-optical elements and their characterization. *J. Opt.*, **12**, 124010-1–124010-8.

76. Malinauskas, M., Žukauskas, A., Gabija, B., Gadonas, R., and Juodkazis, S. (2010) Mechanisms of three-dimensional structuring of photo-polymers by tightly focussed femtosecond laser pulses. *Opt. Express*, **18**, 10209–10221.

77. Ingrosso, C., Sardella, E., Keller, S., Dohn, S., Striccoli, M., Agostiano, A., Boisen, A., and Curri, M.L. (2010) Surface functionalization of epoxy-resist- based microcantilevers with iron oxide nanocrystals. *Adv. Mater.*, **22**, 3288–3292.

78. Xia, H., Wang, J., Tian, Y., Chen, Q.-D., Du, X.-B., Zhang, Y.-L., He, Y., and Sun, H.-B. (2010) Ferrofluids for fabrication of remotely controllable micro-nanomachines by two-photon polymerization. *Adv. Mater.*, **22**, 3204–3207.

79. Tian, Y., Zhang, Y.-L., Ku, J.-F., He, Y., Xu, B.-B., Chen, Q.-D., Xia, H., and Sun, H.-B. (2010) High performance

magnetically controllable microturbines. *Lab Chip*, **10**, 2902–2905.

80. Guo, L., Xia, H., Fan, H.T., Zhang, Y.L., Chen, Q.D., Zhang, T., and Sun, H.B. (2010) Femtosecond laser direct patterning of sensing materials toward flexible integration of micronanosensors. *Opt. Lett.*, **35**, 1695–1697.
81. Houbertz, R., Frölich, L., Popall, M., Streppel, U., Dannberg, P., Bräuer, A., Serbin, J., and Chichkov, B.N. (2003) Inorganic-organic hybrid polymers for information technology: from planar technology to 3D nanostructures. *Adv. Eng. Mater.*, **5**, 551–555.
82. Houbertz, R. (2005) Laser interaction in sol-gel based materials - 3-D lithography for photonic applications. *Appl. Surf. Sci.*, **247**, 504–512.
83. Kumpfmueller, J., Stadlmann, K., Li, Z., Satzinger, V., Stampfl, J., and Liska, R. (2014) Two-photon-induced thiol-ene polymerization as a fabrication tool for flexible optical waveguides. *Des. Monomers Polym.*, **17**, 390–400.
84. Hashmi, S. (2014) *Comprehensive Materials Processing*, vol. 4, Elsevier Science, Oxford.
85. Pashley, D.W., Stowell, M.J., Jacobs, M.H., and Law, T.J. (1964) The growth and structure of gold and silver deposits formed by evaporation inside an electron microscope. *Philos. Mag.*, **10**, 127–158.
86. Kelly, P.J. and Arnell, R.D. (2000) Magnetron sputtering: a review of recent developments and applications. *Vacuum*, **56**, 159–172.
87. Xiong, X., Jiang, S.-C., Hu, Y.-H., Peng, R.-W., and Wang, M. (2013) Structured metal film as a perfect absorber. *Adv. Mater.*, **25**, 3994–4000.
88. Xiong, X., Xue, Z.-H., Meng, C., Jiang, S.-C., Hu, Y.-H., Peng, R.-W., and Wang, M. (2013) Polarization-dependent perfect absorbers/reflectors based on a three-dimensional metamaterial. *Phys. Rev. B*, **88**, 115105-1–115105-7.
89. Park, J.-H. and Sudarshan, T.S. (2000) *Chemical Vapor Deposition*, Surface Engineering Series, vol. 2, ASM International, Materials Park, OH.
90. Kriegler, C.E., Rill, M.S., Thiel, M., Müller, E., Essig, S., Frölich, A., von Freymann, G., Linden, S., Gerthsen, D., Hahn, H., Busch, K., and Wegener, M. (2009) Transition between corrugated metal films and split-ring-resonator arrays. *Appl. Phys. B*, **96**, 749–755.
91. Aekbote, B.L., Jacak, J., Schütz, G.J., Csányi, E., Szegletes, Z., Ormos, P., and Kelemen, L. (2012) Aminosilane-based functionalization of two-photon polymerized 3D SU-8 microstructures. *Eur. Polym. J.*, **48**, 1745–1754.
92. Turkevich, J., Stevenson, P.C., and Hillier, J. (1953) The formation of colloidal gold. *J. Phys. Chem.*, **57**, 670–673.
93. Reinhardt, C., Evlyukhin, A.B., Cheng, W., Gaidukeviciute, A., Kuznetsov, A.I., Zywietz, U., and Chichkov, B.N. (2012) Optical sensing elements based on ordered semiconductor and metal nanoparticle arrays and surface plasmons. Smart Sensor Phenomena, Technology, Networks, and System Integration, San Diego, CA, March 11–15, 2012, SPIE, Vol. 8346, pp. 83460O-1-83460O-8.
94. Chen, Y.-S., Tal, A., and Kuebler, S.M. (2007) Route to three-dimensional metallized microstructures using cross-linkable epoxide SU-8. *Chem. Mater.*, **19**, 3858–3860.
95. Terzaki, K., Vasilantonakis, N., Gaidukeviciute, A., Reinhardt, C., Fotakis, C., Vamvakaki, M., and Farsari, M. (2011) 3D conducting nanostructures fabricated using direct laser writing. *Opt. Mater. Express*, **1**, 586–597.
96. Formanek, F., Takeyasu, N., Tanaka, T., Chiyoda, K., Ishikawa, A., and Kawata, S. (2006) Three-dimensional fabrication of metallic nanostructures over large areas by two-photon polymerization. *Opt. Express*, **14**, 800–809.
97. LaFratta, C.N., Lim, D., O'Malley, K., Baldacchini, T., and Fourkas, J.T. (2006) Direct laser patterning of conductive wires on three-dimensional polymeric microstructures. *Chem. Mater.*, **18**, 2038–2042.

98. Loudon, G.M. (1988) *Organic Chemistry*, 2nd edn, Benjamin/Cummings, Menlo Park, CA.
99. Radke, A., Gissibl, T., Klotzbücher, T., Braun, P.V., and Giessen, H. (2011) Three-dimensional bichiral plasmonic crystals fabricated by direct laser writing and electroless silver plating. *Adv. Mater.*, **23**, 3018–3021.
100. Radke, A., Wolff, F., Giessen, H., and Klotzbücher, T. (2010) Fabrication of dielectric and metallo-dielectric 3D nanostructures by direct laser writing and electroless plating. Laser-Based Micro- and Nanopackaging and Assembly IV, San Fransisco, CA, January 23–28, 2010, SPIE, Vol. 7585, pp. 75850M-1–75850M-6.
101. Malureanu, R., Alabastri, A., Cheng, W., Kiyan, R., Chichkov, B., Andryieuski, A., and Lavrinenko, A. (2011) Enhanced broadband optical transmission in metallized woodpiles. *Appl. Phys. A-Mater.*, **103**, 749–753.
102. Tal, A., Chen, Y.-S., Williams, H.E., Rumpf, R.C., and Kuebler, S.M. (2007) Fabrication and characterization of three-dimensional copper metallodielectric photonic crystals. *Opt. Express*, **15**, 18283–18293.
103. Takeyasu, N., Formanek, F., Chiyoda, K., Tanaka, T., Ishikawa, A., and Kawata, S. (2006) Site-selective metal deposition on 3D micro/nanostructures fabricated by two-photon polymerization. Plasmonics: Nanoimaging, Nanofabrication, and their Applications II, San Diego, CA, August 13–17, 2006, SPIE, Vol. 6324, pp. 63240W-1–63240W-9.
104. Hidber, P.C., Helbig, W., Kim, E., and Whitesides, G.M. (1996) Microcontact printing of palladium colloids: Micron-scale patterning by electroless deposition of copper. *Langmuir*, **12**, 1375–1380.
105. Ikegami, T., Stocker, M.P., Monaco, K., Fourkas, J.T., and Maruo, S. (2012) Fabrication of three-dimensional metalized movable microstructures by the combination of two-photon microfabrication and electroless plating. *Jpn. J. Appl. Phys.*, **51**, 06FL17-1–06FL17-4.
106. Ikegami, T., Ozawa, R., Stocker, M.P., Monaco, K., Fourkas, J.T., and Maruo, S. (2013) Development of optically-driven metallic micromotors using two-photon microfabrication. *J Laser Micro/ Nanoeng.*, **8**, 6–10.
107. Danscher, G. (1981) Histochemical demonstration of heavy metals. *Histochemistry*, **71**, 1–16.
108. Hacker, G.W., Grimelius, L., Danscher, G., Bernatzky, G., Muss, W., Adam, H., and Thurner, J. (1988) Silver acetate autometallography: an alternative enhancement technique for immunogold-silver staining (IGSS) and silver amplification of gold, silver, mercury and zinc in tissues. *J. Histotechnol.*, **11**, 213–221.
109. Henry, A.C. and McCarley, R.L. (2001) Selective deposition of metals on plastics used in the construction of microanalytical devices: Photo-directed formation of metal features on PMMA. *J. Phys. Chem. B*, **105**, 8755–8761.
110. Henry, A.C., Tutt, T.J., Galloway, M., Davidson, Y.Y., McWhorter, C.S., Soper, S.A., and McCarley, R.L. (2000) Surface modification of poly(methyl methacrylate) used in the fabrication of microanalytical devices. *Anal. Chem.*, **72**, 5331–5337.
111. Mizeikis, V., Juodkazis, S., Tarozaitė, R., Juodkazytė, J., Juodkazis, K., and Misawa, H. (2007) Fabrication and properties of metalo-dielectric photonic crystal structures for infrared spectral region. *Opt. Express*, **15**, 8454–8464.
112. Mizeikis, V., Tarozaite, R., Juodkazis, S., and Misawa, H. (2009) Metalo-dielectric 3D photonic crystals prepared by metalization of opticaly microfabricated polymeric templates. Advanced Fabrication Technologies for Micro/Nano Optics and Photonics II, San Jose, CA, January 24–29, 2009, SPIE, pp. 72050I-1–72050I-8.
113. Ren, L., Wang, D.-G., Niu, L.-G., Xu, B.-B., Song, J.-F., Chen, Q.-D., and Sun, H.-B. (2013) Complex three-dimensional polymer-metal core-shell structures towards emission control. *Phys. Chem. Chem. Phys.*, **15**, 9590–9593.

114. Wang, W.-K., Sun, Z.-B., Zheng, M.-L., Dong, X.-Z., Zhao, Z.-S., and Duan, X.-M. (2011) Magnetic nickel–phosphorus/polymer composite and remotely driven three-dimensional micromachine fabricated by nanoplating and two-photon polymerization. *J. Phys. Chem. C*, **115**, 11275–11281.

115. Shukla, S., Furlani, E.P., Vidal, X., Swihart, M.T., and Prasad, P.N. (2010) Two-photon lithography of sub-wavelength metallic structures in a polymer matrix. *Adv. Mater.*, **22**, 3695–3699.

116. Shukla, S., Vidal, X., Furlani, E.P., Swihart, M.T., Kim, K.-T., Yoon, Y.-K., Urbas, A., and Prasad, P.N. (2011) Subwavelength direct laser patterning of conductive gold nanostructures by simultaneous photopolymerization and photoreduction. *ACS Nano*, **5**, 1947–1957.

117. Vora, K., Kang, S.Y., Shukla, S., and Mazur, E. (2012) Fabrication of disconnected three-dimensional silver nanostructures in a polymer matrix. *Appl. Phys. Lett.*, **100**, 063120-1–063120-3.

118. Prabhakaran, P., Son, Y., Jang, K.K., Yang, D.-Y., and Lee, K.-S. (2011) Realizing metal and quantum dot containing patterns by two-photon lithography. Linear and Nonlinear Optics of Organic Materials XI, San Diego, CA, August 21–22, 2011, SPIE, Vol. 8113, pp. 81130B-1-81130B-9.

119. Tribuzi, V., Corrêa, D.S., Avansi, W., Ribeiro, C., Longo, E., and Mendonça, C.R. (2012) Indirect doping of microstructures fabricated by two-photon polymerization with gold nanoparticles. *Opt. Express*, **20**, 21107–21113.

120. Tanaka, T., Yamaguchi, K., and Yamamoto, S. (2002) Rhodamine-B-doped and Au(III)-doped PMMA film for three-dimensional multi-layered optical memory. *Opt. Commun.*, **212**, 45–50.

121. Maruo, S. and Saeki, T. (2008) Femtosecond laser direct writing of metallicmicrostructures by photoreduction of silvernitrate in a polymer matrix. *Opt. Express*, **16**, 1174–1179.

122. Ameloot, R., Roeffaers, M.B.J., De Cremer, G., Vermoortele, F., Hofkens, J., Sels, B.F., and De Vos, D.E. (2011) Metal-organic framework single crystals as photoactive matrices for the generation of metallic microstructures. *Adv. Mater.*, **23**, 1788–1791.

123. Ishikawa, A., Tanaka, T., and Kawata, S. (2006) Improvement in the reduction of silver ions in aqueous solution using two-photon sensitive dye. *Appl. Phys. Lett.*, **89**, 113102-1–113102-3.

124. Lu, W.-E., Zhang, Y.-L., Zheng, M.-L., Jia, Y.-P., Liu, J., Dong, X.-Z., Zhao, Z.-S., Li, C.-B., Xia, Y., Ye, T.-C., and Duan, X.-M. (2013) Femtosecond direct laser writing of gold nanostructures by ionic liquid assisted multiphoton photoreduction. *Opt. Mater. Express*, **3**, 1660–1673.

125. Baldacchini, T., Pons, A., Pons, J., LaFratta, C., Fourkas, J., Sun, Y., and Naughton, M. (2005) Multiphoton laser direct writing of two-dimensional silver structures. *Opt. Express*, **13**, 1275–1280.

126. Reiss, P., Protiere, M., and Li, L. (2009) Core/shell semiconductor nanocrystals. *Small*, **5**, 154–168.

127. Jia, B., Buso, D., Li, J., and Gu, M. (2009) Active three-dimensional photonic crystals with high third-order nonlinearity in telecommunication. 8th Pacific Rim Conference on Lasers and Electro-Optics (CLEO/Pacific Rim 2009), Shanghai, China, August 30 – September 3, 2009, IEEE, pp. 978-1–978-2.

128. Li, J., Serbin, J., Bullen, C., and Gu, M. (2009) Incorporation of quantum dots into 3D photonic crystals for emission control. Nanoscience and Nanotechnology, Queensland, July 3–7, 2009, IEEE, pp. 439-441.

129. Jia, B., Buso, D., Embden, J.V., Li, J., and Gu, M. (2010) Highly non-linear quantum dot doped nanocomposites for functional three-dimensional structures generated by two-photon polymerization. *Adv. Mater.*, **22**, 2463–2467.

130. Jia, B., Buso, D., Li, J., and Gu, M. (2010) Nonlinear nanocomposites for three-dimensional photonic crystals using two-photon polymerization.

Quantum Electronics and Laser Science Conference, San Jose, CA, May 16–21, 2010, IEEE, pp. JWA18-1-JWA18-2.

131. Li, J., Jia, B., Zhou, G., and Gu, M. (2006) Fabrication of three-dimensional woodpile photonic crystals in a PbSe quantum dot composite material. *Opt. Express*, **14**, 10740–10745.

132. Ventura, M.J., Bullen, C., and Gu, M. (2007) Direct laser writing of three-dimensional photonic crystal lattices within a PbS quantum dot-doped polymer material. *Opt. Express*, **15**, 1817–1822.

133. Gan, Z., Cao, Y., and Gu, M. (2013) Direct laser writing of three-dimensional narrow bandgap and high refractive-index PbSe structures in a solution. *Opt. Express*, **21**, 11202–11208.

134. Sun, Z.-B., Dong, X.-Z., Chen, W.-Q., Nakanishi, S., Duan, X.-M., and Kawata, S. (2008) Multicolor polymer nanocomposites: in situ synthesis and fabrication of 3D microstructures. *Adv. Mater.*, **20**, 914–919.

135. Li, J., Jia, B., Zhou, G., Bullen, C., Serbin, J., and Gu, M. (2007) Spectral redistribution in spontaneous emission from quantum-dot-infiltrated 3D woodpile photonic crystals for telecommunications. *Adv. Mater.*, **19**, 3276–3280.

136. Yu, W.W., Falkner, J.C., Shih, B.S., and Colvin, V.L. (2004) Preparation and characterization of monodisperse PbSe semiconductor nanocrystals in a non-coordinating solvent. *Chem. Mater.*, **16**, 3318–3322.

137. Gan, Z., Jia, B., Liu, J.-F., and Gu, M. (2012) Enhancement of spontaneous emission in three-dimensional low refractive-index photonic crystals with designed defects. *Appl. Phys. Lett.*, **101**, 1–4.

138. Buso, D., Nicoletti, E., Li, J., and Gu, M. (2010) Engineering the refractive index of threedimensional photonic crystals through multilayer deposition of CdS films. *Opt. Express*, **18**, 1033–1040.

139. Gratson, G.M., Garcia-Santamaria, F., Lousse, V., Xu, M., Fan, S., Lewis, J.A., and Braun, P.V. (2006) Direct-write assembly of three-dimensional photonic crystals: conversion of polymer scaffolds to silicon hollow-woodpile structures. *Adv. Mater.*, **18**, 461–465.

140. Staude, I., Thiel, M., Essig, S., Wolff, C., Busch, K., von Freymann, G., and Wegener, M. (2010) Fabrication and characterization of silicon woodpile photonic crystals with a complete bandgap at telecom wavelengths. *Opt. Lett.*, **35**, 1094–1096.

141. McMillen, B., Zhang, B., Chem, K.P., Benayas, A., and Jaque, D. (2012) Ultrafast laser fabrication of low-loss waveguides in chalcogenide glass with 0.65 dB/cm loss. *Opt. Lett.*, **37**, 1418–1420.

142. Florea, C., Sanghera, J.S., and Aggarwal, I.D. (2008) Direct-write gratings in chalcogenide bulk glasses and fibers using a femtosecond laser. *Opt. Mater.*, **30**, 1603–1606.

143. Efimov, O.M., Glebov, L.B., Richardson, K.A., Van Stryland, E., Cardinal, T., Park, S.H., Couzi, M., and Bruneel, J.L. (2001) Waveguide writing in chalcogenide glasses by a train of femtosecond laser pulses. *Opt. Mater.*, **17**, 379–386.

144. Ródenas, A., Martin, G., Arezki, B., Psaila, N., Jose, G., Jha, A., Labadie, L., Kern, P., Kar, A., and Thomson, R. (2012) Three-dimensional mid-infrared photonic circuits in chalcogenide glass. *Opt. Lett.*, **37**, 392–394.

145. Zoubir, A., Richardson, M., Rivero, C., Schulte, A., Lopez, C., Richardson, K., Hô, N., and Vallée, R. (2004) Direct femtosecond laser writing of waveguides in As_2S_3 thin films. *Opt. Lett.*, **29**, 748–750.

146. von Freymann, G., Wong, S., Ozin, G.A., John, S., Perez-Willard, F., Deubel, M., and Wegener, M. (2005) Direct laser writing of three-dimensional photonic crystals in high index of refraction chalcogenide glasses. Lasers and Electro-Optics, 3-D Photonic Crystals: Noval Fabrication, Baltimore, MD, May 22–27, 2005, Vol. 2, pp. CTuU5-1–CTuU5-3.

147. Wong, S., von Freymann, G., Fenske, D., Kiowski, O., Kappes,M., Peiris, F.C., Lindner, J.K.N. , Ozin, G.A., Thiel, M.,

Braun,M., Ledermann,A., and Wegener, M. (2008) Er doped As_2S_3 photoresist for 3-D direct laser fabrication of 3-D nanostructures. Quantum Electronics and Laser Science CLEO/QELS, San Jose, CA, May 4–9, 2008, IEEE, pp. CWO6-1–CWO6-2.

148. Wong, S., Kiowski, O., Kappes, M., Lindner, J.K.N., Mandal, N., Peiris, F.C., Ozin, G.A., Thiel, M., Braun, M., Wegener, M., and von Freymann, G. (2008) Spatially localized photoluminescence at 1.5 micrometers wavelength in direct laser written optical nanostructures. *Adv. Mater.*, **20**, 4097–4102.

149. Wong, S.H., Thiel, M., Brodersen, P., Fenske, D., Ozin, G.A., Wegener, M., and von Freymann, G. (2007) Highly selective wet etch for high-resolution three-dimensional nanostructures in arsenic sulfide all-inorganic photoresist. *Chem. Mater.*, **19**, 4213–4221.

150. Nicoletti, E., Bulla, D., Luther-Davies, B., and Gu, M. (2011) Wide-angle stopgap chalcogenide photonic crystals generated by direct multiple-line laser writing. *Appl. Phys. B*, **105**, 847–850.

151. Nicoletti, E., Bulla, D., Luther-Davies, B., and Gu, M. (2011) Generation of $\lambda/12$ nanowires in chalcogenide glasses. *Nano Lett.*, **11**, 4218–4221.

152. Nicoletti, E., Bulla, D., Luther-Davies, B., and Gu, M. (2011) High resolution fabrication in chalcogenide glasses. International Quantum Electronics Conference, Sydney, Australia, September 28 – October 1, 2011, IEEE, pp. 1103-1105.

153. Nicoletti, E., Buso, D., Guangyong, Z., Jia, B., Bulla, D., Luther-Davis, B., and Gu, M. (2009) As_2S_3 photonic crystals for spontaneous emission control of PbSe\CdSe core-shell quantum dots. 9th IEEE Conference on Nanotechnology, Genoa, Italy, July 26–30, 2009, IEEE, pp. 717-718.

154. Nicoletti, E., Zhou, G., Jia, B., Ventura, M.J., Bulla, D., Luther-Davies, B., and Gu, M. (2008) Observation of multiple higher-order stopgaps from three-dimensional chalcogenide glass photonic crystals. *Opt. Lett.*, **33**, 2311–2313.

155. Cumming, B.P., Turner, M.D., Schröder-Turk, G.E., Debbarma, S., Luther-Davies, B., and Gu, M. (2014) Adaptive optics enhanced direct laser writing of high refractive index gyroid photonic crystals in chalcogenide glass. *Opt. Express*, **22**, 689–698.

156. Cumming, B.P., Jesacher, A., Booth, M.J., Wilson, T., and Gu, M. (2011) Adaptive aberration compensation for three-dimensional micro-fabrication of photonic crystals in lithium niobate. *Opt. Express*, **19**, 9419–9425.

Part IV
Applications

Multiphoton Lithography: Techniques, Materials and Applications, First Edition.
Edited by Jürgen Stampfl, Robert Liska, and Aleksandr Ovsianikov.

10
Fabrication of Waveguides and Other Optical Elements by Multiphoton Lithography

Samuel Clark Ligon, Josef Kumpfmüller, Niklas Pucher, Jürgen Stampfl, and Robert Liska

10.1
Introduction

The demands for higher bandwidth data transmission and faster data processing encourage scientists to develop improved techniques for building ever smaller electrical and electrooptical components [1, 2]. Photonics has for many years provided solutions in long-distance broadband data transmission and has become an increasingly viable option for shorter distances at transfer rates of 10 GHz and beyond [3]. Where board-to-board optical interconnects are commercially common today, chip-to-chip and on-chip optical integration are less widespread but hold great promise for the future [4]. Research in photonic integrated circuits (ICs) has been an active field for more than 30 years that has advanced borrowing from the complementary metal oxide semiconductor (CMOS) fabrication platform. CMOS coating processes include chemical vapor deposition (CVD), sputtering, and spin coating, while subtractive processes include chemical etching and focused-ion-beam milling. Photolithography is pervasive in CMOS and has evolved with specialized techniques including interferometric lithography, nanoimprint lithography, nanosphere lithography, and multiphoton lithography (MPL) [5].

Photonic ICs, which use photons instead of electrons to mediate a function, consist of circuit elements including lenses, attenuators, switches, and wavelength filters. Photonic circuit elements are constructed from optical materials including silicon, silica, III–V and II–VI semiconductors, and organic polymers. While chemists use the term *hybrid* for materials with both organic and inorganic molecular moieties, the term *hybrid photonics* refers to circuits or circuit elements constructed from two or more classes of material (i.e., waveguides with silica cores and polymer cladding) [3]. Organic polymers are increasingly becoming common in hybrid photonic ICs because of their versatility in function, ease of processing, and their high thermo-optic coefficients ($\Delta n/\Delta T$), which are exploited in tunable lenses and switches [6]. However, polymers tend to be not very thermally stable, which can put constraints on vapor deposition methods

Multiphoton Lithography: Techniques, Materials and Applications, First Edition.
Edited by Jürgen Stampfl, Robert Liska, and Aleksandr Ovsianikov.

[7]. Lithography is a common technique for the fabrication of photonic circuits and circuit elements. In CMOS fabrication, negative photoresists are transient. By comparison, photo-cured polymers in photonic ICs are meant to last for many years and to retain the optical and mechanical properties required for their application. Thus, new photo-curable polymeric materials, and particularly those with tailorable refractive indices, are investigated for optical applications. We cover some more recent advances in photo-curable polymers for photonics with emphasis on the use of MPL, as this technique offers unique solutions for constructing microscopic optical elements [8].

10.2 Acrylate Monomers for Multiphoton Lithography

Acrylate monomers are the principal components of resins for traditional stereolithography (SLA) and, unsurprisingly, are widely used for MPL as well. A wide variety of (meth)acrylates ranging in molecular weights, structural moieties, and degree of functionality are commercially available. The popularity of acrylates for lithography arises principally from the generally rapid rate of propagation with an efficient termination in the presence of oxygen, which allows excellent spatial and temporal control of the polymerization process [9, 10]. With viscosity, reactivity, and shrinkage all contributing to the performance of a resin, monomers must be selected carefully. Some common and commercially available acrylate monomers are shown in Figure 10.1.

Low molecular weight reactive diluents such as TTA (trimethylpropane triacrylate) or hexane-1,6-diol diacrylate (HDDA) form tightly crosslinked networks with good resistance against solvent swelling. These monomers have generally low viscosity and high reactivity, but, due to the higher number of reactive groups per gram, greater polymerization shrinkage is exhibited. To account for this, higher molecular weight monomers such as ETA (Figure 10.1) can be applied. While such monomers do increase the overall viscosity of the resin significantly, reduction in shrinkage is perhaps more critical particularly for layer-by-layer lithographic techniques where delamination and curling must be avoided [11]. Aside from the molecular weight of the monomers, the selection of functional groups in the backbone is also important. Urethane groups provide interchain

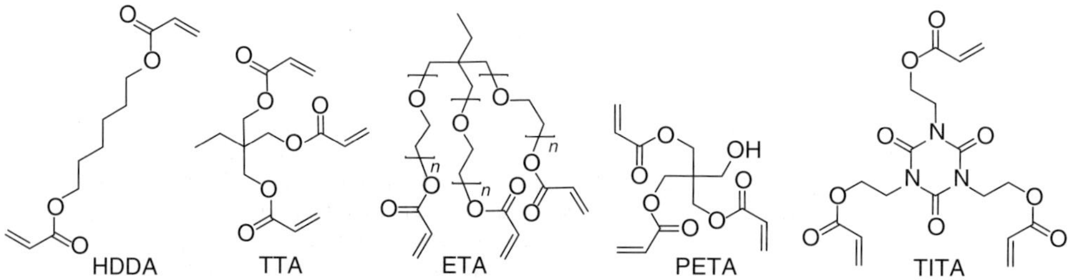

Figure 10.1 Common acrylate monomers.

hydrogen bridges, allowing improved mechanical properties, but also an increase in viscosity. Linear aliphatic chains and polyether units (poly(ethylene glycol) and poly(propylene glycol)) or polycaprolactone units between the acrylate functionalities tend to decrease the glass transition temperature of the cured polymer while monomers with aromatic spacers give stiff materials. By choosing the right set of monomers and reactive diluents, the mechanical properties can be tuned over a broad range from elastomer-like polymers to hard and stiff materials.

SCR500 from Japan Synthetic Rubber is an acrylate-based formulation described in the literature for MPL that consists of different urethane acrylate oligomers with molecular weights from 480 to 1200 Da. While the exact composition of SCR500 is not disclosed, the molecular weight of the monomers and the diluting agents is sufficiently low to allow a viscosity of about 850 mPa s. SCR500 contains either Irgacure 369 or Irgacure 184 (BASF) as radical photoinitiators. The cured material exhibits a modulus of 160 MPa and a tensile strength of 6.0 MPa, along with a strain at failure of 10% and a glass transition temperature of 140 °C [12]. The refractive index before and after polymerization is given as 1.52 and 1.53, respectively [13]. Spin coating is used to provide thin, uniform films for MPL. Covering the film with glass slides prevents oxygen inhibition and maintains film thickness during the building process [14]. The glass slide provides the additional benefit of excluding moisture, which can hydrolyze Ti^{4+} complexes that are sometimes added [15].

In 2004, Duan *et al.* reported the MPL fabrication of a diamond photonic crystal (PhC) using SCR500 as the building material (Figure 10.2) [15]. In order to increase the refractive index, TiO_2 nanoparticles were integrated in the photopolymer. To maximize the dispersion of titania within the system, titanium ethoxide was reacted with methacrylic acid and mixed with SCR500

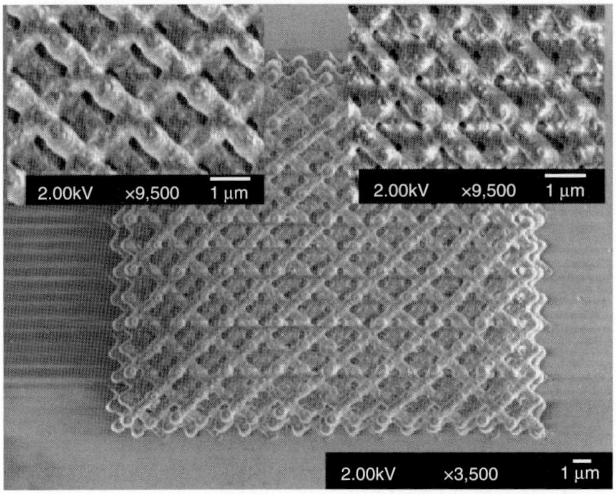

Figure 10.2 Scanning electron microscopy (SEM) image of 3D diamond PhCs structure of inorganic nanoparticle–polymer composite materials produced by MPL. (Duan 2004 [15]. Reprinted with permission of Elsevier.)

using 2,2-diethoxy-acetophenone as initiator. After MPL processing of this novel resin, TiO_2 was formed via the hydrolysis of the Ti^{4+} – methacrylate complexes. Finally, development with ethanol and thermal post-treatment were carried out. The existence of a high-quality photonic bandgap (PBG) was confirmed using transmission spectroscopy.

Sun *et al.* studied the temperature dependence of PhC writing in SCR500 [16]. They predicted that the structuring resolution could be greatly improved by decreasing the process temperature. This hypothesis was based on the suppression of local photo-induced radicals and the reduction of Brownian motion. On the contrary, the voxel size at −60 °C, which is significantly below the freezing point of SCR500, was found to be only slightly smaller than at ambient temperature. While the hypothesis was not proven, a phenomenon was discovered, in which the previously formed radicals in the frozen state did not polymerize until the resin was heated above 10 °C. This indicates a possible method for minimizing unwanted effects such as floating or distortion during laser scanning. Conducting MPL at elevated temperature gave less crosslinked voxels because of enhanced chain termination and thus a lower polymerization rate. Finally, the PBGs of the PhCs were tailored via thermal processing. In some other studies, Xia and coworkers were able to manufacture 15-μm-scale micro-spherical and Fresnel lenses with SCR500 [13].

An innovative approach to fabricate a feedback laser resonator was realized by the group of Duan, who used a carbosilane dendrimer functionalized with an allyl derivative of fluorescein to structure SCR500 [17]. Every Si–H terminated branch of the dendrimer was linked to the allyl group of the dye via hydrosilylation. The novel dye was then mixed with SCR500 in chloroform, enabling a higher doping concentration of fluorescein without self-aggregation, as was the case with conventional fluorescein doping. SCR500 was used more recently with 2,7-dibromo-3,6-bis(3,7-dimethyloctyloxy)9,9-spirobifluorenone as a two-photon sensitizer to test an autofocus method, which allowed voxel position precision of ±0.045 μm [18].

Another acrylate resin that is frequently employed for MPL of photonic elements is tris(2-hydroxyethyl)isocyanurate triacrylate (TITA in Figure 10.1). TITA is a solid at room temperature but exhibits a rather low viscosity above its melting temperature of 52–54 °C [19]. The polar isocyanurate ring of this monomer provides films with excellent adhesion, low brittleness, high strength, and good weathering properties. Other benefits include high reactivity, endowing polymers with scratch, chemical, and heat resistance [20]. Thin uniform films of TITA are formed by casting followed by solvent evaporation or by using calibrated spacers between glass plates [21]. Fourkas *et al.* pretreated the glass substrate with methacryloxypropyltrimethoxysilane (MEMO) to enhance the adhesion with the photopolymer formulation [22]. The need for regulation of the refractive index contrast between the core and clad of a waveguide can make it necessary to prepolymerize the matrix with white light [21]. Dimethylformamide (DMF) [22], ethanol [23], and methanol [24] are all potential solvents for the development of TITA-based resins. Rinsing duration is on a time scale of seconds to minutes. TITA has been used in various formulations, which were modified for MPL with the intention of fabricating optical elements. The monomer and other components such as binder and photoinitiator can be mixed in chloroform, in which case the solvent is subsequently removed after casting [21]. This method was applied by Dorkenoo *et al.*, who succeeded in connecting optical fibers with waveguide structures. Furthermore, they were able to fabricate Y-splitters and a March–Zehnder structure. In this case, poly(styrene-*co*-acrylonitrile) served as the binder and 1,13-bis-[4-(diethylamino)phenyl]-tri-deca-1,3,5,6,8,10,12-hexaen-7-one as initiator [25]. Fourkas *et al.* used TITA with dipentaerythritol pentaacrylate as co-monomer and Lucirin-TPO-L as initiator to create a high-performance micro-ring resonator (Figure 10.3) [23].

TITA was used by Perry *et al.* in a 50 : 50 wt% blend with the alkoxylated difunctional acrylate Sartomer 9008 along with 4,4′-bis(di-*n*-butylamino)biphenyl or *E,E*-1,4-bis[4-(di-*n*-butylamino)styryl]-2,5-dimethoxybenzene as initiator. Woodpile PhCs with 60 nm line widths were constructed (Figure 10.4), which exhibited stop bandgaps in the near-IR spectral region [24].

Another frequently used monomer for MPL is pentaerythritol triacrylate (PETA in Figure 10.1), a trifunctional monomer that is well known from applications including binder adhesives, coatings, and inks. PETA gives highly crosslinked networks because of its three acrylate moieties. The free hydroxyl group allows better adhesion, lower volatility, and the possibility for subsequent chemical modification [26]. The monomer is a clear liquid with a viscosity of 520 cP at 25 °C exhibiting low volatility and fast cure response in free-radical polymerization [20]. The glass transition temperature of the homopolymer of PETA is 103 °C. The refractive index of the uncured monomer (1.4801) is comparatively low for optical applications. Mixed acrylate films for MPL containing PETA are prepared by first dissolving the components in chloroform. For thicker films, polymers such as polymethylmethacrylate (PMMA) can be added to increase the formulation's viscosity [27]. Recently de Miguel *et al.* used a two-step lithographic process with PETA resin and isopropylthioxanthone (ITX) as photoinitiator to achieve $\lambda/20$ axial

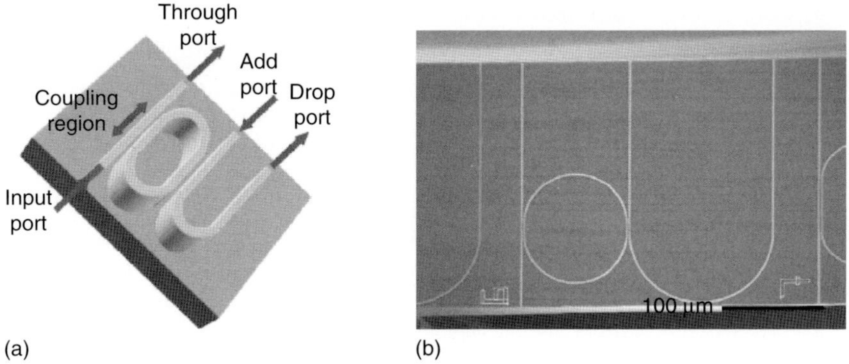

Figure 10.3 (a) Principle of the microring resonator and (b) MPL fabricated waveguide structures. (Li 2008 [23]. Reproduced with permission of Wiley.)

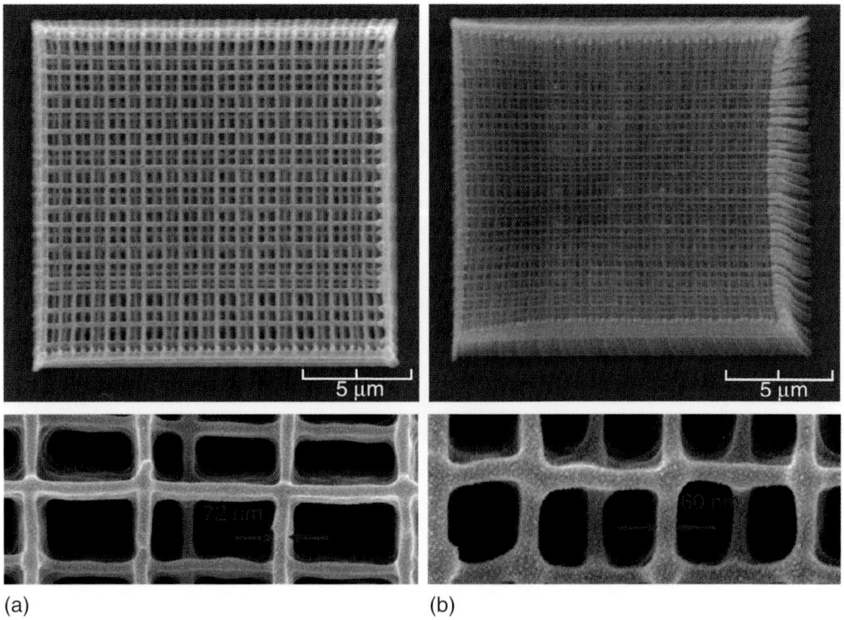

Figure 10.4 Woodpile PhCs with (a) 72 nm (b) and 60 nm feature widths. (Haske 2007 [24]. Reproduced with permission of Optical Society of America.)

control of features. The first step was MPL with high laser power, while the second step was to induce well-defined thermal expansion with lower laser power [28].

Infiltration of existing PhCs with the resin followed by MPL has been used to create well-controlled defects within the PhC [29]. Kaino *et al.* [27] used a resin with both radical and cationic monomers (PETA as crosslinking agent with *tert*-butyl-calix-8-arene acrylate) crosslinked via MPL to form an optical waveguide core. 3,2-Benzimidazolyl-7-diethylamino coumarin was used as a two-photon absorber and sensitizer for the initiator lauryl peroxide. The cationic component of the resin, 3-ethyl-3-ethyloxetane-3-yl methoxy methyl oxetane, remained stable throughout this step and was excluded from the core volume. The refractive index of the core material was in this way increased relative to the surrounding unreacted material. Radical-induced cationic polymerization was found to play only a minor role. The authors then used an unconventional route for postprocessing: After MPL structuring of the waveguide cores, the surrounding unreacted resin, which consisted of oxetanes (cationic curable monomers) and residual uncured acrylate (radical monomers), was photopolymerized with UV light. Only under these conditions could the cationic initiator be activated efficiently. In other studies, PETA was applied by Ober and coworkers in a resin with the multiphoton initiator NOPD and the fluorescent dye FITC1 (Figure 10.5) [29]. NOPD, a novel two-photon photoacid generator (PAG), is able to initiate both radical and cationic polymerization reactions [30].

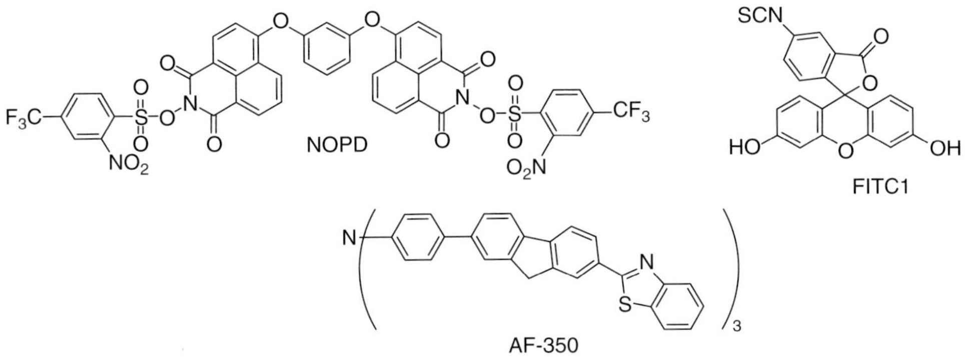

Figure 10.5 Two-photon acid generators NOPD, fluorescent dye FITC1, and MPL initiator AF-350.

The same group also applied MPL to create 3D defects within resin-filled organic modified ceramic (ORMOCER) PhCs with inverse opal features. Unreacted resin was removed from the sample by rinsing it with ethyl acetate for 30 s. To fabricate the PhC, PMMA colloidal particles were self-assembled to give an opal template, and then infiltrated with ORMOCER. In the next step, the ORMOCER was UV-polymerized, and then the PMMA template was removed with tetrahydrofuran (THF) [29]. This approach makes use of colloidal self-assembly to form large-scale periodic structures, which is a method that may not be so well suited for MPL due to its lower throughput. Nevertheless, the high resolution and 3D flexibility of MPL allow well-defined defects in the size range of the lattice period. This was detected by means of the fluorescent dye FITC1, which emits in the region of the PBG of the PhC. Scattering was avoided by matching the refractive index of the resin (1.5) with that of the ORMOCER material of the inverse opal PhC. Figure 10.6 gives an overview of the process starting from the ORMOCER inverse opal.

Another monomer that is applied for the introduction of defects in PhC structures is TTA, shown in Figure 10.1. This trifunctional monomer has a low viscosity of 106 cP at 25 °C and cures to give polymers with good resistance to water, chemicals, abrasion, and heat. Volatility of the monomer is low, and curing proceeds rapidly. The homopolymer has a glass transition temperature of 62 °C and a relatively low refractive index of 1.4723. Braun *et al.* used TTA to structure waveguides within a PhC (Figure 10.7) [31–35]. First, a silicon–air PhC was created from an SU-8 template, which had been formed beforehand via interference lithography. The advantage of this method is that periodic structures are formed in a relatively high-throughput manner. In the next step, the PhC structures were infiltrated with a MPL-sensitive TTA-based resin. The authors used 0.1 wt% of AF-350 (tris[4-(7-benzothiazol-2-yl-9,9-diethylfluoren-2-yl)phenyl]amine) as initiator and a boron dipyromethene fluorescence dye for imaging. MPL was then used to introduce nonperiodic defects. Removal of unreacted resin was unproblematic with ethanol [32].

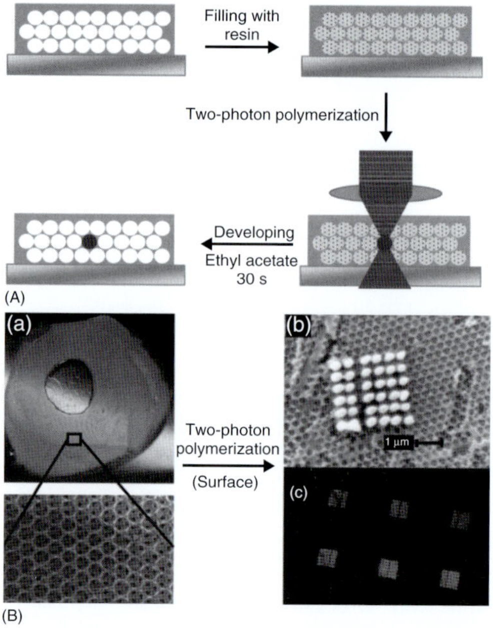

Figure 10.6 (A) Procedure for introducing defects in an inverse opal via MPL. (B) (a) Digital photograph of an ORMOCER replica (2 cm^2) and an SEM image showing a magnification of the inverse opal structure. (b) SEM image of defects introduced on the surface of the replica with two-photon lithography. (c) Fluorescent defect structures on a replica structure, seen under a fluorescence microscope. (Lange 2007 [29]. Reproduced with permission of Wiley.)

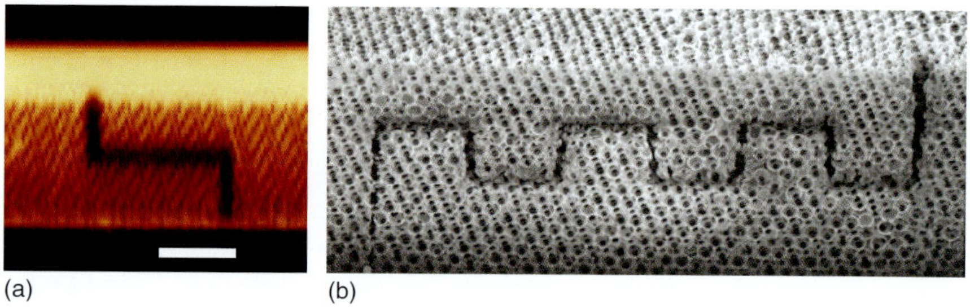

Figure 10.7 (a) Double-bend waveguides embedded in a 3D PBG material (scale bar 10 μm). (b) Vertical cross section of air features embedded within silicon–air inverse opals after removing the polymer. (Rinne 2008 [31]. Reproduced with permission of Nature Publishing Group.)

In another paper from Braun and coworkers, 3D-defects were created within a silica colloidal crystal by means of MPL. A preliminary high-index replication step followed by removal of the opal template allows well-defined defects within a 3D silicon PhC. This allows waveguiding of NIR radiation around sharp bends within a complete PBG material [31]. Another example of using MPL to create flexible waveguide structures is given by Krivec *et al.* In this case, low-refractive-index, organically modified silica monoliths were prepared, which were mechanically flexible and served as a host for polymeric, acrylate-based waveguides [36]. The porous monoliths were fabricated using a surfactant-based template (Figure 10.8).

After removal of the template, the host material was infiltrated with high-refractive-index 2-phenoxyethyl acrylate (POEA) and the initiator NDPD (Figure 10.8) [37]. MPL yielded waveguide structures with good refractive index contrast (Figure 10.9).

Another concept is derived from photolocking: A low-refractive-index polydimethyl siloxane (PDMS) film was swollen with an acrylic mixture containing the initiator B3K [8, 38, 39]. The extent of swelling was controlled via the ratio of acrylic acid isobornyl ester, which is compatible with PDMS, and 1,4-butanediol diacrylate crosslinker, which has poor PDMS compatibility. MPL structuring of waveguides was followed by removal of unreacted monomers under reduced pressure. More recently, swelling has been minimized by mixing the monomer formulation and the liquid polysiloxane matrix material with orthogonal reactivity prior to selective matrix curing [40]. Contrast is improved with high-refractive-index monomers such as for 2-phenylthioethyl acrylate [41].

An analogous process for fabricating optical waveguides was developed by the group of Kern [42]. They used phenyl and benzyl methacrylate as high-refractive-index monomers, which do not interfere with hydrosilylation. This allows a process in which curing of the silicone rubber can be accomplished within the

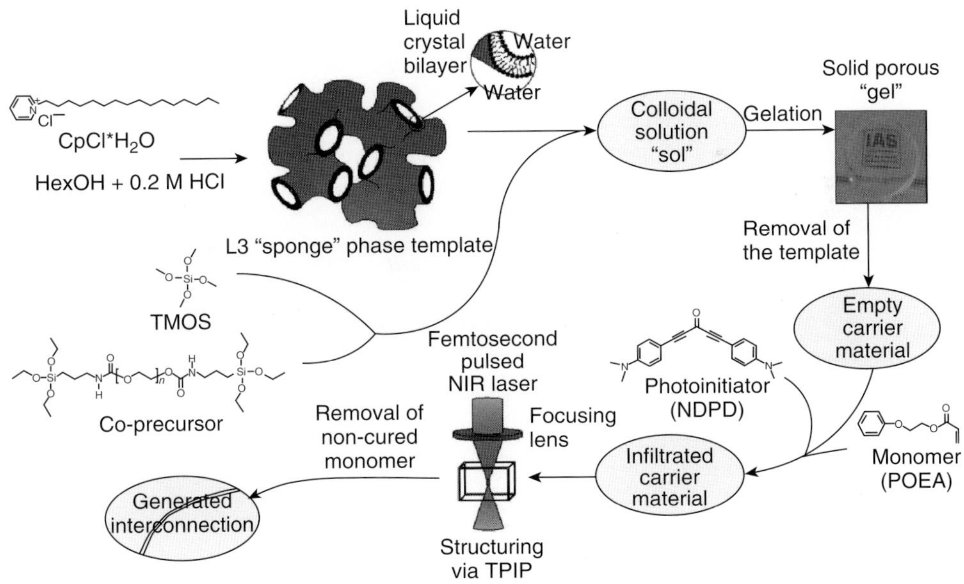

Figure 10.8 Flow diagram for patterning waveguides by MPL in a porous silica host material. The porosity is generated by employing a liquid L3 "sponge" phase, to which appropriate silica precursors are added. (Krivec 2010 [36]. Reproduced with permission of Wiley.)

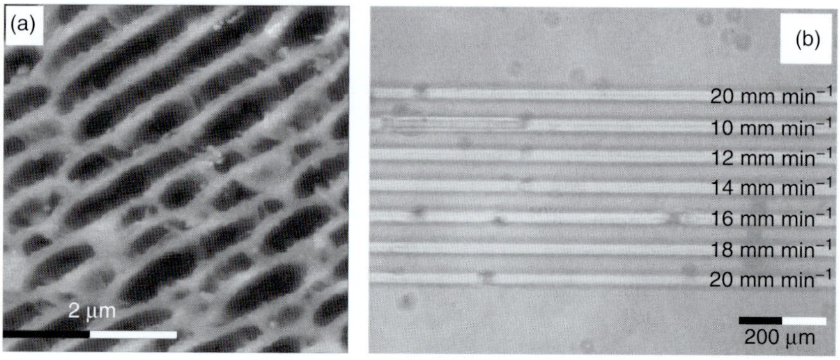

Figure 10.9 (a) Atomic force microscopy (AFM) image of a porous monolith and (b) light microscopy image showing waveguide bundles with a spacing of 75 mm structured in a monolithic sample at writing speeds from 10 to 20 mm min^{-1}. (Krivec 2010 [36]. Reproduced with permission of Wiley.)

MPL reactive monomer formulation. The method was used to produce optical interconnects for printed circuit boards. A refractive index difference of 0.02 between the illuminated core and the nonilluminated cladding material was achieved. The authors suggest that the optical damping value of 1.69 dB cm^{-1} obtained from cut-back measurements can be improved by changing the shape of the waveguides. The same group has more recently used a related technique to photocrosslink acrylate-terminated oligosiloxanes giving waveguides with low optical loss (0.65 dB cm^{-1} at 1310 nm) [43]. Waveguide patterns were produced with different laser powers, and it was found that contrast and optical loss of the waveguides were improved at higher power, although burning was observed beyond 260 μW. With optimized process conditions, a bundle of seven waveguides with core diameters of approximately 45 μm was produced (Figure 10.10). Crosstalk between waveguide channels within the bundle was not observed.

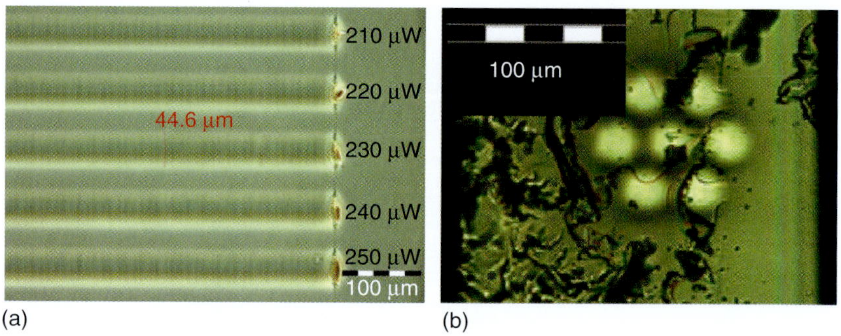

Figure 10.10 (a) Waveguide patterns in polysiloxane polymer produced with increasing laser power (210–250 μW). (b) Illuminated cross sections of a bundle of seven waveguides written with a laser power of 200 μW. (Woods 2014 [43]. Reproduced with permission of Optical Society of America.)

10.3
Thiol–Ene Resins

Another interesting class of material for MPL is thiol–ene resins. In contrast to acrylates, polymerization follows a step-growth mechanism (Figure 10.11), which has great bearing on the architecture of the polymer network and thus on material properties. In addition to having higher refractive indices (due to the presence of sulfur), thiol–ene polymers tend to have higher impact strength, which arises from a more homogeneous network structure. Another advantage is lower sensitivity to oxygen inhibition, since thiols can transfer hydrogens to peroxy radicals to give highly reactive thiyl radicals. Yield and specificity of thiol–ene chemistry are high, which are hallmarks of "click" methodology and allow polymers with a range of structural moieties. Excellent reviews have been written by Hoyle and Bowman [44–46].

UV-curable thiol–ene resins are commercially available from Norland Optical Adhesives (NOA). As an example, NOA 61 resin consists of a mixture of isocyanate-derived thiols [47]. NOA 63, 72, and 81 are other resins based on polyurethane mercapto esters. Although details are proprietary, tetrahydrofurfuryl methacrylate is disclosed as a component of NOA 72 [48]. The viscosity of these resins ranges from 155 to 2000 cP, which make them amenable to many processing techniques. The curing of NOA resins is recommended in two steps: precure in the time range of seconds, followed by a final cure of a few minutes. This brings about full crosslinking and solvent resistance. Surface adhesion can be improved during the time after cure, when chemical bonds are formed with the substrate. The thiol–ene polymerization exhibits lower shrinkage (e.g., 1.5% for NOA 61) than other resin types. Moreover, comparatively high refractive indices of around 1.56 at 635 nm can be achieved. Their mechanical properties are given in Table 10.1.

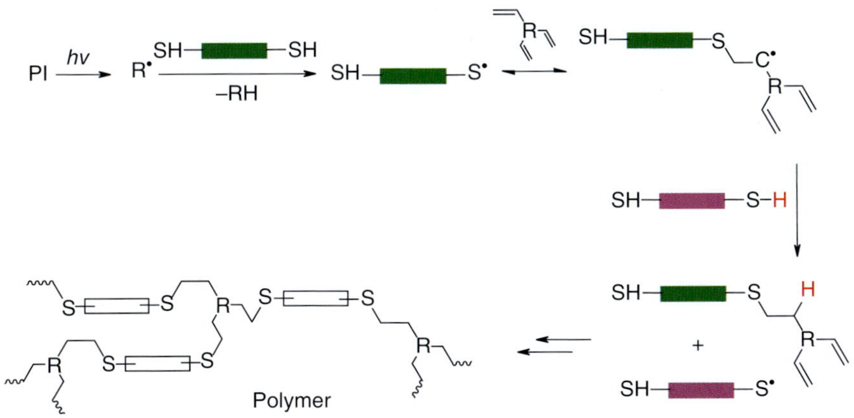

Figure 10.11 Network formation in thiol–ene-based formulations.

Table 10.1 Some important mechanical properties of NOA resins for stereolithography.

	NOA 61	NOA 63	NOA 72	NOA 81
Modulus (MPa)	930–1100	1650	—[a]	1380
Tensile (MPa)	19–24	34	—[a]	28
Elongation at failure (%)	35	6	—[a]	25
Shore D hardness	85	90	75	90

a) No data available.

All NOA thiol–ene resins exhibit very good transparency in the visible and NIR regions. Multiphoton polymerization was generally initiated by an undisclosed initiator, which is also applicable to MPL. In one case, curing was induced with a 465 nm laser using an additional NIR two-photon absorbing fluorophore AF183 (6-benzothiazol-2-yl-2-naphtyl diphenylamine) to sensitize crosslinking of the NOA 72 resin [49]. Following MPL, the development of the microstructures is performed with standard solvents including ethanol [50], ethyl acetate [29], and methanol. Thiol–ene resins have been used for several MPL experiments to build optical elements. Already in 1999, Prasad *et al.* used a mixture of NOA 72, the epoxy glue EPO-TEK 301 (Polytek PT), and the 2P initiator AF 183 to build optical elements such as beam splitters, grating couplers, and waveguides in bulk [49]. EPO-TEK 301 here acts as a hardener, which causes a rubber-like consistency of the NOA 72 film prior to MPL writing after thermal crosslinking. A refractive index difference of 0.0234 between the nonilluminated clad material and the illuminated waveguide core was sufficient for waveguiding without an additional development step. NOA 72 was also used by Kirkpatrick *et al.* to construct periodic grid structures with a holographic MPL technique [51]. In this case, the 2P initiator AF 380 was dissolved in NOA 72 in a benzene solution and spin-coated to give a film for MPL. Also using NOA 61, Sun *et al.* fabricated a Fresnel zone plate and multistep diffractive phase lenses (Figure 10.12) [50].

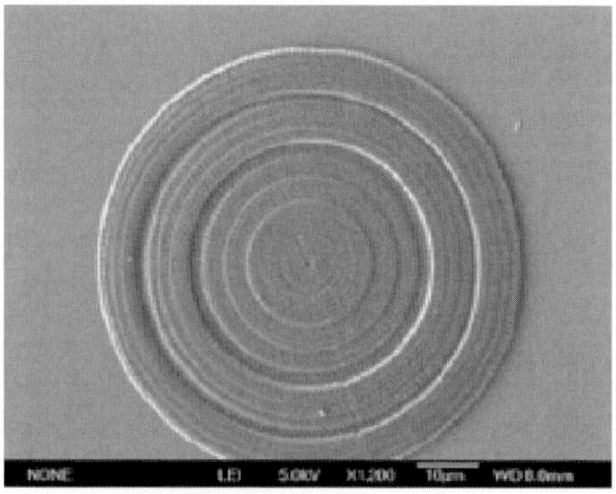

Figure 10.12 Multistep diffractive phase lenses fabricated via MPL. (Chen 2007 [50]. Reprinted with permission of American Institute of Physics.)

In another report, NOA 63 was used to place line and point defects within ORMOCER replica PhC structures using isopropyl thioxanthone (ITX) as photoinitiator [29]. The PhC was first infiltrated with NOA 63 thinned with acetone. Coumarin 6 was also added to the system to improve the contrast for fluorescence microscopy. An interesting application of MPL was pioneered by the group of Cojoc, who fabricated axicon lenses, convergent lenses, and ring-shaped phase masks on top of an optical fiber (Figure 10.13) [52]. NOA 63 was used as the building material because of its good adhesion to glass, easy processing, a suitable refractive index of 1.56, and low cost. NOA 63 has also been used to fabricate optical rotors, which can be manipulated with the assistance of laser tweezers [53]. In a different paper, microscopic diffractive optical elements, amenable to laser-based optical trapping, were fabricated with NOA 81 [54].

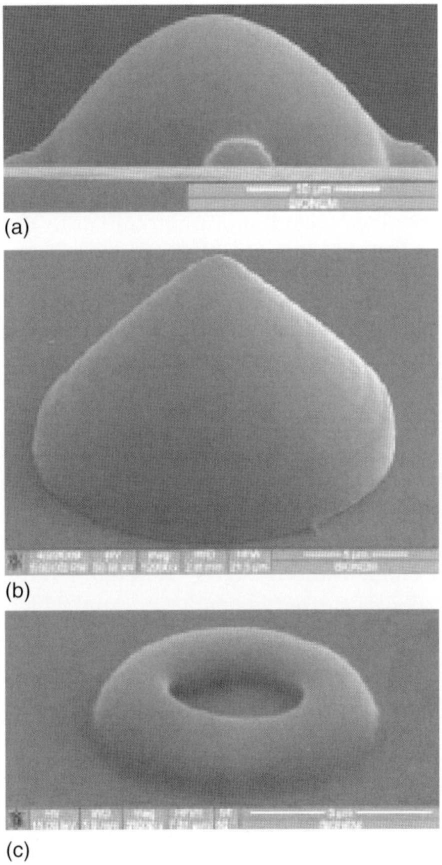

Figure 10.13 Different micro-optical elements fabricated by MPL on top of an optical fiber: (a) convergent lens, (b) axicon lens, and (c) ring-shaped phase mask. (Cojoc 2010 [52]. Reprinted with permission of Elsevier.)

Thiol–ene chemistry has been applied by our own research group as a tool for selectively templating waveguide structures within a siloxane prepolymer [55]. To avoid phase separation, a siloxane dithiol monomer was synthesized and used with a commercial siloxane triallyl monomer. The siloxane thiol–ene system could be soaked in PDMS to give a swollen gel and then selectively crosslinked via MPL. An alternate strategy was to gel triacetoxy vinyl silane in the presence of thiol. In this case, the thiol and ene groups survived the tin-catalyzed condensation and could later be selectively patterned by MPL. This caused a refractive index increase of 0.013, which was sufficient for waveguiding [56]. In another work, Griesser *et al.* selectively photocrosslinked ring-opening metathesis polymerization (ROMP)-derived polyenes with thiol monomers via MPL to fabricate lenses [57].

10.4 Sol–Gel-Derived Resins

Composites, in particular nanocomposites, are intensively studied for optical applications because of the great breadth in material properties. ORMOCERs are a popular class of nanocomposites, known both for their excellent material properties and for ease of processing [58]. ORMOCER photosensitive resins consist of sol–gel-erived oligomers containing organic and inorganic groups, purely organic monomers, and other additives including a photoinitiator. The inorganic element of the ORMOCER is most commonly silicon but may also be Al, Zr, or Ti, which tend to be bound via alkoxyl bridges to organic moieties including acrylates. Thanks to the hydrolytic stability of the Si–C bond, it is possible to introduce various functional groups to modify properties including flexibility, refractive index, and so on. If the organic moiety is amenable to polymerization, it will contribute to network formation. Other silanes, such as, for example, diphenylsilanediol (DPD), are employed as network modifiers [58]. According to the literature, low light absorption – a key property for photonic materials – can be achieved by lowering the content of C–H and silanol moieties. A synthetic method for modifying silanol groups is depicted in Figure 10.14.

DPD is reacted with MEMO with a condensation catalyst at temperatures between 60 and 100 °C. This sol–gel process is carried out without water, which improves the yield [60]. Alkoxylation is favored here because of the bulky phenyl rings which sterically prevent self-condensation of DPD. Thus silanol groups are hardly present in the final polysiloxane network. The byproduct methanol or ethanol can be removed by evaporation [61]. Typical ORMOCER formulations for optical applications include Ormocore (DPD + MEMO), Ormoclad (DPD + MEMO + TTS (3,3,3-trifluoropropyl trimethoxysilane)), and Ormocomp (TTS + TTASS) [62]. For many applications, it is desirable to integrate highly refractive elements such as sulfur (Ormocomp) or titanium in the resins. Typical Ti-containing formulations consist of 17 mol% MEMO, 50 mol% DPD, and 33 mol% $Ti(OEt)_4$. Processing viscosity can be lowered with solvents such as THF or cyclopentanone. ORMOCER resins can be cured with UV radiation, in

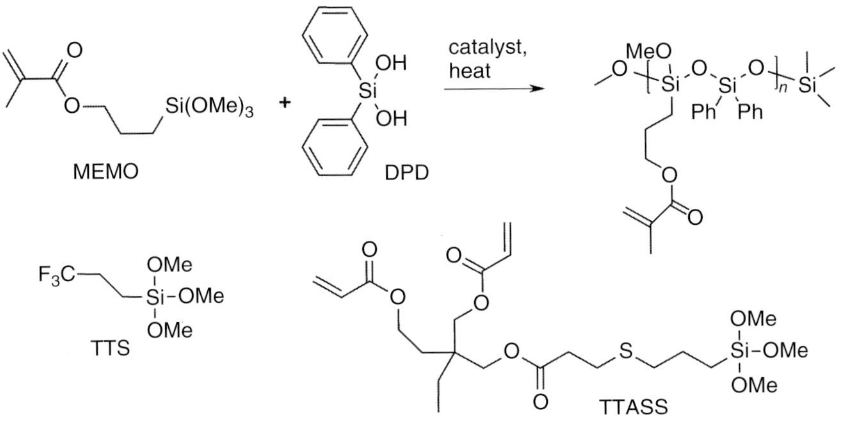

Figure 10.14 Exemplary ORMOCER sol–gel process and components (TTS and TTASS) of Ormocomp [59].

which case two striking benefits distinguish this technology from conventional UV-curable resins. Precursor siloxane oligomers organize with sizes of 1–5 nm to allow very high resolution in UV lithography. Resolution is further enhanced by high crosslink density, which leads to rapid termination and minimizes unwanted polymerization in nonilluminated regions [58].

ORMOCERs gain hardness, chemical resistance, and transparency from their glassy component. Durability, ease of processing, and range of functionality arise from their partially organic character. Silicone portions ensure flexibility and adequate surface properties [63]. By applying trialkoxysilanes and trifunctional metal alkoxides, which form tight crosslinks, Young's moduli greater than 10 GPa are obtained. While some of these materials have ambient-temperature glass transitions, in other materials no transition is observed at all. Difunctional monomers form linear or cyclic oligomers with the behavior of elastomers: low Young's moduli of around 5–100 MPa and glass transitions below room temperature. Mixing different precursors such as tri- or difunctional alkoxysilanes, metaloxides, and macromonomers enables control of the mechanical properties from the range of soft and flexible polymers to rigid and sometimes brittle glasses [58]. A crucial optical property, namely optical attenuation, can be as low as 0.2–0.3 dB cm^{-1} at 1320 nm. Refractive indices can be tuned from 1.47 to 1.58 (at 633 nm), which allows sufficient contrast for waveguides. As photo-curable ORMOCER resins do not require any solvents, shrinkage is comparatively moderate. High thermal stability up to 270 °C is necessary for mass fabrication processes of, for example, printed circuit boards, which are laminated at a similar temperature. The very low root-mean-squared (RMS) roughness indicates an almost perfect planarization behavior, which is important for substrates and makes the material interesting for highly sophisticated optical applications (Table 10.2) [61].

Table 10.2 Properties of three ORMOCER resins for optical applications.

Property [62, 64]	Ormocore	Ormoclad	Ormocomp
Thermal stability (rate 5 K min^{-1})	Weight loss <5% up to 270 °C	—	—
Young's modulus (MPa)	860 ± 120	—	925 ± 100
Shrinkage (vol%)	2–5		—
RMS roughness (nm)	2–4		—
Refractive index (635 nm)	1.553	1.534	1.519
Optical attenuation (633 nm) (dB cm^{-1})	<0.1	≤0.1	—
Optical attenuation (1310 nm) (dB cm^{-1})	0.23	0.26	—
Optical attenuation (1550 nm) (dB cm^{-1})	0.5–0.6	0.48	—

Prior to MPL microfabrication of optical elements, ORMOCER resins may be prebaked for a few minutes at 80 °C to increase resin viscosity. It is also useful to carefully clean the (glass) substrate and thus ensure good adhesion of the resin. Films can be prepared by spin coating but also by simply casting the resin into the cavity between two cover slips [65–67]. As radical initiator, Irgacure 369 is frequently used [68–70]. The refractive index of the resin will change upon polymerization, which has to be taken into account in the optical setup. Online monitoring of the fabrication process is thus simplified, which is an advantage over cationic resins such as SU-8, where polymerization occurs only during the post-baking step.

MPL methodology tends to be streamlined relative to traditional UV methods: pre- and post-exposure baking steps, as well as solvent dilution, are not necessary as with conventional UV curing, because the polymerization takes place within the resin volume [61]. Post-baking can, however, be useful because the adhesion of the microstructures to the substrate is improved [66]. As shown in a study on PhCs [71], it is possible to enhance the resolution of PhC microstructures via post-baking at temperatures between 150 and 300 °C after the solvent development step [68]. Final curing ensures that the maximum crosslinking degree is reached.

In the literature, there are several different applications where ORMOCERs were successfully employed. A distributed-feedback resonator, which might be used for optical lab-on-a-chip applications, was fabricated via MPL by Woggon *et al.* using the Ormocomp resin [66]. The resonator was based on a 400-nm periodic grating with 40 nm height modulation. Partial MPL structuring within the substrate, which was caused accidently by slight tilting of the cover slip, was circumvented by fabricating the resonator on a stage with a thickness of a voxel. For the active material layer, the organic semiconductor tris-8-hydroxyquinoline aluminum doped with the laser dye 4-(dicyanomethylene)-2-methyl-6-(4-dimethylaminostyryl)-4*H*-pyrane was applied.

Leyrer *et al.* introduced waveguide defects via MPL into polymer opals using ORMOCER B59. This well demonstrates the applicability of MPL for creating nonperiodic structures, which are more complicated with other nanostructuring techniques (e.g., interference lithography) [68, 72]. Although the post-processing of ORMOCERs is only incidentally mentioned in most MPL-related literature, post thermal treatment can provide significant improvement in the spatial resolution of the fabricated structures, as demonstrated by Gu and coworkers [71]. During the post thermal treatment step, further polymerization of methacrylate groups and partial decomposition of organic parts were detected. In addition to increased feature resolution, an enhancement in refractive index was also observed. Reinhardt *et al.* presented MPL-fabricated nanostructures showing effective excitation and focusing of surface plasmon polaritons (SPPs). In this case, the ORMOCER with a sufficiently high dielectric constant was covered with a 50-nm gold film by electron sputtering. In Figure 10.15, an SPP excitation event at a 3D ramp structure is shown [67].

Finally, the extensive work at the Fraunhofer Institute, where ORMOCER materials were actually developed, must be mentioned. In particular, Houbertz and coworkers have successfully used ORMOCER resins for MPL structuring of optical elements [59, 61, 69, 73, 74]. In one work, waveguides were fabricated to connect a photodiode and a laser diode on a printed circuit board using only one material. The MPL structured areas served as the core, while the surrounding material served as the clad. During polymerization, organic polymerizable moieties diffuse toward the polymerizing region and cause the desired refractive index change. Importantly, the solvent development step is eliminated. Instead, only a post-curing thermal step is required. PhCs have also been fabricated by the Fraunhofer group using ORMOCERs. Shrinkage problems are addressed by scaling computer-aided design (CAD) structures with a shrinkage constant prior to building [61]. Addressing the problem of not being able to create a full PBG with existent resins (ORMOCER, SU-8, acrylates, etc.), Houbertz *et al.* invented a Ti-containing ORMOCER-based MPL resin named ORMOCER PD92. This was then used to fabricate PhCs with a PBG in the NIR region. Shrinkage during MPL, which is typical to Ti-containing resins, remains a problem [73].

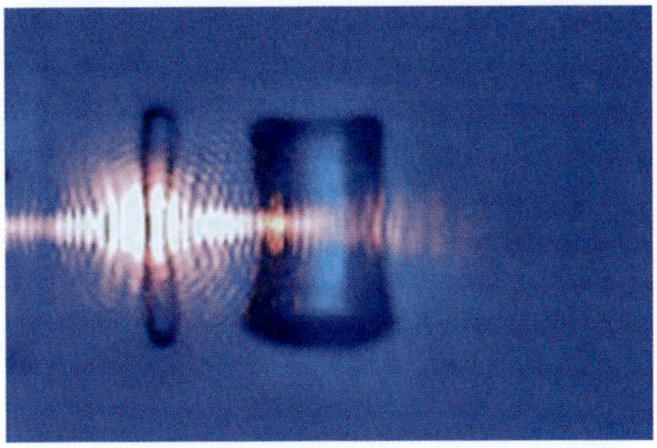

Figure 10.15 Leakage radiation microscopy (LRM) image of SPPs excited on the dielectric line and propagating over an MPL-fabricated 3D ramp structure. (Reinhardt 2007 [67]. Reprinted with permission of Springer.)

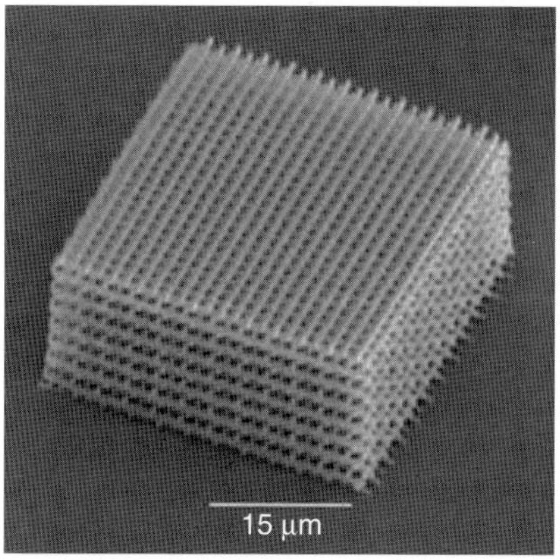

Figure 10.16 SEM image of Zr-containing PhC based on MEMO/ZPO (8 : 2) [77]. (Ovsianikov http://www.hindawi.com/archive/2008/493059/. Used under CC BY 3.0 license http://creativecommons.org/licenses/by/3.0/.)

In addition to ORMOCERs, there are other sol–gel materials that are used in MPL technology for photonics. As an example, Zr-based sol–gel resins exhibit advantages such as good control of refractive index via the Zr content, improved mechanical stability and – perhaps most importantly – very low volume shrinkage [75]. With sufficient laser power, PhC structures from Zr resins with no measurable shrinkage are even possible [76]. This fundamental benefit is thought to arise from the combined inorganic and organic polymerization reactions occurring in the resin. Synthesis of these Zr resins is carried out by chelating zirconium *n*-propoxide (ZPO) with methyl methacrylic acid and adding a hydrolyzed product of MEMO [77]. Afterward, water is added and an interconnected porous cluster structure is obtained. 4,4-Bis(diethylamino)benzophenone is used as the initiator. The gelation step involves the removal of solvents and is typically carried out at 100 °C, where the formation of Si–O–Zr linkages is kinetically favored. Thus condensation is accelerated compared to other sol–gel-derived resins, which is another advantage of this technology. Films for MPL are formed via drop casting or spin coating. 1-Propanol is a suitable solvent to remove the unreacted resin after MPL structuring. The described Zr-based resin has been applied to fabricate PhC with a clear bandgap in the NIR region (Figure 10.16) [76–78]. More recently, Emons *et al.* used a 4 : 1 MEMO/ZPO resin with a dipentaerythritol hexa acrylate crosslinker to create PhCs with feature widths down to 45 nm [79].

10.5 Cationic Polymerization and Stereolithography

Photo-induced cationic polymerization is not as rapid as radical polymerization but has a number of advantages that make it popular in the field of SLA. Notably, cationic processes are not susceptible to oxygen inhibition, and polymerization shrinkage tends to be fairly low [80]. The most commonly used monomers for photo-induced cationic polymerization are epoxies (in particular 3,4-epoxycyclohexylmethyl-3′4′-epoxycyclohexane carboxylate (ECC) and bisphenol-A diglycidylether (BADGE) in Figure 10.17), although oxetanes are

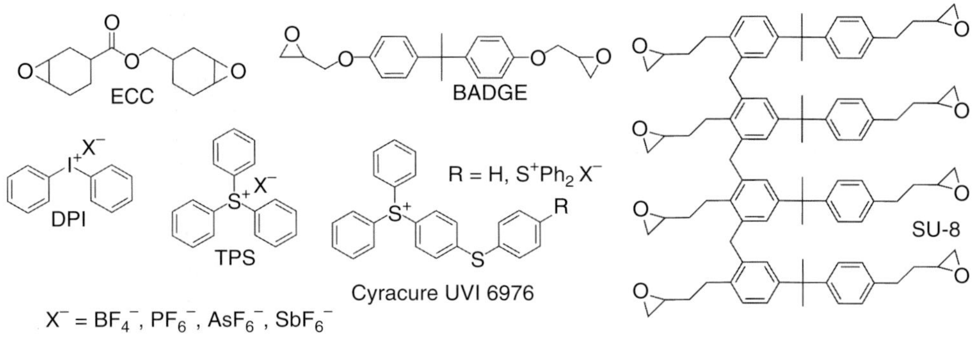

Figure 10.17 Commonly used epoxy monomers and photoinitiators for cationic polymerization.

also found in commercial SLA resins [81]. Photo-induced cationic polymerization is induced with PAGs, which are typically aryl-iodonium or sulfonium salts with fluorinated counterions. As the absorption of these initiators generally tails out at ~300 nm, extended conjugation derivatives (such as Cyracure UVI-6976) or sensitizers can be employed to allow polymerization at higher wavelengths. Iodonium PAGs tend to be more easily sensitized (by compounds such as anthracenes, perylenes, phenothiazines, xanthones, or thioxanthones). To reduce the brittleness of the cured material, chain-transfer agents based on polyols can be included at concentrations up to 20 wt%.

The epoxy-based monomer SU-8 (Figure 10.17), originally developed as a UV-curable negative photoresist by IBM, is well established in the field of MPL microfabrication, as it is readily available and inexpensive [82]. SU-8 formulations typically contain 10 wt% triphenylsulfonium hexafluoroantimonate as PAG with a variable ratio of SU-8 resin to organic solvents. The ratio of solvent is used to control viscosity, which is a key process term for producing films of defined thickness. Gamma butyrolactone was initially employed as solvent for SU-8, but it has been recently replaced with cyclopentanone because of its improved wettability of low-surface-energy substrates and faster processing times [83].

Upon exposure to light of an appropriate wavelength, a PAG will decompose and react with the solvent or monomer to give a Bronsted acid, which catalyzes the ring-opening polymerization of epoxide monomers. In a resin consisting of multifunctional epoxy monomers (ECC, BADGE, and SU-8), a network will rapidly form with propagation of the cationic chain. While ring-opening polymerization is slower than radical chain growth, termination is also slower and conversion can be significantly advanced by heating the material after lithography. In case of SU-8, the photocuring step itself may be performed at an elevated temperature [84]. SU-8 has good optical transparency and a high refractive index [85]. With eight epoxide moieties per molecule, very high crosslink density is possible, and chemical and thermal stability of the cured system are good. Additionally, high aspect ratios as well as straight sidewall profiles can be achieved. Another important advantage is the comparatively high glass transition temperature of the non-cross-linked resist of about 50 °C. This turns out to be critical to the attainment of high-resolution features, since diffusion of the PAG prior to post-exposure baking is limited [86]. Some of the more important properties of SU-8 are summarized in Table 10.3 [87].

While SU-8 has very good material properties, there are some disadvantages to the resin. For one, processing of SU-8 is slightly complicated. Substrate pretreatment is followed by coating, soft baking, and exposure. Postprocessing involves postbaking, development, rinsing, drying, and hard baking. Another shortcoming is shrinkage of ~7% depending on the thickness and area size of the structure [88]. Film thicknesses from 2 to 300 μm are achievable by controlling the spin-coating parameters (time, speed, viscosity). In order to remove the solvent and densify the film, Microchem Corp. recommends a softbake step at 95 °C prior to lithography. The spectral range for single photon exposure is given from 350 to 400 nm.

The exact mechanism of multiphoton-induced crosslinking of SU-8 has not been fully clarified yet. As a result of pulse irradiances near the dielectric breakdown (around 1 TW cm^{-2}) and thus a highly ionized environment, the cationic photopolymerization is likely to proceed in a different manner from the single-photon-induced reaction. It can be explained by the absorption of the spectrally broad emission of thermally nonequilibrated electrons when high-irradiance femtosecond pulses are applied to the resist [89]. Ti:sapphire (800 nm) femtosecond lasers have proven to be suitable for effective MPL structuring. Compared to other polymeric materials for MPL microfabrication, SU-8 formulations do not undergo a liquid–solid phase transition under illumination, which turns out to be advantageous for feature resolution. In a conventional UV lithographic process and the majority of MPL patterning experiments reported up to now, the subsequent post-bake process provides the activation energy for cationic crosslinking, which is inhibited at room temperature. Recently, however, certain MPL processing parameters were found that allow sufficient crosslinking

Table 10.3 Material properties of SU-8.

Property	Value
Young's modulus E (95 °C postbake)	4.02 GPa
Young's modulus E (200 °C hardbake)	4.9 ± 0.4 GPa
Biaxial modulus of elasticity $E/(1-\nu)$	5.2 ± 0.9 GPa
Film stress (95 °C postbake)	16–19 MPa
Maximum stress (200 °C hardbake)	34 MPa
Friction coefficient μ_F (95 °C postbake)	0.19
Glass transition T_g (unexposed)	~50 °C
Glass transition T_g (fully crosslinked)	>200 °C
Decomposition T_d (fully crosslinked)	~380 °C
Thermal expansion coefficient α (95 °C postbake)	52 ± 5 ppm K^{-1}
Volumetric shrinkage	7.5%

without postbaking. Here, the laser itself provides sufficient heat to fully crosslink the resin [89]. The sample is then developed with solvents such as ethyl lactate.

Because of the limited success with MPL using conventional PAGs, Marder *et al.* synthesized the cationic photoinitiator BSB-S_2 based on a bis[(diarylamino)styryl]benzene core. The PAG is quadrupolar, with dialkyl aryl sulfonium moieties covalently attached to the donor amines [90]. The initiator was then used to fabricate a "stack-of-logs" photonic bandgap structure with the SU-8 resin. Despite its numerous advantages, SU-8 does not allow the production of PhCs with spectrally broad PBGs. This is due to its relatively low refractive index of 1.6, which is well below that of many inorganic optical materials. While this is a shortcoming for direct fabrication of optical elements, SU-8 is well suited as a template material for PhCs. Deubel *et al.* [91] produced PhC templates via MPL with SU-8, stating that the small refractive index change upon polymerization made it possible to structure behind already polymerized regions. As further advantages, they mention a lack of oxygen inhibition and the fact that SU-8 is solid during MPL, which eliminated the need to interconnect written features immediately. As shown by Serbin *et al.*, SU-8 scaffolds for PhCs can be readily removed after infiltration and subsequent sol–gel crosslinking with titanium(IV)-isopropoxide [85]. Seet *et al.* presented an SU-8 double inversion process, which provided structures with feature sizes below 100 nm [92]. Another interesting experiment was conducted by Sun *et al.* using MPL with SU-8 to fabricate a microlens with 100% filling factor [93]. SU-8 was used here because of its high transmission of light from the visible to the NIR, low volumetric shrinkage, and its good mechanical properties. Transmission measurement revealed that the optical absorption by the microlens array was negligible: for example, a 1.5-μm-thick polymerized SU-8 film had a transmittance of 99% in the wavelength range 360–1100 nm. The group of Chanda fabricated a three-level optical diffractive element using SU-8 [94]. In another paper, Turberfield *et al.* were able to place a correctly aligned Mach–Zehnder interferometer structure consisting of linear waveguide motives via MPL within a PhC that was created by means of holographic lithography (Figure 10.18) [95].

Lee *et al.* applied a two-photon laser-scanning microscope to produce 2D PhC structures in SU-8 polymer films with circular, elliptical, rectangular, or diamond-shaped unit cells within a hexagonal or square lattice [96]. An aspect ratio of 6.9 with 250 nm line width was possible. Specific patterns, for example, lines, dots, and Y-splitters, were placed in the otherwise perfect PhCs.

10.6 Materials Based on Multiphoton Photochromism

Multiphoton absorption (MPA) can induce changes in the optical properties of a material by numerous mechanisms. Change in fluorescence, absorption, or refractive index can be induced not just by polymerization but also via chemical events such as photochromic transformations (isomerism, dimerizations, cyclizations,

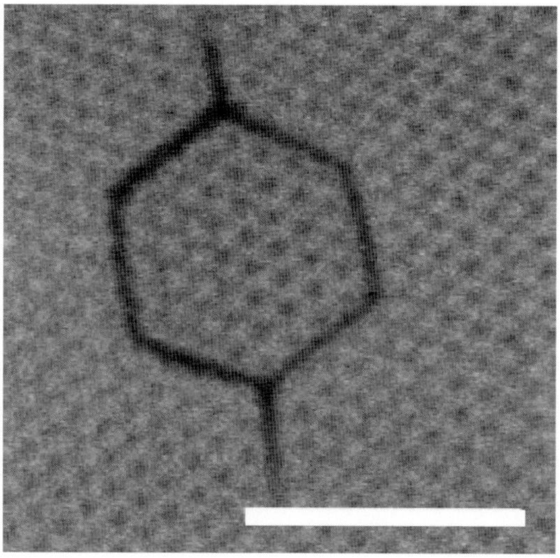

Figure 10.18 PhC pattern with inscribed Mach–Zehnder structure (confocal section through fluorescence of infiltrated dye). Scale bar: 5 μm. (Scrimgeour 2006 [95]. Reproduced with permission of Wiley.)

etc.) [97–101], photobleaching [102–104], or protonation by means of a PAG [105, 106]. These single-molecule transformations are not catalytic processes such as the case with MPL, where one photoinitiation event causes a chain reaction. Reactions are thus more volumetrically confined, making photochromism highly appropriate for 3D optical data storage. Multiphoton photochromism has another benefit in contrast to the more established technique of MPL: higher concentrations of the active ingredients within resins are possible. This allows a better response of functional nanodevices to external stimuli [107]. Furthermore, elaborate processing procedures including post-baking or development steps are avoided. Some of the more commonly studied photochromic moieties include fulgides, spiropyranes, diarylethenes, polycylic aromatic compounds, and azo-based chromophores [99]. Generally, the two-photon absorbing molecules are embedded in transparent polymer matrices. Prior to two-photon illumination, the chromophores are introduced into a carrier polymer – usually PMMA – via uniform doping in organic solvent, copolymerization, or grafting reactions. Films were created from polymer solutions via drop coating, spin coating, or drop casting.

In an interesting contribution, Zhang *et al.* developed an optical polarization storage technology based on two-photon isomerization (TPI) of a bisazobenzene-containing MMA copolymer [101]. Under MPA conditions, azogroups undergo a cis–trans isomerization, which causes anisotropy and birefringence within the material. This allows the encoding of two different patterns in the same layer and region, which can be read out separately by changing the polarization of the read-out beam by 45° (Figure 10.19).

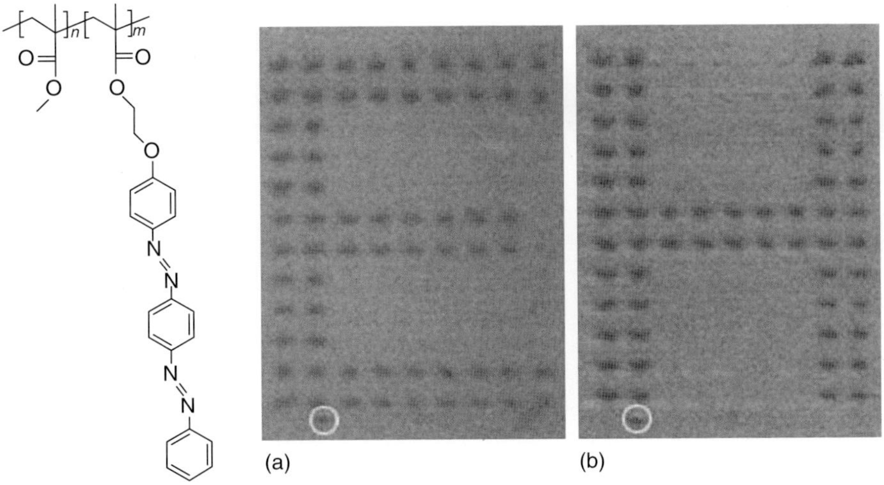

Figure 10.19 (a) Bisazo chromophore-containing copolymer and (b,c) polarization encoded patterns in the same region of a given layer (circled defects represent the same point in each case) with the polarization direction of the recording beam oriented (b) at 0° and (c) at 45°. (Zhang 2009 [101]. Reproduced with permission of Elsevier.)

Bisazobenzene systems, in contrast to other azo-containing polymers, have some advantages including better nonlinear optical and photoresponsive properties and enhanced birefringence as well as thermal stability. Forming a copolymer and thus attaching the dye covalently to the polymer matrix is an effective measure against dye aggregation, which is generally problematic in guest–host systems.

Azo isomerism is not only useful a in case of 3D data storage. Ambrosio and coworkers presented a method for MPA patterning based on photo-induced mass migration [108]. Using a polyurethane, which was obtained via condensation reaction with an azo-containing, Y-shaped chromophore, they were able to create structures including a diffraction gratings. Rentzepis *et al.* solved the problem of nondestructive read out of 3D optical stored data in an elegant way [99]. Most previously proposed photochromic molecules had the same absorption bands for the read-out and erasing states, meaning that some information would be erased during each read-out cycle. To solve this problem, "read," "write," and "erase" operations were each assigned their own separate wavelength. The idea of this invention is shown in Figure 10.20. The molecule consists of a driver part and a dye part, in which the driver part is a fulgimide, the ring moiety (write form) of which can be opened with light of 530 nm to the read form. The reaction is reversible, with light of 400 nm inducing the backward reaction (erasing). Regardless of the state of the driver, the benzophenoxazine dye's absorption wavelength remains unchanged at 660 nm. Importantly, however, the driver is able to switch the dye's fluorescence intensity, which depends on whether the driver is in the "read" or "write" state. The reading wavelength of 660 nm is sufficiently different from the

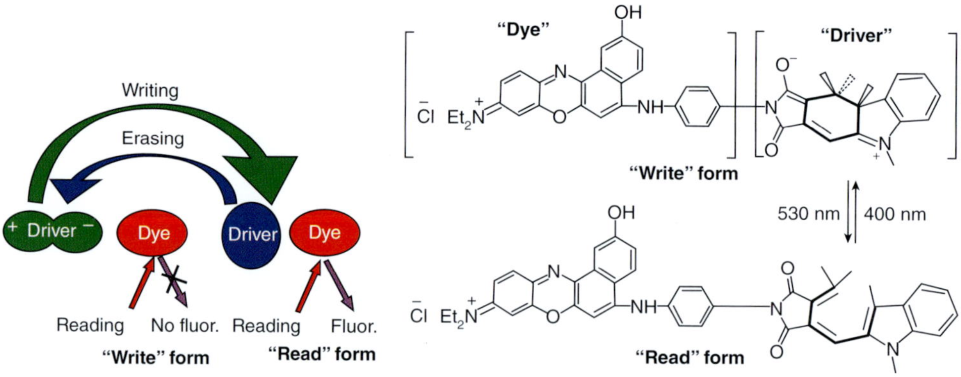

Figure 10.20 Mode of action of novel MPA data storage molecule. (a) Schematic. (b) Chemical. (Dvornikov 2009 [99]. Reproduced with permission of American Chemical Society.)

erasing wavelength of 400 nm, so unwanted erasing does not occur during the reading mode of a typical device.

Another creative approach for optical data storage was developed by Belfield *et al.* [98]. In this system, a fluorinated diarylethene derivative was employed, which reversibly ring-closed under UV exposure and opened with visible light. The photo-reversible molecule served exclusively in its closed conformation as an energy-transfer quencher for a the fluorescent dye 2,7-bis[4-(9,9-didecylfluoren-2-yl)vinyl]phenylbenzothiazole. Based on this principle, the fluorescence of the dye could be modulated. This system allowed a 10 000-fold readout, with only a small extent of erasing. In 2007, Kern and coworkers showed that at the laser wavelength of 610 nm, refractive index change can be accomplished via two-photon-induced photo-Fries rearrangement and decarboxylation (Figure 10.21) [109].

The latter photoreaction can also be triggered by MPA, which has been exploited by the same group to write waveguide structures that could potentially serve as optical interconnects in flexible printed circuit boards (Figure 10.22) [110]. To realize multiphoton fabrication, 4-(diphenylamino)phenylbenzoate was attached to polymer networks such as polystyrene and PDMS. The polymer underwent two-photon-induced decarboxylation with a yield of 40%.

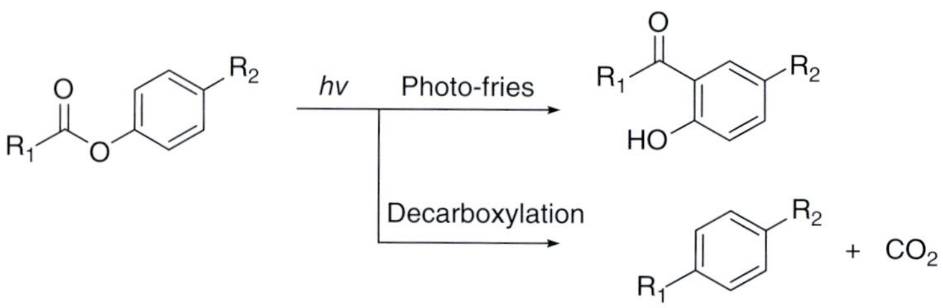

Figure 10.21 Two major photoreactions of aromatic esters triggered by UV irradiation: Photo–Fries reaction and photoextrusion decarboxylation.

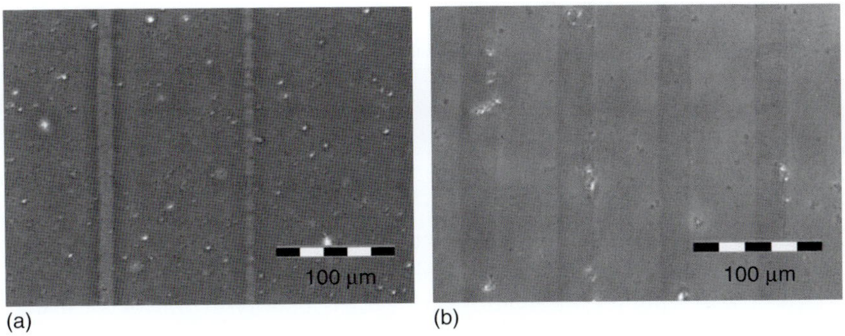

Figure 10.22 Waveguide structures in (a) polystyrene and (b) PDMS formed via two-photon-induced decarboxylation. (Daschiel 2009 [110]. Reproduced with permission of SPIE.)

Sun *et al.* recently demonstrated two-photon structuring via cis–trans isomerization within a viscous fluorescent polymer DPO-PPV (poly(*p*-phenylene vinylene)) [107]. In contrast to the cis conformation, the trans isomer is not or only weakly soluble in organic solvent, which made selective removal of unreacted polymer after TPI possible (Figure 10.23). In this manner, a PhC and a Fresnel

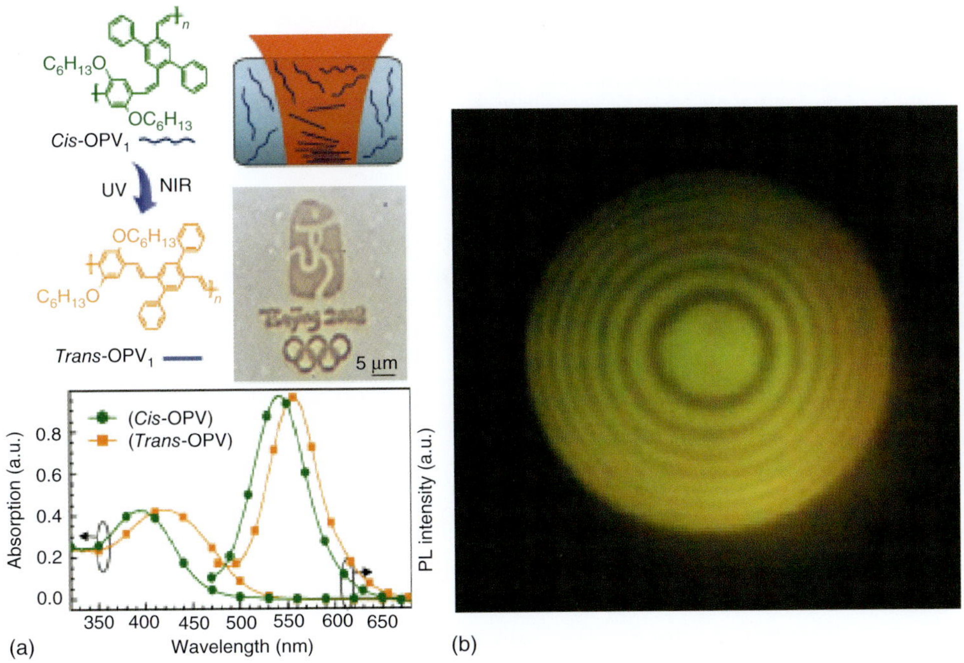

Figure 10.23 (a) Cis–trans isomerization causing insolubility of chromophore and (b) fluorescing Fresnel zone plate. (Xia 2009 [107]. Reproduced with permission of American Institute of Physics.)

zone plate were written with a femtosecond laser. Moreover, the structured areas could be loaded by nearly 100% of fluorescent material.

10.7 Conclusions

Photons have some inherent advantages to electrons in data transmission, storage, and processing, although the manipulation of light at length scales below the diffraction limit is not very straightforward. MPL is a powerful tool for constructing well-defined 3D microstructures with feature widths well below 100 nm. Some of the more common optical elements fabricated by MPL include PhCs, waveguides, lenses, and thermooptic switches. Photo-curable polymers based on acrylates and epoxides are the most commonly applied materials, although thiol–ene resins are also commercially available and increasingly popular. Photo-curable nanocomposite resins (principally monomers with Si, Ti, or Zr and covalently bound acrylate moieties) permit direct structuring of microscopic structures with lower optical loss and higher refractive indices than allowed by traditional organic resins. Multiphoton-induced chemical transformations, aside from polymerization, are also applied in the field of photonics. In particular, MP photochromism is intensively studied for ultra-compact optical data storage. In the coming years, the use of microphotonics is expected to become more widespread. If so, MPL will serve as an enabling technology certainly in the development and possibly the production stages of these next-generation photonic devices.

Acknowledgments

We thank the Christian Doppler Society for establishing the Laboratory for Digital and Restorative Dentistry and the Austrian Science Foundation FWF (project P27059) for additional financial support.

References

1. Dou, X., Wang, X., Lin, X., Ding, D., Pan, D.Z., and Chen, R.T. (2010) *Opt. Express*, **18**, 16227–16233.
2. Schröder, M., Bülters, M., von Kopylow, C., and Bergmann, R.B. (2012) *J. Eur. Opt. Soc. Rap Public*, **7**, 12027.
3. Bettotti, P. (2014) *Adv. Opt.*, **2014**, 24.
4. Izuhara, T., Fujita, J., Gerhardt, R., Sui, B., Lin, W., and Grek, B. (2013) in *Optoelectronic Integrated Circuits XV*, Proceedings of SPIE, vol. **8628** (eds L.A. Eldada and E.-H. Lee), SPIE, p. 862807.
5. Baev, A., Prasad, P.N., Ågren, H., Samoć, M., and Wegener, M. (2015) *Phys. Rep.*, **594**, 1–60.
6. Paquet, C. and Kumacheva, E. (2008) *Mater. Today*, **11**, 48–56.
7. Raghunathan, V., Izuhara, T., Michel, J., and Kimerling, L. (2012) *Opt. Express*, **20**, 16059–16066.
8. Pucher, N., Li, Z., Ligon, S.C., and Liska, R. (2012) *Rad News*, **82**, 33–46.

9. Husár, B., Ligon, S.C., Wutzel, H., Hoffmann, H., and Liska, R. (2014) *Prog. Org. Coat.*, **77**, 1789–1798.
10. Ligon, S.C., Husar, B., Wutzel, H., Holman, R., and Liska, R. (2014) *Chem. Rev.*, **114**, 557–589.
11. LaFratta, C.N., Fourkas, J.T., Baldacchini, T., and Farrer, R.A. (2007) *Angew. Chem. Int. Ed.*, **46**, 6238–6258.
12. Prinz, F.B., Atwood, C.L., Aubin, R.F., Beaman, J.J., Brown, R.L., Fussell, P.S., Lightman, A.J., Sachs, E.J., Weiss, L.E., and Wozny, M.J. (1997) *JTEC/WTEC Panel Report on Rapid Prototyping*, Rapid Prototyping Association of the Society of Manufacturing Engineers, Baltimore, MD.
13. Guo, R., Xiao, S., Zhai, X., Li, J., Xia, A., and Huang, W. (2006) *Opt. Express*, **14**, 810–816.
14. Duan, X.-M., Sun, H.-B., and Kawata, S. (2004) *J. Photopolym. Sci. Technol.*, **17**, 393–396.
15. Duan, X.-M., Sun, H.-B., Kaneko, K., and Kawata, S. (2004) *Thin Solid Films*, **453-454**, 518–521.
16. Takada, K., Kaneko, K., Li, Y.-D., Kawata, S., Chen, Q.-D., and Sun, H.-B. (2008) *Appl. Phys. Lett.*, **92**, 041902/041901–041902/041903.
17. Li, C.-F., Dong, X.-Z., Jin, F., Jin, W., Chen, W.-Q., Zhao, Z.-S., and Duan, X.-M. (2007) *Appl. Phys. A: Mater. Sci. Process.*, **89**, 145–148.
18. Jeon, B.G., Jung, B.J., Kong, H.J., and Cho, Y.H. (2015) *Appl. Opt.*, **54**, 7323–7329.
19. Dubosclard-Gottardi, C., Caubere, P., and Fort, Y. (1995) *Tetrahedron*, **51**, 2561–2572.
20. Sartomer Arkema Group www.sartomer.com. (accessed 15 September 2015).
21. Klein, S., Barsella, A., Taupier, G., Stortz, V., Fort, A., and Dorkenoo, K.D. (2006) *Appl. Surf. Sci.*, **252**, 4919–4922.
22. Farrer, R.A., LaFratta, C.N., Li, L., Praino, J., Naughton, M.J., Saleh, B.E.A., Teich, M.C., and Fourkas, J.T. (2006) *J. Am. Chem. Soc.*, **128**, 1796–1797.
23. Li, L., Gershgoren, E., Kumi, G., Chen, W.-Y., Ho, P.T., Herman, W.N., and Fourkas, J.T. (2008) *Adv. Mater.*, **20**, 3668–3671.
24. Haske, W., Chen, V.W., Hales, J.M., Dong, W., Barlow, S., Marder, S.R., and Perry, J.W. (2007) *Opt. Express*, **15**, 3426–3436.
25. Dorkenoo, K., Klein, S., Bombenger, J.-P., Barsella, A., Mager, L., and Fort, A. (2006) *Mol. Cryst. Liq. Cryst.*, **446**, 151–160.
26. Wang, J., Ward, M.D., Ebersole, R.C., and Foss, R.P. (1993) *Anal. Chem.*, **65**, 2553–2562.
27. Ishihara, J., Komatsu, K., Sugihara, O., and Kaino, T. (2007) *Appl. Phys. Lett.*, **90**, 033511/033511–033511/033513.
28. de Miguel, G., Duocastella, M., Vicidomini, G., and Diaspro, A. (2015) *Opt. Express*, **23**, 24850–24858.
29. Lange, B., Jhaveri, S.J., Steidl, L., Ayothi, R., Ober, C.K., and Zentel, R. (2007) *Macromol. Rapid Commun.*, **28**, 922–926.
30. Steidl, L., Jhaveri, S.J., Ayothi, R., Sha, J., McMullen, J.D., Ng, S.Y.C., Zipfel, W.R., Zentel, R., and Ober, C.K. (2009) *J. Mater. Chem.*, **19**, 505–513.
31. Rinne, S.A., Garcia-Santamaria, F., and Braun, P.V. (2008) *Natl. Photonics*, **2**, 52–56.
32. Ramanan, V., Nelson, E., Brzezinski, A., Braun, P.V., and Wiltzius, P. (2008) *Appl. Phys. Lett.*, **92**, 173304/173301–173304/173303.
33. Pruzinsky, S.A. and Braun, P.V. (2005) *Adv. Funct. Mater.*, **15**, 1995–2004.
34. Nelson, E.C., Garcia-Santamaria, F., and Braun, P.V. (2008) *Adv. Funct. Mater.*, **18**, 1983–1989.
35. Nelson, E.C. and Braun, P.V. (2007) *Los Alamos Nat. Lab. Prepr. Arch. Condens. Matter*, **0710**, 0851, 1–6.
36. Krivec, S., Matsko, N., Satzinger, V., Pucher, N., Galler, N., Koch, T., Schmidt, V., Grogger, W., Liska, R., and Lichtenegger, H. (2010) *Adv. Funct. Mater.*, **20**, 811–819.
37. Pucher, N., Rosspeintner, A., Satzinger, V., Schmidt, V., Gescheidt, G., Stampfl, J., and Liska, R. (2009) *Macromolecules*, **42**, 6519–6528.
38. Infuehr, R., Pucher, N., Heller, C., Lichtenegger, H., Liska, R., Schmidt, V., Kuna, L., Haase, A., and Stampfl, J. (2007) *Appl. Surf. Sci.*, **254**, 836–840.

39. Chandross, E.A., Pryde, C.A., Tomlinson, W.J., and Weber, H.P. (1974) *Appl. Phys. Lett.*, **24**, 72–74.
40. Kumpfmueller, J., Stadlmann, K., Satzinger, V., Li, Z., Stampfl, J., and Liska, R. (2011) *J. Laser Micro/Nanoeng.*, **6**, 195–198.
41. Kumpfmueller, J., Stadlmann, K., Li, Z., Satzinger, V., Stampfl, J., and Liska, R. (2012) *MRS Online Proc. Lib.*, **1438**, 01–08.
42. Bichler, S., Feldbacher, S., Woods, R., Satzinger, V., Schmidt, V., Jakopic, G., Langer, G., and Kern, W. (2009) *Proc. SPIE*, **7413**, 74130W/1–74130W/11.
43. Woods, R., Feldbacher, S., Zidar, D., Langer, G., Satzinger, V., Schmidt, V., Pucher, N., Liska, R., and Kern, W. (2014) *Opt. Mater. Express*, **4**, 486–498.
44. Hoyle, C.E., Lowe, A.B., and Bowman, C.N. (2010) *Chem. Soc. Rev.*, **39**, 1355–1387.
45. Hoyle, C.E., Lee, T.Y., and Roper, T. (2004) *J. Polym. Sci., Part A: Polym. Chem.*, **42**, 5301–5338.
46. Hoyle, C.E. and Bowman, C.N. (2010) *Angew. Chem. Int. Ed.*, **49**, 1540–1573.
47. Gong, X., Wen, W., and Sheng, P. (2009) *Langmuir*, **25**, 7072–7077.
48. Norland Products. www.norlandprod.com (accessed 15 September 2015).
49. Joshi, M.P., Pudavar, H.E., Swiatkiewicz, J., Prasad, P.N., and Reinhardt, B.A. (1999) *Appl. Phys. Lett.*, **74**, 170–172.
50. Chen, Q.-D., Wu, D., Niu, L.-G., Wang, J., Lin, X.-F., Xia, H., and Sun, H.-B. (2007) *Appl. Phys. Lett.*, **91**, 171105/171101–171105/171103.
51. Kirkpatrick, S.M., Baur, J.W., Clark, C.M., Denny, L.R., Tomlin, D.W., Reinhardt, B.R., Kannan, R., and Stone, M.O. (1999) *Appl. Phys. A: Mater. Sci. Process.*, **69**, 461–464.
52. Cojoc, G., Liberale, C., Candeloro, P., Gentile, F., Das, G., De Angelis, F., and Di Fabrizio, E. (2010) *Microelectron. Eng.*, **87**, 876–879.
53. Galajda, P. and Ormos, P. (2001) *Appl. Phys. Lett.*, **78**, 249–251.
54. Asavei, T.A., Nieminen, T.A., Heckenberg, N.R., and Rubinsztein-Dunlop, H. (2009) *J. Opt. A*, **11**, 034001/1–034001/7.
55. Kumpfmueller, J., Stadlmann, K., Li, Z., Satzinger, V., Stampfl, J., and Liska, R. (2013) *Des. Monomers Polym.*, **17**, 390–400.
56. Kumpfmueller, J. (2013) Fabrication of flexible optical waveguides by means of two-photon-induced thiol-ene polymerization. PhD thesis. Vienna University of Technology.
57. Griesser, T., Wolfberger, A., Daschiel, U., Schmidt, V., Fian, A., Jerrar, A., Teichert, C., and Kern, W. (2013) *Polym. Chem.*, **4**, 1708–1714.
58. Sanchez, C., Julian, B., Belleville, P., and Popall, M. (2005) *J. Mater. Chem.*, **15**, 3559–3592.
59. Houbertz, R., Domann, G., Cronauer, C., Schmitt, A., Martin, H., Park, J.U., Frohlich, L., Buestrich, R., Popall, M., Streppel, U., Dannberg, P., Wachter, C., and Brauer, A. (2003) *Thin Solid Films*, **442**, 194–200.
60. Buestrich, R., Kahlenberg, F., Popall, M., Dannberg, P., Muller-Fiedler, R., and Rosch, O. (2001) *J. Sol-Gel Sci. Technol.*, **20**, 181–186.
61. Ovsianikov, A., Passinger, S., Houbertz, R., and Chichkov, B.N. (2007) in *Laser Ablation and its Applications*, Springer Series in Optical Sciences, vol. **129** (ed. C. Phipps), Springer, Berlin, pp. 121–157.
62. Obi, S. (2006) Replicated Optical Microstructures in Hybrid Polymers: Process Technology and Applications. PhD thesis. University of Neuchâtel.
63. ORMOCER®s www.ormocere.de (accessed 15 September 2015).
64. Micro Resist Technology www.microresist.de (accessed 15 September 2015).
65. Houbertz, R., Satzinger, V., Schmid, V., Leeb, W., and Langer, G. (2008) SPIE Photonics 2008 in San Diego, CA, Symposium-OP08P.
66. Woggon, T., Kleiner, T., Punke, M., and Lemmer, U. (2009) *Opt. Express*, **17**, 2500–2507.
67. Reinhardt, C., Kiyan, R., Passinger, S., Stepanov, A.L., Ostendorf, A., and Chichkov, B.N. (2007) *Appl. Phys. A: Mater. Sci. Process.*, **89**, 321–325.

68. Wohlleben, W., Bartels, F.W., Boyle, M., and Leyrer, R.J. (2008) *Langmuir*, **24**, 5627–5635.
69. Houbertz, R. (2005) *Appl. Surf. Sci.*, **247**, 504–512.
70. Declerck, P., Houbertz, R., Jakopic, G., Passinger, S., and Chichkov, B. (2007) *Mater. Res. Soc. Symp. Proc.*, **1007**, 1007-S1001-1002.
71. Li, J., Jia, B., and Gu, M. (2008) *Opt. Express*, **16**, 20073–20080.
72. Boyle, M., Neumeister, A., Kiyan, R., Reinhardt, C., Stute, U., Chichkov, B., Wohlleben, W., and Leyrer, R.J. (2007) *Proc. SPIE-Int. Soc. Opt. Eng.*, **6462**, 646212/646211–646212/646219.
73. Houbertz, R., Declerck, P., Passinger, S., Ovsianikov, A., Serbin, J., and Chichkov, B.N. (2007) *Phys. Status Solidi A: Appl. Mater. Sci.*, **204**, 3662–3675.
74. Serbin, J., Egbert, A., Ostendorf, A., Chichkov, B.N., Houbertz, R., Domann, G., Schulz, J., Cronauer, C., Froehlich, L., and Popall, M. (2003) *Opt. Lett.*, **28**, 301–303.
75. Bhuian, B., Winfield, R.J., O'Brien, S., and Crean, G.M. (2006) *Appl. Surf. Sci.*, **252**, 4845–4849.
76. Ovsianikov, A., Shizhou, X., Farsari, M., Vamvakaki, M., Fotakis, C., and Chichkov, B.N. (2009) *Opt. Express*, **17**, 2143–2148.
77. Ovsianikov, A., Gaidukeviciute, A., Chichkov, B.N., Oubaha, M., MacCraith, B.D., Sakellari, I., Giakoumaki, A., Gray, D., Vamvakaki, M., Farsari, M., and Fotakis, C. (2008) *Laser Chem.*, **2008**, 493059/493051–493059/493057.
78. Ovsianikov, A., Viertl, J., Chichkov, B., Oubaha, M., MacCraith, B., Sakellari, I., Giakoumaki, A., Gray, D., Vamvakaki, M., Farsari, M., and Fotakis, C. (2008) *ACS Nano*, **2**, 2257–2262.
79. Emons, M., Obata, K., Binhammer, T., Ovsianikov, A., Chichkov, B.N., and Morgner, U. (2012) *Opt. Mater. Express*, **2**, 942–947.
80. Crivello, J.V. (1999) *J. Polym. Sci., Part A: Polym. Chem.*, **37**, 4241–4254.
81. Somos® (2015) *Stereolithography Materials Selector Guide*, DSM Desotech, Elgin, IL.
82. Lee, K.Y., LaBianca, N., Rishton, S.A., Zolghamain, S., Gelorme, J.D., Shaw, J., and Chang, T.H.P. (1995) *J. Vac. Sci. Technol., B*, **13**, 3012–3016.
83. Shaw, M., Nawrocki, D., Hurditch, R., and Johnson, D. (2003) *Microsyst. Technol.*, **10**, 1–6.
84. Genolet, G. (2001) New photoplastic fabrication techniques and devices based on high aspect ratio photoresist. PhD thesis. Ecole Polytechnique Federale de Lausanne.
85. Serbin, J., Ovsianikov, A., and Chichkov, B. (2004) *Opt. Express*, **12**, 5221–5228.
86. Moon, J.H. and Yang, S. (2010) *Chem. Rev.*, **110**, 547–574.
87. Teh, W.H., Duerig, U., Drechsler, U., Smith, C.G., and Guentherodt, H.J. (2005) *J. Appl. Phys.*, **97**, 054907/054901–054907/054911.
88. Juodkazis, S., Mizeikis, V., and Misawa, H. (2009) *J. Appl. Phys.*, **106**, 051101/051101–051101/051114.
89. Seet, K.K., Juodkazis, S., Jarutis, V., and Misawa, H. (2006) *Appl. Phys. Lett.*, **89**, 024106/024101–024106/024103.
90. Kuebler, S.M., Braun, K.L., Zhou, W., Cammack, J.K., Yu, T., Ober, C.K., Marder, S.R., and Perry, J.W. (2003) *J. Photochem. Photobiol., A*, **158**, 163–170.
91. Deubel, M., von Freymann, G., Wegener, M., Pereira, S., Busch, K., and Soukoulis, C.M. (2004) *Nat. Mater.*, **3**, 444–447.
92. Seet, K.K., Mizeikis, V., Kannari, K., Juodkazis, S., Misawa, H., Tetreault, N., and John, S. (2008) *IEEE J. Sel. Top. Quantum Electron.*, **14**, 1064–1073.
93. Wu, D., Chen, Q.-D., Niu, L.-G., Jiao, J., Xia, H., Song, J.-F., and Sun, H.-B. (2009) *IEEE Photonics Technol. Lett.*, **21**, 1535–1537.
94. Chanda, D., Abolghasemi Ladan, E., Haque, M., Ng Mi, L., and Herman Peter, R. (2008) *Opt. Express*, **16**, 15402–15414.
95. Scrimgeour, J., Sharp, D.N., Blanford, C.F., Roche, O.M., Denning, R.G., and Turberfield, A.J. (2006) *Adv. Mater.*, **18**, 1557–1560.

96. Lee, T.-w., Mitrofanov, O., White, C.A., and Hsu, J.W.P. (2003) *Mater. Res. Soc. Symp. Proc.*, **776**, 143–148.

97. Belfield, K.D., Liu, Y., Negres, R.A., Fan, M., Pan, G., Hagan, D.J., and Hernandez, F.E. (2002) *Chem. Mater.*, **14**, 3663–3667.

98. Corredor, C.C., Huang, Z.-L., and Belfield, K.D. (2006) *Adv. Mater.*, **18**, 2910–2914.

99. Dvornikov, A.S., Walker, E.P., and Rentzepis, P.M. (2009) *J. Phys. Chem. A*, **113**, 13633–13644.

100. Parthenopoulos, D.A. and Rentzepis, P.M. (1989) *Science*, **245**, 843–845.

101. Zhang, Z., Hu, Y., Luo, Y., Zhang, Q., Huang, W., and Zou, G. (2009) *Opt. Commun.*, **282**, 3282–3285.

102. Polyzos, I., Tsigaridas, G., Fakis, M., Giannetas, V., Persephonis, P., and Mikroyannidis, J. (2003) *Chem. Phys. Lett.*, **369**, 264–268.

103. Gu, M. and Day, D. (1999) *Opt. Lett.*, **24**, 288–290.

104. Day, D. and Gu, M. (1998) *Appl. Opt.*, **37**, 6299–6304.

105. Yanez, C.O., Andrade, C.D., Yao, S., Luchita, G., Bondar, M.V., and Belfield, K.D. (2009) *ACS Appl. Mater. Interfaces*, **1**, 2219–2229.

106. Dvornikov, A.S. and Rentzepis, P.M. (1997) *Opt. Commun.*, **136**, 1–6.

107. Xia, H., Zhang, W.-Y., Wang, F.-F., Wu, D., Liu, X.-W., Chen, L., Chen, Q.-D., Ma, Y.-G., and Sun, H.-B. (2009) *Appl. Phys. Lett.*, **95**, 083118/083111–083118/083113.

108. Ambrosio, A., Orabona, E., Maddalena, P., Camposeo, A., Polo, M., Neves, A.A.R., Pisignano, D., Carella, A., Borbone, F., and Roviello, A. (2009) *Appl. Phys. Lett.*, **94**, 011115/011111–011115/011113.

109. Daschiel, U., Hoefler, T., Jakopic, G., Schmidt, V., and Kern, W. (2007) *Macromol. Chem. Phys.*, **208**, 1190–1201.

110. Daschiel, U., Fuehrer, R., Bauer, T., Jakopic, G., Schmidt, V., Satzinger, V., Riess, G., and Kern, W. (2009) *Proc. SPIE*, **7413**, 74130V/74131–74130V/74111.

11
Fabricating Nano and Microstructures Made by Narrow Bandgap Semiconductors and Metals using Multiphoton Lithography

Min Gu, Zongsong Gan, and Yaoyu Cao

11.1
Introduction[1)]

Multiphoton lithography has become a widely implemented and enabling technique for the fabrication of three-dimensional (3D) complex structures at the micrometer or nanometer scale [1–6]. It has the exceptional advantages of noninvasive and maskless features based on a nonlinear process induced by a focused laser beam. A number of important advances made in optics and photonics are based on this unique technique. This technique has enabled a wide range of applications [1–8] from photonic crystals [1] to metamaterials [2], and from micromechanics [3] to all-on-chip optical processing [7] in the recent decade.

To fabricate functional nano and microstructures with unique new features via multiphoton lithography technique, different materials have been developed with intrinsic properties including high refractive index, high nonlinearity [9, 10], and high mechanical strength [6, 11]. However, as conventional multiphoton lithography is based on the top-down approach, the materials suitable for multiphoton lithography are mainly wide electronic bandgap materials, which can be categorized into inorganic material such as As_2S_3 chalcogenide glass [9, 10] and $LiNbO_3$ crystal [12], and organic photoresins such as SU8 [4], Ormocer [5], and IP-L 780 (Nanoscribe, GmbH) [6] which absorb the incident laser energy via a two-photon or multiphoton process.

When multiphoton lithography is applied to fabricate 3D structures inside narrow electronic bandgap semiconductors via the top-down approach, it encounters the problem of single-photon absorption, which disables the 3D

1) Reproduced with permission from $\lambda/26$ silver nanodots fabricated by direct laser writing through highly sensitive two-photon photoreduction, Cao, Yaoyu and Gu, Min, *Applied Physics Letters*, **103**, 213104 (2013), DOI: http://dx.doi.org/10.1063/1.4830221. Copyright 2013, AIP Publishing LLC; and from Zongsong Gan, Yaoyu Cao and Min Gu, Direct laser writing of three-dimensional narrow bandgap and high refractive-index PbSe structures in a solution, *Opt. Express* **21**, 11202–11208 (2013), DOI: 10.1364/OE.21.011202.

Multiphoton Lithography: Techniques, Materials and Applications, First Edition.
Edited by Jürgen Stampfl, Robert Liska, and Aleksandr Ovsianikov.

fabrication ability. Although a long-wavelength lithography laser can be used for narrow electronic bandgap semiconductors, it obviously lowers the lithography resolution. In addition, compared with wide electronic bandgap materials, narrow electronic bandgap semiconductors normally have much higher refractive indices [13]. This also challenges multiphoton lithography technique with narrow electronic bandgap semiconductors because of the refractive index mismatch. For metals, it is fundamentally impossible to fabricate 3D nano and microstructures in metals using multiphoton lithography technique with the top-down approach because a laser beam cannot be focused inside a metal.

Even though narrow electronic bandgap semiconductors and metals have unique optical and electrical properties that are highly desired for a wide range of applications, the fabrication of their 3D structures by multiphoton lithography technique has not been well demonstrated. In this chapter, we show the fabrication of 3D structures made of lead selenide (PbSe) as a typical example of the narrow electronic bandgap semiconductors and silver as a typical example of metals by conducting multiphoton lithography technique in a solution based on the bottom-up approach [14]. This avoids the problems arising from the refractive index mismatch and single-photon absorption within the top-down fabrication approach for narrow bandgap semiconductors and also the problem of focusing laser into a metal in the top-down fabrication approach [15].

11.2 Fabrication of 3D Structures Made by PbSe with Multiphoton Lithography

As a typical narrow electronic bandgap semiconductor, bulk rock-salt PbSe has a bandgap as narrow as 0.27 eV. It has a refractive index as high as 4.82 at 6 μm [13], which is the largest refractive index value at 6 μm among the current materials used in the multiphoton lithography technique. The third-order nonlinear refractive index of PbSe is 10^{-13} cm^2 W^{-1} [16], which is about one order of magnitude larger than that of As_2S_3 chalcogenide glass [17]. In addition, PbSe exhibits unique optical and electrical properties such as multi-exciton generation [18], high carrier mobility, and the unique thermoelectric property with a thermoelectric figure of merit (ZT) value larger than 1 [19], which makes it a highly promising material applicable in photonics, nanoelectronics, and nano-thermoelectrics.

11.2.1 Challenges of Multiphoton Lithography with Top-Down Approach for Narrow Electronic Bandgap Semiconductors

The challenges of multiphoton lithography with top-down approach to get metal 3D nano and microstructures are obvious because light cannot be focused inside a metal. But the challenges of multiphoton lithography with the top-down approach to get 3D nano and microstructures made by narrow electronic bandgap materials are not straightforward. The main challenges of 3D fabrication with multiphoton lithography in bulk semiconductor PbSe for the top-down strategy come from the narrow electronic bandgap and the high refractive index of PbSe.

Two-photon lithography is performed in a material that absorbs the excitation photon via a two-photon process. However, two-photon lithography in bulk PbSe requires the excitation wavelength to be longer than 5 μm, as the bandgap of bulk PbSe is 0.27 eV. This excitation wavelength of longer than 5 μm does not facilitate the manipulation of the multiphoton lithography system and is the hurdle to sub-micrometer lithography resolution. Furthermore, the refractive index value of 4.82 is also a serious problem for 3D multiphoton lithography in bulk PbSe. As multiphoton lithography uses a high numerical aperture (NA) objective lens to focus the excitation laser beam into the fabricated material, the refractive index of the objective immersion oil (normally about 1.52 for a 1.4 NA objective) is much smaller than 4.82. For 3D multiphoton lithography in a material with refractive index much larger than that of the immersion oil, when the excitation focal spot is moved deeply into the material, the excitation focal spot is more defocused, more elongated, and more distorted [20]. This hinders a symmetric, high-resolution, and depth-unaffected fabrication, which is clearly shown in Figure 11.1a. The large refractive index mismatch requires strong wavefront aberration compensation, which is a big challenge for 3D multiphoton lithography in bulk PbSe.

11.2.2 Photoresin Development

As the top-down strategy encounters the problems in narrow bandgap and high refractive index materials, the bottom-up strategy is utilized to fabricate 3D PbSe nano and microstructures by two-photon lithography, the basic concept of which is as follows: First, the basic chemical components to form PbSe need to be put into the fabrication photoresin beforehand. Then, under the irradiation of the fabrication laser beam, the basic chemical components in the photoresin can produce PbSe to form the designed PbSe micro/nanostructures. To avoid any significant refractive index mismatch, the developed photoresin should have a low refractive index (Figure 11.1b). To enable 3D fabrication, the photoresin is required to have no single-photon absorption but should have two-photon absorption at the excitation wavelength.

The photoresin for fabrication was prepared by dissolving Pb in the form of lead oleate (PbOA) and Se in the form of TOPSe (Se dissolved in trioctylphosphine) in the organic solvent 1-octadecene (ODE). This organic solution had a high solubility of PbOA and TOPSe, which allowed the formation of PbSe with sufficient amount of elemental Pb and Se in the local position of the excitation laser beam focal spot. The refractive index of ODE is 1.44. After dissolving of PbOA and TOPSe, the final mixture had a refractive index of ~1.50, which served as a low refractive index platform for the two-photon lithography of PbSe without aberration compensation. ODE is a colorless solution and has no single-photon absorption at the laser wavelength range 500–800 nm. This provided the possibility of fabricating 3D PbSe structures in ODE. TOPSe in ODE was also a colorless

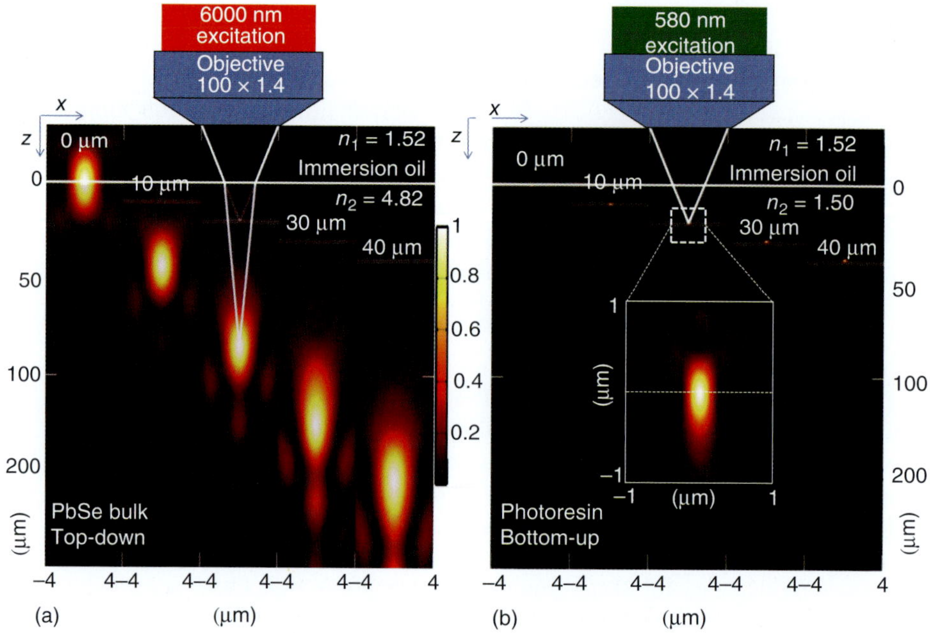

Figure 11.1 A comparison of the top-down strategy and the bottom-up strategy for the fabrication of PbSe material by two-photon lithography. From left to right, the excitation focal spot was moved deeper with an increment of 10 μm. (a) Top-down strategy in PbSe bulk material. (b) Bottom-up strategy in a photoresin. The inset shows a detailed view of the focal spot. (Gan 2013 [15]. Reproduced with permission of the Optical Society of America.)

solution. Pure PbOA was a wax-like solid. After dissolving PbOA and TOPSe in ODE, the final photoresin was a transparent and colorless solution (Figure 11.2a).

For synthesizing PbOA as the precursor of the fabrication photoresin with maximum dissolved of Pb elements, a mixture of PbO (0.506 g), oleic acid (1.280 g), and technological-grade ODE (2.300 g) was stirred and heated to 120 °C for 1 h under vacuum. After that, the mixture was kept under nitrogen gas and the temperature of the mixture was allowed to increase to 180 °C. The mixture was kept at this temperature for 15 min in a three-necked flask under nitrogen gas. Then, the mixture was quickly cooled down to the room temperature under the protection of nitrogen gas. The mixture of PbOA in ODE was centrifuged to remove the small amount of undissolved PbO. The obtained solution after the centrifugation was PbOA dissolved in ODE, which was a colorless solution. The concentration of PbOA was ~0.5 mol l^{-1}. The TOPSe solution was prepared by dissolving an excess amount of selenium (0.358 g) in trioctylphosphine (3.220 g). The mixture of Se and TOP (trioctylphosphine) was centrifuged to remove the undissolved Se. The concentration of TOPSe was ~1.0 mol l^{-1}. At room temperature, the Pb mixture and the Se mixture were mixed together at a volume ratio 1 : 1, which corresponds to a mole ratio of 1 : 2. This mole ratio (rather than 1 : 1) contributed to the fast

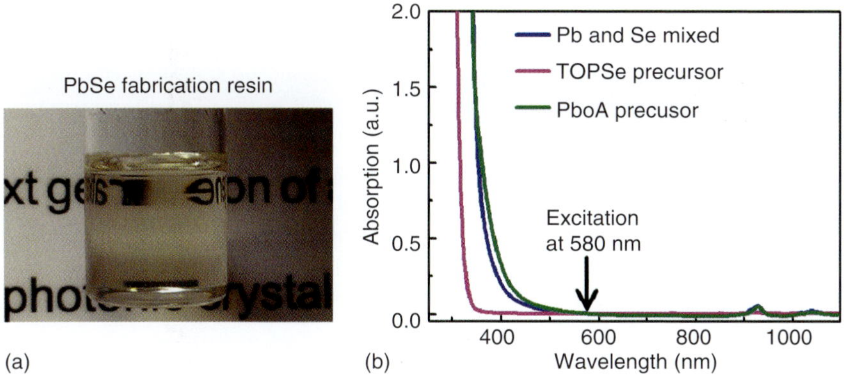

Figure 11.2 Synthesized photoresin for the PbSe material fabrication based on the bottom-up strategy via two-photon lithography. (a) The photograph shows that the photoresin is transparent and colorless. (b) The absorption spectrum of the photoresin, TOPSe precursor, and PbOA precursor. (Gan 2013 [15]. Reproduced with permission of the Optical Society of America.)

formation of PbSe with high chemical efficiency [21]. To ensure complete mixing of these two precursors, the final mixture was vigorously shaken in a sealed glass vial. The final mixture was used as the fabrication photoresin.

Figure 11.2b shows the absorption spectrum of the photoresin, TOPSe precursor, and PbOA precursor. The absorption spectra of both PbOA and TOPSe precursors in Figure 11.2b show an absorption band edge at 410 and 340 nm, respectively. The photoresin shows an absorption band edge at 400 nm, which confirms no absorption in the wavelength range 500–800 nm. A laser beam with a wavelength of 580 nm was used as the lithography excitation source for two-photon absorption. Under excitation with the laser, the photoresin can absorb 580 nm light via two-photon absorption and trigger the formation of PbSe within the focal spot volume.

The sample for fabrication was prepared by sandwiching a drop of the photoresin between two cover slips separated by about 60 μm with thick sticky tapes. The sample was then used for the two-photon lithography. A femtosecond pulsed beam of wavelength 800 nm, generated from a Ti:sapphire laser (Spectra Physics Mai Tai), was set to go through an optical parametric oscillator (Coherent Mira (OPO) optical parametric oscillator). The output beam with a wavelength of 580 nm, a repetition rate of 76 MHz, and a pulse duration of 200 fs was first expanded and then focused into the photoresin with a high-NA oil immersion objective lens (Olympus, NA = 1.4, 100×) to induce PbSe formation. The sandwiched sample was affixed to a 200 × 200 × 200 μm *xyz* piezoelectric scanning stage (P562, Physik Instrumente) for fabrication. A mechanical shutter, synchronized with the scanning stage, was used to control the on and off status of the laser exposure. Both the mechanical shutter and the scanning stage were controlled by a computer program with the designed structure. The fabrication process was monitored in real time by a charge-coupled device (CCD) camera.

A red lamp was used as an illumination source for the CCD camera. The 3D structure was fabricated on the top surface of the bottom cover slip by controlling the movement of the scanning stage. After fabrication, the sample was annealed at 120 °C for 5 min. The cover slip without the stick of the fabricated structure was mechanically taken off with care, and then the whole bottom cover slip with the fabricated structure attached was immersed into a pure toluene solution for 5 min to remove the unsolidified photoresin and to reveal the fabricated structure. The fabricated structure was finally dried in air at room temperature.

The fabricated structures were measured with a ZEISS Supra 40 VP field-emission scanning electron microscope (SEM). Before measurement, the sample was coated with a gold layer of thickness 2–3 nm.

11.2.3
Two-Photon Lithography of PbSe Structures

At room temperature, the mixed PbOA and TOPSe in ODE can form PbSe at a slow reaction rate. The formed PbSe has a dark color, therefore the color of the mixture changes from colorless to dark brown. This is due to the formation of PbSe nanoparticles in the photoresin. However, as the reaction rate is slow, this can be observed only 12 h after the mixing of PbOA and TOPSe in ODE, which leaves a long enough time interval to use two-photon lithography in this photoresin.

Under the irradiation of a femtosecond pulsed laser beam with wavelength 580 nm, PbSe can be formed instantly. This laser-induced instant formation of PbSe in the photoresin enabled the fabrication of PbSe structures in a solution and facilitated the successful fabrication of narrow bandgap semiconductor materials by two-photon lithography. By moving the relative position of the photoresin with a scanning stage, two-dimensional structures could be fabricated. A typical image obtained from the CCD shows the fabrication of lines with dark color (Figure 11.3a). The significant refractive index change at the position of the fabricated lines indicates the formation of the high-refractive-index material. The experimentally achieved line width in Figure 11.3b is about 900 nm, which confirms the sub-micrometer fabrication resolution. To show the 3D fabrication ability, pillars with a height of $\sim$12 μm were fabricated by moving the scanning stage vertically away from the objective (Figure 11.3c).

We also fabricated 3D structures with different layers (Figure 11.3d). As we used a linearly polarized beam, lines at different layers fabricated along perpendicular directions (top layer and the second layer in Figure 11.3d) show significant line width differences (the top layer line width is about 900 nm, the second layer line width is about 500 nm) because of the nonsymmetrical property of the laser beam focal spot. However, as can be seen from the SEM images (Figure 11.3b–d), the surface of the fabricated structure is rough. This originates from the fast reaction rate of PbOA with TOPSe under the irradiation of the femtosecond pulsed laser beam, which leads to the uncontrolled growth of PbSe within the volume of the focal spot. To make it smoother, the sample was annealed at 120 °C for 5 min after the fabrication and before washing. The structure after annealing (Figure 11.3e)

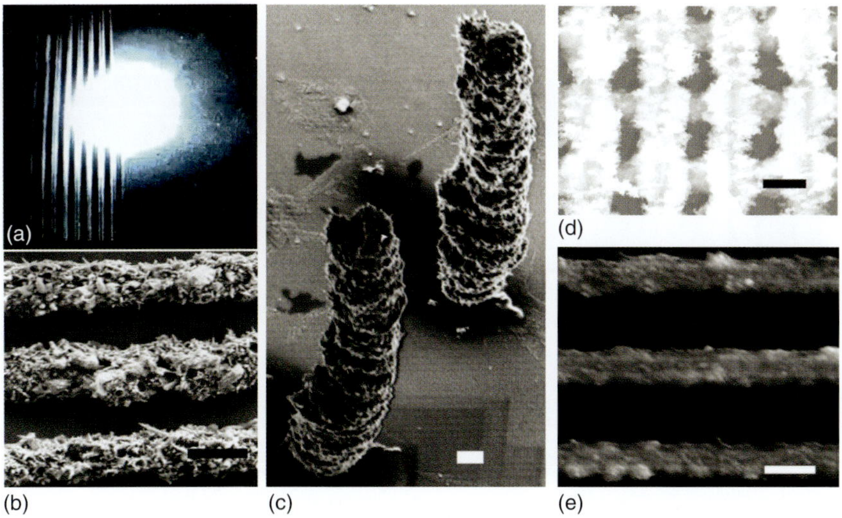

Figure 11.3 Fabrication of the PbSe structures by two-photon direct laser writing (DLW) in the photoresin. (a) A typical picture taken during the fabrication process from the CCD camera. (b) The SEM images of two-dimensional lines. (c) The SEM images of pillars with a height of about 12 μm. (d) The SEM images of 3D structure with different layers. (e) The SEM images of the two-dimensional structures after fabrication and annealing. All scale bars are 1000 nm. (Gan 2013 [15]. Reproduced with permission of the Optical Society of America.)

is much smoother, and the feature size of the line is reduced as well compared to Figure 11.3b, which shows an effective way to improve the quality of the fabricated structure.

11.2.4 Confirmation of PbSe Formation

It is important to confirm the formation of the PbSe structure by two-photon lithography. The energy-dispersive spectrum (EDS) of the fabricated structure was measured, which is shown in Figure 11.4a. From the EDS measurement, Pb and Se elements can be identified with an atomic ratio of Pb/Se = 1.07 : 1. This almost 1 : 1 ratio of Pb and Se confirms that the formed structure is mainly constituted with nearly equal amounts of Pb and Se elements. The slightly larger amount of Pb can be explained as a Pb-rich structure surface due to the low utilization rate of Se in the photoresin [22]. There are also tiny amounts of carbon and oxygen elements identified. These are residual organic molecules attached to the surface of the fabricated PbSe structure. To show the crystal structure of the fabricated PbSe structures, a high-resolution transmission electron microscopy (HRTEM) image is shown in Figure 11.4b. The image confirms the formation of PbSe with a rocksalt phase.

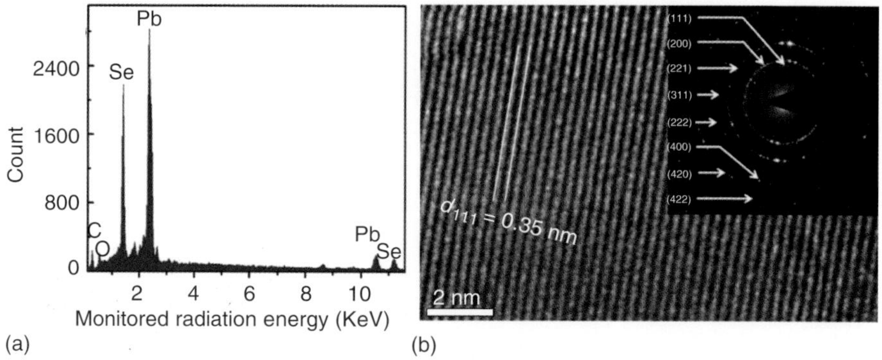

Figure 11.4 (a) Energy-dispersive spectrum of the fabricated PbSe structures. (b) HRTEM image of the fabricated PbSe structures. (Gan 2013 [15]. Reproduced with permission of the Optical Society of America.)

The EDS measurement was performed at a fabricated PbSe block at an area of $20 \times 20\,\mu m^2$ to reduce random error. Because Pb and Se are the main heavy elements of the fabricated structure, the atomic ratio measured by the EDS is quantitative. HRTEM was performed on a FEI Tecnai F30 transmission electron microscope.

11.3 Fabrication of Silver Structures with Multiphoton Lithography

Metals have unique properties compared to dielectric materials. Metals with large dielectric permittivity allow the fabrication of structures of high contrast in refractive index, which is beneficial to achieve a complete photonic bandgap [23]. Metals with surface plasmon resonance enable effective manipulation of light in the nanoscale, which contributes to miniature devices for photonic chip integration. Metals with higher light damage threshold can stand high temperatures for specified applications.

Bottom-up strategy is the only method currently used for multiphoton lithography to fabricate 3D metal nano and microstructures. Photoreduction provides a unique mechanism for multiphoton lithography to fabricate metal nano and microstructures. The mechanism is similar to PbSe fabrication in Section 11.2, but using photoreduction to change metal ions to metal. Two-photon photoreduction was first used to fabricate silver nanostructures in 2006 [24]. However, compared with the top-down strategy, there has been little work exploring ways to improve the lithography resolution for this bottom-up strategy. Using the super-resolution photo-induced inhibition nanolithography (SPIN) approach, multiphoton lithography technique based on bottom-up strategy can achieve feature size down to 9 nm and resolution about 50 nm [11, 24]. In comparison, the technique can only realize feature size and resolution for metal fabrication of more than 100 nm and several hundreds of nanometers, respectively, due to the diffraction limit of light

[25, 26]. For the PbSe structures mentioned in Section 11.2, the minimum feature size is realized at ∼500 nm. This is a significant limitation for the application of structures fabricated by multiphoton lithography and made of narrow electronic bandgap materials as well as metals. In this section, we will show the fabrication of silver nanodots as small as 22 nm by employing photoreduction-based nonlinear ultra-high photosensitivity to reduce the feature size.

11.3.1
Principle of Resolution Improvement by Increasing Photosensitivity in Photoreduction

Photosensitivity is an intrinsic property of a material to evaluate the utilization rate of light in fabricating a structure. Tuning photosensitivity can not only provide an energy-efficient way for the fabrication but also improve fabrication resolution. For photoreduction-based multiphoton fabrication, as the metal structure is realized through sequential three-step reactions, namely, nucleation, nuclei growth, and particle aggregation, the photoreduction photosensitivity differs from each other at different reaction step. Photoreduction of metal ions can show ultrahigh photosensitivity with a photon flux dependence, as the reduced metal enhances the light absorption and promotes the electron transfer from donors to acceptors. Consequently, this type of the photosensitivity can provide significant improvement in the resolution during fabrication.

For reducing the silver ions and producing a silver dot, the laser energy should exceed two thresholds (Figure 11.5). One is the nucleation threshold, and the other is set up by the formation of silver dots through the aggregation of growing silver nanoparticles accompanied by an accelerated photoreduction of silver ions (Figure 11.5a,b). The nucleation threshold determines the minimum laser energy to initiate the photoreduction reactions, while the laser energy below the threshold cannot reduce the silver ions to generate silver nuclei. Above the nucleation threshold, the silver nuclei require more laser energy to grow and aggregate to form a silver dot, which imposes a higher fabrication threshold. Meanwhile, the photosensitivity of the photoreduction solution dramatically increases with increasing the laser energy, as the nuclei increase the chemical reaction on their surface [26]. In this case, the minimum feature size of the silver dot can be realized by carefully tuning the laser energy to achieve the fabrication threshold (Figure 11.5d). The effective focal spot in terms of the size of the photoreduction area can be adjusted smaller than the focal spot of the laser beam. As the photosensitivity of the photoreduction solution increases, the size of the effective focal spot can be further squeezed (Figure 11.5a) at the threshold level of the laser energy.

11.3.2
Photosensitivity Enhancement by Tuning Laser Wavelength

Silver dots were fabricated with a standard two-photon laser writing setup [10, 27]. An ultrafast Chameleon OPO laser system (Coherent) with a repetition rate

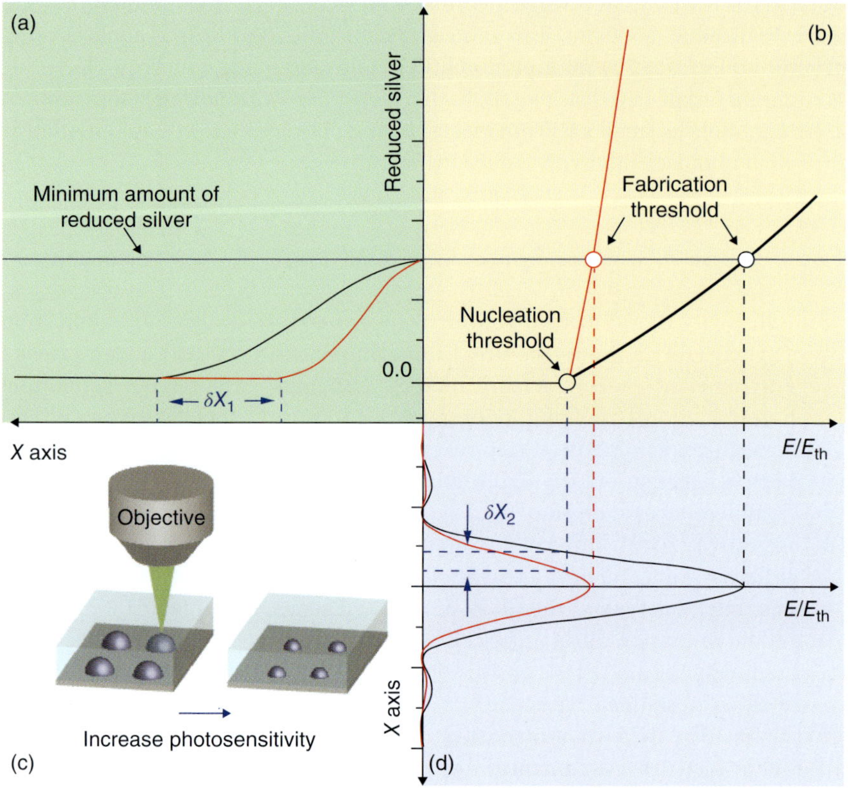

Figure 11.5 Schematic of the resolution improvement by increasing photosensitivity in the photoreduction process. (a) The distribution of the reduced silver dots. The red curve and the black curve are the distribution profiles of the reduced silver dot in case of high photosensitivity and low photosensitivity, respectively. δX_1 is the decreased size of the silver dot with increasing the photosensitivity. (b) The amount of reduced silver plotted as a function of the fabrication laser energy. E_{th} is the laser energy threshold required for the nucleation of silver ions. The red curve with a higher gradient indicates high photosensitivity for the photoreduction solution. The black curve with a lower gradient indicates low photosensitivity for the photoreduction solution. When the laser energy level is below the nucleation threshold, the photoreduction reaction cannot be initiated, which results in the reduced silver equal to zero. When the laser energy level is in between the nucleation threshold and the fabrication threshold, the reduced silver cannot form a silver dot. (c) Illustration of the direct laser writing with high photosensitivity and low photosensitivity, respectively, under the laser irradiation at the threshold level. (d) Distribution of the laser energy in the focal region of the fabrication laser beam. The red curve and the black curve are the energy distribution at the fabrication threshold levels in case of the high photosensitivity and the low photosensitivity, respectively. δX_2 is the decreased size of the exposure zone with the laser energy above the nucleation threshold as the photosensitivity increases. (Cao 2013 [14]. Reproduced with permission of American Institute of Physics.)

of 80 MHz and a pulse width of 140 fs was used as the light source. The laser beam passing through a telescope-based optical setup was focused onto the surface of a cover slip with an oil immersion objective (100×, NA = 1.4). The cover slip was fixed on a high-precision 3D piezo-scanning stage controlled by a computer. The photoreduction solution was formulated by mixing silver nitrate, ammonium, and the surfactant *n*-decanoylsarcosine sodium with the concentrations of 0.07, 0.035, and 0.08 M, respectively. To prepare the sample, the photoreduction solution was sandwiched between two cover slips and a sticky tape with the thickness of 45 μm as the spacer. After the fabrication, the sample was rinsed by deionized water and dried in the open air.

In practice, the photosensitivity for the photoreduction solution can be increased by tuning the wavelength of the femtosecond laser. As shown in Figure 11.6, the dot size increased from 200 to 290 nm as the wavelength was tuned from 620 to 580 nm with a fixed laser energy of 4 μJ. In this regard, the weight of the reduced silver dot, calculated based on its diameter (the weight: $w = \rho_{\text{silver}} \times 2/3 \times \pi \times (d/2)^3$, d is the diameter of the silver dot and ρ_{silver} is the silver density: $10.5\,\text{g cm}^{-3}$), increased by 3 times from 0.022×10^{-12} to 0.069×10^{-12} g with the blue shift of the laser wavelength and assuming the silver dot as a hemisphere [25]. As the wavelength of the laser beam is tuned to 800 nm, the fabrication threshold was >4 μJ (Figure 11.6c), which implies that more laser energy is required to form a silver dot at a longer wavelength. In addition, Figure 11.6b,c show the weight variation of the photoreduced silver dot as a function of the laser energy under exposure with laser beams of different wavelengths. While the wavelength is tuned from 800 to 580 nm, the weight of the reduced silver dot fabricated with the laser energy in the threshold region increased faster with increasing the laser energy, which suggests that the shorter wavelength rendered higher photosensitivity. Moreover, as the diffusion of the silver ions into the focal region sets up a limit for the photoreduction [24], for a given wavelength the photosensitivity drops with increasing laser energy. At the wavelength of 580 nm, the photosensitivity decreased from 0.025×10^{-12} to $0.010 \times 10^{-12}\,\text{g}\,\mu\text{J}^{-1}$, which is calculated by the gradient of the curve for the weight change of the photoreduced silver dot (the photosensitivity $\text{Ph} = [w(E_2) - w(E_1)]/(E_2 - E_1)$, where $w(E)$ is the weight of the silver dot produced with the laser energy of E) while the laser energy increased from 1 to 5 μJ.

As the fabrication with the laser energy in the threshold region results in higher resolution, the corresponding photosensitivity plays a critical role in producing small features. Figure 11.7 shows the variation of the photosensitivity for the photoreduction solution in the threshold region and the minimum feature size of the silver dots as the laser beam wavelength was tuned from 800 to 580 nm. The photosensitivity in the threshold region was calculated based on the gradients of the curves plotted in Figure 11.6 between the first two dots fabricated with low laser energies. As the wavelength is shifted from 800 to 580 nm, the photosensitivity increased from 0.006×10^{-12} to $0.025 \times 10^{-12}\,\text{g}\,\mu\text{J}^{-1}$ (Figure 11.7a). Eventually, the increase of the photosensitivity results in the feature size of the silver dot decreasing from 220 to 130 nm (Figure 11.7b), which implies improved resolution for

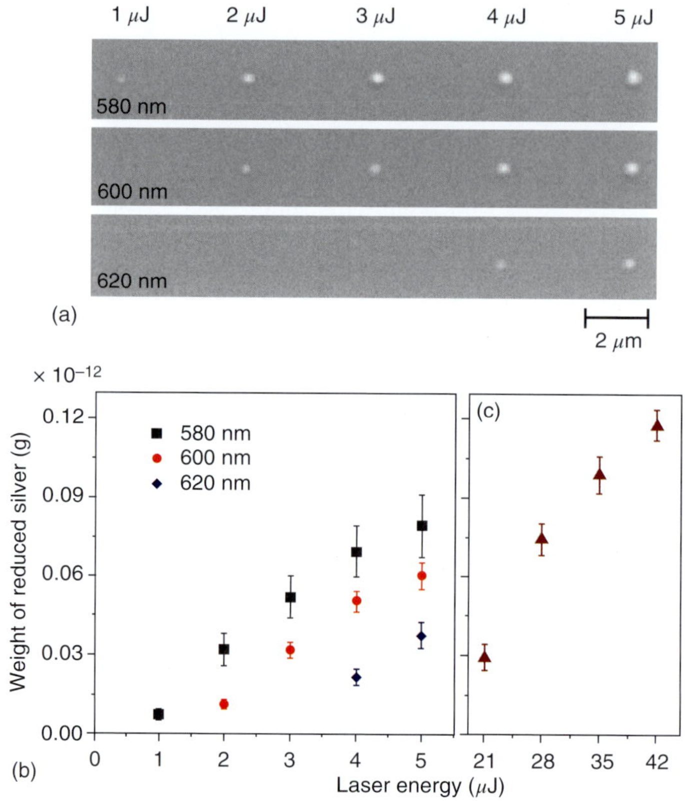

Figure 11.6 (a) SEM images of the silver dots fabricated with different laser energy levels under the exposure of the fabrication laser beam at different wavelengths of 580, 600, and 620 nm, respectively. The scale bar is 2 μm. (b) Weight of the reduced silver dots as a function of the laser energy under the exposure of the fabrication laser beam at different wavelengths of 580, 600, and 620 nm, respectively. (c) Weight of the reduced silver dots as a function of the laser energy under the exposure of the fabrication laser beam at the wavelengths of 800 nm. (Cao 2013 [14]. Reproduced with permission of American Institute of Physics.)

fabrication, while the laser energies used in the fabrication are 7 μJ for 800 nm laser beam, 4 μJ for 620 nm laser beam, 2 μJ for 600 nm laser beam, and 1 μJ for 580 nm laser beam, respectively.

11.3.3
Dot Size Model Based on Photosensitivity

While the photosensitivity significantly affects the silver dot size, the smallest dot size can be predicted by a model based on the distribution of photon flux for the photoreduction process. For a focused laser beam with a Gaussian distribution, the photon flux $N(r, t)$ at a point at a distance r to the center position of the beam

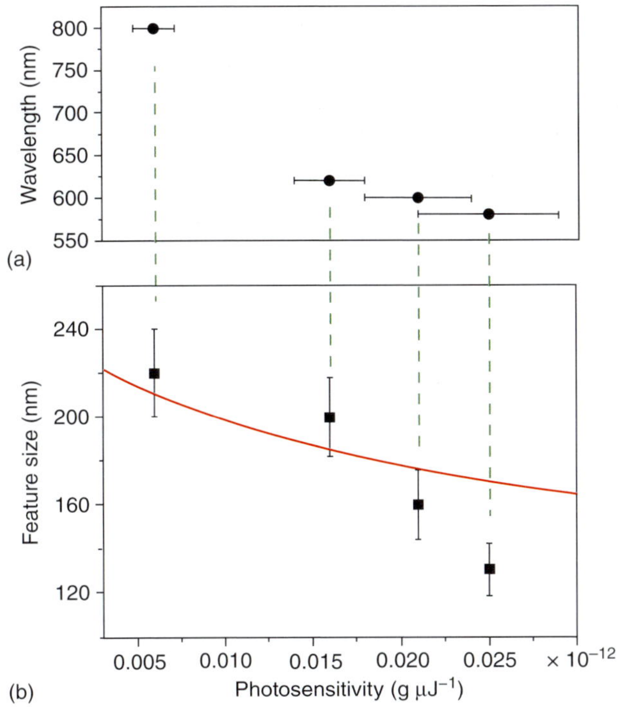

Figure 11.7 (a) Photosensitivity for the photoreduction solution as a function of the wavelength of the fabrication laser beam with laser energy at the fabrication threshold level. (b) Minimum feature size of the silver dot fabricated as a function of the photosensitivity. The red curve is the simulation result. (Cao 2013 [14]. Reproduced with permission of American Institute of Physics.)

in the focal plane ($z = 0$) is $N(r,t) = N_0(t)\exp(-2r^2/{r_0}^2)$, where $N_0(t)$ is the photon flux at the center of the beam, r is the distance of the position to the center of the beam, and r_0 is the waist of the laser beam. For a given exposure time, we can assume that the amount of the photoreduced silver at the position r is $S = k_{\text{eff}} \times \sigma_{\text{eff}} N^2(r,t)$, where $\sigma_{\text{eff}} N^2(r,t)$ is the two-photon absorbance of the photoreduction solution,σ_{eff} is the effective absorbance cross section, and k_{eff} represents the photosensitivity, when the photo flux is above the nucleation threshold [28].

For the photon flux at the fabrication threshold level, which represents the minimum photon flux required for fabricating silver dots, the size of the reduced silver dot (size $d = 2r$), based on the photon flux distribution, is given by $d = \sqrt{2}r_0\sqrt{\ln(1 + \sqrt{S_{\text{th}}/(k_{\text{eff}} \times \sigma_{\text{eff}})}{N_{\text{th}}}^{-1})}$, where S_{th} represents the minimum amount of the reduced silver (in Figure 11.1) required for the producing a silver dot, and N_{th} is the nucleation threshold for the photon flux. Thus, the diameter of the silver dot depending on the photosensitivity can be expressed as: $d \propto r_0\sqrt{\ln(1 + \gamma\sqrt{{K_{\text{eff}}}^{-1}})}$, where the γ is $\sqrt{S_{\text{th}}/\sigma_{\text{eff}}}{N_{\text{th}}}^{-1}$ and r_0 is $0.61 \times \lambda/\text{NA}$.

The calculated results can fit the experimental results (Figure 11.7b), which indicates that the high photosensitivity for the photoreduction process can reduce the feature size of the silver dot to a deep sub-diffraction level.

11.3.4
Further Increase the Photosensitivity with an Electron Donor

For further increasing the photosensitivity, we have added a water-soluble dye (2-hydroxy-4′-(2-hydroxyethoxy)-2-methylpropiophenone, 0.6 wt%) as an electron donor, which has the absorption band up to 380 nm. Figure 11.4a shows the SEM images of silver dots fabricated under the exposure of the laser power from 0.05 mW (equivalent to the laser intensity of 2.23 GW cm^{-2}, given the laser spot size of 505 nm) to 0.4 mW (equivalent to the laser intensity of 17.84 GW cm^{-2}) with various exposure times between 5 and 50 ms by using the 580 nm laser beam. It can be observed that the photosensitivity of the material improved to 0.054×10^{-12} g μJ^{-1} under the exposure of the laser power of 0.1 mW, which could contribute to a larger K_{eff}, as shown in Figure 11.8a,b. Meanwhile, the feature size of the silver dot dramatically increased with increasing the laser power above the threshold. This could be explained by the increased chemical reactivity and photoreduction sensitivity due to the photoabsorption of the reduced silver [26]. As the laser power is increased from 0.05 to 0.4 mW, the feature size increased from sub-100 nm up to around 400 nm, which approached the size of the diffraction-limited focal spot at 505 nm. In addition, the laser energy for the fabrication threshold was reduced to 0.25 μJ, which can also weaken the thermal effect to produce small silver dots. With the exposure time of 5 ms and a laser power of 0.05 mW, we achieved the minimum silver dot of 22 nm (Figure 11.8c), equivalent to $\lambda/26$ of the wavelength of the laser beam.

11.4
Conclusions

In this chapter, we introduced the progress in the fabrication of nano and microstructures made of narrow electronic bandgap semiconductors and metals via multiphoton lithography. Compared with the top-down strategy, the current

Figure 11.8 (a) SEM image of the silver dot array fabricated with different exposure times under the irradiation of the laser beam of varied laser power. The arrow indicates the direction of the increase of the exposure time from 5 to 50 ms and the laser power from 0.05 to 0.4 mW. (b) Feature size of the silver dots plotted as a function of the exposure time and the laser power. (c) Left: SEM image of the silver dot fabricated with the exposure time of 5 ms under the exposure of the laser beam of 0.05 mW. The scale bar is 50 nm. Right: Graph for the plot of the cross-sectional profile along the blue line in left SEM image. (Cao 2013 [14]. Reproduced with permission of American Institute of Physics.)

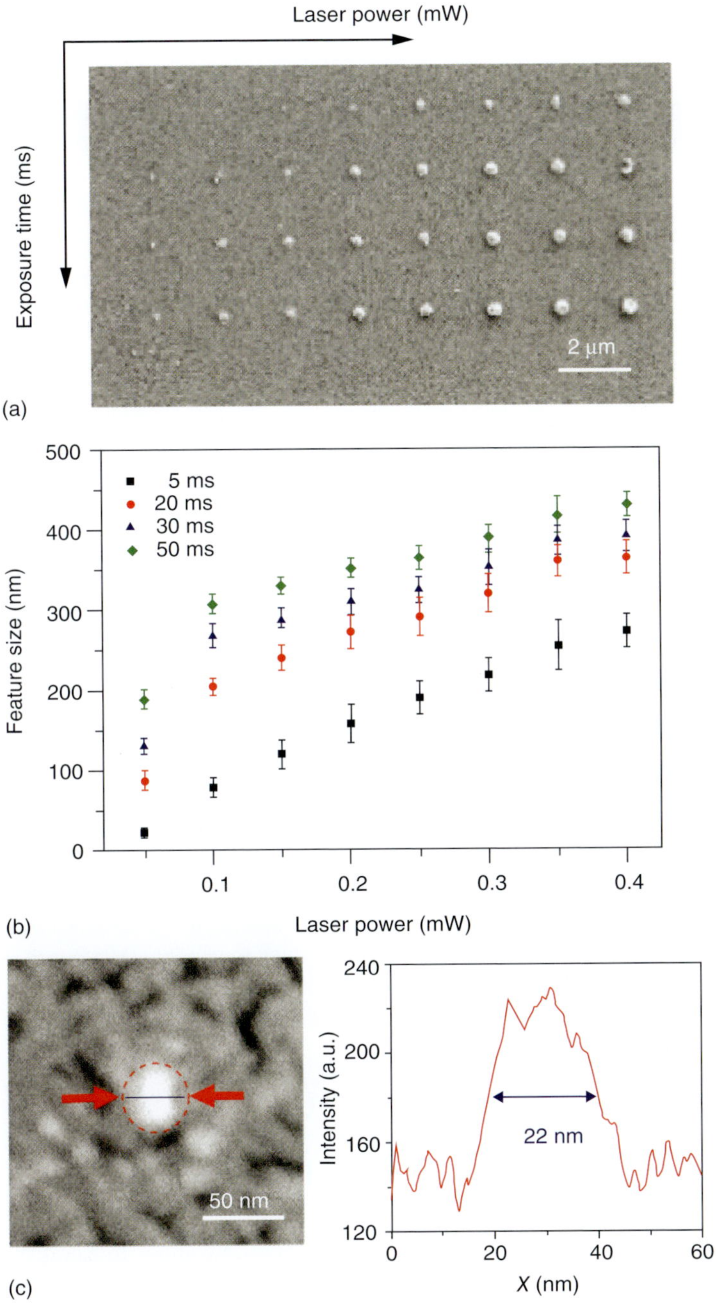
Laser power (mW)
Exposure time (ms)
2 μm
(a)
500
400
300
200
100
0
Feature size (nm)
5 ms
20 ms
30 ms
50 ms
0.1
0.2
0.3
0.4
(b)
Laser power (mW)
50 nm
240
200
160
120
Intensity (a.u.)
22 nm
0
20
40
60
X (nm)
(c)

development of multiphoton lithography technique based on the bottom-up strategy is still at its infancy, especially for the fabrication of nano and microstructures made of narrow electronic bandgap semiconductors.

The demonstration of 3D nano and microstructures made of PbSe and silver enables the implementation of materials with unique intrinsic properties for 3D device applications, which are otherwise unavailable for wide electronic bandgap materials. With a similar lithography principle, PbSe and silver used for fabrication can be extended to other kinds of narrow bandgap semiconductors and metals (such as lead telluride, cadmium sulfide, cadmium selenide, and gold). We expect that this approach can provide a simple and feasible nanofabrication technique based on far-field optics for producing nanophotonic and nanoelectronic devices made from a wide range of unique materials including narrow electronic bandgap semiconductors and metals.

Acknowledgments

This work was supported in part by the Australian Research Council (ARC) Centre for Ultrahigh-bandwidth Devices for Optical Systems (CUDOS) (project number CE110001018) and by the Laureate Fellowship scheme (FL100100099). This work was also supported in part by the Science and Industry Endowment Fund (SIEF) under the John Stocker Postdoctoral Fellowship (project number PF14-100).

References

1. Straub, M. and Gu, M. (2002) Near-infrared photonic crystals with higher-order bandgaps generated by two-photon photopolymerization. *Opt. Lett.*, **27** (20), 1824–1826.
2. Rill, M.S., Plet, C., Thiel, M., Staude, I., von Freymann, G., Linden, S., and Wegener, M. (2008) Photonic metamaterials by direct laser writing and silver chemical vapor deposition. *Nat. Mater.*, **7**, 543–546.
3. Galajda, P. and Ormos, P. (2001) Complex micromechines produced and driven by light. *Appl. Phys. Lett.*, **78**, 249–251.
4. Ohlinger, K., Lin, Y., Poole, Z., and Chen, K.P. (2011) Undistorted 3D microstructures in SU8 formed through two-photon polymerization. *AIP Adv.*, **1**, 032163.
5. Serbin, J. and Gu, M. (2006) Superprism phenomena in waveguide-coupled woodpile structures fabricated by two-photon polymerization. *Opt. Express*, **14** (8), 3563–3568.
6. Turner, M.D., Schröder-Turk, G.E., and Gu, M. (2011) Fabrication and characterization of three-dimensional biomimetic chiral composites. *Opt. Express*, **19** (10), 10001–10008.
7. Swinburne. http://www.swinburne.edu.au/engineering/cmp/cudos.php (accessed 20 April 2016)
8. Gu, M., Li, X., and Cao, Y. (2014) Optical storage arrays: a perspective for future big data storage. *Light: Sci. Appl.*, **3**, e177.
9. Wong, S., Deubel, M., Pérez-Willard, F., John, S., Ozin, G.A., Wegener, M., and von Freymann, G. (2006) Direct laser writing of three-dimensional photonic crystals with a complete photonic bandgap in chalcogenide glasses. *Adv. Mater.*, **18**, 265–269.
10. Nicoletti, E., Bulla, D., Luther-Davies, B., and Gu, M. (2011) Generation of $\lambda/12$

nanowires in chalcogenide glasses. *Nano Lett.*, **11** (10), 4218–4221.
11. Gan, Z., Cao, Y., Evans, R.A., and Gu, M. (2013) Three-dimensional deep sub-diffraction optical beam lithography with 9 nm feature size. *Nat. Commun.*, **4**, 2061.
12. Rodenas, A., Zhou, G., Jaque, D., and Gu, M. (2009) Rare-earth spontaneous emission control in three-dimensional lithium niobate photonic crystals. *Adv. Mater.*, **21** (34), 3526–3530.
13. Bass, M. (1994) *Handbook of Optics*, 2nd edn, vol. **2**, McGraw-Hill. http://refractiveindex.info/?group=CRYSTALS&material (accessed 20 April 2016).
14. Cao, Y.Y. and Gu, M. (2013) $\lambda/26$ silver nanodots fabricated by direct laser writing through highly sensitive two-photon photoreduction. *Appl. Phys. Lett.*, **103** (21), 213104.
15. Gan, Z., Cao, Y.Y., and Gu, M. (2013) Direct laser writing of three-dimensional narrow bandgap and high refractive-index PbSe structures in a solution. *Opt. Express*, **21** (9), 11202–11208.
16. Moreels, I., Hens, Z., Kockaert, P., Loicq, J., and Van Thourhout, D. (2006) Spectroscopy of the nonlinear refractive index of colloidal PbSe nanocrystals. *Appl. Phys. Lett.*, **89**, 193106.
17. Kobayashi, H., Kanbara, H., Koga, M., and Kubodera, K. (1993) Third-order nonlinear optical properties of As_2S_3 chalcogenide glass. *J. Appl. Phys.*, **74**, 3683–3687.
18. Ellingson, R.J., Beard, M.C., Johnson, J.C., Yu, P., Micic, O.I., Nozik, A.J., Shabaev, A., and Efros, A.L. (2005) Highly efficient multiple exciton generation in colloidal PbSe and PbS quantum dots. *Nano Lett.*, **5**, 865–871.
19. Parker, D. and Singh, D.J. (2010) High-temperature thermoelectric performance of heavily doped PbSe. *Phys. Rev. B*, **82**, 035204.
20. Cumming, B.P., Jesacher, A., Booth, M.J., Wilson, T., and Gu, M. (2011) Adaptive aberration compensation for three-dimensional micro-fabrication of photonic crystals in lithium niobate. *Opt. Express*, **19**, 9419–9425.
21. De Lasa, H.I. and Rosales, B.S. (2009) *Advances in Chemical Engineering*, Academic Press, pp. 37–67.
22. Ouyang, J., Schuurmans, C., Zhang, Y., Nagelkerke, R., Wu, X., Kingston, D., Wang, Z.Y., Wilkinson, D., Li, C., Leek, D.M., Tao, Y., and Yu, K. (2011) Low-temperature approach to high-yield and reproducible syntheses of high-quality small-sized PbSe colloidal nanocrystals for photovoltaic applications. *ACS Appl. Mater. Interfaces*, **3**, 553–565.
23. Fleming, J.G., Lin, S.Y., El-Kady, I., Biswas, R., and Ho, K.M. (2002) All-metallic three-dimensional photonic crystals with a large infrared bandgap. *Nature*, **417**, 52–55.
24. Tanaka, T., Ishikawa, A., and Kawata, S. (2006) Two-photon-induced reduction of metal ions for fabricating three-dimensional electrically conductive metallic microstructure. *Appl. Phys. Lett.*, **88**, 081107.
25. Cao, Y.Y., Takeyasu, N., Tanaka, T., Duan, X.M., and Kawata, S. (2009) 3D metallic nano-structure fabrication by surfactant-assisted multi-photon-induced reduction. *Small*, **5**, 1144.
26. Maillard, M., Huang, P., and Brus, L. (2003) Silver nanodisk growth by surface plasmon enhanced photoreduction of adsorbed [Ag+]. *Nano Lett.*, **3**, 1611.
27. Turner, M.D., Matthias, S., Zhang, Q.M., Benjamin, P.C., Gerd, E.S., and Gu, M. (2013) Miniature chiral beamsplitter based on gyroid photonic crystals. *Nat. Photonics*, **7**, 801–805.
28. Xing, J.F., Dong, X.Z., Chen, W.Q., Duan, X.M., Takeyasu, N., Tanaka, T., and Kawata, S. (2007) Improving spatial resolution of two-photon microfabrication by using photoinitiator with high initiating efficiency. *Appl. Phys. Lett.*, **90**, 131106.

12
Microfluidic Devices Produced by Two-Photon-Induced Polymerization

Shoji Maruo

12.1
Introduction

Three-dimensional (3D) printing technologies have attracted much attention because of their ability to produce complex 3D structures with a wide variety of materials. There are several approaches for 3D printing, including fused deposition modeling, inkjet methods, laser sintering, and stereolithography [1–4]. In particular, stereolithography is one of the most precise 3D printing techniques using photopolymers as materials for 3D objects. The original concept of stereolithography was first proposed and demonstrated as a form of stereolithography by Kodama in 1981 [5]. In his system, layer-by-layer photopolymerization according to the sliced data of the desired 3D structure was utilized to produce complex 3D objects. By using the layer-by-layer process, Ikuta and Hirowatari first developed a microstereolithography system that could produce sophisticated 3D microstructures whose sizes were in the range of 10–100 μm in 1993 [6]. For example, they succeeded in making 3D microstructures such as a microcoil of 70 μm diameter with sub-10-μm resolution by the optimization of both the optical system and the sensitivity of the photopolymer. In addition, they utilized the microstereolithography for the production of 3D microfluidic devices called *chemical IC chips* [7]. At that time, most of microfluidic devices were limited to two-dimensional microchannels, because they were made by photolithographic patterning on glass or silicon substrates and the subsequent bonding process. On the other hand, the chemical IC chips were constructed of 3D microchannels made by stereolithography. The 3D microchannels provide several advantages such as minimization of the total size of the microfluidic devices and improvement of mixing efficiency. Additionally, several types of modular functional microfluidic devices such as micropumps, microvalves, and micromixers were developed and assembled into a micro total-analysis system on demand [8]. Although the concept of chemical IC chips is an innovative and efficient strategy, each modular microfluidic chip is made by the layer-by-layer process of stereolithography. Therefore, both the lateral and depth resolutions

Multiphoton Lithography: Techniques, Materials and Applications, First Edition.
Edited by Jürgen Stampfl, Robert Liska, and Aleksandr Ovsianikov.

of stereolithography are still restricted to several tens of micrometers due to the layer-by-layer nature of the technique.

As an alternative ultrahigh-precision microstereolithography, Maruo *et al.* utilized two-photon-induced polymerization for producing 3D microstructures using a photopolymer [9]. This fabrication process is called *two-photon microfabrication* or *two-photon microstereolithography*. In the two-photon process, a photopolymer absorbs two near-infrared photons simultaneously in a single quantum event whose collective energy corresponds to the UV region of the spectrum. The rate of two-photon absorption is proportional to the square of the light intensity, so that the near-infrared light is strongly absorbed only at the focal point within the photopolymer. The quadratic intensity dependence of two-photon absorption allows the confinement of the solidification to submicrometer dimensions, beyond the diffraction limit of light [10]. Two-photon microfabrication has widely been used for many applications including photonics, biology, lab-on-a-chip systems, and microelectromechanical systems (MEMS), as described elsewhere in this book [11]. Our group has proposed and developed several types of optically driven microfluidic devices. In this chapter, we introduce optically driven microfluidic devices including micropumps and micromanipulators using movable microcomponents fabricated by two-photon microfabrication.

12.2
Fabrication of Movable Micromachines

Two-photon microfabrication allows the creation of 3D microstructures by scanning the laser focus inside a photopolymer, unlike conventional stereolithography. Fabrication is accomplished either by scanning the sample relative to the laser focus or the laser focus relative to the sample. Since a stage scanning system can provide an ideal focused spot without aberrations and can be used to fabricate over a large area, this approach has been used to produce ultraprecise 3D microstructures. Figure 12.1 shows some example of 3D microstructures made by two-photon microfabrication. This process allows the fabrication of 3D microstructures not only on flat substrates but also on curved surfaces. Additionally, fragile micromodels, such as lattice models, can be also produced without deformation due to the surface tension of a rinse during the washing process by supercritical drying technique [12].

On the other hand, a mirror scanning system using galvanometric scanners is suitable for making freely movable microstructures such as microgears [13] and nanotweezers [14]. Although such movable microstructures have been produced by a sacrificial layer technique using micromachining in silicon based on photolithography, two-photon microfabrication makes it possible to produce 3D movable microstructures by direct writing using a femtosecond pulsed laser beam inside a photopolymer.

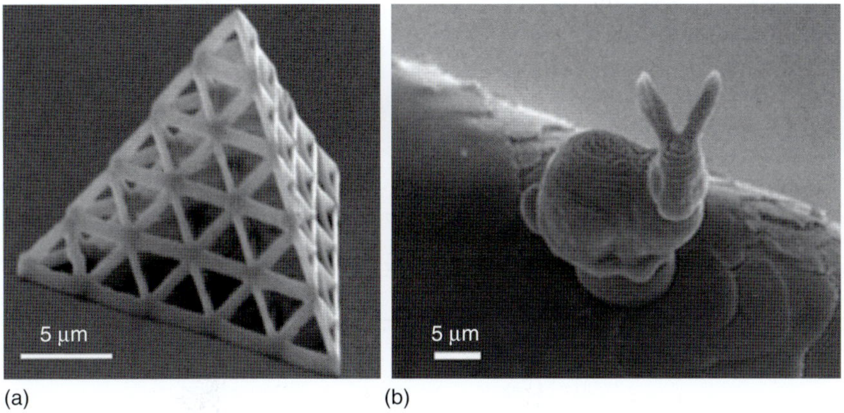

Figure 12.1 Three-dimensional microstructures made by two-photon microfabrication. (a) Micro lattice model on a glass substrate. (b) Micro bunny model on a human hair.

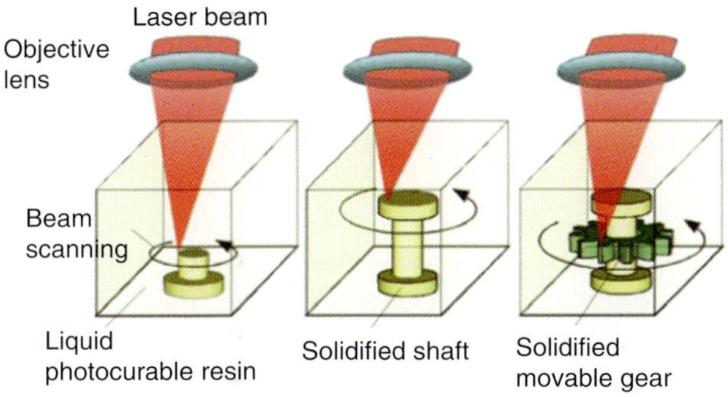

Figure 12.2 Assembly-free, single-step fabrication process based on direct laser writing. A microgear with a shaft is fabricated without any supporting parts [13].

The process of fabricating a microgear is illustrated in Figure 12.2 [13]. In conventional microstereolithography, although simple raster scanning of a laser beam is used to make a microstructure fixed to a substrate, it is not suitable for the assembly-free fabrication of freestanding microparts because of the shrinkage of the photocurable resin during polymerization. The problem is overcome by optimizing the scanning pattern of a laser beam to reduce the deformation of the solidified microstructure. For example, circular scanning of a laser beam is used to make a rotation-symmetric movable microstructure such as a microgear [13]. An attached shaft and stopper are first fabricated by scanning a laser beam circularly at each layer, while the stage is lowered step by step at a constant interval along a central axis. Next, the stage is moved to the position to make a movable gear wheel. The circular part of a gear wheel is fabricated by scanning the laser beam circularly with increasing

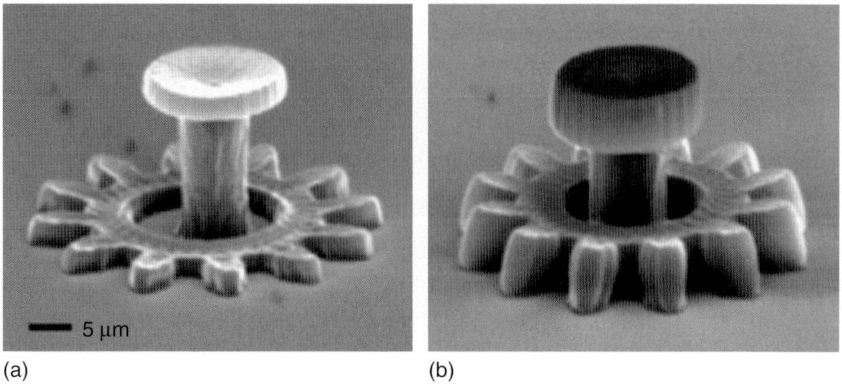

Figure 12.3 SEM images of microgears fabricated with objective lenses of different numerical apertures (NAs). (a) NA: 1.25, laser power: 100 mW, thickness: 0.59 μm. (b) NA: 0.95, laser power: 250 mW, thickness: 1.57 μm. (Maruo 2008 [15]. Reproduced with permission of Springer.)

radius. Finally, the teeth of the gear wheel are fabricated by circularly scanning the laser beam by increasing the radius while the scanning speed is changed periodically. In this case, the nonlinear response of two-photon absorption enables the solidification of the resin only in the region where the scanning speed is low.

Not only thin but also thick movable microparts can be fabricated by the direct laser writing method. Figure 12.3 shows scanning electron microscopy (SEM) images of microgears with different thicknesses [15, 16]. The microgears were fabricated by changing both the numerical aperture of an objective lens and the exposure conditions. As the numerical aperture is reduced, a thicker microgear is fabricated. This is because the focal depth is enlarged by reducing the numerical aperture of the objective lens. Using an objective lens of low numerical aperture, the input laser power must be increased to generate two-photon-induced polymerization. If the laser power exceeds a critical value, the solidified microparts will be destroyed owing to laser-induced breakdown. For this reason, the thickness of the movable microstructure is limited by the critical laser power. To make thicker microparts, the layer-by-layer process is introduced into the laser direct writing method. Figure 12.4a–c shows the SEM images of microgears fabricated by the layer-by-layer process without any supporting parts. In addition, microturbines with curved blades were fabricated by the layer-by-layer process as shown in Figure 12.5. The sophisticated movable parts are useful for functional microfluidic devices.

Several types of manipulators could be also fabricated by optimizing the scanning pattern of a laser beam. Figure 12.6 shows the SEM images of nanotweezers [14]. The arm parts of the tweezers were fabricated by circular scanning, and then the probe tips were fabricated by point-by-point exposure to the sharp tips. The diameter of the probe was 250 nm.

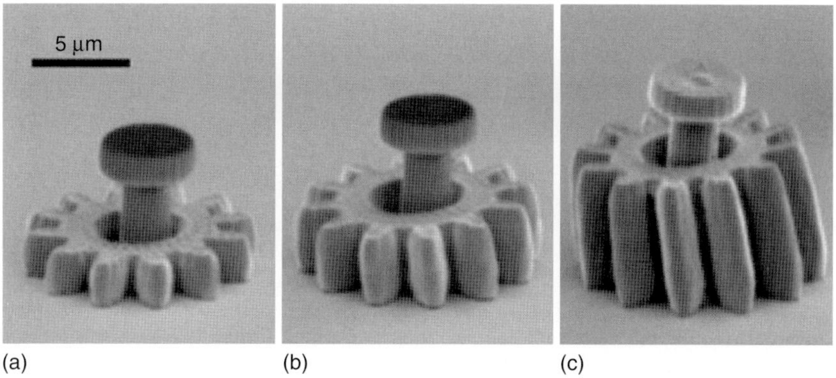

Figure 12.4 SEM images of microgears fabricated by the layer-by-layer process based on direct laser writing. (a) Number of stacking layers: 2, thickness: 1.66 μm. (b) Number of stacking layers: 6, thickness: 2.71 μm. (c) Number of stacking layers: 10, thickness: 4.83 μm. (Maruo 2008 [15]. Reproduced with permission of Springer.)

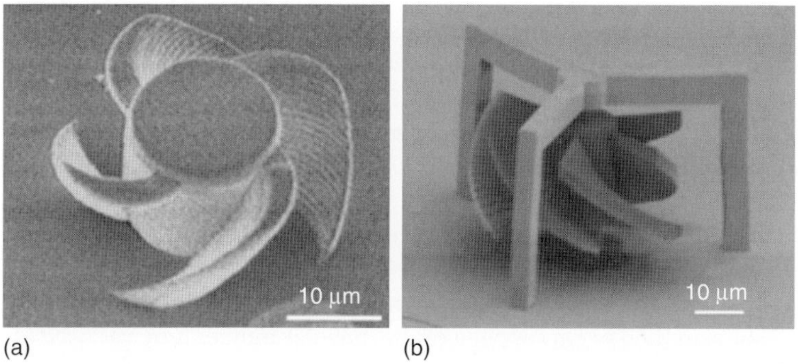

Figure 12.5 SEM images of microturbines (a) Microturbine with a shaft. (b) Spinning-top-type microturbine.

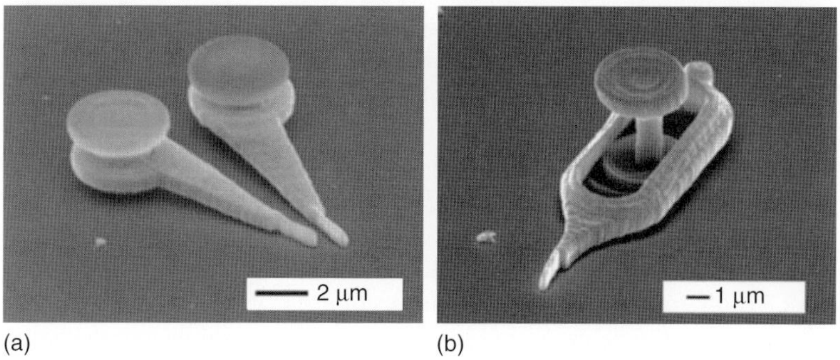

Figure 12.6 SEM images of micro/nano manipulators. (a) Nanotweezers. (b) Nanoneedle. (Maruo 2003 [14]. Reproduced with permission of American Institute of Physics.)

12.3 Optically Driven Micromachines

In the late 1980s, MEMS such as microgears and microactuators were developed by using silicon-based micromachining techniques [17]. Most MEMS utilize electrostatic actuators as the major driving source. Electrostatic actuators have been used in practical microdevices, including digital micro-mirror devices (DMD)s and RF MEMS. However, it is difficult to utilize electrostatic force in liquids such as electrolytes. For this reason, electrostatic actuators are not suitable for biological applications such as microfluidic devices and micromanipulation tools for cells.

On the other hand, radiation pressure from a tightly focused laser beam can be used as optical tweezers to confine, position, and transport microparticles in liquid. Ashkin's group first demonstrated this technique in 1986 [18]. Optical tweezers provide unique features such as remote manipulation of micro/nanoparticles in liquid, noninvasive manipulation of biological samples, precise manipulation in a sealed environment, and extremely small torque on the order of 10^{-12} Nm. For these reasons, optical tweezers and the related techniques have been widely applied to studies on biological cells and DNA molecules, and microchemistry with microdroplets and microbeads. The history and previous work on optical tweezers are available in earlier reviews [19–22].

Recently, optical manipulation techniques have been applied to remotely controlled microfluidic devices. For example, synchronized manipulation of microparticles inside a microchannel was applied to create microfluidic components such as micropumps and microvalves [23, 24]. Use of the spin angular momentum carried by circularly polarized optical tweezers was also made to construct a micropump with birefringent microparticles [25]. In addition, a Laguerre–Gaussian (LG) beam carrying orbital angular momentum was used for revolving multiple microparticles along its ring-like intensity distribution, called *optical vortex* [26, 27]. The rotation of microparticles using optical vortices was applied to microfluidic devices [28]. However, in the above examples, the target samples manipulated by optical tweezers are merely microparticles. Therefore, the structure and performance of the microfluidic devices are limited by the shape and size of the microparticles.

By contrast, we have utilized sophisticated 3D microstructures produced by two-photon microfabrication instead of target microparticles. By controlling the 3D microstructures according to the desired motions, several types of micromachines could be produced [13–16, 29–33]. The optically driven micromachines offer unique, remotely controlled microdevices such as micropumps and micromanipulators applicable for microfluidic devices.

Figure 12.7 shows optical microscope images of optically driven micromanipulators. The micromanipulators were synchronously controlled by a time-shared optical trapping technique [15]. By scanning the laser beam at high speed, multiple micromanipulators can be driven simultaneously. In this demonstration, a glass microbead whose diameter was 5 μm was grasped by the manipulator in a

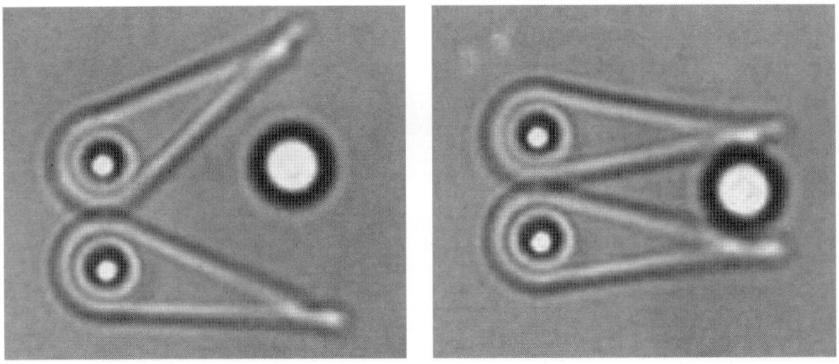

(a) (b)

Figure 12.7 Manipulation of a microparticle using optically controlled micromanipulators. (Maruo 2008 [15]. Reproduced with permission of Springer.)

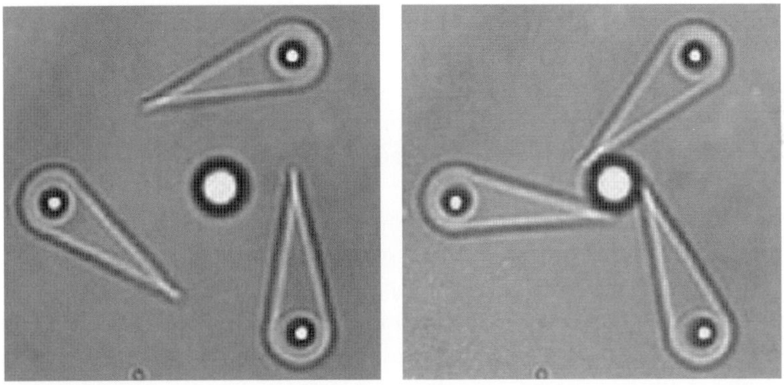

(a) (b)

Figure 12.8 Holding of a microparticle using three-hand micromanipulators. (Maruo 2008 [15]. Reproduced with permission of Springer.)

liquid. Using three micromanipulators, another type of micromanipulators could be constructed, as shown in Figure 12.8. In this case, a glass microbead could be tightly fixed at the center of the manipulators.

In addition, we demonstrated a novel type of 3D micromanipulators for holding of a microobject in three dimensions [29]. Figure 12.9a shows a schematic diagram and the driving principle of optically controlled micromanipulators with rotating arms. The micromanipulator arm has a handle to rotate the arm about its long axis. A microparticle is attached to the tip of the handle as a trapping point. Figure 12.9b shows a cross section of the micromanipulators. In Figure 12.9b, when a laser beam is focused on the handle below the attached microparticle and scanned perpendicular to the arm in the focal plane, the handle is inclined toward the focus because of the net radiation pressure exerted on the handle and the particle. At that time, since the base of the manipulator arm is restricted to the housing, the manipulator arm is rotated on its long axis. The rotational mechanism allows

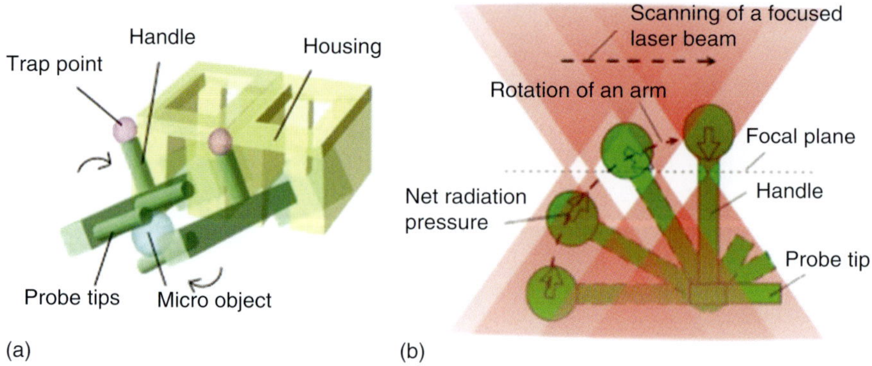

Figure 12.9 Schematic design and driving method of optically controlled micromanipulators using rotating arms. (a) Schematic diagram of micromanipulators. (b) Optical driving of the rotating arm by scanning a focused laser beam linearly in the focal plane. (Maruo 2007 [29]. Reproduced with permission of Fuji Technology Press Ltd.)

the manipulatation of a microobject in three dimensions. If the probes attached to the manipulator arms are designed to grasp a microobject tightly, it can be caught between the probes while the arms are synchronously rotated by use of time-shared scanning of a laser beam.

A unique feature of the 3D micromanipulators is the rotational motion of the manipulator arm. In most optical trapping techniques using galvano scanners, since a single laser beam is scanned in the focal plane, the motion of the trapped particle is restricted to translational or rotational motions in the focal plane. For this reason, the optically controlled manipulators that we have already developed [14] were driven in the focal plane. Therefore, the targeted microparticle is merely held in two dimensions on the substrate. On the other hand, the micromanipulators with rotating arms allow the handling of a microobject separated from the substrate. The 3D manipulation of microobjects will be useful for the manipulation of floating biological samples such as blood cells and microbes and for holding a biological sample released from a substrate.

In 3D micromanipulators, the manipulator arm is rotated about its long axis because of the radiation pressure exerted on both the handle and the particle attached to the handle, as shown in Figure 12.9b. The net radiation pressure depends on some parameters such as the length of the handle, diameter of the particle, and focusing position along the handle. To evaluate the driving performance of the manipulators, several manipulator arms with handles of different lengths were fabricated through the direct writing process of a femtosecond laser beam. Figure 12.10 shows the top and the side views of the manipulator arms (length: 22.5 μm, width: 2 μm) with a handle (length: 5 μm, diameter: 2 μm). For the simplification of the experiments, the manipulator arms do not have probe tips for grasping a microobject. A microparticle (diameter: 2.4 μm) is attached to the top of each handle. The handle with the microparticle was fabricated along the optical axis, while the supporting stage is lowered stepwise at a constant pitch

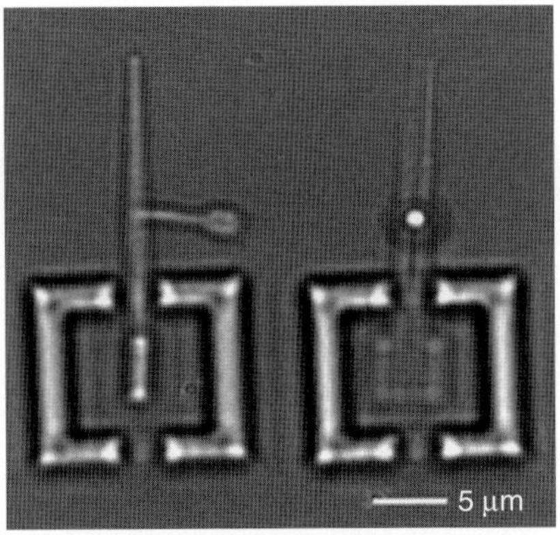

Figure 12.10 Optical microscope image of micromanipulators with handles for the evaluation of driving performance. (Maruo 2007 [29]. Reproduced with permission of Fuji Technology Press Ltd.)

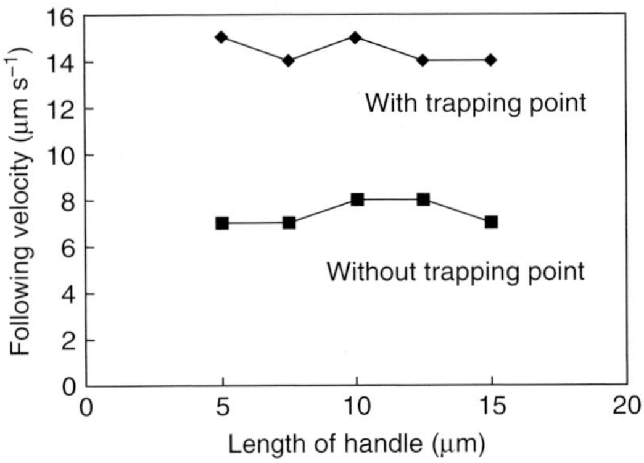

Figure 12.11 Dependence of the following velocity of the handle attached to the arm on the length of the handle. (Maruo 2007 [29]. Reproduced with permission of Fuji Technology Press Ltd.)

of 0.5 μm. The length is changed from 5 to 15 μm. The power and the scanning speed of the laser beam are 200 mW and 50 μm s^{-1}, respectively.

As shown in Figure 12.9b, the manipulator arm is rolled by slanting the handle by scanning the focused laser beam linearly in the direction perpendicular to the long axis of the arm. In our experiments, the magnitude of the net radiation pressure to roll the arm is evaluated by measuring the maximum following velocity of the handle while the scanning speed of the laser beam is increased. When the handle is rotated at the maximum following velocity, the net radiation pressure is equal to the viscous drag exerted on the handle and the attached microparticle.

Figure 12.11 shows the dependence of the following velocity of the handle on its length. When the microparticle as a trapping point is not attached to the handle, the following velocities are almost constant at ~7 μm s^{-1}. On the other hand, the following velocities of the handles with trapping points are twice as fast as those of the cases without trapping points. This result indicates that the trapping point plays an important role to tilt the handle.

Manipulation of a microparticle using the micromanipulators was demonstrated. Figure 12.12a shows the handling of a 5-μm-diameter microparticle with two manipulator arms. The targeted microparticle was also fabricated by two-photon microfabrication for the demonstration of the performance of the manipulators. It was demonstrated that the targeted microparticle could be held away from the substrate. In the above experiments on the evaluation of the following velocity of the arm, the trapping point was 2.4 μm in diameter. However, the holding force at the probe tips was not sufficient to grasp a 5-μm-diameter microparticle. Therefore, in the experiments, the diameter of the

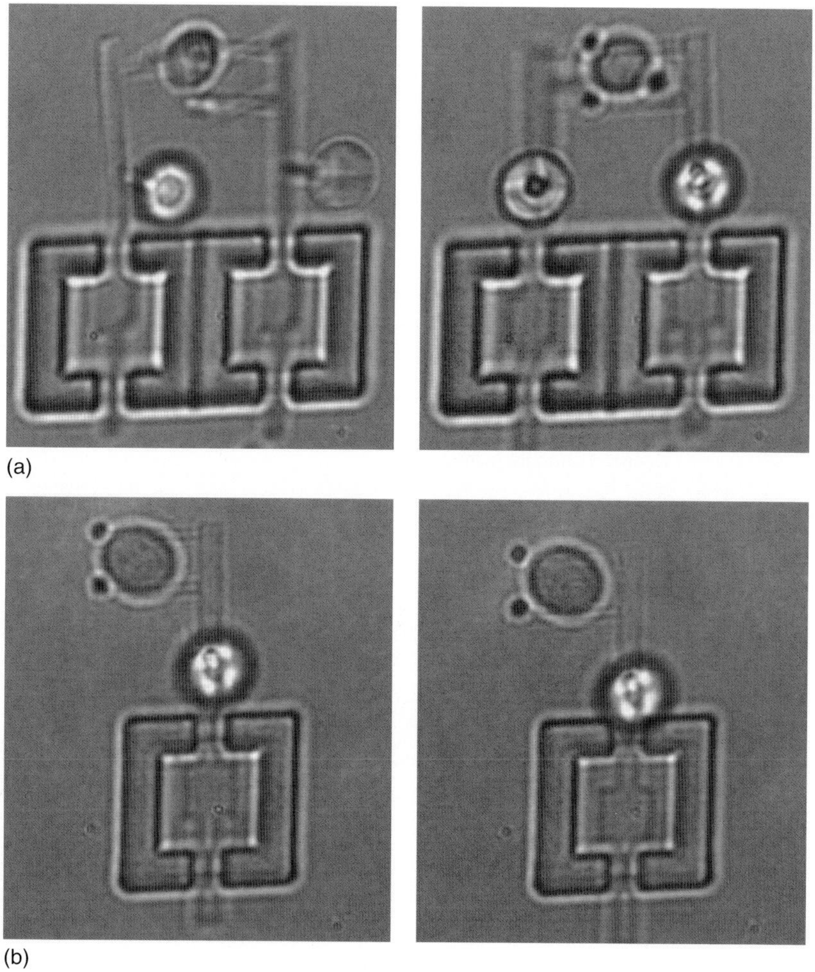

Figure 12.12 Manipulation of a microparticle using optically driven micromanipulators with rotating arms. (a) Grasping a microobject using two arms. (b) Holding and sliding of a microparticle on a cover glass. (Maruo 2007 [29]. Reproduced with permission of Fuji Technology Press Ltd.)

trapping point was increased to 6 μm in order to increase the holding force. As a result, the targeted microparticle was grasped tightly with the two rotating arms. In the experiment, the two arms with probes were cooperatively controlled by time-shared scanning of a laser beam. The power and scanning speed of the laser beam were 500 mW and 3 $\mu m\ s^{-1}$, respectively. The repetition rate of time-shared laser scanning for trapping each handle is 200 Hz. Figure 12.12b shows the transportation of a microparticle using a single arm with two probes. In this case, the targeted microparticle is pushed and slid on the glass substrate.

Optically controlled micromanipulators will be applied for noninvasive manipulation of living cells.

12.4 Microfluidic Devices Driven by a Scanning Laser Beam

In the development of functional lab-on-a-chip devices, built-in micropumps based on pressure-driven flow are necessary for the further extension of application fields. Although various kinds of pressure-driven micropumps have been developed using a range of micromachining techniques [30, 31], most of the previously developed micropumps are diaphragm types, in which piezoelectric actuators or pneumatic actuation are utilized to deform an elastic membrane that drives fluid transport. To obtain adequate displacement of the membrane, the diameter of the membrane must range from 100 to 500 μm. As a result, the overall dimensions of the micropumps are much larger than the microchannels. This prevents the further miniaturization and integration of microfluidic components. In addition, the use of built-in high-precision microactuators makes biochips expensive. For these reasons, most current disposable microfluidic chips for chemical synthesis and cell analysis still utilize external pumps such as syringe types rather than built-in micropumps.

As a promising method of addressing the above issues, we have developed an optically driven micropump by using two-photon microstereolithography [32]. The micropump consist of two lobed rotors that are incorporated into a microchannel. The lobed rotors are individually confined to a microchannel by their own shaft, thus preventing the rotors from moving unless continuously irradiated with light. This allows the micropump to be easily integrated into a mass-produced biochips made from plastics and glass.

Figure 12.13a shows a schematic diagram of the optically driven lobed micropump. This micropump is driven by means of radiation pressure generated by focusing a laser beam. When the laser beam is focused on the side of the rotor, the net radiation pressure applied to the rotor points toward the focus, allowing the rotor to be controlled by scanning the laser beam along two circular trajectories alternately. Figure 12.13b shows the time-divided laser scanning method to control the two rotors simultaneously. The black points shown in Figure 12.13b are trapping points for each status. The laser beam is divided and scanned along two circular trajectories in opposite directions. In this case, on–off control of

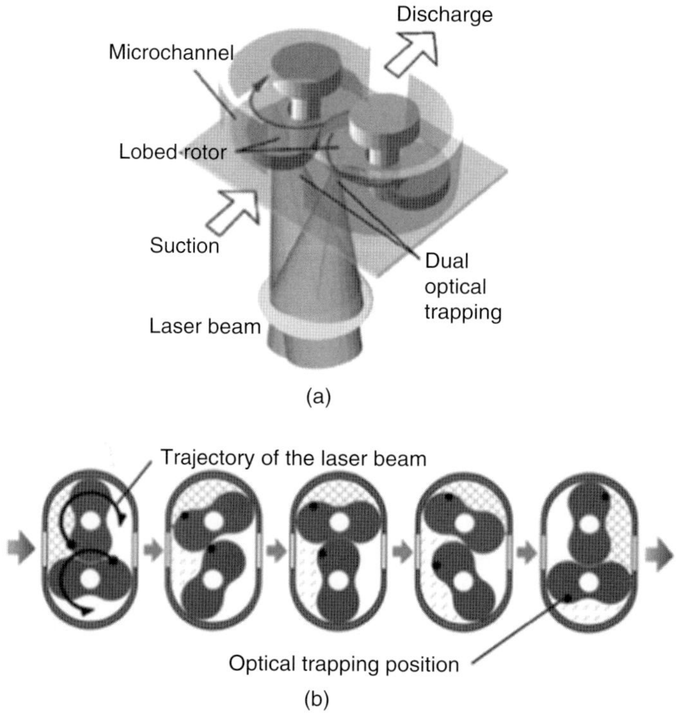

Figure 12.13 Optically driven lobed micropump. (a) Schematic diagram of the lobed micropump driven by the time-divided laser scanning of a single laser beam. (b) Fluid transport by optically rotating two rotors. (Maruo 2006 [32]. Reproduced with permission of American Institute of Physics.)

the laser beam is not performed when the laser beam alternates between the two trajectories. While the rotors are tightly engaged and rotated, fluid is trapped in the spaces between the rotors and the cage, and carried around in them as shown in Figure 12.13b.

Figure 12.14 shows an SEM image of a prototype of the lobed micropump. The length of the major axis of each rotor is 9 μm. The thickness of the rotor is 2.5 μm. The lobed rotors were incorporated into a microchannel whose inlet and outlet are in 5 μm wide and 7 μm high. The microchannel was also fabricated by using two-photon microstereolithography.

We demonstrated that fluid was transported with the optically driven micropump. Figure 12.15a,b shows sequential images of the micropump being driven. A tracer particle was also fabricated in the microchannel using two-photon microstereolithography. The tracer particle was optically trapped to prevent it from moving around while unsolidified photocurable resin was being washed out. After the washing process, the tracer particle was released, after which the rotors were interlocked using the time-divided laser scanning technique. As a result, the tracer particle was successfully brought into the microchamber by the

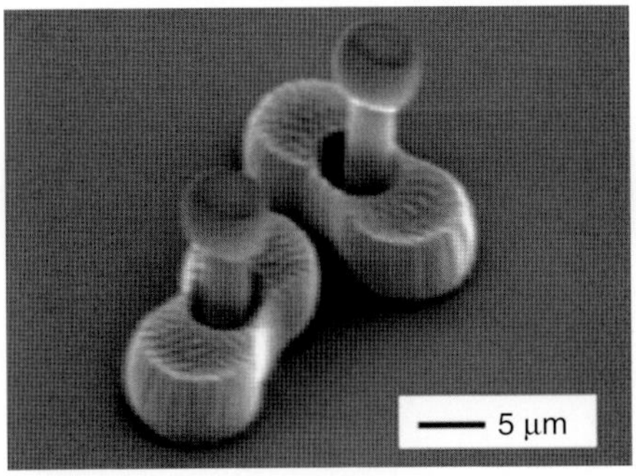

Figure 12.14 SEM image of a prototype of the lobed micropump. (Maruo 2006 [32]. Reproduced with permission of American Institute of Physics.)

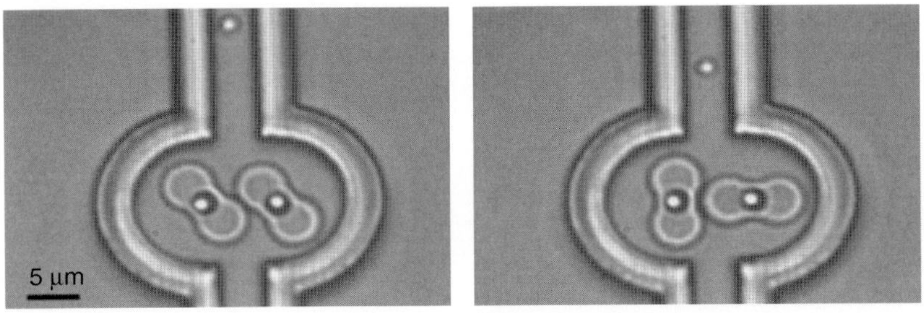

Figure 12.15 Sequential images taken while driving the optically driven micropump. (Maruo 2006 [32]. Reproduced with permission of American Institute of Physics.)

pumped flow. When the rotation direction was reversed, the tracer particle was pushed to the outlet. Figure 12.16 shows the dependence of the velocity of the tracer particle on the rotational speed of the rotors. It is clear that the velocity of the tracer particle is proportional to the rotational speed of the rotors. The flow rate was estimated at less than 1 pl min^{-1}. Ultralow flow rates such as this give our mechanism the potential for use in micro/nanofluidic devices.

12.5 Microfluidic Devices Driven by a Focused Laser Beam

In the microscale, viscous drag force is dominant compared to inertia force due to the scale effect. Using this unique property, we have proposed and developed a viscous micropump using a single-disk microrotor, as shown in Figure 12.17 [33]. In this case, we designed a U-shaped microchannel to generate laminar flow around the single-disk microrotor. Although we have successfully demonstrated that a microparticle was transported along the U-shaped channel, the rotational speed of the disk is low. This is because the disk is rotated by time-shared scanning of a

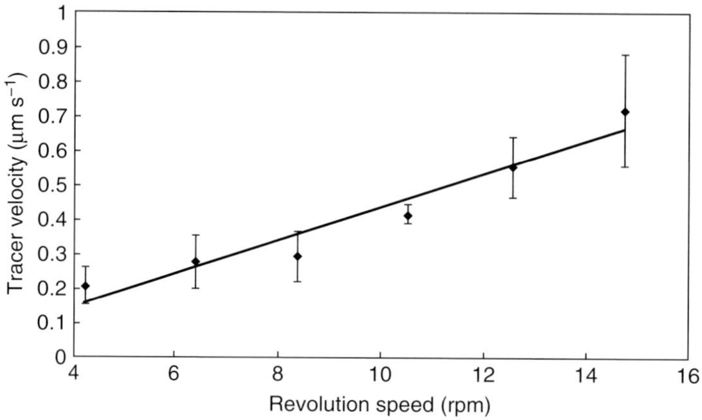

Figure 12.16 Dependence of the velocity of the tracer particle on the rotational speed of the rotors. (Maruo 2006 [32]. Reproduced with permission of American Institute of Physics.)

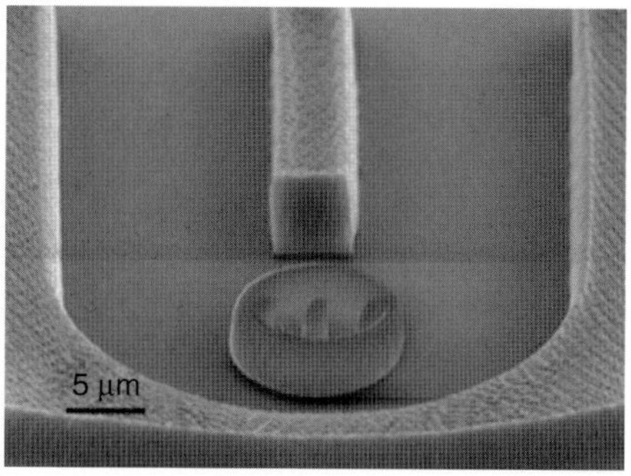

Figure 12.17 Viscous micropump using a single-disk micropump. (Maruo 2007 [33].)

single laser beam. To overcome the limitation of the rotational speed, a spinning microrotor with twin spiral blades was utilized as a microrotor for the viscous micropump. Figure 12.18 shows an optically driven viscous micropump using a spinning microrotor with twin spiral blades [34]. The outer cylinder attached to the twin spiral microrotor plays an important role in that it generates a continuous unidirectional flow without flow separation. The twin spiral microrotor is confined to a U-shaped microchannel to generate a laminar flow like the previous viscous micropump using a single-disk microrotor. When the twin spiral microrotor is optically trapped by focusing a laser beam at the corner of the U-shaped microchannel, it is stably rotated at a high speed of several hundred micrometers per minute while the microrotor is trapped in three dimensions at the focus. As a result, the microrotor rotation provides a continuous, unidirectional flow without pulsation like the micropump using a rotating single disk.

The optically induced rotation of a microrotor with a single spiral blade was first reported by Galajda and Ormos [35]. In their experiments, they demonstrated that the rotation direction of the spiral microrotor is reversed when the trapped

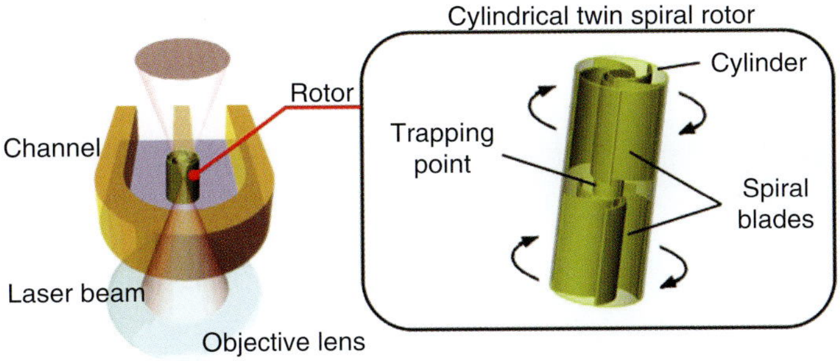

Figure 12.18 Viscous miropump using a twin spiral microrotor. (Maruo 2009 [34].)

position of the microrotor is changed from below the focus to beyond the focus along the optical axis. We employ the phenomenon of the reversed rotation of the single spiral microrotor to create a novel microrotor that has higher rotational speed than a single spiral microrotor. As shown in Figure 12.18, in our twin spiral microrotor, two spiral blades with reversed directions are connected along the optical axis inside an outer cylinder. When a laser beam is focused on the center of the twin spiral microrotor, the optical torque induced by the radiation pressure and exerted on the spiral blades is in the same direction, because the spiral microrotors are mounted in opposite directions. Therefore, the net optical torque generated by the twin spiral microrotor can exceed that of a single spiral microrotor.

Figure 12.19 shows the SEM images of the twin spiral microrotor with an outer cylinder. The diameter and height of the microrotor were 4 and 9 μm, respectively. The opposite facing spiral blades were fabricated by two-photon microfabrication. When extracting the microrotor from unsolidified photopolymer, we used a supercritical CO_2 drying process to reduce harmful deformation caused by the surface tension of the rinse solution [12].

To drive the twin spiral microrotor, we also used the optical system of two-photon microfabrication, while operating the Ti:sapphire laser in the continuous wave (CW) mode. The surrounding liquid was glycol ether ester, which we used for washing out the liquid photopolymer. The rotation speed of the twin spiral

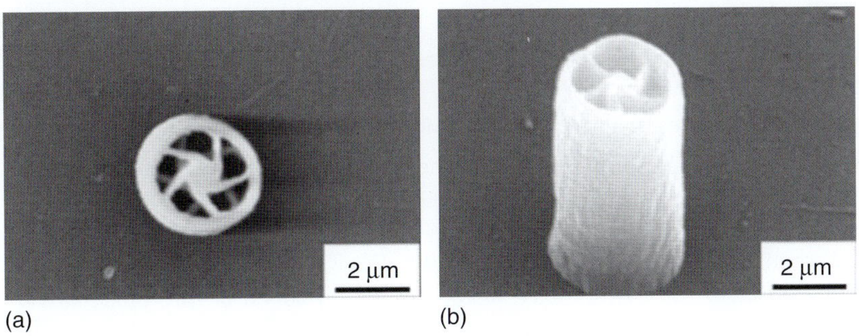

Figure 12.19 SEM images of the twin spiral microrotor with an outer cylinder. (Maruo 2009 [34].)

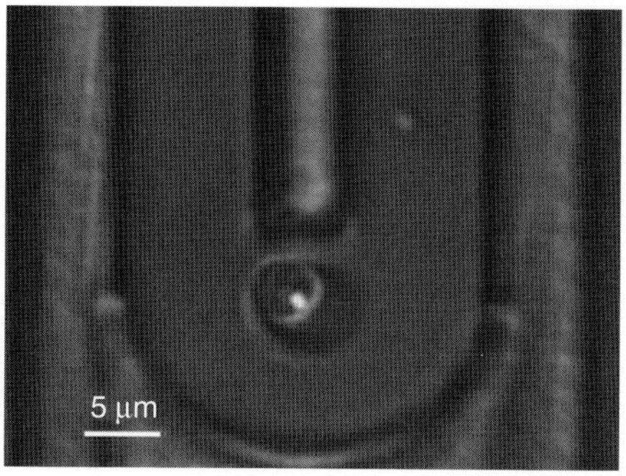

Figure 12.20 Demonstration of microparticle transportation in the viscous micropump using a twin spiral microrotor. (Maruo 2009 [34].)

microrotor was proportional to the intensity of the induced laser beam. The maximum speed reached 300 rpm at a laser power of 500 mW.

The pumping performance of the viscous micropump with the twin spiral microrotor depends on several parameters including the width of the U-shaped microchannel and the diameter and position of the microrotor, like a viscous micropump using a single rotating disk. We examined the pressure field and streamlines of the micropump by changing the width of the microchannel via simulation using COMSOL Multiphysics software. As a result, a steady laminar flow was generated in the linear regions of the U-shaped microchannels by selecting proper parameters. In this simulation, the flow rate was saturated at a microchannel width of 10 μm with a 4-μm-diameter microrotor.

A prototype of the viscous micropump containing a twin spiral microrotor (diameter: 4 μm) was fabricated by two-photon microfabrication. Figure 12.20 shows that the viscous micropump with a twin spiral microrotor can be used for pumping a fluid and transporting microobjects. The sequence of images was taken at 2-s intervals, showing a tracer particle being pumped through the channel when the microrotor was rotating at a speed of 300 rpm.

12.6 Microfluidic Devices Driven by an Optical Vortex

As another promising way to rotate a microrotor, we have utilized an optical vortex. In our experiments, we generated a Laguerre–Gaussian beam by using a spatial light modulator [36]. Since a Laguerre–Gaussian beam has orbital angular momentum, microparticles can be rotated along the doughnut-ring-like optical intensity distribution at a focal plane [26, 27]. To design the microrotor, we first examined the dependence of the rotational speed of microparticles by changing the diameter of doughnut-ring-like optical intensity distribution. The diameter of the doughnut beam could be controlled by changing the topological charge (TC) of the Laguerre–Gaussian beam with the spatial light modulator.

Figure 12.21 shows the polystyrene microparticles (diameter: 1 μm) driven by Laguerre–Gaussian beams with different TCs. The rotational speed was proportional to TC. In the experiments, it was found that the rotational speed attained the maximum value at TC = 16. According to the results, we designed a microrotor whose diameter was same as the diameter of the doughnut beam with TC = 16. Figure 12.22 shows the SEM images of the microrotor fabricated by two-photon microfabrication. To construct a viscous micropump, the microrotor was built in a U-shaped microchannel as shown in Figure 12.23. By illuminating the Laguerre–Gaussian beam with TC = 16, the microrotor was successfully rotated and a tracer particle was transported in the U-shaped microchannel. In this experiment, the rotational speed of the microrotor was 38.5 rpm, and the velocity of the tracer particle was 3.1 $\mu m\ s^{-1}$.

12.7 Future Prospects

We have developed various kinds of optically driven microfluidic devices by using two-photon microfabrication. The microfluidic devices have unique features such

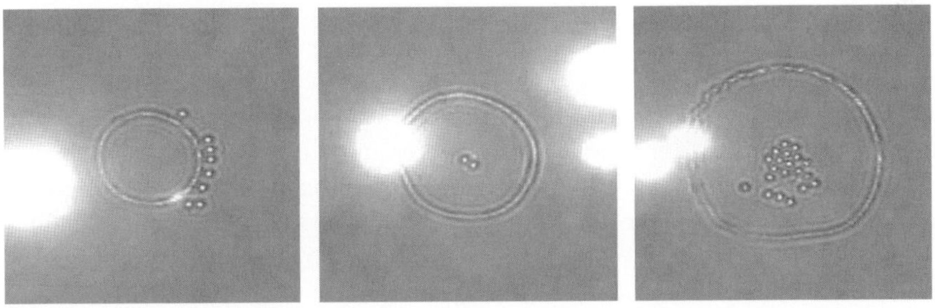

(a) (b) (c)

Figure 12.21 Microparticles driven by Laguerre–Gaussian beams with different TCs [36]. (a) TC = 16, (b) TC = 32, and (c) TC = 64.

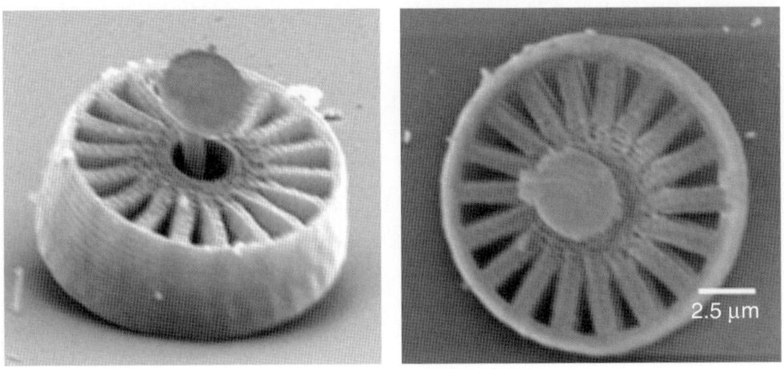

(a) (b)

Figure 12.22 Microrotor driven by a Laguerre–Gaussian beam [36].

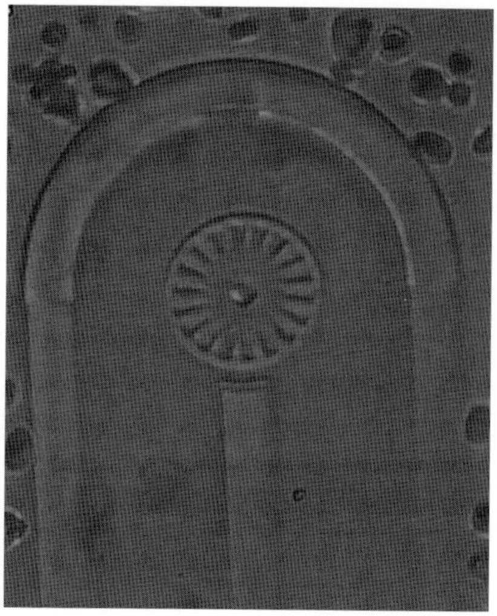

Figure 12.23 Viscous micropump using a microrotor driven by a Laguerre–Gaussian beam [36].

as synchronous remote drive of multiple components, high integration of functional elements, and ultralow flow rate. The optical-driven microfluidic devices can be used for low-cost, all-polymer lab-on-a-chip devices, because all of built-in components can be remotely controlled without any wiring and power supply. To produce practical analytical systems based on optically driven microfluidic devices, there remain some issues such as the improvement of the optical driving efficiency and chip-to-world connections. As a promising method to improve the optical driving efficiency, we have recently utilized metalized micromachines [37, 38]. The metalized micromachines are fabricated by the combination of two-photon microfabrication and electroless plating. Since the metalized micromachines are be driven by repulsive force due to reflection of the incident laser beam at the metalized surface, the optical radiation pressure is drastically improved. We have already demonstrated that a metalized microrotor could be driven by a laser beam whose laser power was less than 1 mW. Such metalized micromachines will be useful for high-efficiency, optically driven microfluidic devices. The reduction of the required laser power to drive micromachines enables the reduction of both the size and cost of the total driving system. Therefore, portable lab-on-a-chip systems will be realized for medical diagnostic applications in the near future.

References

1. Sachs, E., Cima, M., Williams, P., Brancazio, D., and Cornie, J. (1992) 3-dimensional printing – rapid tooling and prototypes directly from a CAD model. *J. Eng. Ind. Trans. ASME*, **114**, 481–488.
2. Pham, D.T. and Gault, R.S. (1998) A comparison of rapid prototyping technologies. *Int. J. Mach. Tools Manuf.*, **38**, 1257–1287.
3. Melchels, F.P.W., Feijen, J., and Grijpma, D.W. (2010) A review on

stereolithography and its applications in biomedical engineering. *Biomaterials*, **31**, 6121–6130.
4. Kruth, J.P., Levy, G., Klocke, F., and Childs, T.H.C. (2007) Consolidation phenomena in laser and powder-bed based layered manufacturing. *CIRP Ann. Manuf. Technol.*, **56**, 730–759.
5. Kodama, H. (1981) Automatic method for fabricating a 3-dimensional plastic model with photo-hardening polymer. *Rev. Sci. Instrum.*, **52**, 1770–1773.
6. Ikuta, K. and Hirowatari, K. (1993) Real three dimensional microfabrication using stereo lithography and metal molding. Proceedings of IEEE Micro Electro Mechanical Systems, 1993, IEEE, pp. 42–47.
7. Ikuta, K. (1996) 3D micro integrated fluid system toward living LSI - Challenge for artificial cellular devices. Proceedings of the Fifth International Workshop on the Synthesis and Simulation of Living Systems (Complex Adaptive Systems), pp. 17–24.
8. Ikuta, K. (2000) Biochemical IC using micro stereolithography-towards artificial cellular device. Proceedings of Microprocesses and Nanotechnology Conference 2000, IEEE, pp. 26–27.
9. Maruo, S., Nakamura, O., and Kawata, S. (1997) Three-dimensional microfabrication with two-photon-absorbed photopolymerization. *Opt. Lett.*, **22**, 132–134.
10. Kawata, S., Sun, H.B., Tanaka, T., and Takada, K. (2001) Finer features for functional microdevices – micromachines can be created with higher resolution using two-photon absorption. *Nature*, **412**, 697–698.
11. Maruo, S. and Fourkas, J.T. (2008) Recent progress in multiphoton microfabrication. *Laser Photonics Rev.*, **2**, 100–111.
12. Maruo, S., Hasegawa, T., and Yoshimura, N. (2009) Single-anchor support and supercritical CO_2 drying enable high-precision microfabrication of three-dimensional structures. *Opt. Express*, **17**, 20945–20951.
13. Maruo, S., Ikuta, K., and Korogi, H. (2003) Force-controllable, optically driven micromachines fabricated by single-step two-photon micro stereolithography. *J. Microelectromech. Syst.*, **12**, 533–539.
14. Maruo, S., Ikuta, K., and Korogi, H. (2003) Submicron manipulation tools driven by light in a liquid. *Appl. Phys. Lett.*, **82**, 133–135.
15. Maruo, S. (2008) in *Optically Driven Micromachines for Biochip Application*, Nano- and Micromaterials SeriesAdvances in Materials Research, vol. **9** (eds K. Ohno, M. Tanaka, J. Takeda, and Y. Kawazoe), Springer.
16. Inoue, H., Haga, S., and Maruo, S. (2006) Development of optically driven microgears by using two-photon microstereolithography. *IEEJ Trans. Sens. Micromach.*, **126**, 216–221.
17. Bustillo, J.M., Howe, R.T., and Muller, R.S. (1998) Surface micromachining for microelectromechanical systems. *Proc. IEEE*, **86**, 1552–1574.
18. Ashkin, A., Dziedzic, J.M., Bjorkholm, J.E., and Chu, S. (1986) Observation of a single-beam gradient force optical trap for dielectric particles. *Opt. Lett.*, **11**, 288–290.
19. Ashkin, A. (2000) History of optical trapping and manipulation of small-neutral particle, atoms, and molecules. *IEEE J. Sel. Top. Quantum Electron.*, **6**, 841–856.
20. Molloy, J.E. and Padgett, M.J. (2002) Lights, action: optical tweezers. *Contemp. Phys.*, **43**, 241–258.
21. Neuman, K.C. and Block, S.M. (2004) Optical trapping. *Rev. Sci. Instrum.*, **75**, 2787–2809.
22. Grier, D.G. (2003) A revolution in optical manipulation. *Nature*, **424**, 810–816.
23. Terray, A., Oakey, J., and Marr, D.W.M. (2002) Microfluidic control using colloidal devices. *Science*, **296**, 1841–1844.
24. Terray, A., Oakey, J., and Marr, D.W.M. (2002) Fabrication of linear colloidal structures for microfluidic applications. *Appl. Phys. Lett.*, **81**, 1555–1557.
25. Leach, J., Mushfique, H., di Leonardo, R., Padgett, M., and Cooper, J. (2006) An optically driven pump for microfluidics. *Lab Chip*, **6**, 735–739.
26. Curtis, J.E. and Grier, D.G. (2003) Structure of optical vortices. *Phys. Rev. Lett.*, **90**, 133901.

27. Gahagan, K.T. and Swartzlander, G.A. (1996) Optical vortex trapping of particles. *Opt. Lett.*, **21**, 827–829.
28. Ladavac, K. and Grier, D.G. (2004) Microoptomechanical pumps assembled and driven by holographic optical vortex arrays. *Opt. Express*, **12**, 1144–1149.
29. Maruo, S. and Hiratsuka, Y. (2007) Optically driven micromanipulators with rotating arms. *J. Rob. Mechatron.*, **19**, 565–568.
30. Zhang, C., Xing, D., and Li, Y. (2007) Micropumps, microvalves, and micromixers within PCR microfluidic chips: advances and trends. *Biotechnol. Adv.*, **25**, 483–514.
31. Shoji, S. and Esashi, M. (1994) Microflow devices and systems. *J. Micromech. Microeng.*, **4**, 157–171.
32. Maruo, S. and Inoue, H. (2006) Optically driven micropump produced by three-dimensional two-photon microfabrication. *Appl. Phys. Lett.*, **89**, 144101.
33. Maruo, S. and Inoue, H. (2007) Optically driven viscous micropump using a rotating microdisk. *Appl. Phys. Lett.*, **91**, 084101.
34. Maruo, S., Takaura, A., and Saito, Y. (2009) Optically driven micropump with a twin spiral microrotor. *Opt. Express*, **17**, 18525–18532.
35. Galajda, P. and Ormos, P. (2002) Rotors produced and driven in laser tweezers with reversed direction of rotation. *Appl. Phys. Lett.*, **80**, 4653–4655.
36. Maruo, S. and Saito, Y. (2009) A viscous micropump using a spinning microrotor driven by a Laguerre-Gaussian beam in Proceedings of International Symposium on Micro-Nanomechatronics and Human Science (MHS) 2009, pp. 316–319.
37. Ikegami, T., Stocker, M.P., Monaco, K., Fourkas, J.T., and Maruo, S. (2012) Fabrication of three-dimensional metalized movable microstructures by the combination of two-photon microfabrication and electroless plating. *Jpn. J. Appl. Phys.*, **51**, 06FL17-1–06FL17-4.
38. Ikegami, T., Ozawa, R., Stocker, M.P., Monaco, K., Fourkas, J.T., and Maruo, S. (2013) Development of optically-driven metallic microrotors using two-photon microfabrication. *J. Laser Micro Nanoeng.*, **8**, 6–10.

13
Nanoreplication Printing and Nanosurface Processing

Christopher N. LaFratta

13.1
Introduction: Limitations of Multiphoton Lithography

Multiphoton lithography is unique in its ability to create nanoscale objects with any arbitrary 3D geometry, but suffers from two limitations: slow fabrication speed and limited options for polymerizable materials [1]. Since the patterning process is serial, where each device is fabricated one at a time, it is necessarily time consuming to scale-up for the mass production of parts. Also, since two-photon induced polymerization (TPIP) can only take place in specific kinds of materials, such as acrylates or epoxies, the choice of materials and, therefore, the mechanical, magnetic, electronic, and optical properties of these materials are quite limited.

One solution to these issues can be found by using TPIP to create master structures and then to mold them using micro-transfer molding (μTM) to create replicas [2]. Since the replicas can be created much more quickly than the masters, this method lends itself to mass production. Also, since the crosslinking reaction used in the mold is not TPIP, different materials can be used to cast devices that cannot be created directly from TPIP. Another solution to the limited material availability is to chemically alter the microstructure post fabrication and give it a functional coating: for example, by depositing metal on the outside of a structure, thereby creating electrically conductive devices [3]. Yet another solution is to use laser ablation to remove material, which may enable some structures to be fabricated faster and can generate a more reactive surface on which to perform reactions.

This chapter will summarize the recent work where various molding techniques, such as μTM, have been used to nano-replicate structures made by TPIP. It will also discuss nanosurface processes where added functionality has been introduced to the device by controlling its surface chemistry.

Multiphoton Lithography: Techniques, Materials and Applications, First Edition.
Edited by Jürgen Stampfl, Robert Liska, and Aleksandr Ovsianikov.

13.2
Micro-transfer Molding (μTM)

TPIP is a slow way to create a device not only because each object is made one at a time but also because each point within that object is individually created and connected to the previous one. This means that, despite fabrication speeds now approaching centimeter per second, it can still be rather time consuming to create a single device [4]. Alternative methods have been shown to increase fabrication speed, such as projection lithography [5, 6] and beam splitting [7, 8], but these optical setups can be somewhat complicated. Conventional photolithography, which uses masks, does not suffer from this serial writing problem because millions of features are patterned simultaneously in parallel. So while the mask used in conventional lithography is expensive and slow to make itself, it can be used over and over again to quickly produce millions of devices. This ability to replicate devices is at the core of the inexpensive electronic devices available today. A similar parallel strategy can be applied to multiphoton lithography, but instead of using a 2D mask to reproduce features, a 3D mold can be used.

The technique of μTM was popularized by Prof. George Whitesides as part of a suite of tools known as *Soft Lithography* [9]. In Soft Lithography, an elastomeric polymer, polydimethylsiloxane (PDMS), is poured over a lithographed master and cured. The cured PDMS can then be used as a stamp, a mold, or as a functional device itself. When it is used as a mold, the technique is called *micro-transfer molding* (Figure 13.1). The low surface energy of PDMS enables it to wet surfaces very well, resulting in feature resolutions that are less than 100 nm. The

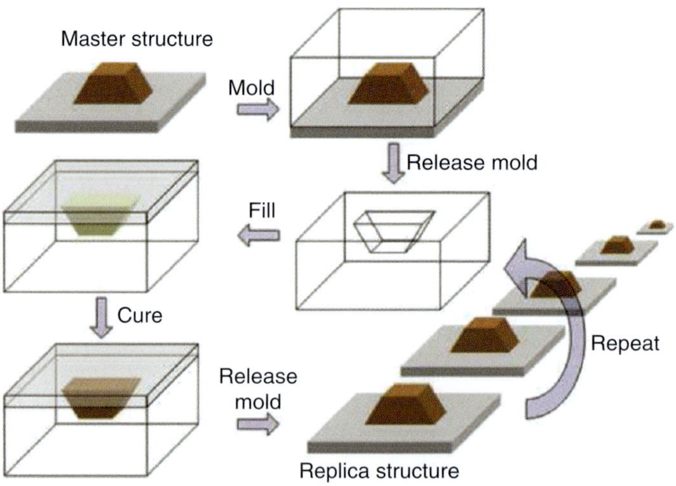

Figure 13.1 The process of μTM. A master structure, which can be made by TPIP, is molded in PDMS, and the mold is then filled and cured before releasing the replica structure. The mold can then be reused [1].

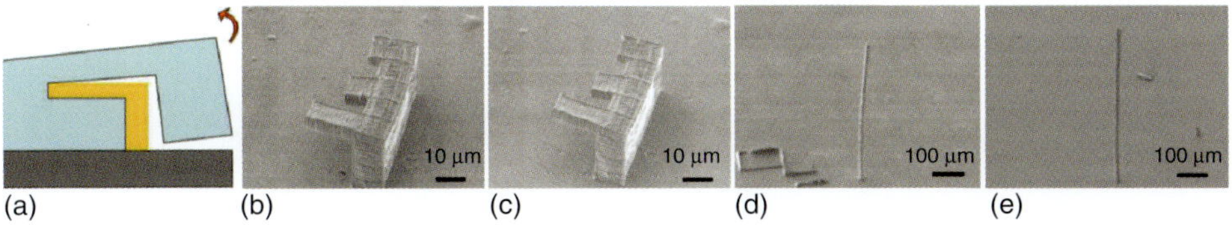

Figure 13.2 (a) Schematic of how a PDMS mold can flex around an undercut structure. (b) Master structure made by TPIP and (c) its replica. High-aspect-ratio tower (d) master and (e) replicas that are $10 \times 10\,\mu m$ in cross section and $500\,\mu m$ tall [1].

early examples of this technique predated multiphoton lithography and focused on conventionally lithographed masters, but it was not long after that masters made by TPIP were used with μTM.

Since structures made by TPIP can have any arbitrary geometry, some of the first demonstrations of nanoreplication using μTM showed the ability to reproduce undercut and high-aspect-ratio features, which are nontrivial geometries to fabricate by conventional photolithography. For example, the set of cantilevers shown in Figure 13.2 demonstrates that large overhanging features can be reproducibly molded and released using PDMS molds. PDMS is able to mold such geometries because of its low modulus, $2\,N\,mm^{-2}$, which makes it very flexible. PDMS has also been used to mold some very high-aspect-ratio structures, such as the tower shown in Figure 13.2. This tower is $10\,\mu m \times 10\,\mu m$ in cross section and $500\,\mu m$ tall. The replicas can also be used to create new molds, and masters can be reused to create new molds repeatedly.[1]

One issue with μTM is that, usually, the molded part is not able to be isolated as a free piece because it is connected to a replica of the substrate. Even though only the molded piece is filled and the mold is cleaned of liquid resin, there is inevitably some "flash" remaining around the new part. Macroscopically molded parts are usually deflashed by tumbling or grinding, but this is not feasible with microscopic parts.

There are two molding techniques: nanoimprint lithography (NIL) [10–12] and particle replication in nonwetting templates (PRINT) [13, 14], which are closely related to μTM but result in parts without flash. These techniques have not yet been used with TPIP-created masters. NIL is similar to μTM except that pressure is applied before curing to ensure the layer of flash is extremely thin, usually ~100 nm. After curing, this thin layer is removed by reactive ion etching, leaving clean parts on a substrate that is not made of the same material. In PRINT, the molding elastomer is not PDMS, but instead a fluorinated silane that is highly nonwetting, which causes only the part to be filled with monomer and results in essentially flash-free particles. Liquidia Technologies Inc. has already scaled PRINT up for the mass production of microscopic particles using a reel-to-reel format for the mold. These two techniques are poised for interesting future application of multiphoton nanoreplication, since both of these methods eliminate flash.

13.3
μTM of Complex Geometries

μTM has been demonstrated to work well for challenging geometries; however, it is still unable to reproduce all of the true 3D features that can be made using TPIP. Closed loops, for example, such as a ring or a coil, can be easily made by TPIP, but cannot be replicated by μTM.

A solution to the closed-loop topological problem was found by Fourkas and coworkers with their development of membrane-assisted micro-transfer molding or (MA-μTM) [15]. This technique takes advantage of a unique characteristic of PDMS: it is self-healing. This self-healing is shown in Figure 13.3, where the seam between two pieces of PDMS nearly disappears when they are pressed together. Upon contact, the two pieces of PDMS form a liquid-tight junction that can be formed reversibly.

In MA-μTM, the master structure is intentionally made to be different. Then the intended daughter structure as the master contains a membrane to prevent the formation of a closed loop during the molding process. As shown in Figure 13.4, if a structure containing a closed loop were fabricated, then upon molding and mold release, the original structure would be destroyed. However, if the original structure is built having a thin membrane (<1 μm thick), then upon molding no closed loop is formed. The mold is readily released from such a structure with moderate undercut features. The mold can then be pressed together, and the area where the membrane was can now be sealed, because PDMS self-heals. This mold can then be filled with a monomer, which is polymerized by exposure to UV light. Upon release from the mold, the daughter structure contains a closed loop. The mold can be then be reused to create more closed-loop structures.

In Figure 13.5 are shown more examples of structures created using MA-μTM, including a table, a coil, and a free-standing ring. Maruo and coworkers have also

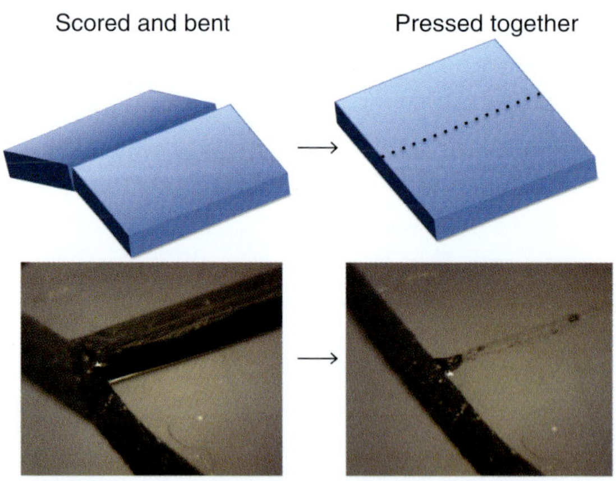

Figure 13.3 Wetting/self-healing of PDMS. After cutting, a piece of PDMS, approximately 2 mm thick, can be reversed and sealed back together.

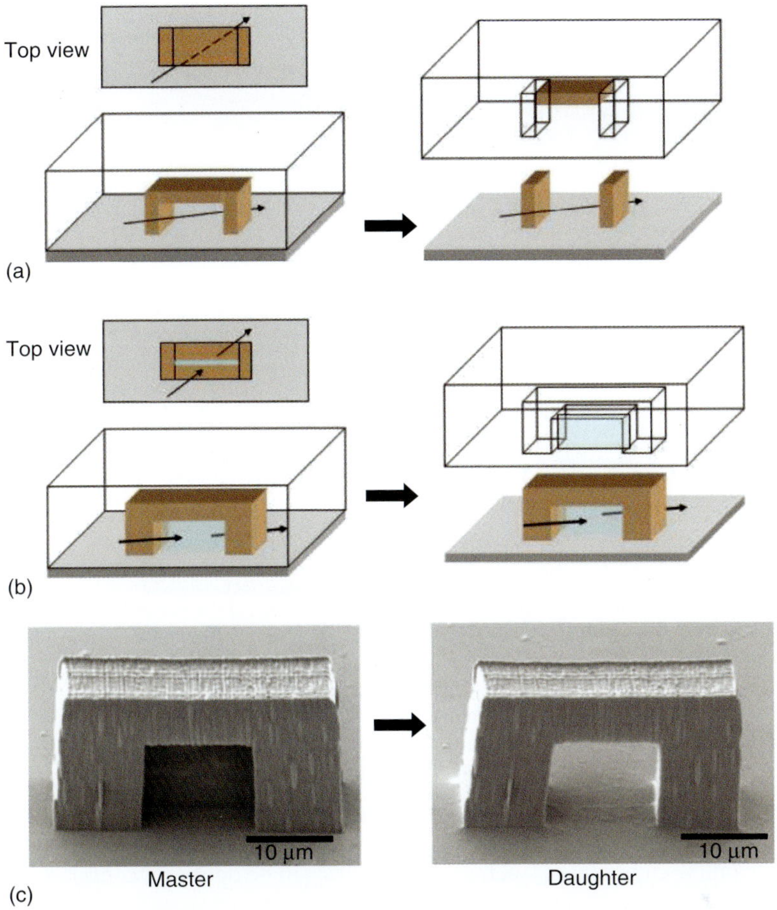

Figure 13.4 (a) Schematic of the closed-loop issue with μTM. (b) The membrane shown in blue prevents the mold from closing around the structure and enables the mold to be released. (c) Electron micrographs of a master structure containing a membrane and its daughter structure having a closed loop [1].

used this technique to create a freely moving gear, a vertical coil, and a micromanipulator [16]. These features were significantly smaller than previously demonstrated objects and were <10 μm in size. The membrane in these structures was ~100 nm wide.

13.4
Nano-replication of Other Materials

The vast majority of objects made using multiphoton lithography are acrylates, epoxides, or organically modified ceramics, ORMOCERS, which are hybrid

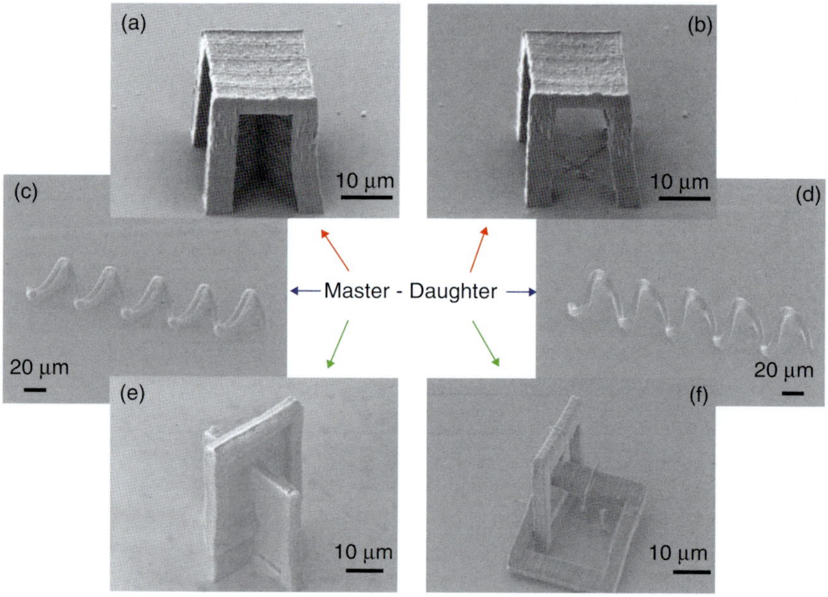

Figure 13.5 Various master structures (a,c,e) used for MA-μTM to create daughter structures (b,d,f) [1].

materials having a –O–Si–O– backbone and acrylic or epoxy groups for crosslinking [17, 18]. Chalcogenide resins [19] and resins containing titanium and zirconium [20, 21] have also been used in TPIP (see Chapter 3 for more examples). But another way to create structures out of different material is by making a master structure out of a photopolymer, molding it, and then filling the mold with a different material, such as silicon. This can be done in a conservative way where the mold can be reused, as in conventional injection molding, or in a sacrificial way where the master and mold are consumed for each part made, as in sand casting.

Wegener and coworkers were the first to create a 3D photonic crystal out of silicon by first creating a pattern by multiphoton lithography and then "double inverting" it into a Si structure [22]. The schematic in Figure 13.6 shows the steps involved. Briefly, the polymer microstructure made by TPIP is coated with SiO_2 using chemical vapor deposition (CVD) to create a mold. The original polymer material is then vaporized at high temperature, and the void is filled with Si, which is also deposited by CVD. Their final log-pile structure showed a bandgap near 2.3 μm.

Chichkov and coworkers also made photonic crystals by inverting an SU-8 master into titania [23]. Following fabrication of their woodpile structure, it was soaked in a solution of Ti(IV) isopropoxide in ethanol and allowed to react under an inert atmosphere. A sol–gel reaction proceeded following exposure to moist air, forming nanoporous TiO_2. The SU-8 master was then removed by calcination at 600 °C. The final photonic crystal showed a bandgap in the near-infrared.

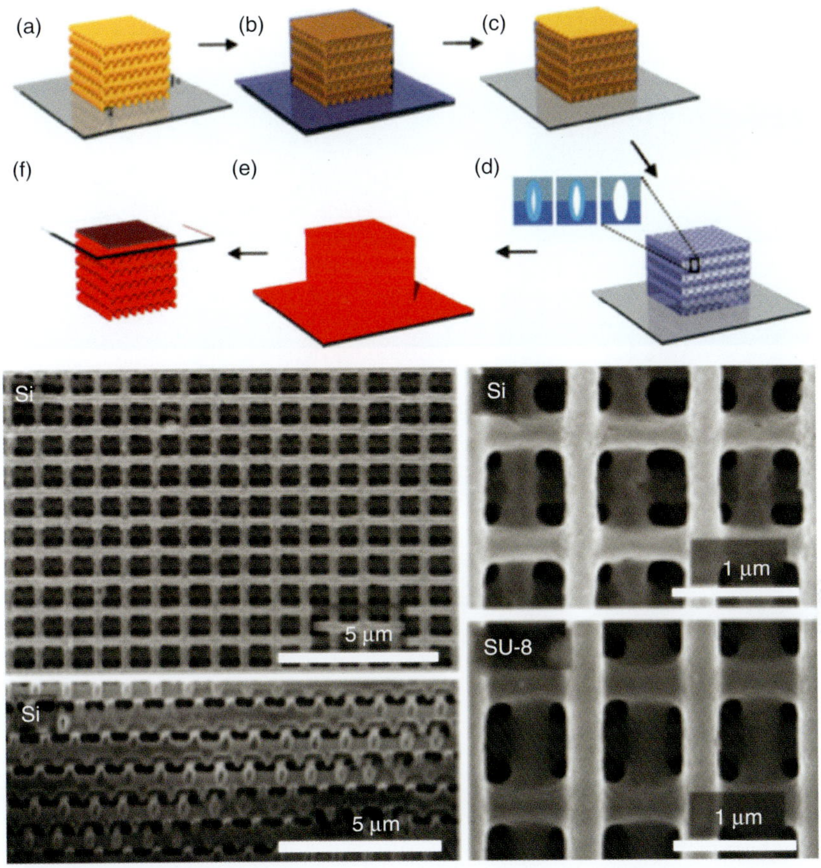

Figure 13.6 Master structure made of SU-8 is infiltrated with SiO_2, then the SU-8 is calcinated and removed, before being re-infiltrated with Si by chemical vapor deposition, and finally the SiO_2 is removed by HF etching. Scanning electron micrographs are shown of the master SU-8 structure and its replica made of Si. (Tétreault 2006 [22]. Reproduced with permission of Wiley.)

A chiral photonic crystal was also made by Gansel *et al.* using multiphoton lithography, but this time by using a positive photoresist on a transparent conductive indium-tin oxide (ITO) substrate [24]. After development of the exposed photoresist, the unexposed resist acted like a mold for the electrodeposition of gold. This method of galvanoforming has been used in the past as part of a set of techniques known by the German acronym LIGA (Lithographi, Galvanoformung, Abformung), but this was the first demonstration of LIGA using a multiphoton process to create the initial pattern. The initial pattern was of spirals having varying numbers of turns. It was found the gold photonic crystal had very different transmission spectra between 3 and 8 µm for left- and right-handed circularly polarized light.

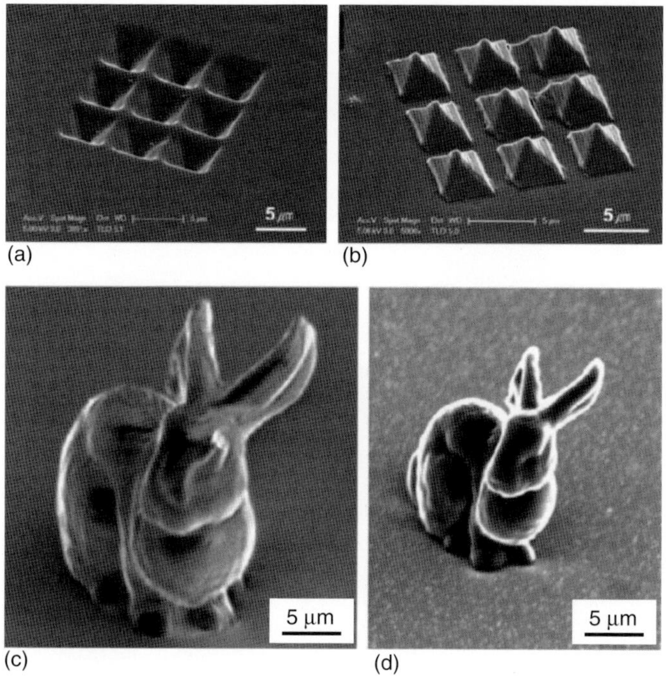

Figure 13.7 (a) Scanning electron micrographs of the mold of a set of pyramids made by TPIP and (b) a replica of the pyramids made from SiCN. (c) Master structure made of epoxy by TPIP and (d) a replica structure made from (c). (Daicho 2013 [26]. Reproduced with permission of Optical Society of America.)

Both these methods to create photonic crystals out of Si and Au are very impressive, but both processes only use the master and mold structure once and are inevitably wasteful and slow. A preferable method would be to reuse the mold for each replica. Unfortunately, this may not be possible given the complex geometry of a photonic crystal, but it has been done with simpler shapes such as a pyramid.

SiC-based ceramics have been patterned by μTM of urethane acrylate masters. The molds were filled with polyvinylsilazane, which, upon heating to 800 °C, becomes amorphous SiC ceramic with >50% shrinkage [25]. Carbon microstructures were also molded by Maruo and coworkers using specially developed resins that carbonize efficiently [26]. These resins contain an epoxy, resorcinol diglycidyl ether, which has a high carbon content, and shrinks <30% upon carbonization. Figure 13.7 shows SiC pyramids and an approximately 5-μm-wide rabbit made of carbon.

13.5 Nanosurface Metallization Processing

Direct multiphoton reduction of silver and gold has been used to create metal microstructures in two dimensions and to a lesser extent in three dimensions. One issue with direct photoreduction is that the resulting structures are granular and often quite rough. This granularity is due to the mechanism of photoreduction, which first creates nanoparticles that grow and eventually coalesce. The necessity of a reducing agent, which is usually an organic molecule, also means that frequently there is an organic impurity in or on the metal that can limit its conductivity.

Another strategy that is used more often is to create 3D metal patterns by coating polymer microstructures with a metal [27]. The first metalized polymer microstructures were made by Kawata and coworkers, who used Tollen's test reagents to electrolessly deposit silver [28, 29]. First, a polymer microstructure was made using a resin having monomers containing phenyl groups, which aided silver adhesion. The sample was fabricated on a hydrophobic glass surface. Next, they sensitized the surface with an aqueous solution of Sn(II). Since the surface was hydrophobic, the Sn(II) did not adhere to the substrate but only to the microstructure. The sample was then soaked in an aqueous solution of ammoniacal silver nitrate and glucose, which resulted in the formation of a silver mirror on the microstructures. Figure 13.8 shows a schematic of this process as well as electron micrographs of a coil microstructured before and after silver deposition.

Using an analogous strategy, Terzaki *et al.* added a tertiary amine-containing monomer, 2-(dimethyl amino) ethylmethacrylate (DMAEMA), to their silicon–zirconium hybrid resin to aid its metal-binding capability [30]. Following fabrication, they demonstrated two silver electroless plating techniques: (i) by forming gold nanoparticle or (ii) silver nanoparticle seeds on the microstructure. These seeds then provided a catalytic surface for the reduction of silver using either Tollen's reagents or a solution containing gum Arabic, citrate buffer, and silver lactate. By increasing the amount of the DMAEMA, the researchers could increase the density of the nanoparticle seeding, but at the expense of feature resolution. The best compromise of smaller features and high seeding density was 30 mol% DMAEMA. They were able to create ohmically conductive, silver-coated microstructures using this method.

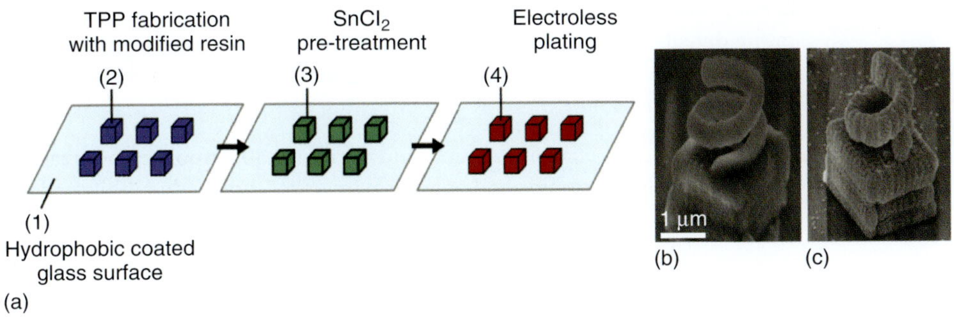

Figure 13.8 (a) Scheme showing the strategy for electroless plating of silver onto TPIP microstructures. Panels (b) and (c) show a microcoil before and after coating with silver. (Formanek 2006 [28]. Reproduced with permission of Optical Society of America.)

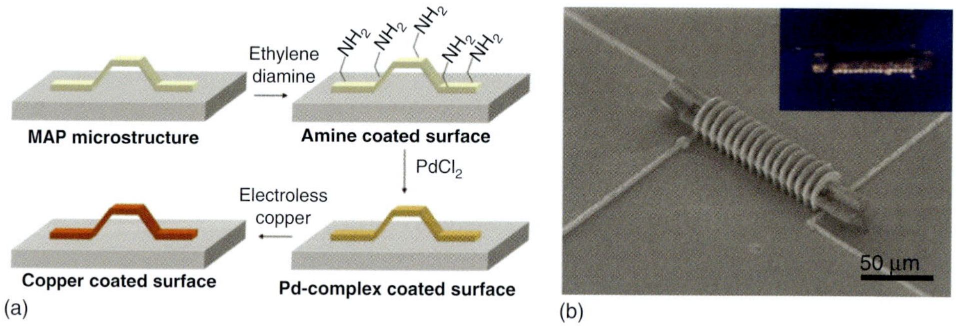

Figure 13.9 (a) Schematic of electroless deposition of copper onto microstructures made by TPIP. (b) Scanning electron micrograph of an inductor coated with copper and having a methacrylate support structure, which was uncoated. The inset shows an optical micrograph of the structure [1].

Nickel-coated photonic crystals were also prepared by Misawa and coworkers using SU-8 photoresist [31]. After fabrication and development, the samples were dipped successively in aqueous solutions of Sn(II) and then Pd(II) before being put in an electroless Ni plating bath. These Ni-coated SU-8 photonic crystals showed a bandgap in the 3–5 µm range.

The work by Farrer *et al.* [3] and also by Chen *et al.* [32] used a slightly different strategy than the previously described reports by taking advantage of the inherent chemistry of the acrylic monomers used for the TPIP resin. Using a Michael reaction, ethylene diamine was covalently attached to the surface of their microstructures [3]. This amine functionality was then used as a handle to immobilize Pd(II) catalysts, which then enabled gold or copper electroless reduction.

One important benefit of this reaction scheme is that the Michael reaction is significantly slower in its reaction with methacrylates compared to acrylates. Therefore, if a single microstructure is made in two steps using a methacrylate resin and an acrylic resin, then only the acrylic parts of the structure will undergo metallization (Figure 13.9). Li *et al.* went on to show that the density of the amine modification was a function of the amount of unpolymerized acrylate groups [33]. This amine density is dependent on the exposure dosage during polymerization, and they were able to show gray-scale patterning of this amine functionality. They demonstrated applications for biopatterning using this gray-scale method, but they did not demonstrate gray-scale metallization. The method used by Chen *et al.* [32] was similar using a diamino compound, but instead of using Pd to catalyze further metal deposition they used gold nanoparticles [32].

13.6 Nanosurface Structuring via Ablation

A relatively new method of fabrication for multiphoton lithography is via 3D laser ablation. Laser ablation is the process whereby a focused laser beam is used to vaporize a target material. Ablation has been used for decades in pulsed laser deposition to create high-quality crystalline thin films of ceramic materials by removing material from a bulk sample and exposing a substrate to the resulting plume. As it relates to multiphoton lithography, this ablation method has only been reported by a few groups, but it is poised to become an important tool in the microfabrication arsenal.

The basic idea of multiphoton ablation is very simple and is shown schematically in Figure 13.10a. After creation of a master structure by TPIP, it is developed, and then exposed to the same ultrafast laser for ablation. By scanning the beam with respect to the structure, parts of the structure can be vaporized. Exactly the same optical system can be used for both fabrication and ablation, but usually the laser power required is approximately 5–10 times higher for ablation than for polymerization.

Unlike multiphoton polymerization, ablation is a subtractive fabrication method. Since the process uses an ultrafast laser, it has the same spatial confinement in three dimensions as TPIP, but it can be used to remove material from a structure rather than add on to it. Why might removing material be useful? There are several practical reasons, such as (i) the ability to overcome structural fragility

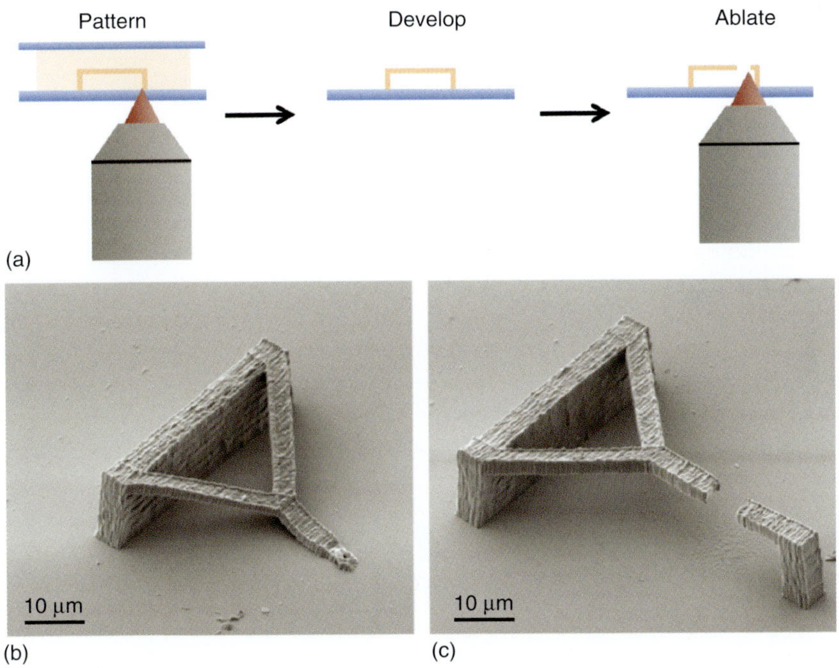

Figure 13.10 (a) Schematic of how ablation can be performed in conjunction with TPIP. The same laser system is used to remove material following polymerization and development. (b) A cantilever that fell during development due to capillary forces of the developing solvent. (c) A bridge-like structure that survived development and was subsequently ablated to create a cantilever like the one shown in (b).

during developing, (ii) reducing feature sizes, (iii) creating patterns faster, and (iv) as a means to generate a novel surface for chemical reactions.

One issue with some microstructures is that they can deform during development due to the surface tension of the developer solvent as it dries. This surface tension can cause some fragile structures to collapse. For instance, the cantilever in Figure 13.10b collapsed during development. To overcome this problem, a modified structure was created that resembled a bridge. After development, this bridge structure did not collapse. A portion of the bridge was then ablated, leaving the intended cantilever behind (Figure 13.10c) [1].

The image in Figure 13.10c shows some remnants of the removed material on the substrate. This micrograph highlights one of the flaws with ablation: while most of the material is vaporized and leaves the sample in the form of a gas, some parts may remain on the substrate or on the microstructure that can be difficult to remove.

Lim *et al.* also reported ablation as a means to create a cantilever by ablating a bridge, and they described another problem of ablation: surface roughness [34]. They ablated lines from thin films of the acrylate resin SCR500, and found that the surface roughness of the lines depended on both the thickness of the film and the location of the voxel relative to the surface. Thinner (300-nm-thick) films gave smoother sidewalls compared to thicker (500-nm-thick) films. They also found that if the focal point was placed at the surface of the film, the sidewalls were smoother than if it was placed at the bottom of the film. These surface roughness results highlight the importance of characterizing the ablation process, which has not been reported in the lithography literature. While ablation of bulk materials from their surface has been performed for years, there is a significant amount of experimental and theoretical work remaining to be done for ablation as it applies to multiphoton lithography.

The paper by Lim *et al.* also demonstrates a novel way to decrease feature size using ablation [34]. By simply ablating a portion of a line, a smaller line-width feature can be created (Figure 13.11). While the line sizes can be made significantly smaller this way, it should be noted that this does not prove that the resolution can be improved. Since resolution is defined by how close two lines can be to one another, making one line narrower is not equivalent to making two lines closer

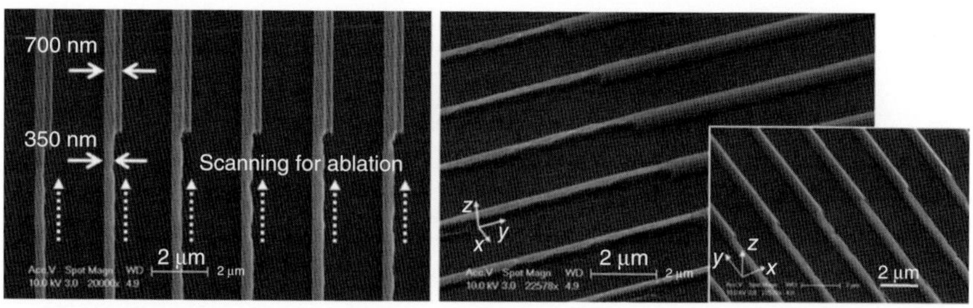

Figure 13.11 Line widths made smaller by ablation. (Lim 2010 [34]. Reproduced with permission of Springer.)

together. This could be an important method, however, in making small structures stronger. Typically, fine features suffer from low crosslinking density since they are made with a low exposure dosage. Small features made by ablation should be mechanically stronger, having been fabricated with more photons, and subsequently made smaller by ablation.

A subtractive technique, such as ablation, also enables some structures to be made significantly faster than additive structures. For example, a 3D microfluidic network can be very difficult and slow to create by traditional TPIP, because great care would be needed to polymerize lines that overlap sufficiently to not leak in some areas but are spaced appropriately apart in other areas to create channels. However, if an entire block is polymerized first, and then a 3D channel is ablated afterward, the structure can be made much more quickly. Xiong *et al.* showed exactly that process [35], a micrograph of which is shown in Figure 13.12. For this sample, the researchers used the IP-L resin from Nanoscribe Inc. They mention that the UV–vis absorbance spectrum of polymerized IP-L showed that it did not significantly absorb 390 nm, but that it did absorb well at 260 nm. They infer from this spectrum that the ablation process must be a three-photon nonlinear process and not a two-photon process, since their laser is of 780 nm. They show in their report that the ablated voxel is approximately 200 nm in diameter. They also

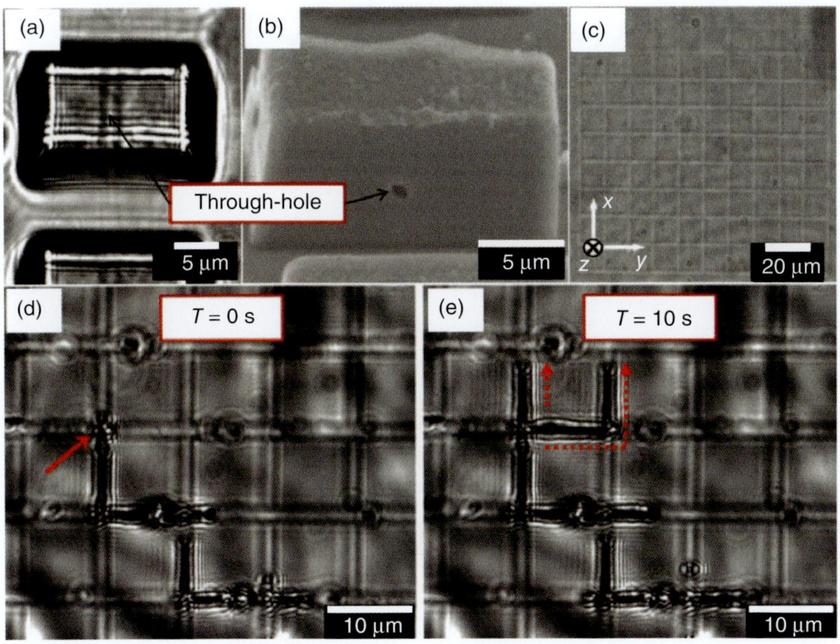

Figure 13.12 (a) Optical micrograph of a microfluidic channel inside a cube of solid IP-L photoresist. (b) Electron micrograph of the same cube. (c) Optical micrograph showing the channels within the cube. (d,e) Liquid flow inside the channel network at $t = 0$ and 10 s. (Xiong 2012 [35]. Reproduced with permission of Springer.)

demonstrate that fiber Bragg gratings can be made by ablating a periodic pattern in a waveguide, but they did not optically characterize the waveguides they made.

Midorikawa and coworkers were also able to create 3D microfluidic channels using femtosecond laser-assisted etching (FLAE) [36, 37]. While this process is not ablation, the results are very similar. They used near-IR pulses ~100 fs in duration with microjoule per pulse energies to ablate both fused silica and the photosensitive glass FORTAN. The photochemical reaction within the glass was multiphoton and could be localized in three dimensions. After patterning the glass, the exposed regions were wet-etched using dilute HF or KOH. Channels with arbitrary cross sections could be created, having length-to-width aspect ratios upwards of 200. These researchers have also pioneered the technique of using multiphoton polymerization inside these FLAE channels to pattern structures such as filters and mixers out of SU-8. They equate this fabrication method to creating a "ship-inside-a-bottle" (Figure 13.13).

Another promising application of ablation is for the chemical activation of surfaces. Work by Chen *et al.* used the fourth harmonic of a Nd:YAG laser at 266 nm

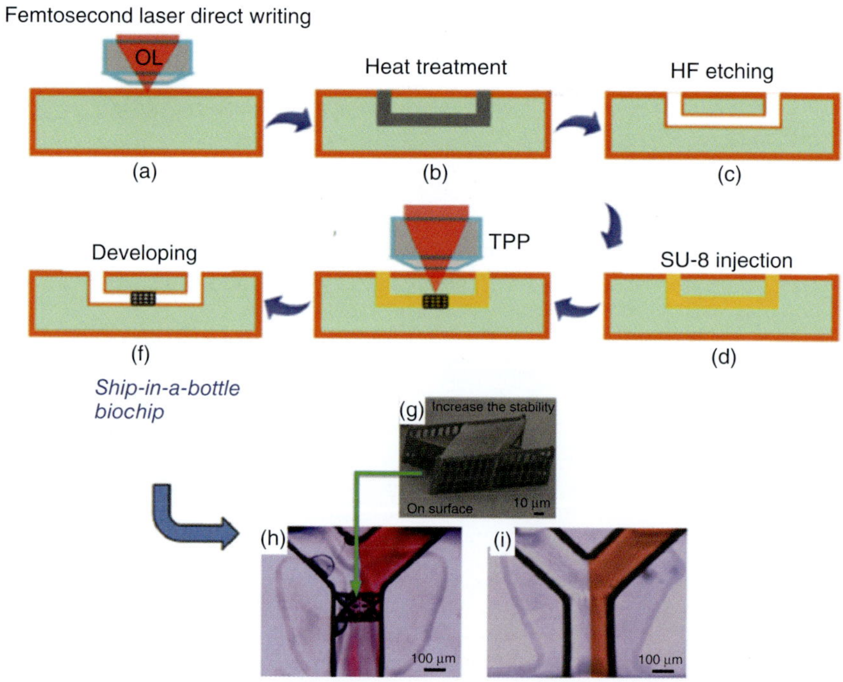

Figure 13.13 "Ship-in-a-bottle" process shown schematically. (a–c) A 3D pattern is created in glass by femtosecond-laser assisted etching. The channel is then filled with SU-8, and a mixer is created within the channel (d–f). Electron and optical micrographs show the mixer alone (g), the mixer in a channel (h), and a channel without a mixer for comparison (i). (Sugioka 2014 [37]. Reproduced with permission of Royal Society of Chemistry.)

to ablate a film of polyimide (Kapton) [38]. The patterned polyimide film was then exposed to aqueous solutions of ammoniacal silver, resulting in silver seed particles that adhered only to the ablated areas. The film was then exposed to a copper electroless plating bath to deposit copper wires. The wires were approximately 100 μm wide and 1 μm thick.

Kordas *et al.* also used laser ablation to enhance the metallization reaction [39]. They ablated low-temperature co-fired ceramics (LTCCs), which are mostly used in electronics packaging, using a Q-switched Nd:YAG laser at 1064 nm, with 150-ns pulses having 2 mJ per pulse. Ablated patterns could then be coated in copper, silver, or gold using electroless plating. They found that both ablation and annealing took place near the focal point and that annealing was more favorable for metal deposition. The ablated ceramics were analyzed by X-ray diffraction and Raman spectroscopy, and it was found that there were minor increases in the calcium aluminum borate peaks and the lanthanum borate peaks. They also found that the Raman spectrum significantly broadened, which suggested the formation of new chemical bonds. The researchers were able to create a miniaturized RLC circuit on the surface of their LTCC using their ablation technique.

Midorikawa and coworkers have also published on ablation-enhanced electroless deposition. They have shown that multiphoton-ablated glass, either fused silica or FORTAN, will selectively grow copper only in the ablated regions. The reason for the selectivity, they explain, is surface roughening due to ablation, which then causes particles of copper to stick and become anchored to the surface. These seed particles then catalyze the growth of a thin film of copper. They have used this technique to metalize the walls of microfluidic channels to create electrodes capable of manipulating Euglena cells via dielectrophoresis.

While these examples of metallization are impressive, it should be noted that they have not yet been used with conventional materials used in multiphoton lithography. This provides an opportunity for the TPIP community. There is reason to believe that acrylates, epoxies, or organically modified ceramics may also have the ability to be activated by laser ablation. If so, ablation-enhanced metallization could become a new route to 3D electronic devices.

13.7 Conclusion and Future Directions

While multiphoton lithography is somewhat limited in the speed with which objects can be made and the materials that can be used, many researchers have made progress to work around these issues. Using μTM, time-consuming devices can be quickly replicated even if they have quite complex geometries. MA-μTM allows the creation of daughter structures that can contain closed loops, such as rings, bridges, or coils.

One possible future direction is the use of ablation in conjunction with μTM to ease mold-filling and the release of parts. Many parts made by conventional injection molding are filled from a central runner that feeds molten polymer to the mold cavities. After the parts have hardened, they are broken free from the runner. This is not possible with microscale parts because of their size, but this could be accomplished via laser ablation, with its pinpoint ability to remove material. This new method would be an adaptation of a soft lithographic method known as microinjection molding in capillaries (MIMIC). Typically, in MIMIC the features are two dimensional, but by using TPIP, 3D structures can be produced in large numbers this way and then isolated from one another by ablation.

Another future direction that is exciting for replication is the use of fluoropolymer molds, such as those used in PRINT, which help eliminate mold flash. At present, the PRINT process has been commercialized and focuses predominately on simple geometries for drug delivery. Typical shapes for particles made by the PRINT process include hexagonal nut-shaped pieces, rectangular prisms, triangular prisms, and toroids, all having dimensions between ~1 and 10 μm. The mold is made using a proprietary process based on conventional photolithography and is made reel to reel, enabling mass production of particles. If masters made by TPIP were used instead to create the mold, more complex particles could be mass-produced using the PRINT process.

PRINT also exemplifies an exciting prospect for nanoreplication technologies – replicating unconventional materials. It may not be possible for a drug or a semiconductor to be patterned by TPIP, but these different materials can be cast. This area of casting functional nanomaterials is still in its infancy, and only a few examples have been shown thus far. Functional materials such as biocompatible polymers, semiconductors, metals, and magnetic materials have not yet been molded from TPIP-created masters. Whitesides and coworkers have developed what they call "microsolidics", in which traditional PDMS-based microfluidic channels are filled with low-melting alloys of In, Ga, and Sn and then solidified, leaving behind complex metal patterns [40]. These patterns may be able to extend one or two orders of magnitude smaller using TPIP to create the molds. Metals can also be cast from nanoparticle-based solutions to create metallic nanoparts. The Grzybowski group showed that a solution of Au nanoparticles having "sticky" capping groups could be cast and polymerized to yield microscopic Au parts [41]. Even alloy nanoparticles could be used to create mesoscopic metallic alloy parts. These are but two examples of the novel ways to cast microscopic metal, but there are numerous applications in optics, electronics, and medicine if semiconductors, ionic conductors, biopolymers, or magnetic nanostructured materials could be made by replication.

Another exciting possibility is the use of ablation to generate chemically reactive surfaces. Several examples were discussed previously, but this could become an important area of research as new materials are discovered to have this property. By combining molding and ablation, one can imagine the use of ablation of replicated parts made out of materials not suitable for TPIP. Ablation of this other material can then be used for patterning new surface chemistry.

After WW II, the plastics industry was revolutionized by the invention of screw injection molding, because it enabled complex parts to be mass-produced inexpensively; nanoreplication by μTM may offer an analogous advancement to nanomanufacturing. At present, TPIP is used to create devices that simply cannot be created any other way, but it is time consuming to create multiple devices. Nanoreplication offers a solution around this problem. These exciting solutions will continue to evolve and are sure to increase the reach of multiphoton lithography into a more widespread technology.

References

1. LaFratta, C.N. (2006) *Multiphoton Polymerization: Issues and Solutions*, University of Maryland College Park, College Park, MD.
2. LaFratta, C., Baldacchini, T., and Farrer, R. (2004) *J. Phys. Chem. B*, **108**, 11256–11258.
3. Farrer, R.A., LaFratta, C.N., Li, L., Praino, J., Naughton, M.J., Saleh, B.E.A., Teich, M.C., and Fourkas, J.T. (2006) *J. Am. Chem. Soc.*, **128**, 1796–1797.
4. Kumi, G., Yanez, C.O., Belfield, K.D., and Fourkas, J.T. (2010) *Lab Chip*, **10**, 1057.
5. Nielson, R., Kaehr, B., and Shear, J.B. (2009) *Small*, **5**, 120–125.
6. Kaehr, B. and Shear, J.B. (2007) *J. Am. Chem. Soc.*, **129**, 1904–1905.
7. Kato, J.-I., Takeyasu, N., Adachi, Y., Sun, H.-B., and Kawata, S. (2005) *Appl. Phys. Lett.*, **86**, 044102.
8. Gittard, S.D., Nguyen, A., Obata, K., Koroleva, A., Narayan, R.J., and Chichkov, B.N. (2011) *Biomed. Opt. Express*, **2**, 3167–3178.
9. Zhao, Y., Li, M., Lu, Q., and Shi, Z. (2008) *Langmuir*, **24**, 12651–12657.
10. Chou, S.Y., Krauss, P.R., and Renstrom, P.J. (1996) *Science*, **272**, 85–87.
11. Kim, Y.S., Park, J., and Lee, H.H. (2002) *Appl. Phys. Lett.*, **81**, 1011.
12. Costner, E.A., Lin, M.W., Jen, W.-L., and Willson, C.G. (2009) *Annu. Rev. Mater. Res.*, **39**, 155–180.
13. Rolland, J.P., Maynor, B.W., Euliss, L.E., Exner, A.E., Denison, G.M., and DeSimone, J.M. (2005) *J. Am. Chem. Soc.*, **127**, 10096–10100.
14. Zhang, H., Nunes, J.K., Gratton, S.E.A., Herlihy, K.P., Pohlhaus, P.D., and DeSimone, J.M. (2009) *New J. Phys.*, **11**, 075018.
15. Lafratta, C.N., Li, L., and Fourkas, J.T. (2006) *Proc. Natl. Acad. Sci. U.S.A.*, **103**, 8589–8594.
16. Maruo, S., Hasegawa, T., and Yoshimura, N. (2009) *Jpn. J. Appl. Phys.*, **48**, 06FH05.
17. Serbin, J., Egbert, A., Ostendorf, A., Chichkov, B.N., Houbertz, R., Domann, G., Schulz, J., Cronauer, C., Fröhlich, L., and Popall, M. (2003) *Opt. Lett.*, **28**, 301–303.
18. Haas, K.-H. and Rose, K. (2003) *Rev. Adv. Mater. Sci.*, **5**, 47–52.
19. Wong, S.H., Thiel, M., Brodersen, P., Fenske, D., Ozin, G.A., Wegener, M., and von Freymann, G. (2007) *Chem. Mater.*, **19**, 4213–4221.
20. Passinger, S., Saifullah, M.S.M., Reinhardt, C., Subramanian, K.R.V., Chichkov, B.N., and Welland, M.E. (2007) *Adv. Mater.*, **19**, 1218–1221.
21. Ovsianikov, A., Viertl, J., Chichkov, B., Oubaha, M., MacCraith, B., Sakellari, I., Giakoumaki, A., Gray, D., Vamvakaki, M., Farsari, M., and Fotakis, C. (2008) *ACS Nano*, **2**, 2257–2262.
22. Tétreault, N., von Freymann, G., Deubel, M., Hermatschweiler, M., Pérez-Willard, F., John, S., Wegener, M., and Ozin, G.A. (2006) *Adv. Mater.*, **18**, 457–460.
23. Serbin, J., Ovsianikov, A., and Chichkov, B. (2004) *Opt. Express*, **12**, 5221–5228.
24. Gansel, J.K., Thiel, M., Rill, M.S., Decker, M., Bade, K., Saile, V., von Freymann, G., Linden, S., and Wegener, M. (2009) *Science*, **325**, 1513–1515.
25. Lim, T.W., Park, S.H., Yang, D.-Y., Pham, T.A., Lee, D.H., Kim, D.-P., Chang, S.-I., and Yoon, J.-B. (2006) *Microelectron. Eng.*, **83**, 2475–2481.
26. Daicho, Y., Murakami, T., Hagiwara, T., and Maruo, S. (2013) *Opt. Mater. Express*, **3**, 875.
27. Tanaka, T., Ishikawa, A., and Kawata, S. (2006) *Appl. Phys. Lett.*, **88**, 081107.
28. Formanek, F., Takeyasu, N., Tanaka, T., Chiyoda, K., Ishikawa, A., and Kawata, S. (2006) *Opt. Express*, **14**, 800–809.

29. Formanek, F., Takeyasu, N., Tanaka, T., Chiyoda, K., Ishikawa, A., and Kawata, S. (2006) *Appl. Phys. Lett.*, **88**, 083110.
30. Terzaki, K., Vasilantonakis, N., Gaidukeviciute, A., Reinhardt, C., Fotakis, C., Vamvakaki, M., and Farsari, M. (2011) *Opt. Mater. Express*, **1**, 586–597.
31. Mizeikis, V., Juodkazis, S., Tarozaite, R., Juodkazyte, J., Juodkazis, K., and Misawa, H. (2007) *Opt. Express*, **15**, 8454–8464.
32. Chen, Y.S., Tal, A., Torrance, D.B., and Kuebler, S.M. (2006) *Adv. Funct. Mater.*, **16**, 1739–1744.
33. Li, L., Driscoll, M., Kumi, G., Hernandez, R., Gaskell, K.J., Losert, W., and Fourkas, J.T. (2008) *J. Am. Chem. Soc.*, **130**, 13512–13513.
34. Lim, T.W., Son, Y., Yang, D.-Y., Kong, H.-J., and Lee, K.-S. (2010) *Appl. Phys. A*, **103**, 1111–1116.
35. Xiong, W., Zhou, Y.S., He, X.N., Gao, Y., Mahjouri-Samani, M., Jiang, L., Baldacchini, T., and Lu, Y.F. (2012) *Light Sci. Appl.*, **1**, e6.
36. Wu, D., Wu, S.-Z., Xu, J., Niu, L.-G., Midorikawa, K., and Sugioka, K. (2014) *Laser Photonics Rev.*, **8**, 458–467.
37. Sugioka, K., Xu, J., Wu, D., Hanada, Y., Wang, Z., Cheng, Y., and Midorikawa, K. (2014) *Lab Chip*, **14**, 3447.
38. Chen, D., Li, Y., Lu, Q., Yin, J., and Zhu, Z. (2005) *Appl. Surf. Sci.*, **246**, 167–173.
39. Kordas, K., Pap, A.E., Saavalainen, J., Jantunen, H., Moilanen, P., Haapaniemi, E., and Leppavuori, S. (2005) *IEEE Trans. Adv. Packag.*, **28**, 259–263.
40. Siegel, A.C., Bruzewicz, D.A., Weibel, D.B., and Whitesides, G.M. (2007) *Adv. Mater.*, **19**, 727–733.
41. Klajn, R., Bishop, K.J.M., Fialkowski, M., Paszewski, M., Campbell, C.J., Gray, T.P., and Grzybowski, B.A. (2007) *Science*, **316**, 261–264.

Part V
Biological Applications

Multiphoton Lithography: Techniques, Materials and Applications, First Edition.
Edited by Jürgen Stampfl, Robert Liska, and Aleksandr Ovsianikov.

14
Three-Dimensional Microstructures for Biological Applications

Adriano J. G. Otuka, Vinicius Tribuzi, Daniel S. Correa, and Cleber R. Mendonça

14.1
Introduction

Several methods have been proposed for the fabrication of three-dimensional (3D) microstructures, aiming at application in distinct fields of science and technology, from microelectronics to biology [1–3]. Among such approaches, two-photon polymerization (2PP) stands out because of its ability to produce complex 3D microstructures with high spatial resolution and virtually no geometrical constraints. Furthermore, different molecules can be conveniently incorporated to obtain functionalized structures, giving special features to the final device. Functionalized structures allow several technological applications, such as optical data storage, reversible birefringence, optically active photonics crystals, and selective metal coating [4–7]. 2PP can be applied to a wide range of polymers, ceramics, and composites, providing high versatility to this fabrication method. An important compound for 2PP is the photoinitiator. This compound is responsible for absorbing two or more photons and starting the polymerization process. Several molecules have been employed as photoinitiator, aiming at improving yield of conversion, and therefore 2PP efficiency. Among the different means to start the photopolymerization, radical polymerization has been the most commonly used for fabrication via 2PP [8].

In 2PP, two near-infrared photons are simultaneously absorbed by the photoinitiator in such a way that their combined energy promotes a transition to a higher excited state, initiating radical formation and, thus, the polymerization reaction. Since two-photon absorption rate scales with the square of light intensity, the photopolymer excitation, and consequently the polymerization, is confined to the vicinity of the focal region, providing high spatial selectivity to this method. Experimentally, fabrication is usually carried out by using either a stage scanning system or a mirror scanning system, depending on specific details of the microstructure to be designed. While in the first method the laser beam is scanned with respect to the sample, in the second one the sample is scanned

Multiphoton Lithography: Techniques, Materials and Applications, First Edition.
Edited by Jürgen Stampfl, Robert Liska, and Aleksandr Ovsianikov.

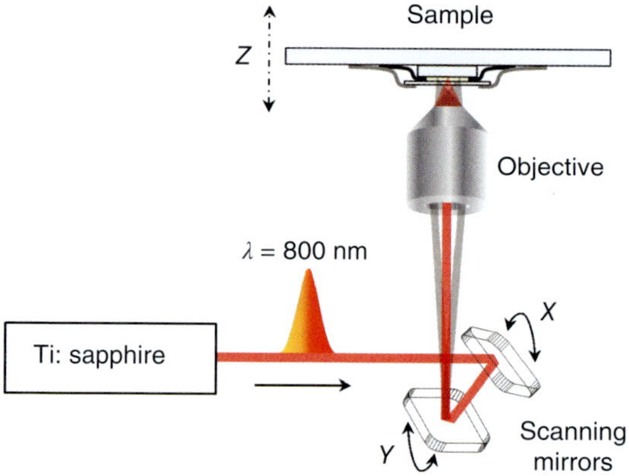

Figure 14.1 Typical experimental setup employed for fabricating 3D microstructures via two-photon polymerization.

relative to the laser beam focus. Figure 14.1 illustrates a typical fabrication setup that uses a scanning mirror system; a mode-locked Ti:sapphire laser oscillator (100 fs, 800 nm, and 80 MHz) is employed to induce the two-photon absorption in the photopolymer. The laser beam is deflected by the scanning mirrors in the $x-y$ direction and then focused into the sample with a microscope objective. The sample is placed on an axial z-stage, which is responsible for its vertical scanning. The desired 3D microstructure is produced by controlling the scanning mirror and the stage using a computer. Upon completion of the microstructure fabrication, the unpolymerized resin is rinsed with an appropriate solvent.

Among the several applications of 2PP, its use as a method to fabricate structures tailored for biological applications has received special attention in the last few years. This approach allows, for instance, precise control of the architecture, material, and doping of the 3D microenvironment in which microorganisms can be cultured, thereby enabling the study of the influence of different parameters on their adhesion, migration, and development. Therefore, the use of 2PP has prompted an interesting approach to obtain crucial information on microorganisms' behavior in controlled 3D environments. For example, structures fabricated by 2PP have been elucidating aspects of cancer cell migration in engineered scaffolds [9] and used to study stem-cell homing [10]. Aspects of bacterial growth and development have also been observed in two-photon fabricated structures [11], even containing antibiotics. Additional applications include areas such as tissue engineering, stem-cell growth and differentiation, and drug delivery. In this chapter we present some recent advances on the use of 2PP for the fabrication of microstructures for biological applications in topics related to (i) scaffolds and structures for cell migration investigations, (ii) structures containing biomaterials, (iii) structures for bacterial investigation, and, finally, (iv) microstructures for drug delivery.

14.2
3D Structures for Cells Studies

Standard approaches to study cell behavior in tailored environments are usually carried out on two-dimensional (2D) patterned substrates [12, 13] fabricated using well-known methods such as soft lithography and photolithography. The use of 2PP as a tool to fabricate scaffolds, however, has allowed the study of cell behavior in controlled 3D microenvironments, which has led to important advances to this area.

One of the first works proposing the use of 2PP to fabricate microstructures for studying cellular growth was conducted by Mazur's group at Harvard University [9]. In such pioneering work, 2PP was employed to produce interconnected structures (woodpile structures) containing square-shaped pores, with sizes varying from 12 to 110 μm, while the microstructures height and width were, respectively, 100 and 350 μm. These structures were fabricated using photopolymerized triacrylate materials. Although relatively simple, the advantage of this scaffold is that it allows cell migration as well as flow of the culture medium throughout the entire structure. Prior to use in cell studies, the microstructures were carefully rinsed in methanol to eliminate the toxic resin that was not polymerized during the fabrication process. Additionally, the authors carried out experiments to evaluate the mechanical properties of the scaffolds and their biocompatibility for cell growth [9].

Human fibrosarcoma cells (cell line HT1080) were labeled with green-fluorescent protein (GFP) by GFP plasmid transfection in order to enable 3D cell monitoring on the microscaffolds by confocal fluorescence microscopy. The labeled cells were seeded on the scaffolds (cell density 500 000 cells per ml), and cultured for a period of 24 h prior to starting the imaging procedure.

A representative bright-field image of cells, obtained 5 h after seeding, is presented in Figure 14.2a [9], for a microstructure with a pore size of 52 μm. It is interesting to point out that, in the beginning, the cells were situated only on the cover slip, but they started to move in three dimensions a few hours after seeding, as illustrated in Figure 14.2a.

As previously mentioned, the scaffolds were imaged 24 h after seeding using a spinning-disk confocal microscope. Images were obtained using a microscope objective with a numerical aperture of 0.75 (20×). The temperature of the microscope stage was kept at 37.8 °C, in such a way that all physiological conditions (temperature of 37.8 °C, 5% CO_2 and humidified atmosphere) were preserved during the experiments. In their work, authors were able to monitor cell movement by assigning it spatial coordinates (x,y,z) as a function of time using the intensity of the image obtained from the confocal microscope. Cell migration was then analyzed using three parameters; (i) the average speed, (ii) probability of movement, and (iii) motion persistence.

Three-dimensional cell movement was promptly observed by the authors [9] 24 h after seeding, as illustrated in Figure 14.2b, which displays a rendered 3D image of the labeled cells. These results reveal that the structures are indeed

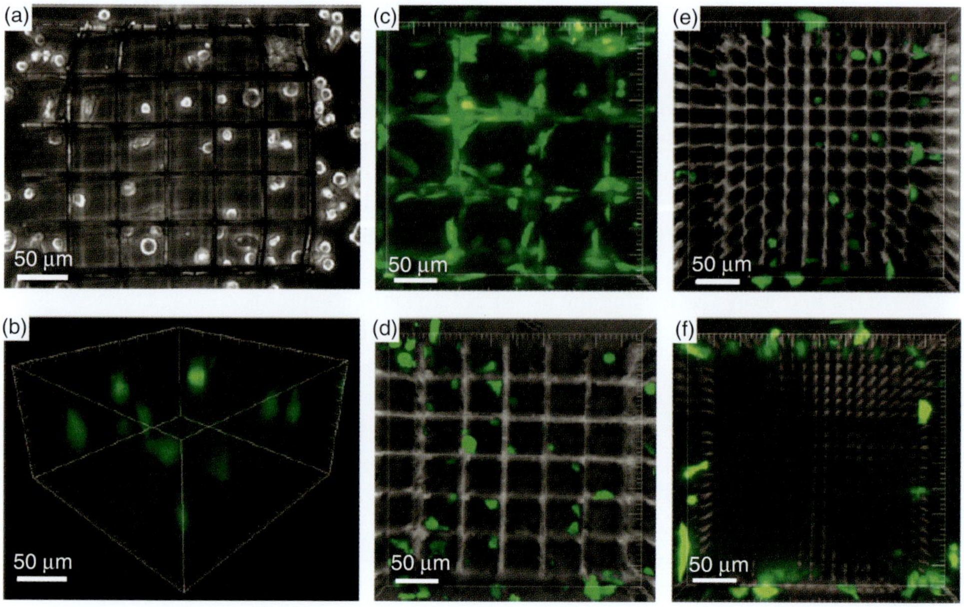

Figure 14.2 (a) Image (brightfield – top view) of HT1080 cells in a 52 μm pore-size scaffold obtained 5 h after seeding. (b) Rendered 3D fluorescent image of the cells inside the scaffold 24 h after seeding. (c–f) Overlay of fluorescence and differential interference contrast images, displaying the cells on the scaffolds with 110, 52, 25, and 12 μm, respectively. (Tayalia 2008. Reproduced with permission of Wiley.)

interconnected and that the cells are free to move in three dimensions in the microenvironment. Their data also reveal that for scaffolds with pore size of 110 μm, cell attachment and movements prevails along the beams of the woodpile structure, leaving the pores mostly unoccupied, as can be seen in Figure 14.2c. On the other hand, for structures with pore sizes of 52 and 25 μm, the cells tend to occupy the pores of the scaffolds. As seen in Figure 14.2d,e, for the 52-μm pore size scaffold the cellular distribution in the central region of scaffold is a little more uniform than for the 25-μm pore size scaffold, since the cells can migrate more easily because of appropriate pore size. For the 12-μm scaffold (Figure 14.2f), the cells remain mostly in the outer part of the structure because the pore size is smaller than the size of a single cell, which is estimated as ~15 μm.

By using image analysis, the authors [9] have tracked cell movement by analyzing images acquired for a period of 8 h at intervals of 10 min. The results demonstrated that, although the lateral speed and persistence of the cells in the $x-y$ plane of the 3D structure were the same as the ones obtained in a 2D substrate, the overall mean speed was higher in the 3D scaffold. On the other hand, the probability of movement in x and y directions for cells in the scaffold was lower than in the 2D substrate, which was be attributed to fewer hindrances encountered by cells on 2D surfaces. Interestingly, an increase in the overall mean speed was observed as the pore size increased, though the probability of movement and persistence remained nearly the same.

These results clearly showed the possibility of using 2PP to develop 3D scaffolds with controlled geometrical parameters that could be employed to either monitor or control cellular growth and adhesion, for instance. Aside from the obtained results, the relevance of this investigation [9] reside on the method described (fabrication and analysis technique), indicating it as a viable approach for cell migration studies in 3D environments.

A further advance in the fabrication of 3D microstructures for cell studies was achieved by using biodegradable and biocompatible polymers, which are important to design improved scaffolds for cellular growth platforms and tissue engineering. Biodegradable polymers of great importance are ε-polycaprolactone (PCL) and its derivatives. PCL is obtained by ring-opening polymerization of ε-caprolactone and has appealing features for biological application, including high degradability under typical physiological conditions (e.g., human plasma). The chemical degradability of PCL occurs by hydrolysis of ester groups, which makes this biomaterial suitable for applications in medical implants [14, 15]. The miniaturization of 3D structures with PCL increases the potential of scaffolds once structures with distinct geometries and high surface area can be tailored.

Fabrication of biodegradable microstructures has already been successfully reported in the literature by Claeyssens *et al.* [16], as depicted in scanning electron microscope (SEM) images displayed in Figure 14.3. In this figure, a structure based on triblock copolymer poly(ε-caprolactone-*co*-trimethylenecarbonate)-*b*-poly(ethyleneglycol)-*b*-poly(ε-caprolactone-*co*-trimethylenecarbonate) consisting of squares and horizontal/vertical lines that work as cell dividers is shown. This structure was fabricated via 2PP, using a power of 30 mW and a scanning speed of 50 $\mu m\ s^{-1}$. Although the authors observed distortion of the structure due to polymer shrinkage during photopolymerization, the shape and porosity was

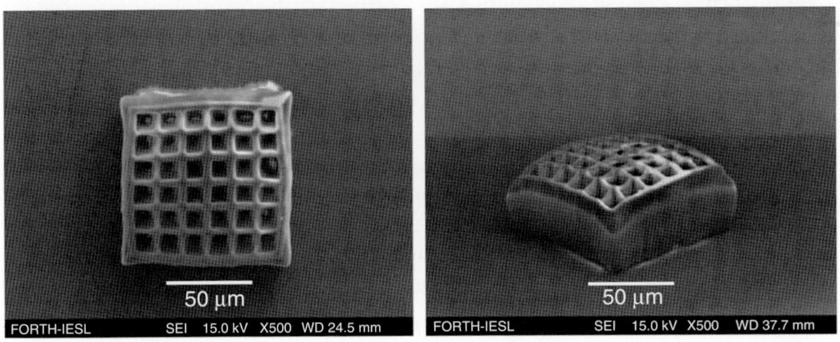

(a) (b)

Figure 14.3 Top view (a) and side view (b) of 3D microstructure fabricated using two-photon polymerization. These microstructures are composed of the triblock copolymer poly (ε-caprolactone-*co*-trimethylenecarbonate)-*b*-poly(ethyleneglycol)-*b*-poly(ε-caprolactone-*co*-trimethylenecarbonate). (Claeyssens 2009. Reproduced with permission of American Chemical Society.)

maintained. Further, cytotoxicity tests revealed that the biodegradable polymer did not affect cell proliferation, which is a crucial feature in tissue engineering and biomedical scaffolds applications.

Three-dimensional microstructures fabricated by 2PP using acrylated poly(ethylene glycol) have also been demonstrated [17]. The authors fabricated tubular scaffolds having a hexagonal outer shape and hollow hexagonal area in the central part, and seeded cells into these structures by using the laser-induced forward transfer (LIFT) technique. This technique uses a propulsive force, arising from laser-induced shock wave, to propel individual cells or groups of cells from a donor substrate toward the receiver substrate. This approach combines accurate control to fabricate structures within the size scale of the natural cell environment with precise deposition of cells on specific structure site. Such advantages are fundamental for precise 3D fabrication of scaffolds for *in vitro* studies as well as tissue engineering applications.

Doraiswamy and coworkers [18] used organically modified ceramic (Ormocer) materials for fabricating 3D microstructured scaffolds for tissue engineering because of their biocompatibility and good mechanical properties. These scaffolds were produced on glass substrates by means of 2PP.

In order to determine cell growth viability on Ormocer, live/dead assays of B35 neuroblast-like cells were performed. Prior to the test, the Ormocer material was placed in a plastic Petri dish sterilized by exposure to ultraviolet-B light. The cells were then incubated in these Petri dishes. Living cells were dyed to present green fluorescence, while dead cells would emit red fluorescence. The cells were imaged after different time intervals aiming at establishing a growth profile. This was made by using an inverted optical microscope and a camera to record the cells' epifluorescence.

After 48 h of incubation, fluorescence images revealed that all cells remained alive. The authors observed an increase in the number of neuroblast-like cells over time. The cell growth rates on polystyrene or Ormocer surfaces presented no noticeable difference. These results led to the conclusion that Ormocer materials do not significantly impair cell growth or viability.

Three-dimensional scaffolds in the form of arrays of cylindrical pillars were produced on both sides of an Ormocer surface. These structures might be interlocked in a lego-like manner to be used as single or stackable tissue engineering scaffolds. Figure 14.4 shows a fluorescence image of neuroblast-like cells on one of these samples.

The bright green shape in Figure 14.4b,d is one of the pillars in the scaffold. The pillars are 20 μm high and have a diameter of 75 μm. In these structures, the incubated cells tend to be oriented along the pillar walls. All the neuroblast-like cells were found alive after 48 h of incubation and their numbers increased with time, indicating that the scaffolds do not impair cell development. Two-photon polymerization of 3D interlocking scaffolds provides a useful method for engineering of multilayered tissues.

The geometric shape of microstructures employed for biological applications is a very interesting feature to be considered before starting the fabrication process.

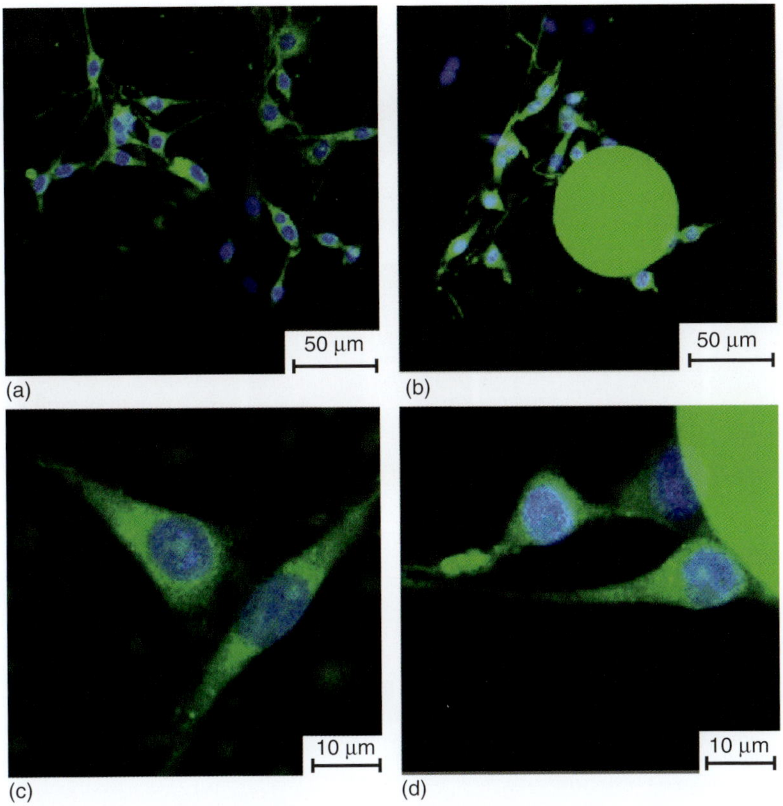

Figure 14.4 Fluorescent micrographs at several magnifications of B35 neuroblast-like cells on an Ormocer pillar 48 h after seeding. The cytoskeleton is denoted by green color and the nuclei are denoted by blue color. (Doraiswamy 2007. Reproduced with permission of Elsevier.)

Different height and pore dimensions (regions where bacteria or cells can enter the microenvironment) are interesting factors to explore in microenvironment fabrication.

For instance, Raimondi and coauthors [10] evaluated the biocompatibility, cell migration, and adhesion of cells in microenvironments, or niches, developed using different heights (20 and 80–100 μm) and lattice pore dimensions (10, 20, 30 μm, and graded). They fabricated the niches using a hybrid organic–inorganic photoresist SZ2080 and also used primary rat mesenchymal stem cells (MSCs) to study cell viability, migration, and proliferation in these niches. In their study, cell culture was carried out only up to 6 days. Fluorescent compounds were used as the main analytical tool in their study because cells could not be extracted from the niches independently from the cells adhered to the neighboring flat surface [10]. Furthermore, fluorescent compounds facilitate confocal observation. Figure 14.5 shows a summary of the results presented in the work of Raimondi *et al.* [10],

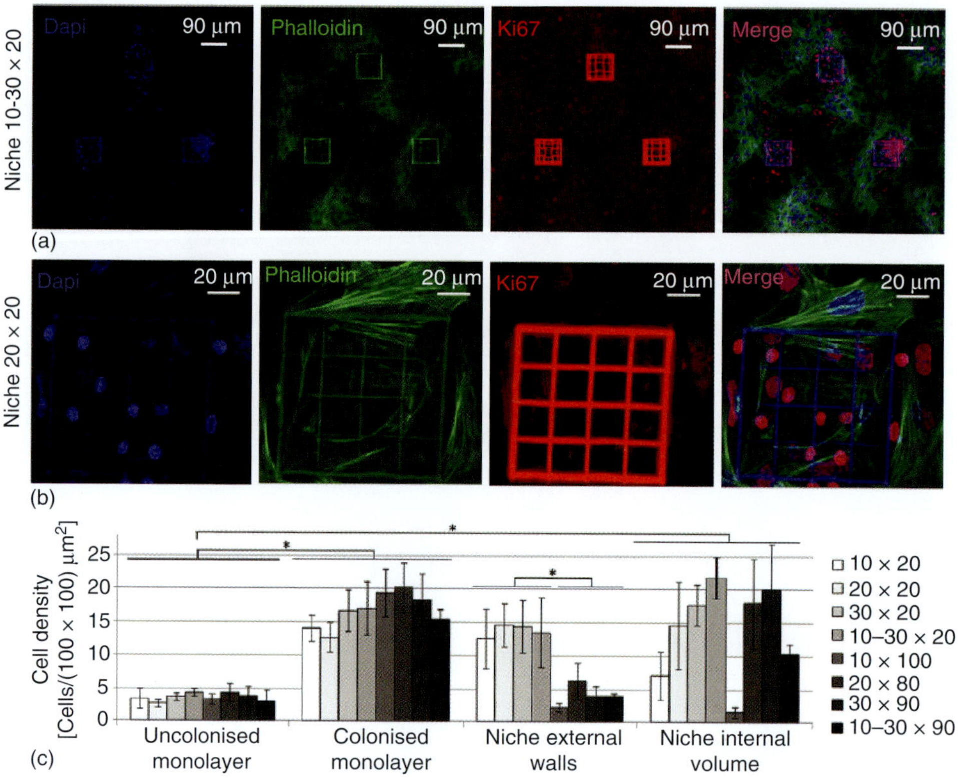

Figure 14.5 *Z*-stack projections of confocal images acquired on MSC-seeded 2PP-engineered niches at 6 culture days. Nuclei are stained with DAPI (blue). Actin filaments are stained with phalloidin FITC (green). Dividing cells are visualized by anti-Ki67 Cy3 (red). Nuclei of dividing cells co-express Ki-67 and DAPI, and appear pink in the merged image. (a) Cells adhere extensively to the glass substrate and form colonies in and around the niches. Proliferating cells are mainly localized in the central areas of the colonies. (b) Cells adhere extensively to the external niche walls and to the internal niche lattice. Proliferating cells are localized in all the niche internal volumes and on the niche external walls. (c) Cell densities measured for each niche architecture on four representative areas of the MSC-seeded culture wells containing the 2PP-engineered niches, at 6 culture days. (Raimondi 2013. Reproduced with permission of Elsevier.)

in which nuclei were labeled with DAPI (blue); actin filaments are labeled with phalloidin FITC (green), and dividing cells are visualized by anti-Ki67 Cy3 (red).

Raimondi *et al.* considered the niche with the smallest 10 μm pore size unviable because they had incompatible dimensions for cell development. However, they observed evidence of cell proliferation in the other niches (Figure 14.5a,b). Nuclei of dividing cells co-express Ki-67 and DAPI and appear pink in the merged image.

The authors found no significant difference in the average cell density between short and tall niches (Figure 14.5c). This result showed that the cells could physically migrate through the niche internal lattice and fill the niche up to a density primarily dependent on their colony-forming capacity and not on the available space [10]. In short niches, the authors concluded that the cell density increased with pore size, suggesting that high external walls represented an obstacle to cell invasion.

In another study performed by Malinauskas *et al.* [19], 3D scaffolds were fabricated with AKRE37 and SR2080 photopolymers, which are recognized materials for 3D micro and nanostructuring. In their work, typical parameters for fabrication, such as the average laser power and sample scanning speed, were determined, where they tested the scaffold's biocompatibility using adult rabbit myogenic stem cells. Furthermore, a cell growth comparison was performed in scaffolds made with Ormocer and PEG-DA (Poly(ethylene glycol) diacrylate), which are biocompatible materials commonly used as the reference photopolymers for determining biocompatibility for cell growth [20, 21]. The scaffolds produced had a truncated octahedral form and exhibited high mechanical strength. The typical pores sizes of the scaffold were engineered with twice the size of a single cell, and for this study it corresponded to tens of micrometers [22]. The results obtained by Malinauskas *et al.* [19] showed that all scaffolds had similar cell density, proving that the scaffolds made with AKRE37 and SZ2080 are potential materials for producing biocompatible microenvironments to evaluate cell growth.

14.3 Biocompatible Materials

As previously mentioned, the use of biocompatible materials is of utmost importance for fabricating structures with improved features, aiming at biological applications. In this direction, 3D microstructures containing biomolecules have been proposed by Drakakis *et al.* [23]. In their work, they constructed 3D microstructures using the commercial and biocompatible hybrid Ormocer. This material is highly transparent, and allows three-photon polymerization using an infrared femtosecond laser source, yielding optical resolution surpassing those attained by 2PP. Biotin was subsequently attached to the hybrid structure using a photosensitive compound called *photobiotin*, which can be activated by UV light. The integrity of the photolytically immobilized biotin was confirmed by fluorescence microscopy, which could detect the binding of avidin to it (see Figure 14.6a). Three-dimensional samples treated only with avidin but not with photobiotin showed no fluorescence (see Figure 14.6b), as avidin could not be retained in the porous matrices. This is due to the fact that avidin incorporation occurs only by a chemical bound to the biotin present in the structure. When irradiated by UV light, photobiotin released reactive biotin, which reacted with dangling Si–O bonds at the surface of the polymerized structure [23]. The results attained by Drakakis *et al.* [23] demonstrate the possibility of using this approach to build complex 3D structures with several types of biomaterials.

Aiming at designing biocompatible scaffolds, 3D microstructures containing chitosan fabricated by 2PP have been proposed by Correa *et al.* [24]. Chitosan [(1→4)-2-amino-2-deoxy-β-D-glucan] is a cationic polysaccharide usually

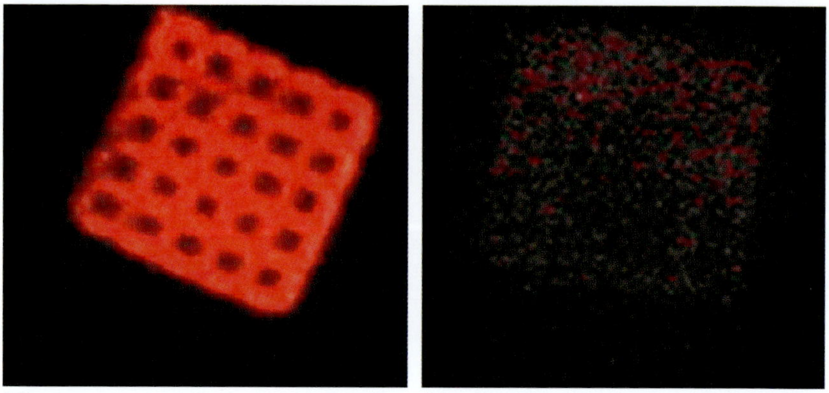

(a) (b)

Figure 14.6 Three-dimensional structure with photolytically attached biotin and streptavidin. (a) The 3D structure presents a strong fluorescence when exposed to green light. (b) A 3D structure with the photobiotin attachment step not accomplished. As streptavidin is not attached to the structure, no fluorescence is observed when the structure is shone with green light. (Drakakis 2006. Reproduced with permission of American Institute of Physics.)

obtained by deacetylation of chitin [(1→4)-2-acetamide-2-deoxy-β-D-glucan], which is a structural polysaccharide found in shrimps and other crustaceans shells. Chitosan presents appealing features including biodegradability and biocompatibility, and has been employed for biomedical applications, including blood coagulation, taste sensors, tissue and bone regeneration, antibacterial action, and drug delivery systems [25–27]. The authors incorporated chitosan in triacrylate-resin-based structures. SEM images of 3D structures containing chitosan are displayed in Figure 14.7, which demonstrate the high spatial resolution and structural integrity attained by the fabrication method.

To verify whether the chitosan was retained in the polymer structure, Raman scattering spectra were obtained from neat and chitosan-containing structures.

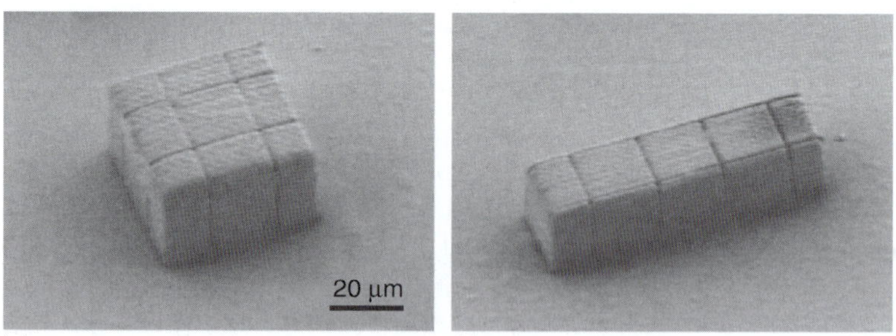

(a) (b)

Figure 14.7 Scanning electron micrographs of microstructures containing chitosan fabricated by two-photon polymerization. Panels (a) and (b) have the same scale bar. (Correa 2009. Reproduced with permission of American Scientific Publishers.)

The Raman spectrum of the neat 3D structure is displayed in Figure 14.8a, which shows vibrational lines at 1637 and 1736 cm^{-1}. The left two images in the inset display Raman maps obtained from the intensity of the vibrational modes at 1637 and 1736 cm^{-1}. The rightmost image of the inset was obtained from the intensity of the vibrational mode at 1858 cm^{-1}, where the resin presents no Raman peak. Figure 14.8b displays the Raman spectrum of a structure containing chitosan, using the same excitation wavelength. Again, Raman measurements did not reveal any specific chitosan peak, but a monotonic increase in background signal, characteristic of the chitosan fluorescence. These data indicate that the structures fabricated retained chitosan in their bulk. Another remarkable feature is that hardness measurement showed that the microstructures containing chitosan retained

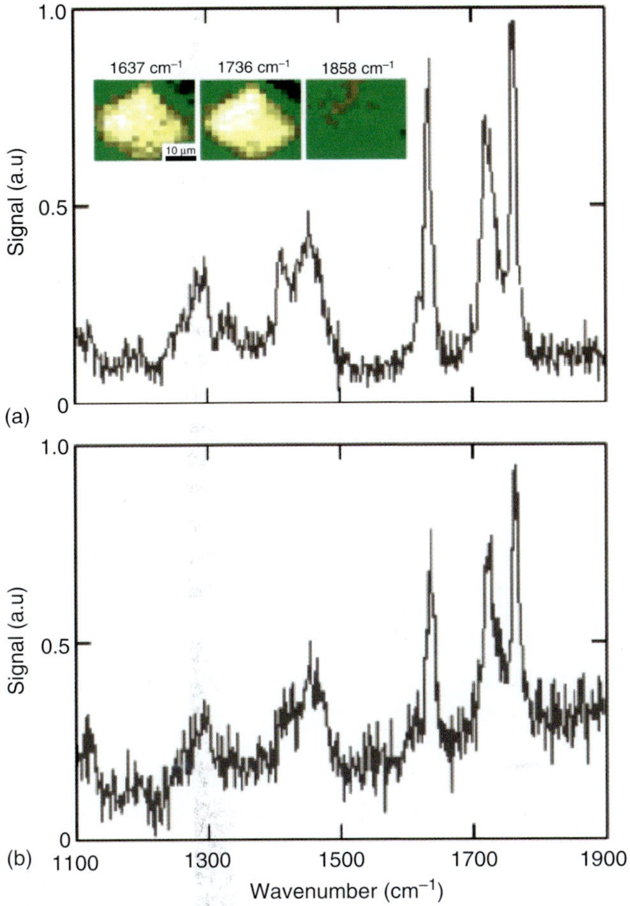

Figure 14.8 Raman spectra of a microstructure (a) without and (b) with chitosan. The inset in (a) shows the images built from the Raman scattering signal at 1637, 1736, and 1858 cm^{-1}, using excitation at 514 nm. (Correa 2009. Reproduced with permission of American Scientific Publishers.)

the same properties as those of neat microstructures, which is important for biomedical applications [24].

Although 2PP is usually applied to photosensitive materials composed mostly of polymeric resins mixed with a suitable photoinitiator, 2PP can as well be used to produce solid microstructures from protein photocrosslinking [28, 29]. In this case, the final structures display high porosity, permeability, and tunable elasticity [30, 31].

Several soluble proteins can be photocrosslinked by means of 2PP, and those are capable of retaining their native molecular activity so that structures with unique chemical and biochemical properties can be fabricated. Also, conjugation of protein-based hydrogels can be used to tailor their chemical functionality, resulting in structures with various bioactive species. Therefore, the production of protein microstructures has great potential for biological and biomedical applications.

The production of microparticles with complex three-dimensional shapes can be of great importance for characterizing and tailoring phagocytotic uptake [32, 33] in order, for example, to optimize drug delivery processes. But in most 2PP systems it is not possible to produce free microparticles. Because of the drifting of the structure during the fabrication process, which results in shape distortion, the fabricated objects must be attached to a fixed surface. Therefore, it is not possible to produce microparticles capable of retaining translational and rotational degrees of freedom.

In order to address this problem, Shear and coworkers [34] have developed a protein-based material with high viscosity that allows the production of free-floating microstructures with arbitrary shapes. This is achieved through a controlled evaporation process of a mixture of the desired soluble protein, a photosensitizer, and DMSO (dimethyl sulfoxide) in aqueous buffer, which is used to demonstrate the production of a series of unconstrained microstructures. Incorporation of DMSO in the mixture is necessary to achieve a homogeneous material.

To demonstrate the possibility of creating functional mechanical devices, a paddlewheel inside a cylindrical chamber was produced from a bovine serum albumin (BSA) protogel. The paddlewheel had a diameter of 12 μm and spun around a vertical post that extended from the bottom to the top of the cylindrical chamber. This device had an opening at the cylinder wall, which allowed the introduction of motile bacteria. For this experiment *Pseudomonas aeruginosa*, a 2-μm-long and 0.5-μm-wide bacterium, was introduced into the structure. These bacteria are propelled by a single flagellum. Rotation of the paddlewheel, due to its interaction with moving bacteria, could be observed even at low bacterial density.

As a first structure to study the interactions between living microorganisms and free-floating microparticles, a trap containing an unconstrained probe was built. The trap was designed to capture and retain motile bacteria; a funneled entry-way led to a small entrance which minimized the escape of the bacteria. This funnel led to a dome, attached to a surface. Inside this compartment an 18-point star was fabricated, which would interact with trapped bacteria.

The trap worked well in accumulating bacteria inside the interaction chamber. Star movement was observed as a result of direct collision with bacteria and also because of the flagellum-generated convective flow inside the trap. Figure 14.9 shows the resulting three-dimensional movement (translation and rotation) of the microprobe [34].

These results demonstrate the great potential of protein-based microstructures with translational and rotational degrees of freedom to study the interaction with motile microorganisms. Fabrication of protein-based microstructures is important for studying drug delivery mechanisms since recent investigations have indicated that phagocytosis might greatly depend on microparticle's size and shape. The versatility of this type of material paves the way for the production of new functional and biocompatible microdevices with improved flexibility and environmental sensitivity.

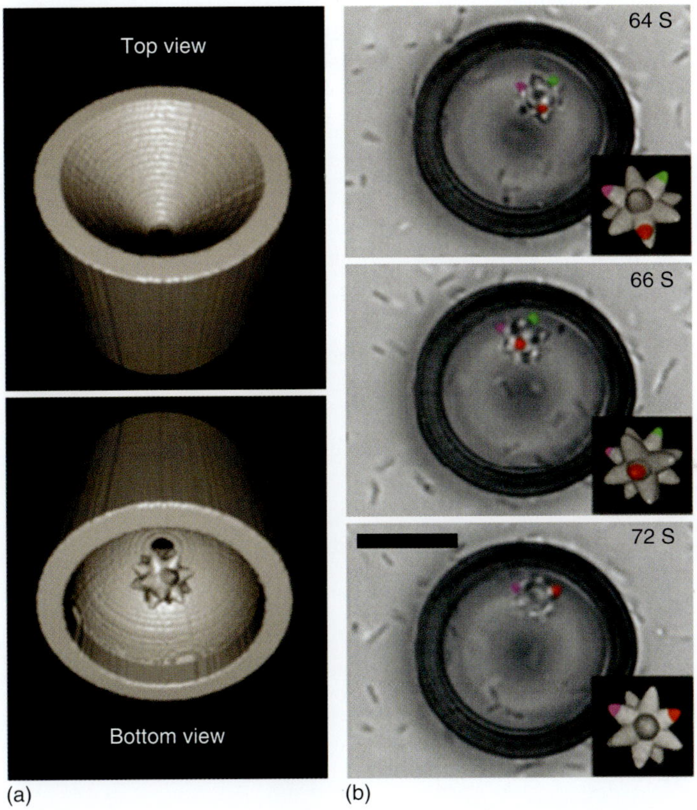

Figure 14.9 Probing bacterial motility using unconstrained microstructure. (a) Reconstruction of the bacterial trap containing unconstrained 18-point stars. The top frame shows the funneled surface used to guide bacteria toward a capture chamber containing a single unconstrained star. The bottom frame provides a view into the capture chamber. (b) Time sequence showing the tumbling movement of an 18-point BSA star caused by *P. aeruginosa*. The scale bar is 15 μm. (Spivey 2013. Reproduced with permission of Wiley.)

14.4 Scaffolds for Bacterial Investigation

In general, the microenvironments used to evaluate bacterial growth require special conditions. For instance, a microenvironment could be used to perform a quantitative analysis of the action of antibiotics in drug delivery systems. For such application, these microenvironments should be doped at specific sites with biological agents, such as antibiotics. In 2014, Otuka *et al.* [11] demonstrated the production of microenvironments prepared to evaluate *Escherichia coli* (*E. coli* in Mueller–Hinton broth) growth. The base resin used in that study was composed by two triacrylate resins and the photoinitiator *2,4,6-trimethylbenzoylphenyl phosphinate* (Lucirin TPO-L) [8]. These environments were doped at specific sites with ciprofloxacin (a typical drug used in the treatment of *E. coli* infection).

The microenvironment was composed of an array of 25 solid cylinders (5×5 matrix), in which the central cylinder was solely doped with 12.5×10^{-2} μg ml^{-1} of ciprofloxacin. The antibiotic was easily mixed with the polymeric resin. The microenvironment was fabricated in steps: first, only the doped portion was polymerized, and after a washing procedure, the undoped parts were polymerized. Figure 14.10 shows an SEM image (tilted view) of the fabricated microenvironment.

The SEM image in Figure 14.10 illustrates that, even for the doped cylinder, the structures had good integrity and definition, indicating that the presence of

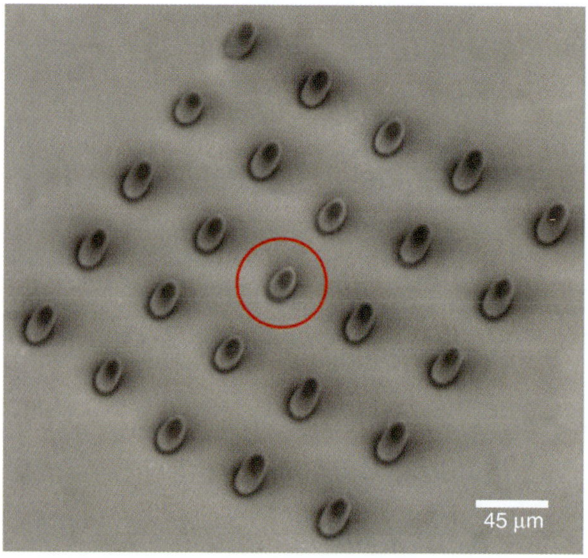

Figure 14.10 SEM micrography (tilted view) of the microenvironment composed of solid cylinders fabricated by 2PP. In the center of the microenvironment, the cylinder is doped with ciprofloxacin. (Otuka 2014. Reproduced with permission of Elsevier.)

the antibiotic in the resin formulation does not affect the 2PP process. Additionally, the microstructures maintain their structural features even after the repeated washing procedures used to remove out the unpolymerized resin before the inoculation of the bacteria into the microenvironment.

E. coli was inoculated in the microenvironment with all the necessary conditions for their development. Bacterial growth was monitored hourly, using an optical microscope. Figure 14.11a shows an optical microscopy image (top view) of the microenvironment after 1 h of incubation. An inhibition halo around the central cylinder can be observed, which is due to the presence of the antibiotic ciprofloxacin in this structure.

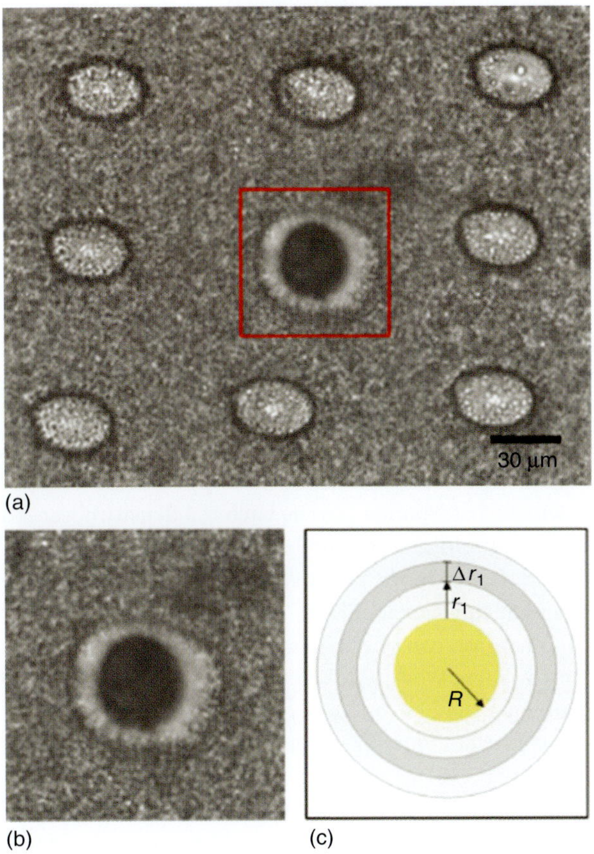

Figure 14.11 (a) Optical microscopy image (top view) of microenvironment after 1 h of incubation with *E. coli*. In detail, it is possible to observe the inhibition zone around the cylinder containing ciprofloxacin. (b) Magnified view of the region where the inhibition of bacterial growth was observed. (c) Illustration of the scheme used to analyze the inhibition zone. (Otuka 2014. Reproduced with permission of Elsevier.)

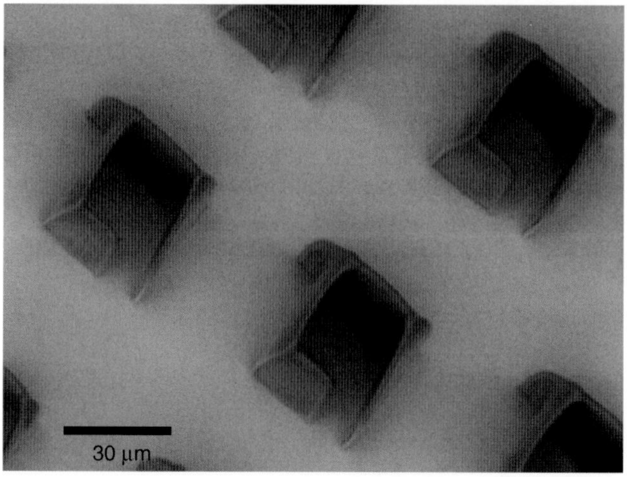

Figure 14.12 SEM micrography of microtraps fabricated by 2PP. (Otuka 2014. Reproduced with permission of Elsevier.)

In Figure 14.11b a magnification of the region where the inhibition of bacterial growth was observed. To evaluate the range of antibiotic action in the microenvironment, the authors [11] used the method illustrated in Figure 14.11c. Simply, the inhibition zone was divided into concentric rings with an inner radius r_i, starting from the edge of the doped microstructure (displayed in yellow color in Figure 14.11c), and thickness Δr_i. The number of bacteria in each ring was counted with the aid of an image software, allowing the determination of the density of bacteria per unit area. In this situation, r_i varies from 0 to 25 μm, and Δr_i was kept constant at 2.5 μm. This study showed that the density of bacteria grew monotonically with r_i, and saturated when r_i reached approximately 12 μm in about 0.7 bacteria μm^{-2}, which corresponded to the bacterial density observed in the areas not affected by the antibiotics. The inhibition zone had a maximum range of approximately 10 μm, being more effective on getting closer to the microstructure impregnated with ciprofloxacin [11].

The dynamics of bacterial migration can also be evaluated using microenvironments fabricated by 2PP. In the same work, Otuka *et al.* [11] demonstrated the fabrication of microenvironments, named microtraps, by which they analyzed the motion and entrapment of *E. coli.* Figure 14.12 shows an SEM micrograph of the fabricated microtraps.

The traps displayed in Figure 14.12 are 35 μm long, and present an aperture of 7 μm, which were designed without a ceiling to facilitate the real-time visualization of the migration and trapping of bacteria. Again, in this case *E. coli* was incubated in the microenvironment with the necessary conditions for their development, and monitoring was done by optical microscopy over the course of 1 h, in intervals of 5 min. The resulting images were analyzed, and the number of bacteria in each microtrap was counted as a function of time. Figure 14.13a shows

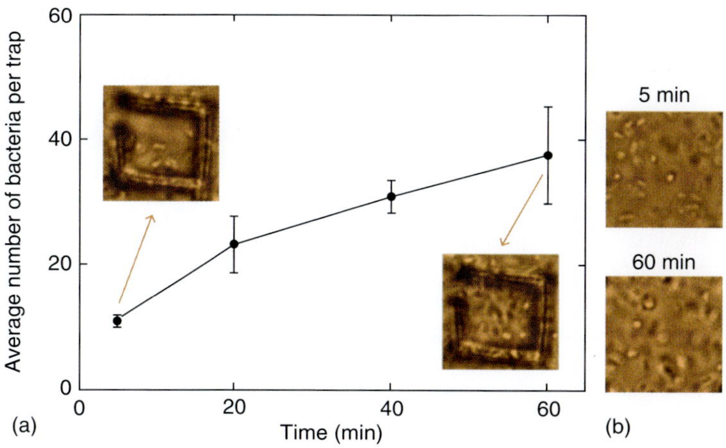

Figure 14.13 (a) Average number of trapped bacteria as a function of time. The error bars represent the standard deviation. The insets display images of the bacterial trapping for 5 and 60 min. (b) Images of the bacteria in regions outside the microtrap for 5 and 60 min, respectively. (Otuka 2014. Reproduced with permission of Elsevier.)

the average number of bacteria trapped per microtrap as function of time. These results correspond to an evaluation of five microtraps [11].

Figure 14.13a shows that the number of bacteria trapped in the micro-fences increases as function of time. The error bars in this figure represent the standard deviation of the average number of bacteria trapped. As the traps have no ceiling, some of the entrapped bacteria can leave the microtrap through the top, increasing the error bars for longer times. The inset in Figure 14.13a displays the images of bacterial trapping for 5 and 60 min. In Figure 14.13b, the images of bacteria in regions outside the microtraps are shown, for incubation times of 5 min (top image) and 60 min (bottom image). The average number of bacteria observed, in this case, was about 35. These values remained approximately constant during the time of the experiment (60 min). The number of bacteria lying outside the microtraps is higher than inside for times smaller than 40 min. The average number of entrapped bacteria approaches that found in regions outside the microtraps for long times. The results showed that particularly designed microenvironments could be used to study bacterial migration and motion, opening new opportunities for microbiological studies.

14.5 Microstructures for Drug Delivery

In many treatments, the drug or vaccine of interest cannot be administered by oral means because it might be metabolized by the liver, kidneys, or intestines before reaching the circulatory system [35]. Therefore, in these cases intravenous administration can be used by applying hypodermic needles. However, the use of a needle cause some unwanted effects, such as pain to the patient and local damage at the injection site. In addition, this kind of application also requires medical expertise and it is not suited for cases that demand sustained delivery of the medicine.

A good alternative for hypodermic needles is an array of microneedles [36]. By using needles with reduced dimensions, it is possible to greatly reduce both pain to the patient [37] and damage at the injection site. Also, applications using microneedles require no expertise, and hollow devices may allow for diffusion-driven drug administration over an extended period. In this way, microneedles provide a unique method for transdermal or intravenous drug delivery of many kinds of pharmacologic agents with sustained or controlled administration. In addition, an array of microneedles may allow injection or extraction over a larger area and at higher rates than with a single hypodermic needle. The production of microneedle arrays may allow the development of novel, portable medical devices and patches.

The development of effective microneedles has to fulfill some requirements: (i) they must penetrate at least 15 μm [38] into the skin; (ii) microneedles must not fracture during skin penetration; and (iii) the medicine should diffuse into the dermis blood vessels. The minimum penetration depth requirement refers to the stratum corneum layer, which contains dead cells and represents a barrier for drug transportation through the skin. Another requirement is that the needles be made of a biocompatible material.

2PP fabrication can be applied to organically modified ceramics, also known by its trade name Ormocer. This organic/inorganic hybrid material presents interesting properties and is suited for the production of microneedles. During the fabrication process, the crosslinking of the organic and inorganic components yields a 3D network that prevents the phase separation of the material. This results in a material with good chemical and thermal stability. In addition, Ormocer has passed through biological assays, including ISO 10933-5 cytotoxicity assay, showing that this material is nontoxic and biologically inert [39].

Because of its versatility, 2PP process allows the production of microneedles with different designs. In addition, in the production of microneedle arrays, 2PP can achieve higher microneedle density than other conventional fabrications methods. Ovsianikov *et al.* [40] have fabricated hollow microneedle arrays with centered and out-of-center channels with different geometrical aspects. Figure 14.14 shows the SEM images of these structures.

All microneedles in Figure 14.14 are 800 μm long, so they can be used for drug delivery and/or drawing blood. The aspect ratio of microneedles was modified by changing the size of the structure base, which ranged from 150 to 300 μm. Another geometry modification was made by adjusting the position of the channel relative to the center of the structure. Channel diameter was also the same for all structures.

Microneedles are subjected to many kinds of forces during insertion, such as compressive forces, bending forces, and shear forces. Microneedle arrays are less prone to fracture compared to single microneedles when subjected to shear force

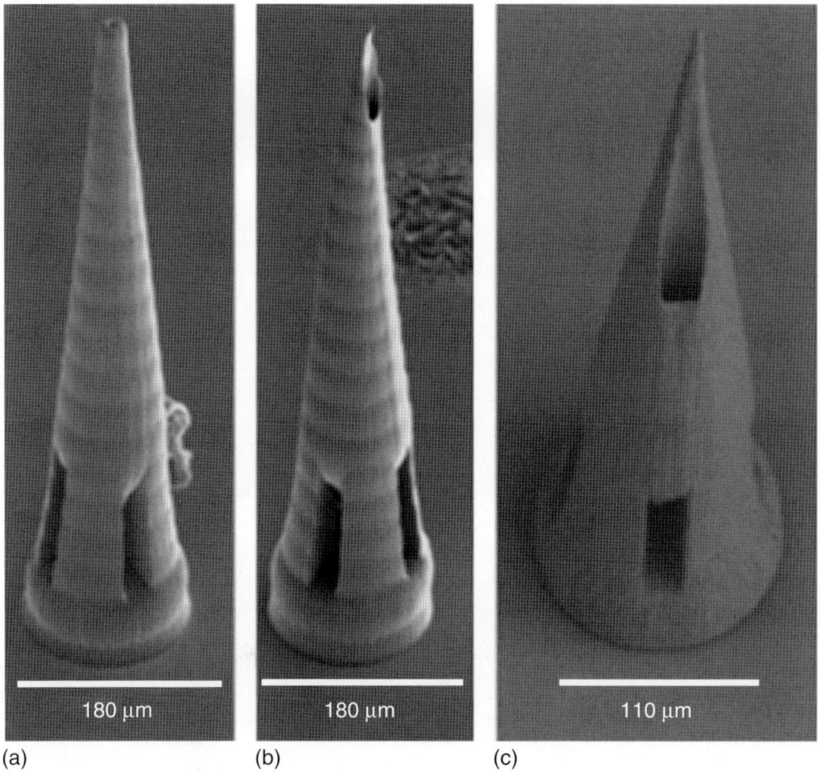

Figure 14.14 Scanning electron micrographs of Ormocer microneedles with (a) 0 μm, (b) 1.4 μm, and (c) 20.4 μm pore–needle center displacement values. (Ovsianikov 2007. Reproduced with permission of Wiley.)

because the force will be distributed over a wider area [41]. Also, this kind of configuration would allow higher rates of blood drawal and drug administration.

Arrays containing needles of each type of geometry were produced and tested against *polytetrafluoroethylene* and cadaveric porcine adipose tissue for penetration and fracture properties [40]. During these tests, the tip of the microneedles showed bending, which might increase the resistance to skin penetration. Also, needles with higher base diameter resisted higher load values. This result suggests that structures with larger bases have greater mechanical stability. The needles with a 20.4-μm off-center channel presented the sharpest tip and the lowest penetration resistance. Therefore, microneedles with higher aspect ratios and smaller tip radii are best suited for applications since they will meet with lower resistance during skin penetration [42]. However, a compromise must be made since microneedles with higher base diameter can resist high loads.

2PP allows the production of microneedles with a wide range of geometrical features, which allows the tailoring of needles depending on the application. Ormocer microneedle arrays were able to penetrate porcine tissues without fracture, reveling that these devices might be suited for medical applications [18, 40, 43].

14.6 Final Remarks

Two-photon polymerization has proven to be a powerful technique for designing 3D micro and nanostructures for biological applications. Several polymers and modified ceramics have been demonstrated biocompatible, allowing applications in biological environments. Additionally, a range of molecules have been investigated to play the role of photoinitiators in 2PP. The applications of these unmodified or (chemically, biologically, etc.) modified 3D structures span from cell studies to scaffolds for tissue regeneration, platforms for bacterial and cell growth investigation, as well as platforms for drug delivery, with great impact in areas such as medicine, biology, and chemistry. Although such microstructures are usually fabricated with femtosecond laser apparatus, by chemically designing specific resins and photoinitiators, less expensive laser systems can be employed in this fabrication process, which is a fundamental aspect for making this fabrication technology available to a wider community.

References

1. Aekbote, B.L., Jacak, J., Schütz, G.J., Csányi, E., Szegletes, Z., Ormos, P., and Kelemen, L. (2012) Aminosilane-based functionalization of two-photon polymerized 3D SU-8 microstructures. *Eur. Polym. J.*, **48** (10), 1745–1754.
2. Maruo, S. and Fourkas, J.T. (2008) Recent progress in multiphoton microfabrication. *Laser Photonics Rev.*, **2** (1–2), 100–111.
3. Ovsianikov, A., Ostendorf, A., and Chichkov, B.N. (2007) Three-dimensional photofabrication with femtosecond lasers for applications in photonics and biomedicine. *Appl. Surf. Sci.*, **253** (15), 6599–6602.
4. Cumpston, B.H., Ananthavel, S.P., Barlow, S., Dyer, D.L., Ehrlich, J.E., Erskine, L.L., Heikal, A.A., Kuebler, S.M., Lee, I.-Y.S., McCord-Maughon, D., Qin, J., Röckel, H., Rumi, M., Wu, X.-L., Marder, S.R., and Perry, J.W. (1999) Two-photon polymerization initiators for three-dimensional optical data storage and microfabrication. *Nature*, **398** (6722), 51–54.
5. Farsari, M., Ovsianikov, A., Vamvakaki, M., Sakellari, I., Gray, D., Chichkov, B.N., and Fotakis, C. (2008) Fabrication of three-dimensional photonic crystal structures containing an active nonlinear optical chromophore. *Appl. Phys. A-Mater. Sci. Process.*, **93** (1), 11–15.
6. Mendonca, C.R., Baldacchini, T., Tayalia, P., and Mazur, E. (2007) Reversible birefringence in microstructures fabricated by two-photon absorption polymerization. *J. Appl. Phys.*, **102** (1), 013109.
7. Takeyasu, N., Tanaka, T., and Kawata, S. (2008) Fabrication of 3D metal/polymer microstructures by site-selective metal coating. *Appl. Phys. A-Mater. Sci. Process.*, **90** (2), 205–209.
8. Baldacchini, T., LaFratta, C.N., Farrer, R.A., Teich, M.C., Saleh, B.E.A., Naughton, M.J., and Fourkas, J.T. (2004) Acrylic-based resin with favorable properties for three-dimensional two-photon polymerization. *J. Appl. Phys.*, **95** (11), 6072–6076.

9. Tayalia, P., Mendonca, C.R., Baldacchini, T., Mooney, D.J., and Mazur, E. (2008) 3D cell-migration studies using two-photon engineered polymer scaffolds. *Adv. Mater.*, **20** (23), 4494–4498.
10. Raimondi, M.T., Eaton, S.M., Laganà, M., Aprile, V., Nava, M.M., Cerullo, G., and Osellame, R. (2013) Three-dimensional structural niches engineered via two-photon laser polymerization promote stem cell homing. *Acta Biomater.*, **9** (1), 4579–4584.
11. Otuka, A.J.G., Correa, D.S., Fontana, C.R., and Mendonca, C.R. (2014) Direct laser writing by two-photon polymerization as a tool for developing microenvironments for evaluation of bacterial growth. *Mater. Sci. Eng. C-Mater. Biol. Appl.*, **35**, 185–189.
12. Bhatia, S.N., Balis, U.J., Yarmush, M.L., and Toner, M. (1998) Microfabrication of hepatocyte/fibroblast co-cultures: Role of homotypic cell interactions. *Biotechnol. Progr.*, **14** (3), 378–387.
13. Chen, C.S., Mrksich, M., Huang, S., Whitesides, G.M., and Ingber, D.E. (1997) Geometric control of cell life and death. *Science*, **276** (5317), 1425–1428.
14. McNeil, S.E., Griffiths, H.R., and Perrie, Y. (2011) Polycaprolactone fibres as a potential delivery system for collagen to support bone regeneration. *Curr. Drug Delivery*, **8** (4), 448–455.
15. Woodruff, M.A. and Hutmacher, D.W. (2010) The return of a forgotten polymer-Polycaprolactone in the 21st century. *Prog. Polym. Sci.*, **35** (10), 1217–1256.
16. Claeyssens, F., Hasan, E.A., Gaidukeviciute, A., Achilleos, D.S., Ranella, A., Reinhardt, C., Ovsianikov, A., Shizhou, X., Fotakis, C., Vamvakaki, M., Chichkov, B.N., and Farsari, M. (2009) Three-dimensional biodegradable structures fabricated by two-photon polymerization. *Langmuir*, **25** (5), 3219–3223.
17. Ovsianikov, A., Gruene, M., Pflaum, M., Koch, L., Maiorana, F., Wilhelmi, M., Haverich, A., and Chichkov, B. (2010) Laser printing of cells into 3D scaffolds. *Biofabrication*, **2** (1), 014104.
18. Doraiswamy, A., Jin, C., Narayan, R.J., Mageswaran, P., Mente, P., Modi, R., Auyeung, R., Chrisey, D.B., Ovsianikov, A., and Chichkov, B. (2006) Two photon induced polymerization of organic-inorganic hybrid biomaterials for microstructured medical devices. *Acta Biomater.*, **2** (3), 267–275.
19. Malinauskas, M., Danilevicius, P., Baltriukiene, D., Rutkauskas, M., Zukauskas, A., Kairyte, Z., Bickauskaite, G., Purlys, V., Paipulas, D., Bukelskiene, V., and Gadonas, R. (2010) 3D artificial polymeric scaffolds for stem cell growth fabricated by femtosecond laser. *Lith. J. Phys.*, **50** (1), 75–82.
20. Almany, L. and Seliktar, D. (2005) Biosynthetic hydrogel scaffolds made from fibrinogen and polyethylene glycol for 3D cell cultures. *Biomaterials*, **26** (15), 2467–2477.
21. Schlie, S., Ngezahayo, A., Ovsianikov, A., Fabian, T., Kolb, H.-A., Haferkamp, H., and Chichkov, B.N. (2007) Three-dimensional cell growth on structures fabricated from ORMOCER (R) by two-photon polymerization technique. *J. Biomater. Appl.*, **22** (3), 275–287.
22. Chachques, J.C., Trainini, J.C., Lago, N., Cortes-Morichetti, M., Schussler, O., and Carpentier, A. (2008) Myocardial assistance by grafting a new Bioartificial upgraded myocardium (MAGNUM trial): clinical feasibility study. *Ann. Thoracic Surg.*, **85** (3), 901–908.
23. Drakakis, T.S., Papadakis, G., Sambani, K., Filippidis, G., Georgiou, S., Gizeli, E., Fotakis, C., and Farsari, M. (2006) Construction of three-dimensional biomolecule structures employing femtosecond lasers. *Appl. Phys. Lett.*, **89** (14), 144108.
24. Correa, D.S., Tayalia, P., Cosendey, G., dos Santos, D.S., Aroca, R.F., Mazur, E., and Mendonca, C.R. (2009) Two-photon polymerization for fabricating structures containing the biopolymer chitosan. *J. Nanosci. Nanotechnol.*, **9** (10), 5845–5849.
25. Du, H.L., Yang, X.Y., and Zhai, G. (2014) Design of chitosan-based nanoformulations for efficient intracellular release of active compounds. *Nanomedicine*, **9** (5), 723–740.

26. Kaur, S. and Dhillon, G.S. (2014) The versatile biopolymer chitosan: potential sources, evaluation of extraction methods and applications. *Crit. Rev. Microbiol.*, **40** (2), 155–175.
27. Thadathil, N. and Velappan, S.P. (2014) Recent developments in chitosanase research and its biotechnological applications: a review. *Food Chem.*, **150**, 392–399.
28. Dawood, F., Qin, S., Li, L.J., Lin, E.Y., and Fourkas, J.T. (2012) Simultaneous microscale optical manipulation, fabrication and immobilisation in aqueous media. *Chem. Sci.*, **3** (8), 2449–2456.
29. Pitts, J.D., Howell, A.R., Taboada, R., Banerjee, I., Wang, J., Goodman, S.L., and Campagnola, P.J. (2002) New photoactivators for multiphoton excited three-dimensional submicron cross-linking of proteins: Bovine serum albumin and type 1 collagen. *Photochem. Photobiol.*, **76** (2), 135–144.
30. Connell, J.L., Wessel, A.K., Parsek, M.R., Ellington, A.D., Whiteley, M., and Shear, J.B. (2010) Probing prokaryotic social behaviors with bacterial "Lobster Traps". *MBio*, **1** (4), e00202-10.
31. Kaehr, B. and Shear, J.B. (2008) Multiphoton fabrication of chemically responsive protein hydrogels for microactuation. *Proc. Natl. Acad. Sci. U.S.A.*, **105** (26), 8850–8854.
32. Sharma, G., Valenta, D.T., Altman, Y., Harvey, S., Xie, H., Mitragotri, S., and Smith, J.W. (2010) Polymer particle shape independently influences binding and internalization by macrophages. *J. Controlled Release*, **147** (3), 408–412.
33. Venkataraman, S., Hedrick, J.L., Ong, Z.Y., Yang, C., Ee, P.L.R., Hammond, P.T., and Yang, Y.Y. (2011) The effects of polymeric nanostructure shape on drug delivery. *Adv. Drug Delivery Rev.*, **63** (14–15), 1228–1246.
34. Spivey, E.C., Ritschdorff, E.T., Connell, J.L., McLennon, C.A., Schmidt, C.E., and Shear, J.B. (2013) Multiphoton lithography of unconstrained three-dimensional protein microstructures. *Adv. Funct. Mater.*, **23** (3), 333–339.
35. Potts, R.O. and Lobo, R.A. (2005) Transdermal drug delivery: clinical considerations for the obstetrician-gynecologist. *Obstet. Gynecol.*, **105** (5), 953–961.
36. Prausnitz, M.R. (2004) Microneedles for transdermal drug delivery. *Adv. Drug Delivery Rev.*, **56** (5), 581–587.
37. Kaushik, S., Hord, A.H., Denson, D.D., McAllister, D.V., Smitra, S., Allen, M.G., and Prausnitz, M.R. (2001) Lack of pain associated with microfabricated microneedles. *Anesthesia Analgesia*, **92** (2), 502–504.
38. Sivamani, R.K., Stoeber, B., Wu, G.C., Zhai, H., Liepmann, D., and Maibach, H. (2005) Clinical microneedle injection of methyl nicotinate: stratum corneum penetration. *Skin Res. Technol.*, **11** (2), 152–156.
39. Al-Hiyasat, A.S., Darmani, H., and Milhem, M.M. (2005) Cytotoxicity evaluation of dental resin composites and their flowable derivatives. *Clin. Oral Invest.*, **9** (1), 21–25.
40. Ovsianikov, A., Chichkov, B., Mente, P., Monteiro-Riviere, N.A., Doraiswamy, A., and Narayan, R.J. (2007) Two photon polymerization of polymer-ceramic hybrid materials for transdermal drug delivery. *Int. J. Appl. Ceram. Technol.*, **4** (1), 22–29.
41. Griss, P. and Stemme, G. (2003) Side-opened out-of-plane microneedles for microfluidic transdermal liquid transfer. *J. Microelectromech. Syst.*, **12** (3), 296–301.
42. Davis, S.P., Landis, B.J., Adams, Z.H., Allen, M.G., and Prausnitz, M.R. (2004) Insertion of microneedles into skin: measurement and prediction of insertion force and needle fracture force. *J. Biomech.*, **37** (8), 1155–1163.
43. Gittard, S.D., Chen, B., Xu, H., Ovsianikov, A., Chichkov, B.N., Monteiro-Riviere, N.A., and Narayan, R.J. (2013) The effects of geometry on skin penetration and failure of polymer microneedles. *J. Adhes. Sci. Technol.*, **27** (3), 227–243.

Index

Multiphoton Lithography: Techniques, Materials and Applications, First Edition.
Edited by Jürgen Stampfl, Robert Liska, and Aleksandr Ovsianikov.

n

o

p

u

v